Cover:

NAMARRKON—THE LIGHTNING SPIRIT

Artist: *Jimmy Nakurridjdjilmi Nganjmira*

Namarrkon is the Lightning Spirit. He is often depicted in cave art and on barks and it is believed that he initiates severe tropical electrical storms, which cause widespread damage, destroying camps, and even killing people. Indeed it is widely held that the *marrkidjbu* or 'clever men' have the power to call on the lightning man to strike a particular person whom they wish to have killed.

Namarrkon is depicted with a circuit of lightning connecting his head to his testicles, either enclosing his limbs and torso completely or passing through his wrists and ankles. This representation suggests the sexual connotations often associated with thunder and lightning in Aboriginal mythology. Stone axes are attached to his elbows and knees, and it is by hitting with these that Namarrkon causes lightning to strike.

Reproduced with permission of the Aboriginal Artists Agency Limited, 12 McLaren Street, North Sydney, NSW 2060, Australia, acting on behalf of the artist.

FRACTALS IN PHYSICS

FRACTALS IN PHYSICS

Proceedings of the Sixth Trieste International Symposium on Fractals in Physics, ICTP, Trieste, Italy, July 9-12, 1985

Edited by

Luciano PIETRONERO
Solid State Physics Laboratory
University of Groningen
The Netherlands

and

Erio TOSATTI
International School for Advanced Studies
Trieste
Italy

1986

NORTH-HOLLAND
AMSTERDAM · OXFORD · NEW YORK · TOKYO

© ELSEVIER SCIENCE PUBLISHERS B.V., 1986

All rights reserved. No part of this publication may be reproduced, stored in a retrieval system, or transmitted, in any form or by any means, electronic, mechanical, photocopying, recording or otherwise, without the permission of the Publisher, Elsevier Science Publishers B.V. (North-Holland Physics Publishing Division), P.O. Box 103, 1000 AC Amsterdam, The Netherlands. Special regulations for readers in the USA: This publication has been registered with the Copyright Clearance Center Inc. (CCC), Salem, Massachusetts. Information can be obtained from the CCC about conditions under which photocopies of parts of this publication may be made in the USA. All other copyright questions, including photocopying outside of the USA, should be referred to the publisher.

ISBN: 0 444 86995 6

Published by:

North-Holland Physics Publishing
a division of
Elsevier Science Publishers B.V.
P.O. Box 103
1000 AC Amsterdam
The Netherlands

Sole distributors for the U.S.A. and Canada:

Elsevier Science Publishing Company, Inc.
52 Vanderbilt Avenue
New York, N.Y. 10017
U.S.A.

First edition March 1986
Reprinted September 1986

Library of Congress Cataloging-in-Publication Data

```
International Symposium on Fractals in Physics (6th :
  1985 : ICTP, Trieste, Italy)
  Fractals in physics.

  Sponsored by the International Center for
Theoretical Physics (I.C.T.P.).
  1. Fractals--Congresses.  2. Mathematical physics--
Congresses.  3. Irreversible processes--Congresses.
I. Pietronero, L. (Luciano)  II. Tosatti, E. (Eric)
III. International Centre for Theoretical Physics.
IV. Title.
QC20.7.G44I585  1985       530.1'5        86-2001
ISBN 0-444-86995-6
```

PRINTED IN THE NETHERLANDS

PREFACE

This volume contains the Proceedings of the International Trieste Symposium on "Fractals in Physics" held on July 9-12, 1985, at I.C.T.P., Trieste, Italy. This is the Sixth of a series of I.C.T.P. Condensed Matter Symposia.

The concepts of self-similarity and scale invariance have arisen independently in several areas. One is the study of the critical properties of phase transitions; another is fractal geometry, which involves the concept of (non-integer) fractal dimension. These two areas have now come together, and their methods have extended to various fields of physics. The purpose of this Symposium was to provide an overview of the physical phenomena that manifest scale invariance and fractal properties with the aim of bringing out the common mathematical features. The emphasis was on theoretical and experimental work related to well defined physical phenomena. Dynamical Systems and Chaos were only marginally included in view of the fact that these fields are already covered in specialized meetings.

The majority of the contributions presented are concerned with the study of the physics underlying irreversible growth phenomena that generate fractal structures. This area of problems, that we may call *"Kinetic Critical Phenomena"* represents one of the most challenging fields in today's theoretical physics.

In view of the large number of papers the volume has been divided into sections. This division should be considered only as a loose indication of the topics covered. Many papers have a substantial overlap with more than one section and we would like to apologize for the subjectivity of this partition.

We have been helped in the preparation of the Symposium by the co-organizers: S. Lundqvist, B. Mandelbrot, H.E. Stanley and C. Evertsz (secretary) and we are very grateful to them. We would also like to thank all the participants who made the Symposium so lively and successful.

The Symposium, from which this book originates, was generously sponsored by the International Center for Theoretical Physics (I.C.T.P.) of Trieste. Co-sponsorship and support was obtained also from the following institutions: I.B.M. - Italy; The Office of Naval Research - U.S.A.; Consiglio Nazionale delle Ricerche - Italy; The University of Groningen - The Netherlands; SOHIO, Cleveland - U.S.A. and Brown Boveri - Switzerland. On behalf of the international scientific community we wish to express our gratitude to all these institutions, as well as to those individuals - we mention here in particular Mrs. Deisa Buranello and Miss Carla Carbone - who have contributed so much to making a successful Symposium, and this book as by-product.

L. Pietronero
E. Tosatti

TABLE OF CONTENTS

PREFACE ... v

Part I. General properties of fractals

Self-affine fractal sets, I: The basic fractal dimensions
B.B. MANDELBROT ... 3

Self-affine fractal sets, II: Length and surface dimensions
B.B. MANDELBROT ... 17

Self-affine fractal sets, III: Hausdorff dimension anomalies and their implications
B.B. MANDELBROT ... 21

Random fractals, flow fractals and the renormalisation group
J. MELROSE ... 29

On finitely ramified fractals and their extensions
R. HILFER and A. BLUMEN ... 33

Part II. Analysis of fractal properties of materials

Structure of random silicates: Polymers, colloids and porous solids
D.W. SCHAEFER and K.D. KEEFER ... 39

Interaction of fractals with fractals: Adsorption of polystyrene on porous Al_2O_3
P. PFEIFER ... 47

Scattering by fractals
E. JAKEMAN ... 55

Optical Fourier transforms of fractals
C. ALLAIN and M. CLOITRE ... 61

On the measure of fractal dimensionalities through physical properties
C. TSALLIS ... 65

Part III. Polymer statistics and self-avoiding-walks

Random walks with memory
L. PELITI ... 73

Survival probability and enhancement factor in polymer statistics
L. PIETRONERO and L. PELITI ... 83

The Laplacian random walk
  J.W. LYKLEMA and C. EVERTSZ … 87

Kinetically growing self-avoiding walks
  J.W. LYKLEMA … 93

The coil-globule transition in 2-dimensions
  N. JAN, A. CONIGLIO, I. MAJID and H.E. STANLEY … 97

Self-similarity of mutual and self-intersections of random fractals
  A.L. STELLA, R. DEKEYSER and A. MARITAN … 101

Statistical mechanics of self-avoiding random surfaces
  A. MARITAN and A.L. STELLA … 107

Bethe - like approximation for self-avoiding random walks and surfaces (and frustrations)
  A. CAPPELLI, R. LIVI, A. MARITAN and S. RUFFO … 111

On the self-avoiding walks on disordered lattices
  S. MILOŠEVIĆ and A. CHERNOUTSAN … 115

Proteins in the experiment
  Y.S. YANG … 119

Part IV. Branched polymers, gelation and percolation

Fractal dimension and the synthesis of branched polymers
  Z. ALEXANDROWICZ … 125

Fractal dimensionalities of backbones and clusters in a kinetic gelation model
  A. CHHABRA, H.J. HERRMANN and D.P. LANDAU … 129

A lattice magnetic model for branched polymers and the sol-gel transition
  A.E. GONZALEZ … 133

Percolation in a concentration gradient
  J.F. GOUYET, M. ROSSO and B. SAPOVAL … 137

Surface tension in Potts models and percolation
  C.K. HARRIS … 141

Percolation on the DAP
  R. DEWAR and C.K. HARRIS … 145

Part V (A). Irreversible growth models: Laplacian fractals, dielectric breakdown, fracture propagation and viscous fingers in liquids

Properties of Laplacian fractals for dielectric breakdown in 2 and 3 dimensions
  H.J. WIESMANN and L. PIETRONERO … 151

Scaling properties of growing zone and capacity of Laplacian fractals
L. PIETRONERO, C. EVERTSZ and H.J. WIESMANN   159

An infinite hierarchy of exponents to describe growth phenomena
A. CONIGLIO   165

2-D dielectric breakdown between parallel lines
M. MURAT   169

Dielectric breakdown in three dimensions
S. SATPATHY   173

The fractal nature of fracture
E. LOUIS, F. GUINEA and F. FLORES   177

Pattern formation of dendritic fractals in fracture and electric breakdown
H. TAKAYASU   181

Collapse of loaded fractal trees
S.A. SOLLA   185

Fractals and the fracture of cracked metals
C.W. LUNG   189

When are viscous fingers fractal?
J. NITTMANN, G. DACCORD and H.E. STANLEY   193

Part V (B). Irreversible growth models: Diffusion-limited aggregation, dendritic growth, Eden model and cluster-cluster aggregation

Some recent advances in the simulation of diffusion limited aggregation and related processes
P. MEAKIN   205

Internal anisotropy of diffusion-limited aggregates
P. MEAKIN and T. VICSEK   213

Growing interface in diffusion-limited aggregation and in the Eden process
M. PLISCHKE and Z. RÁCZ   217

Sticking probability scaling in diffusion-limited aggregation
L.A. TURKEVICH and H. SCHER   223

Cone angle picture and anisotropy in DLA cluster growth
G. ROSSI, B.R. THOMPSON, R.C. BALL and R.M. BRADY   231

Growth of anisotropic DLA clusters
B.R. THOMPSON, G. ROSSI, R.C. BALL and R.M. BRADY   237

Continuum DLA: Random fractal growth generated by a deterministic model
L.M. SANDER   241

Formation of solidification patterns in aggregation models
  T. VICSEK — 247

Scaling properties of the surface of the Eden model
  R. JULLIEN and R. BOTET — 251

Cluster aggregation
  R. BOTET, R. JULLIEN and M. KOLB — 255

Anisotropy in cluster and particle aggregation
  M. KOLB — 259

Reversibility in cluster aggregation
  M. KOLB — 263

Film on aggregation processes
  M. KOLB — 267

Field theory approach to the Eden model and diffusion-limited aggregation
  L. PELITI and Y.-C. ZHANG — 269

Spreading of epidemic processes leading to fractal structures
  P. GRASSBERGER — 273

Random rain simulations of dendritic growth
  B. CAPRILE, A.C. LEVI and L. LIGGIERI — 279

Experimental study of two dimensional aggregation
  C. ALLAIN and M. CLOITRE — 283

Part VI. Kinetics of clustering

Kinetics of clustering in irreversible aggregation
  M.H. ERNST — 289

Tail distribution for large clusters in irreversible aggregation
  P.G.J. VAN DONGEN and M.H. ERNST — 303

Scaling generalization of the Smoluchowski equation
  Z. RÁCZ — 309

Clustering in the Universe
  F. LUCCHIN — 313

Stochastic approach to large scale clustering of matter in the Universe
  L. PIETRONERO and R. KUPERS — 319

Part VII. Dynamical properties of fractal structures

Fractal surfaces and the de Gennes termite model for a two-component random material
  H.E. STANLEY — 327

Dynamical properties of random and non-random fractals
  R. STINCHCOMBE — 337

The elastic behavior of fractal structures
  I. WEBMAN — 343

Static and dynamic properties of loopless aggregates
  S. HAVLIN — 351

The fractal dimension of growth perimeters
  A.E. MARGOLINA — 357

Percolation and fractal behavior in disordered lattices
  P. ARGYRAKIS — 361

Hierarchical fractal graphs and walks thereupon
  J. MELROSE — 365

Fractal-like exciton dynamics: Geometrical and energetical disorder
  R. KOPELMAN — 369

Nyquist, diffusion and flicker (1/f) noise in fractals and percolating networks
  R. RAMMAL — 373

A real-space renormalization group approach to electrical and noise properties of percolation clusters
  J.M. LUCK — 379

Theory of the ac response of rough interfaces
  S.H. LIU, T. KAPLAN and L.J. GRAY — 383

Part VIII. Hierarchical and fractal features of disordered systems

The nature of temporal hierarchies underlying relaxation in disordered systems
  M.F. SHLESINGER and J. KLAFTER — 393

Reactions in disordered media modelled by fractals
  A. BLUMEN, J. KLAFTER and G. ZUMOFEN — 399

Self-similar temporal behavior of random walks in one-dimensional random media
  J. BERNASCONI and W.R. SCHNEIDER — 409

An observation of scaling in trapping reactions
  Z.B. DJORDJEVIĆ — 413

Hierarchically constrained thermodynamics in metastable systems and glasses
  L. PIETRONERO — 417

Fractal clusters and scaling in the Ising model
  J.L. CAMBIER and M. NAUENBERG — 421

Devil's staircase and strange attractor in the Ising model with
competing interactions
    M.J. DE OLIVEIRA, S.R. SALINAS and C.S.O. YOKOI   427

Problems about the self-similar structure of wavefunctions in disordered systems
    A.P. SIEBESMA and L. PIETRONERO   431

Part IX. Chaos, turbulence and related topics

Circle maps in the complex plane
    P. CVITANOVIĆ, M.H. JENSEN, L.P. KADANOFF and I. PROCACCIA   439

Fractal models for two- and three- dimensional turbulence
    G. PALADIN and A. VULPIANI   447

Numerical investigation of nonuniform fractals
    R. BADII and A. POLITI   453

Generalised scale invariance and anisotropic inhomogeneous fractals in turbulence
    D. SCHERTZER and S. LOVEJOY   457

Analysis of the fractal shape of severe convective clouds
    F.S. RYS and A. WALDVOGEL   461

Nested cellular automata: Continuous aspects of discrete systems
    U. QUASTHOFF   465

The hyperbolic helix hypothesis: Stapleton's fractal measure on the
hydrophobic free energy mode distributions of allosteric proteins
    A.J. MANDELL   469

AUTHOR INDEX   475

Part I
GENERAL PROPERTIES OF FRACTALS

SELF-AFFINE FRACTAL SETS, I: THE BASIC FRACTAL DIMENSIONS

Benoit B. MANDELBROT

Physics Department, IBM Research Center
Mathematics Department, Harvard University, Cambridge, MA 02138 USA*

The notion of fractal dimension is explored for various fractal curves or dusts that are not self-similar, but are diagonally self-affine. A diagonal self-affinity stretches the coordinates in different ratios. It is showed that, in contrast to the unique fractal dimension of strictly self-similar sets, one needs in general several distinct notions. Most important are the concepts of dimension obtained via the mass in a sphere and via covering by uniform boxes. One finds it does not matter which definition is taken, but it matters greatly whether one interpolates or extrapolates. Thus, one obtains two sharply distinct dimensions: a local one, valid on scales well below, and a global one, valid on scales well above, a certain crossover scale.

1. INTRODUCTION

This paper examines, on three levels corresponding to three parts, what happens to diverse alternative definitions of fractal dimension when they are generalized from self-similar fractals to certain self-affine fractals. The substance of my Scripta paper[1] is incorporated. Self-similar fractals were the original objects on which diverse fractal dimensions had been tested in detail, and their values were found to coincide.[2] When a method works well in one case, it is tempting to apply it under increasingly wide conditions. The more general context of self-affine fractals now deserves systematic attention.

I have coined "self-affine" and "self-similar" in 1964 (the latter is so accepted now, that its age has become hard to believe), but "affine" goes back to Euler. In this paper, no specific knowledge of affine geometry will be required, but it is amusing to quote a characterization of that field by E. Snapper and R.J. Troyer: "Roughly speaking, affine geometry is what remains after practically all ability to measure length, area, angles, etc... has been removed from Euclidean geometry. One might think that affine geometry is a poverty-stricken subject. On the contrary, it is quite rich". I hope to convince the reader that self-affine fractals also prove to be a very surprisingly rich topic.

One well-known but very special example of self-affine fractal is the record of Wiener's scalar Brownian motion, which is the random process with independent and stationary Gaussian increments. This record has a well-known invariance property: setting $B(0) = 0$, the processes $B(t)$ and $b^{-1/2}B(bt)$ are identical in distribution for every ratio $b > 0$. One observes that the rescaling ratios of t and of B are

*Supported in part by the Office of Naval Research, grant N00014-85-K-0188

different, hence the transformation from $B(t)$ to $b^{-1/2}B(bt)$ is not a similitude but a more general "affinity." This is why $B(t)$ was called "statistically self-affine" on page 350 of my book[2].

While a similarity is a linear transformation that shrinks or expands all the vectors implicit in a geometric figure in the same ratio, an affinity is a linear transformation that shrinks different vectors differently according to their directions. More precisely, $B(t)$ is unchanged statistically under "diagonal affinity", a notion explained in Section 2.

In this paper, diagonally self-affine fractal curves, dusts, and other sets, are examined in detail on several successive levels of generality, first in the plane: records of functions (random or not) analogous to $B(t)$, and two levels of more general sets, and then in space. In each case, the fractal-dimensional properties are shown to exhibit new and surprising complications. The different roles of the single "all-purpose" fractal dimension of a fully self-similar set are now shared by a multitude of different "special purpose" fractal dimensions.

This part covers the mass-box dimension, and the gap dimension. Conceptually and mathematically, the most original finding is that a self-affine fractal's mass-box dimension has distinct local and global variants. For example, in a recursive case of middle generality, the single base b is replaced by two bases b' and b'', and the classical expression $\log_b N$ is replaced by the combination of $\log_{b'}(Nb'/b'')$ (local) and $\log_{b''}(Nb''/b')$ (global). In addition, lying between the above, we often have a distinct "gap dimension", with a single value extending over all scales, namely $\log N / \log\sqrt{(b'b'')}$.

Part II covers the dimensions obtained by walking a divider along a curve or "triangulating" a surface. The values one obtains are "doubly anomalous", namely: distinct from those obtained in Part I.

Part III starts with recent mathematical findings[3,4,5] concerning certain cases of the Hausdorff-Besicovitch dimension D_{HB}, extends them, and discusses the meaning of the "double anomaly" found in certain cases.

The richness and complexity of this study are not purely mathematical, but reflect the richness and complexity of nature. Again, increasingly complex fractals are considered, the continuing refinement of our description of their structure demands a continuingly increasing number of fractal dimensions.

2. THE NOTIONS OF AFFINITY, DIAGONAL AFFINITY, AND SELF-AFFINITY

2.1. Background to diagonal affinity

Let us think of the Brownian record $B(t)$ again. In Wiener's original interpretation, t is time and B is a physical particle's location on a spatial axis. The two coordinates play sharply different roles, and the units of B and t (cm. and second?) can be chosen independently. Rotation is not allowable, because it leads to sets that are no longer records of functions. Forming the expression $B(t)-\delta t$ (another application of affinity) introduces a function called "Brownian motion with a drift," which is a very different process from B(t). In an even earlier interpretation by Louis Bachelier (in 1900), t is time and B is a price in francs. The same remarks apply. However, an interpretation that I advanced later is substantially different: B(t) describes the vertical section of one of my Brown landscapes (my book[2], chapter 28); the coordinates still play different roles, since gravity defines the vertical direction, makes overhangs an exception and makes it is useful to represent the relief by a (single-valued) function. However, both B and t are lengths in this example, and their units can no longer be chosen independently. For reasons that will transpire later, the best is to choose as unit of both B and t to be the t_c such that $|B(t+t_c) - B(t)| \sim |t_c|$. This t_c will be called the "crossover scale." This unit's counterpart in the original Brownian motion interpretation depends upon the units that happen to be selected for t and B; therefore, the crossover is in general not intrinsic.

The local and global dimensions that are introduced examined below are separated by this crossover. Without it, the local versus global distinction could not be clearcut.

2.2. Diagonal affinities

Section 2.1. shows that in the case of the record of B(t) — and the same will hold for the other fractals in this paper — a special role is played by affinities whose invariant set is made of straight lines that are parallel to the coordinate axes. Such an affinity, which I propose to call <u>diagonal</u>, operates in the E-dimensional affine space A^E. Each member of a collection of affinities is specified by giving a fixed point of coordinates $\mathfrak{D}_m$ ($0 \leq m \leq E-1$) and an array of reduction ratios r_m ($0 \leq m \leq E-1$), and considering the map

$$x_m \rightarrow \mathfrak{D}_m + r_m(x_m - \mathfrak{D}_m).$$

The ratios r_m need not be positive. And they must not all be equal, because otherwise the transformation would fall be a similitude. The inverses $1/|r_m| = b_m$, called bases, are integers in the simplest examples that are constructed recursively.

Most of the examples will be sets in the affine plane A^2 (E=2). We shall write b'=max b_m, b"=min b_m, and H=log b"/log b'. This H, called affinity exponent, will satisfy 0<H<1. When E>2, there are E(E-1)/2 affinity exponents and crossover scales.

Formally, a linear transformation is the sequence of a translation and a multiplication by a matrix; we only tackle the cases where the matrix is diagonal and its diagonal terms are not identical. The product of two diagonal affinities is a diagonal affinity. Thus, a collection of diagonal affinities can be used as the basis for a group.

The issues to be addressed involve the meaning of "square", "distance", and "circle" in affine geometry. (See second paragraph of Section 1). These notions remain meaningful for relief cross-sections, but for records of noise or of price, the units along the t axis and along the B axis are set up independently of each other. There being no meaning to equal height and width, a square cannot be defined. Similarly, a circle cannot be defined, because its square radius $R^2 = \Delta t^2 + \Delta B^2$ would have to combine the units along both axes. Furthermore, one cannot "walk a divider" along a self-affine noise record, to measure its approximate length, because the distance covered by each step combines a Δt and a ΔB. On the other hand, a noise record is always represented on the same graph paper as a relief section or an isotropic set. This does <u>not</u> cause the distinction between the affine A^E and the Euclidean R^E to disappear, but sometimes it is elusive, and one is tempted to evaluate various "prohibited" dimensions "mechanically". One should not.

2.3. The recursive constructions in a grid of many standard self-similar fractals extend easily to the self-affine case.

For an example of how to generalize the Sierpinski carpet, take the semi-open unit square as initiator. (Semi-open means that the top and the right sides are open and the bottom and left sides are closed. The rectangles to be considered will also be semi-open). As generator, take the array on Figure 1.

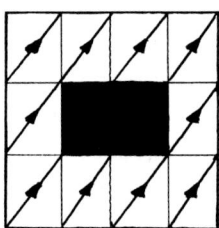

FIGURE 1

Thus, we divide initiator into 3 x 4 = 12 subrectangular parts, and erase the middle two parts, shown in black. Then we erase the middle two of the 12 sub-subrectangular parts, etc... The resulting self-affine carpet is the union of N=10 "tenths". Each tenth is obtained from the whole by a diagonal affinity with $r_n' = 1/3$, $r_n'' = 1/4$ for n=1 to n=N=10. I propose that those signs be summarized by representing the generator in the form of a "stick generator": using arrows along the diagonals of the ten rectangles. In the present example arrows must be placed as marked to insure that the "tenths" of this carpet do not overlap. The fixed points are the four vertices, the midpoints of the left and right sides, and the points 1/3 and 2/3 along the top and bottom sides. Indeed, an affinity's fixed point is the point of intersection of the four straight lines that join the vertices of the whole to these vertices' transforms, each of which is the vertex of the part.

The choice of a unit square as initiator deserves comment. It fixes the units of the coordinates t and B in analogy with the condition $|B(t+t_c) - B(t)| \sim t_c$. The extrapolation that yields scales above this t_c is discussed in section 2.6.

A general fractal generator in a self-affine lattice is obtained by drawing b'x b" subrectangles, and keeping N<b'b" of them. Again, $|r'_n|=1/b'$ and $|r''_n|=1/b''$ for all n, but the orientation of the n-th affinity - as expressed by the signs of the r'_n and r''_n - may depend on n. And I propose that it be represented by a diagonally placed vector. In the example of a construction that will be important in the sequel, the two variants shown in Figures 2 and 3 play especially important roles.

(Surprisingly, Part III will show that D_{HB} depends on which variant is chosen!)

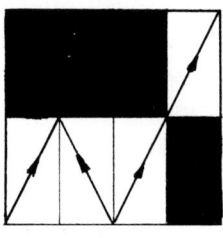
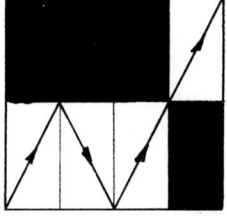

FIGURE 2   FIGURE 3

When one keeps both the rectangles and their diagonals, the resulting fractal is obtained as a limit of nested collections of rectangles, "nested" meaning that each is contained in the preceding one. When one only keeps the diagonals, and these diagonals form a curve, a self-affine fractal curve is obtained as a limit of broken lines. Connectedness of the stick generator imposes a constraint on the retained subrectangles.

However, as in my book,[2] chapter 13, the stick generator may split into several curves, creating "islands" and/or "lakes".

Important special case: When $N=b'$, a stick-generated fractal curve is the record of a (one-valued) continuous function.

2.4. Definition of self-affinity

The preceding examples adequately motivate the following definition.

A set S is self-affine with respect to a list of N diagonal affinities, α_n, if one can write $S = \cup \alpha_n S$, where $\alpha_n S \cap \alpha_m S$ is empty unless $n=m$. That is, S decomposes into N parts (no two of which overlap) each of which results from the whole by an affinity in the list.

2.5. Fractally homogeneous measure

To be able to define a fractal's mass dimension, one must define the mass within a cube. A uniform measure on the fractal is easy to define here.

For records of functions, one takes measure as proportional to time elapsed.

For recursive fractals within lattices, the k-th stage is made up of N^k rectangles; they are given equal measures.

For records of functions constructed recursively, the two methods yield the same result, because the X-projection of the measure according to the second definition is uniform.

2.6 Extrapolation

To understand extrapolation is especially important in the case of self-affine fractals, because (as has been announced and will be proven), the extrapolate is ruled by its own global fractal dimensions. Recall that a Cantor dust's extrapolation is not unique, in fact there is an extrapolate for every infinite sequence of base $N=2$. Since every point in the interpolated dust is determined by a sequence written towards the right, the extrapolative sequence is best written towards the left, as $\ldots a_{-k} \ldots a_{-3}, a_{-2}, a_{-1}, a_0$. If $a_0=0$ (resp., $a_0=1$), our dust is viewed as the left (resp. the right) portion of a dust that has been upsized in a ratio $b=3$. And so on. The same applies to all multi-dimensional recursive constructions, extrapolative sequences being written in the base N. In our self-affine case, the values of n_0 identifies the square initiator with one of the N parts of a super-initiator b' wide and b'' high; therefore, n_0 identifies the super-initiator. Recursively, when the super^k-initiator is known, the values of n_{-k} identifies it within a super^{k+1}-initiator. And so on.

3. THE MULTIPLE FRACTAL DIMENSIONS OF A SELF-AFFINE SET

From either the purely mathematical or the fractal literature, the only very widely known fact on this topic is that the Hausdorff-Besicovitch dimension D_{HB} of the Brown record is 3/2. This result becomes less murky if extended to the more general fractional Brownian motion $B_H(t)$, where $0 < H < 1$. If $B_H(0) = 0$, the random processes $B_H(t)$ and $b^{-H}B_H(bt)$ are identical in distribution. The fractional Brown record satisfies $D_{HB} = 2-H$. The value $H = 1/2$ brings $B(t)$ as a special case of $B_H(t)$.

But what about D_{HB} for self-affine sets that are records of other functions, or are <u>not</u> the records of functions? And what about definitions of fractal dimension other than D_{HB}?

Two words suffice to deal with the "(self-) similarity dimension," D_S. This notion applies most directly to self-similar fractal sets, which are made of N parts, each obtained from the whole by a similitude whose reduction ratio satisfies $|r| = 1/b$. For the self-similar fractal sets, $D_S = \log N/\log b$. For self-affine fractal sets, D_S simply <u>cannot be evaluated</u>. (However, a naturally generalized D_S will enter in Section 4).

As a substitute for D_S, several authors have made guesses, which by and large prove unjustifiable. For example, take $B(t)$. If D_{HB} is written as $1 + 1/2$, its value happens numerically to be $1 + \log_{b'} b"$, where (again) b' is the larger, and $b"$ the smaller base. But in fact, $D_{HB}=2-\log_{b'} b"$, as shown by the study of the fractional Brownian motion $B_H(t)$. Other guesses are less obviously wild.

The basic fact established in this paper (and in part already established earlier[1]) is that different roles that in the self-similar case are taken up by one number called fractal dimension must in the self-affine case be shared among different quantities. Some are local – like D_{HB} – but the newest ones are global. Of particular significance are the global mass dimension D_{MG} and the local box dimension D_{BL}.

Recall that I once defined a fractal, as a set for which $D_{HB} > D_T$ (=topological dimension). This "tentative" definition had looked less and less attractive, and I have long abandoned it (my book's second and later printings, p. 458). An alternative was to view as fractal those sets for which the dimensions in a certain list coincide. This alternative is no longer promising.

4. THE GAP DIMENSION

We begin with a fractal dimension that is simple, but of narrow validity and interest. Since the similarity dimension $\log N/\log b$ is meaningless under self-affinity, it is tempting to "save it" formally, by replacing b by some suitable "effective base" b^*, and then to try and interpret the outcome. Taking for b^* the geometric mean of the b_m, namely $b^* = (b_1, b_2, ... b_E)^{1/E}$, let us show that $D_G = \log N/\log b^*$ is indeed a dimension in case one can define either gaps or islands. The formula for D_G is symmetric with respect to the bases, which is why D_G figures among guesses of what "the" D should be. We shall see that the other and more important D's are <u>not</u> symmetric.

The notion of gap dimension applies to the self-similar fractals in $\mathbb{R}^E$

exemplified by the Cantor dust on the line and by the Sierpinski gasket and carpet in the plane. These shapes have the following properties. Their measure in $\mathbb{R}^E$ vanishes ("fat fractals" - my book[2], chapter 15 - are excluded). Their complement splits into an infinity of <u>gaps</u> (maximal open sets) which are domains in $\mathbb{R}^E$, similar to each other and differing solely by their linear scale. In all these cases, it is known that the following relation holds for all L:

"number of gaps of linear scale > L" $\propto L^{-D_G}$ with $D_G < E$.

The exponent D_G is now called gap fractal dimension, and all other definitions of the fractal dimension of a self-similar fractal give the same value. Now consider self-affine fractals that have gaps. There is good news and there is bad news.

The good news is that it is still true that the number of gaps scales like N^k and volume scales like $b_1 b_2 \ldots b_E$. Define linear size as the (1/E)-th power of volume the geometric mean of the sides. If so, the number-size relation for gaps or islands continues to be a power law valid for all sizes L. The exponent independent of L can continue to be called gap dimension; it is the D_G defined earlier.

The bad news is that D_G bears no direct relation to D_{HB}. For example, consider the generator in Figure 4. Here, $D_G = \log 8/\log\sqrt{4 \times 6} = 1.30$, while Part III will show that $D_{HB} = 1.34$, and we shall soon see that the basic fractal dimensions are $D_{BL} = 1.38$ and $D_{MG} = 1.20$.

For thin fractal dusts (dusts of Lebesgue measure 0) in the special case E=1 (e.g., Cantor dust, my book, p.78),

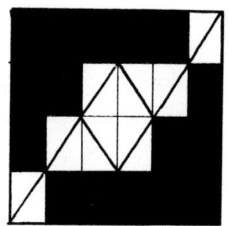

FIGURE 4

the gap dimension is a stronger form (valid for all scales) of an exponent that Besicovitch and Taylor introduced for small scales, and showed to be identical to D_{HB} (my book[2], p.359). The generalization of the identity $D_{BT} = D_{HB}$ to E>1 is known to be correct in some self-similar cases (Sierpinski carpet; my book, p. 134), but now we see that in the self-affine cases, D_G stands alone.

5. SELF-AFFINE PLANAR FRACTAL CURVES DEFINED AS RECORDS OF FUNCTIONS

Sections 5.1. and 5.2. are summary excerpts from my <u>Scripta</u> paper, which includes additional material and more detailed wording. Section 5.3. is new and important.

(One word about the use of the letter H for $\log_b,b"$. It is appropriate in this section, because H is the "Holder exponent" of $B_h(t)$ and other functions we study. However, I originally picked H in homage to H. E. Hurst; my book, Chapter 27).

5.1. The mass dimension's local value for small R is 2-H. For large R the mass dimension is 1.

When a set S is a self-similar fractal, the mass M(R) contained in the intersection of a S with a disc or ball of radius R behaves like $M(R) \propto R^{D_M}$. One can also replace the disc or ball by a square or cube whose sides are parallel to the axes and of length 2R. For some physicists, this property has become almost a definition of the notion of fractal. But this step is not justified. In fact, we must be careful about the meaning of the symbol ∝.

Most important is its meaning to the physicist, who thinks primarily of extrapolation to long radii, of straight log-log plots of mass versus radius, and of $\lim_{R\to\infty} \log M(R)/\log R = D_M$. Let us evaluate this limit for a record of any diagonally self-affine function, such as $B_H(t)$ or a recursively defined function. When $R \gg t_c$, any one of these records is effectively a horizontal interval. Within a square of side 2R, it occupies a very thin horizontal slice. Therefore, if we follow Section 2.5 and weigh our record proportionately to time elapsed, we find that $M(R) \propto R$. That is, we obtain the striking result that $D_{MG} = 1$. The subscript G stands for "global".

The mathematician, however, may also think of the case $R \ll t_c$. There, our record is effectively a collection of vertical intervals, one for each zero. Some algebra yields for the local mass dimension the value $D_{ML} = 2-H$, which is familiar as the D_{HB} of records of $B_H(t)$. The subscript L stands for "local".

Conclusion: We discover that two limits that are identical for self-similar fractals can differ. Besides, $1 < 2-H$ for all H in $0 < H < 1$.

The rest of the paper extends this discovery.

5.2. The box dimension's local value for large b is 2-H. The global value is 1

After a lattice made of boxes of side r=1/b is made to cover a set, let N(b) denote the number of lattice boxes that intersect the set. "Box dimension" is my present term for an exponent that characterizes sets for which N(b) behaves like $N(b) \propto b^{D_B}$. Box dimension is short for "box counting dimension". Its virues are to be short and to have no other meaning, while the alternatives do ("metric dimension" applies to all fractal dimensions, and "capacity dimension" has long been reserved for the Frostman dimension).

Again, what does this ∝ mean?

Most important is its meaning to the mathematician, who thinks of local behavior of the form $\lim_{b\to\infty} \log N(b)/\log b = D_{BL}$. In the case of recursively defined records, try to cover a piece $b'^{-k}=b^{-1}$ wide and b''^{-k} high, using boxes of side b^{-1}. Clearly, these boxes must be piled into vertical stacks. The number of boxes in each stark is $(b''/b')^{-k}=b^{1-H}$, and the number of stacks is b. Hence, $D_{BL}=2-H$. Similarly, in the case of $B_H(t)$, the heuristic box argument given in my book[2], bottom left of page 237 yields $D_{BL}=2-H$.

The physicist, however, may also think of the global limit $b \to 0$ or $r \to \infty$, which requires an unbounded record. The portion of a self-affine record from 0 to $t \gg 1$ is covered by a single box. Hence, $\lim_{b\to 0} \log N(b)/\log b = D_{BG} = 1$.

5.3. Behavior of the dimensions of $B_H(t)$ under section. Bounded locally self-similar fractals

As a rule (my book, p. 135), when a self-similar fractal in the plane is cut by a line, the fractal dimension goes down by one. The rule is fundamental but has numerous exceptions, To apply it to the record of $B_H(t)$, which is not self-similar, care is both needed and well rewarded. Assume that $B_H(t)$ has an intrinsic scale equal to one.

Horizontal cuts. They are self-similar and their local dimension also applies globally. It is 1-H=(2-H)-1, thus, it is obtained from D_{BL}, and the fact that $D_{BG}=1$ does not matter.

Vertical cuts. They reduce to 1 point, whose dimension used to count as exception to the rule.

Skew sections by Y=σt, with 0<σ<∞. When a cell of size $b'^{-k} \times b''^{-k}$ is upsized to a unit square, Y=σt is replaced by $Y=\sigma(b''/b')^{k}t$. Thus, Y=σt is locally indistinguishable from Y=0. The cut is locally self-similar, indeed locally identical to the horizontal cut.

BUT! Self-affinity has sensitized us to also check the global properties and the intrinsic scale.

Globally, a skew cut is <u>not</u> self-similar. As a matter of fact, it is bounded, therefore $D_{MG}=D_{BG}=0$. This is confirmed by observing that, globally, Y=σt scales up to a vertical line. Indeed, $D_{MG}=D_{BG}=0$ are the values obtained from the basic rule, if it is applied to the original curve's $D_{MG}=D_{BG}=1$.

I confess having been insensitive until now to the special status of bounded locally self-similar fractals. They are extremely common in nature, the best examples being individual island coastlines and DLA.

Now to the cut's intrinsic scale. For a bounded dust, the scale is the length of the smallest interval that contains it. As σ→∞, intrinsic scale → 0, which is why a vertical cut reduces to one point and the only thing that matters in the limit is the global dimension, which is 0. In other words, vertical cuts have now ceased to count as exceptions. As σ→0, intrinsic scale →∞, which is why the horizontal cut is unbounded self-similar. This explains why horizontal cuts are not at all affected by global quantities.

5.4. Behavior of the dimensions of the recursive self-affine fractals under section.

This topic, surprisingly complex and interesting, must be postponed to Part III.

6. SELF-AFFINE RECURSIVE PLANAR FRACTALS WHOSE PROJECTIONS FILL THE AXES

6.1. Projections of a self-affine recursive fractal on the axes

When a projection of the generator fills the corresponding side of the initiator, the limit fractal's projection fills the axis. When the projection of the generator fails to fill the side of the initiator, the limit fractal's projection is a dust that leaves uncovered gaps. In particular, when $r_n'>0$ and $r_n''>0$ for all the basic affinities, the limit fractal's projection is a Cantor dust. The above two cases are examined in Sections 6 and 7. Cases when $r_n'<0$ and/or $r_n''<0$ for some of the basic affinities cannot be discussed here.

6.2. The global mass dimension is $1-1/H+\log_{b''}N=\log_{b''}(Nb'/b'')$. The local mass dimension is $1-H+\log_{b'}N = \log_{b'}(Nb''/b')$

Again, we begin with the notion of greatest interest to physics: the extrapolative global mass dimension. On the advice of Section 2, we use a unit square as initiator, and attach a unit mass to it. The extrapolation (Section 2.6) is uniquely specified in a sequence of increasing boxes of mass N^k and area $(b'b'')^k$. This seems to point towards the gap dimension $D_G = \log N/\log b^*$ as mass dimension. But things are more complex, because the mass-radius relation requires the mass to be evaluated within square boxes, not specially adapted rectangular boxes. A square box of side $b"^k$, if chosen at random within a rectangular box of sides b'^k and $b"^k$, contains on the average the mass $N^k(b"/b')^k$. Hence, the surprising new result

$D_{MG} = \log(b"N/b')^k/\log b"^k$
$= 1+\log_{b"}(N/b') = 1-1/H+\log_{b"}N$

In the case of function records, Section 5, $N = b'$, hence this D_{MG} duly yields the already known value $D_{MG} = 1$.

A similar argument applied to local behavior yields

$D_{ML} = 1 - H + \log_{b'} N$.

Again, in the case of function records, Section 5, $N = b'$, hence this D_{MG} duly yields the already known value $2-H$.

The above formulas would be hard to guess, and their most striking feature lies in their being asymmetric in b' and $b"$, and symmetric of each other.

6.3. The global and local box dimensions take the same values as the corresponding mass dimensions

The formula for D_{BL} was obtained (implicitly) long ago by Kline[6]

6.4. One has $\underline{D}_{MG}=\underline{D}_{BG}<\underline{D}_G<\underline{D}_{BL}=D_{ML}$.

Proof: $D_{MG}<D_{BL}$ follows from $b'>b"$. Hence,

$\log(Nb'/b")/\log b' < \log(Nb"/b')/\log b"$

Averaging the numerators and the denominators separately yields a quantity that is strictly contained between D_{MG} and D_{BL}, and that is $\log N/\log\sqrt{(b'b")} = D_G$.

7. SELF-AFFINE PLANAR RECURSIVE FRACTALS AT LEAST ONE OF WHOSE PROJECTIONS IS A SIMPLE CANTOR DUST

7.1. Definitions and example

What is meant in this section's title is that, a) $r_n' > 0$ and $r_n" > 0$ for all n, and b), the X- and/or Y- projections of the limit fractal are Cantor dusts, made (respectively) of N' parts with $r' = 1/b'$ and of N" parts with $r" = 1/b"$.

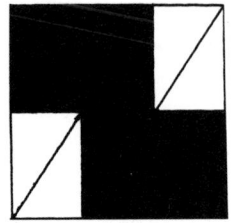

FIGURE 5

The very simplest example uses the generator in figure 5. It is clear that, globally, the resulting fractal is the original dust obtained by "midthirds removal". The global dimensions are, therefore, $\log_3 2$. Locally, on the other hand, the resulting fractal is the Devil's staircase, minus its flat steps. The local dimensions are known to be 1.

7.2. The global mass dimension is $D_{MG}= (1-1/H)\log_{b'}N'+\log_{b"}N=\log_{b"}(NN'^{H-1})$. The local mass dimension is $D_{ML}=(1-H)\log_{b'}N"+\log_{b'}N=\log_{b'}(NN"^{1/H-1})$

The argument concerning D_{MG} runs exactly like in Section 6.2., up to the point where the average mass in a box of side $b"^k$ is evaluated. The new feature is that this mass is now allowed to vanish, but massless squares cannot be part of our hierarchy of boxes. Therefore, it is necessary to exclude the massless boxes, and take a conditional average mass, which is larger than the average mass. Observe that $b'^{kH} = b"^k$. The strip of width b'^k and height $b"^k$ decomposes into $(b'/b")^k$ boxes of side $b"^k$, of which $N'^{k(1-H)}$ are not empty. Hence, the conditional average mass is $N^k N'^{1-k(1-H)}$, and

$$D_{MG} = -(1-H)\log_{b''} N' + \log_{b''} N$$
$$= (1-1/H)\log_{b'} N' + \log_{b''} N.$$

A similar argument applied to local behaviour yields

$$D_{ML} = (1-H)\log_{b''} N'' + \log_{b'} N$$
$$= (1/H-1)\log_{b'} N'' + \log_{b'} N.$$

The first (second) expression for D_{ML} is the exact symmetric of the second (first) expression for D_{MG}.

7.3. Except when $N=N'N''$, one has $D_{ML} > D_{MG}$
Writing $1/\log b'' - 1/\log b' = F$,

$$D_{MG} - D_{ML} = \log_{b''} N - (\log_{b'} N' - \log_{b''} N')$$
$$- \log_{b'} N - (\log_{b'} N'' - \log_{b''} N'')$$
$$= F(\log N - \log N' - \log N'')$$
$$= F\log (N/N'N'').$$

Since $F>0$ and $N<N'N''$, $D_{MG} < D_{ML}$, with equality only if $N=N'N''$, in which case our planar Cantor dust is the Cartesian product of two linear Cantor dusts. A linear Cantor dust is self-similar, thus the simplicity of the self-similar situation carries on to the self-affine case, when it is obtained as Cartesian product.

7.4. The global and local box dimensions take the same values as the mass dimensions

8. SELF-AFFINE SURFACES

This section comments briefly on two functions $Z(x,y)$, where the (x,y) plane is isotropic and there is one scaling parameter H, and on a function $T(x,y,z)$, when the (x,y,z) space is itself affine and there are two scaling parameters G and H.

8.1. Fractal functions of a variable in an isotropic plane. Relief.

My simplest model of the Earth's relief (my book, chapter 28) is a fractional Brown surface $B_H(x,y)$, the point of coordinates x and y being in an isotropic plane. Everything concerning this surface depends on the single parameter H. It is easily seen that $D_{BG} = D_{MG} = 2$, while $D_{BL} = D_{ML} = 3-H$. Also, $D_{HB} = 3-H$.

Dimensions' behavior under vertical or horizontal plane sections. Come back to the rule that, when a fractal is intersected by a plane, the dimension goes down by 1. This rule, again, is fundamental but with the reputation of having many exceptions; let us show how some of these can be ironned away.

The vertical sections of $B_H(x,y)$ have both local and global properties, and the rule applies both to D_{BL} and to D_{BG}, with no fuss.

The horizontal sections are the coastlines of all the islands taken together. They are self-similar and have only one dimension, which is the local dimension of vertical sections. Horizontal sections' intrinsic scale is infinite. Thus, a dimension that tells a lot about the horizontal sections, tells only half of the story about the vertical sections.

Dimensions' behavior under skew plane sections. As for the skew lines in Section 5.3., a skew plane $Z=\sigma x$ downsizes locally to a horizontal plane, and upsizes globally to a vertical plane. Both the local and the global dimensions are decreased by 1.

8.2. Fractal functions of a variable in an isotropic plane. Clouds/rain.

My fractal model of coastlines has been empirically shown by S. Lovejoy to extend to cloud boundaries' vertical projection on the Earth's surface. This has in turn led Lovejoy and Mandelbrot[7] to a two-dimensional model of rain areas or clouds. It is based on "fractal sums of pulses", a self-explanatory new term for a family of self-affine surfaces that I had introduced for other purposes. In the FSP model, some quantity (like temperature, opaqueness or rain intensity) is ruled by a self-affine function $Z_H(x,y)$, where the plane of the (x,y) is isotropic. The main mathematical contrast, compared to $B_H(x,y)$ as applied to relief, lies in the contrasting common experiences that a mountain's altitude is by and large a continuous function, while rainfall intensities are sharply discontinuous in time and space. In the simplest case[7], there is no parameter other than H.

8.3. Fractal functions of a variable in an affine plane. Clouds.

As everyone is aware, large clouds tend to be like pancakes parallel to the Earth's surface, and the conventional argument of the meteorologists is that the atmosphere is three-dimensional on small scales and two-dimensional on large scales, with a crossover scale in between. In a counter-argument, Schertzer and Lovejoy[8] argue from available empirical evidence that the atmosphere itself is self-similar in x and y, but self-affine in x (or y) and z. I view this suggestion as excellent, and I like the way op. cit. adapts various of my models to make them self-affine, or more fully completely self-affine.

The dimensional properties of the corresponding fractals are therefore worthy of exploration. Unfortunately, op. cit.[8] quotes numbers with scant motivation, or none, and shows no awareness of the interesting complications the topic presents.

With little cost, one can immediately consider self-affine functions $T(x,y,z)$, where the horizontal variables (x, y, and z) are isotropic. The basic self-affinity property is invariance under a map whose diagonal terms can be written as r, r, r^G, and r^{GH} with $G<1$. In addition, using the awkward but self-explanatory notation of op. cit., one has $\Delta T(\Delta x) \sim (\Delta x)^{GH}$, $\Delta T(\Delta y) \sim (\Delta y)^{GH}$, and $\Delta T(\Delta z) \sim (\Delta z)^H$, we find that $H<1$. It is easily found that for the record of T, $D_{MG}=D_{BG}=3$ irrespectively of H and G. However, other dimens easily found that for the record of T, $D_{MG}=D_{BG}=3$ irrespectively of H and G. However, other dimensions of the record, and the dimensions of other objects related to T, usually depend on the object itself, and on H and/or on G.

For example, consider D_{BL}. In a box of side $\Delta x=\Delta y=\Delta z \ll 1$, $\Delta T(\Delta x)$ is dominated by $\Delta T(\Delta z) \sim (\Delta z)^{GH}$. Therefore, covering the record of T by boxes of side $\Delta x=\Delta y=\Delta z$ requires $(\Delta x)^{-3}$ stacks with $\sim (\Delta z)^{GH-1}$ boxes in each stack. Conclusion: $D_{BL}=4-GH$.

8.4 Coverings by rectangles, and the × "elliptical dimension".

In the case $E=3$, $b_1=b_2$ and $b_3=b_1^H$, op. cit.[8] gives prominence to the quantity 2+H, which it calls "elliptical dimension of space", D_{el}. The motivation is that in the isotropic 3-D case, $D_{el}=3$, and in the isotropic 2-D case, $D_{el}=2$; "it is therefore natural to regard 2+H as the fractal dimension of this" self-affine space.

This attractive motivation does not suffice. It is, moreover, weakened by the second supporting argument, which notes that the case $E=2$ and $b_2=b_1^H$ with $H=1/2$ yields "$D_{el}=1.5$, which is the same as the fractal dimension suggested" [sic] by me for the record of B(t). We know, however, that for $B_H(t)$, one has $D_{HB}=2-H$ rather than $1+H$; these two formulas take the same value for $H=1/2$ because of a numerical coincidence.

The first supporting argument is that "the number of eddies of horizontal scale λ may be written as λ^{-D}" with $D=D_{el}$. But considering the vertical scale would give $D=1+2/H$; why choose the horizontal scale?

A search for a clearcut interpretation of $2+H$ as dimension has involved private conversations with J. P. Kahane and J. Peyriere, who suggest checking on "intrinsic" coverings that do not use cubes but affine rectangles b''^{-k} high and b'^{-k} wide, the "radius" of a rectangle being its longer side. Local Hausdorff-like dimensions of this kind are discussed by Peyriere[9].

The concrete physical meaning of covering by rectangles is as yet unproven. Its adoption would of course involve additional local and global dimensions, many of which are found to take on very questionable values. For example, the mass dimensions in the case studied in Section 6 become $\log_{b'} N$ globally and $\log_{b''} N$ locally. Both values yield a very biased and incomplete view of these fractals' structure. Specifically, take a self-affine Sierpinski carpet with $b'=9$, and $b''=3$ and one big gap leaving in $N=20$. Its global mean dimension based on intrinsic rectangles is $\log_{b'} N = 1.36$. The same values continue to apply if b'' is replaced by any integer from 3 to 9 (inclusive). The carpets obtained in this fashion greatly differ from each other, except from this peculiar viewpoint. To put this dimension in perspective, note that $D_G=1.81$, $D_{ML}=D_{BL}=\log_b, N+1-H=1.86$ and $D_{MG}=D_{BG}=\log_{b''} N+1-1/H=1.72$. In op. cit.[8], however, $\log_b, N$ is given (without explanation) as the only fractal dimension of this carpet.

In the limit case $N = b'b''$, the main dimensions based on rectangles simplify to $1+2/H$ locally and $2+H$ globally. This last value might provide, after-the-fact, an element of motivation for "the elliptic dimension". But this motivation is no better than the motivation for the unbounded "twin" value $1+2/H$.

Observe also that, in order to verify the standing of this global $2+H$ and this local $1+2/H$ as dimensions, one must know in advance which boxes one should use in the covering, which requires advance knowledge of H. Every dimension based on the common square boxes can be measured by direct algorithms.

9. COMMENT ON "PHYSICAL" EXTRAPOLATION VERSUS "MATHEMATICAL" INTERPOLATION

The constructions and procedures that fractal geometry has borrowed from mathematics all involved infinite interpolation. In physics, on the other hand, interpolation cannot proceed without end, and constructions tend to proceed by extrapolation. At the time of the first uses of fractals in physics, in 1980, this contrast was often brought to my attention by physicists. In the self-similar case, and to the mathematicians' and physicists' surprise, the infinitesimal techniques extend via power law relationships valid uniformly at all scales. In the self-affine case, the two basic procedures involve different tools.

FRACTALS IN PHYSICS
L. Pietronero, E. Tosatti (editors)
© Elsevier Science Publishers B.V., 1986

SELF-AFFINE FRACTAL SETS, II: LENGTH AND SURFACE DIMENSIONS

Benoit B. MANDELBROT

Physics Department, IBM Research Center
Mathematics Department, Harvard University, Cambridge, MA 02138 USA

For a self-similar curve, one is able to estimate the fractal dimension by "walking a divider. It is shown that for self-affine curves, this procedure yields a local and a global value, both doubly anomalous. Other problems raised by length and area measurement for fractals are investigated.

1. INTRODUCTION

It is tempting to measure a curve's length by "walking" an increasingly narrowly opened divider along it, or a surface's area by triangulating it increasingly finely. On standard curves, the procedure works well. On standard surfaces (e.g., a cylinder's) anomalies are known to arise; the main one is the Schwarz area paradox, which deserves to be widely known and is recalled below. On self-similar curves, this procedure leads back to the fractal dimension. Let us know explore it for self-affine fractals, and show that the exponents it yields differ from the mass/box dimension.

2. MEASURING THE LENGTH OF SELF-AFFINE FRACTAL CURVES OBTAINED AS RECORDS OF FUNCTIONS

2.1. The measurement of length using the Minkowski sausage yields local and global dimensions identical to D_{ML} and D_{MG}

In the spirit of Minkowski and Bouligand, one defines a curve's approximate length $B(\eta)$ via the "Minkowski sausage": all the points within a distance η of a point on the curve. For a standard rectifiable curve and $\eta \ll 1$, $B(\eta)=(2\eta)^{-1}$ (area of the sausage). For a self-similar curve (my book[2], p. 36), $B(\eta) \sim \eta^{1-D}$. For a self-affine curve, the area of the sausage behaves for small η as $N(\eta)\eta^{-2} \sim \eta^H$, hence the local sausage dimension is 2-H. The global dimension is 1. Both values are familiar from Part I.

2.2. The measurement of length by walking a divider and monitoring last exits yields local and global dimensions identical to D_{ML} and D_{MG}

One of the many methods of measuring the length of a rectifiable curve is "to walk a divider" along it. The curve may have knots, i.e., multiple points of any order; it suffices that its points be ordered, say "in time". Pick a starting point, P_0. The first point P_1 will be the first exit of the curve from the circle of center P_0 and radius η, etc... If $L(\eta)$ denotes the length of the resulting broken line approximation, the curve's length is $\lim_{\eta \to 0} L(\eta)$.

One may as well select for P_1 the last, rather than the first exit along the curve. And one may move backwards.

For a self-similar curve, one finds $L(\eta) \propto \eta^{1-D}$, and again one can pick either the first or the last exit at will.

For our self-affine curves, the situation is quite different. In addition to the local dimension for $\eta \to 0$, there is a global dimension, which we shall see is equal to 1. And the local divider dimension has two totally different versions: one for last and one for first exits.

Before we proceed, note that arguments on self-similar records are made simpler, and the results are unchanged, if the circle around P_k is replaced by a square.

If so, last exits become a simple matter. Cover our curve with $(b"^k)^{2-H}$ square boxes of side $b"^{-k} \ll 1$; this yields $D \geq 2-H$. Next, add a ring of 8 like boxes all around each cell to multiply the side by three. Clearly, $(b")^{2-H}$ "divider steps" of size $3b"^{-k}$ are sufficient to walk along the curve. Therefore, the divider dimension is $\leq 2-H$. Consequently, it is $= 2-H$.

2.3. The measurement of length by walking a divider and monitoring first exits gives "anomalous" dimensions. The divider dimension's local value for small η is $1/H$. This is the fractal dimension of a fractal trail related to the record. For large η, the divider dimension is 1.

This section is summarized from my Scripta paper, which unfortunately uses the gallicism compass for divider.

When $\eta \gg t_c$ (e.g., when the unit of B_H is very small), the record is effectively a horizontal line. The divider walked along the curve remains mostly parallel to the t-axis, and $L(\eta)$ varies little with η. If one insists on writing $L(\eta) \propto \eta^{1-D}$, the fact that $L(\eta)$ is constant yields 1 as global dimension, irrespective of H.

When, to the contrary, $\eta \ll t_c$ (e.g., when the unit of B_H is huge), the situation is different: a divider walked along the curve remains mostly parallel to the B-axis. This yields $1/H$ as dimension.

This extremely strange value can exceed 2 and is doubly anomalous insofar as it contradicts the value $2-H$ of the other local definitions of a fractal dimension. On the other hand, those familiar with the fractional Brownian motion will identify $1/H$ as being the fractal dimension of the trail (as drawn in an E-dimensional Euclidean space R^E, with E satisfying $E > 1/H$) of a motion whose E coordinates are independent realizations of $B_H(t)$.

In this case, an attempt to take an unusual path to measure "the" fractal dimension for one set actually ends up by measuring the value all paths yield for a different set.

2.4. Affine box dimensions

This section relates length measurements to a line of thought started in Section 8 of Part I. In both limit cases, $\eta \gg 1$ or $\eta \ll 1$, the number of divider steps, $L(\eta)/\eta$, is for all practical purposes a number of rectangular cells $\eta = b"^{-k}$ high and b'^{-k} wide, used to cover our fractal. In the customary definitions of dimension, cells are squares and numbers of cells are written as functions of cell diameter. This wording can be extended to $L(\eta)/\eta$ if the diameter of a rectangular cell is taken to be its longer side. Locally, the longer side is vertical, and one finds, as in Section 2.3., the dimension $1/H$. Globally, the longer side is horizontal, and one finds the dimension 1.

3. MEASURING THE LENGTH OF OTHER SELF-AFFINE CURVES, INCLUDING PEANO MOTION TRAILS

The only interesting case is the counterpart of the divider walking argument in Section 2.3.

Local value. Walking a divider of length $b''^{-k} \ll 1$ will take N^k steps, hence the approximate length exponent is $\log_{b''}(b''N^{-1}) = 1 - \log_{b''}N$, and the dimension is $\log_{b''}N$. In particular, the Peano case $N=b'b''$, gives the dimension $1+1/H$.

Global dimension. It is $\log_{b'}N$ and, in the Peano case, it is $1+H$.

4. THE SCHWARZ AREA PARADOX

Triangulating ordinary surfaces is far more difficult than anyone expects. For example, it was shown by Hermann Amandus Schwarz in the late 1800's that in the case of the unit cylinder of radius 1 and height unity, seemingly innocuous triangulation methods can yield any value from the true value 2π to infinity!

Proceed as follows: divide the height into n layers by the planes $z=p/n$ (p being an integer>0), divide the circumferences of odd-numbered levels by the points $\theta=(2q+1)\pi/m$ (q an integer), and divide the circumferences of even-numbered levels by the points $\theta=2q\pi/m$. Join each point (z, θ) to the points $(z\pm1/n, \theta\pm\pi/m)$. In this fashion, the unit cylinder is triangulated by $2mn$ equal triangles. To obtain the "true area", one is tempted to add these triangles' areas and then to let $m\to\infty$ and $n\to\infty$ independently in any fashion whatsoever.

Straight algebra shows that for large m, the approximate area behaves like $2\pi\sqrt{[1+(\pi^4/4)n^2/m^4]}$. If $m\to\infty$, but $n/m^2\to0$, the approximation converges indeed to 2π. However, if $m\to\infty$ and $n=\lambda m^2$ with λ a constant >0, one can obtain any finite limit > 2π that one wishes!

And we may add that, by selecting $\eta\sim m^\beta$, with $\beta>2$, one can make the approximate area increase as any power one pleases of $1/m$, of $1/n$ or of the triangle area $\sim 1/mn$. A cylinder mimicks a fractal! And is infinitely boundable.

Reason: to let $m/n\to\infty$ is to use triangles that a) become increasingly "thin", i.e., have at least one angle that $\to 0$, and b) lie in planes that tend to become orthogonal to the cylinder. The resulting approximation is in fact increasingly "corrugated" and increasingly removed from the actual surface.

The pragmatist's reaction: keep away from thin triangles. The mathematicians' response: the "Schwartz area paradox" has been one of the "triggers" of "modern" mathematics. For example, it has stimulated Minkowski to his safe definition of length and area via the volumes of increasingly thin Minkowski "sausages" of curves and of increasingly thin Minkowski "comforters" of surfaces. These are the sets made of all points within ϵ of a point on the curve or surface. Then, Minkowski defines the area of an ordinary surface as $\lim_{\epsilon\to\infty}(1/2\pi)$ (volume of the ϵ-comforter).

Contrary to triangles, all intervals are alike, hence for an ordinary curve in the plane, the analog to the Schwarz paradox is absent. It is also absent for self-similar fractal curves, since my book points out that length measurements to various precisions ϵ can be performed in many different ways, but all approximations grow at the same rate ϵ^{1-D}. But for self-affine curves, Sections 2.1. to 2.3. have proven that the situation is more complex. While the length grows like ϵ^{1-D}, one has $D=D_{BL}$ via the Minkowski approach, but $D=D_{CL}>D_{BL}$ via the divider-walking approach. Can one make D take values other than the two described above?

5. MEASURING THE AREA OF SELF-AFFINE FRACTAL SURFACES OBTAINED AS RECORDS OF FUNCTIONS

5.1. Area of a fractal relief $B_H(x,y)$ via the Minkowski comforter

One falls back on the dimensions D_{BL} and D_{BG}.

5.2. Area of a fractal relief via triangulation

Begin with square tiles of x's and of y's, with $\Delta x=\Delta y=1/b$. Each cell's 4 vertices define 4 values of B_H and yield two ways of approximating a piece of area by two "twin triangles". Take the average of the two approximations for each cell and add the averages for the b^2 cells.

Rough triangulation. If one neglects the details below the critical values $x_c=y_c$, my Brown model of Earth's relief is close to having a well-defined area, not much higher than the area of its projection over an idealized plane (or sphere). This stands in sharp contrast with all I have argued about island coastlines.

Consider in this light the two nonGaussian landscapes in my book, Plate C13. They are obtained from the same Gaussian landscape by non – linear transformations that are meant to insure that t_c is very small in the valley of C13 top and on the "mesa" of C13 bottom, but very high in the "sierra" of C13 top and the "canyon" of C13 bottom. Also, I like to point out in lectures that good airports' runways are as rough as the Himalayas – but their vertical scale is smaller. We see now that these quantitative differences have qualitative effects. First of all, as suggested by the eye and "common sense", an airport does have a well defined area for most sensible yardsticks. As to the Himalayas, the usual photographs taken from far away suggest that "overall slopes" are about $\pi/4$. This in turn suggests that there is interesting detail in the crossover region; hence, different measures of their area obtained with different yardsticks less than t_c should be expected to fall upon a curve whose doubly logarithmic graph is definitely <u>not</u> straight.

Fine triangulation. The area will be arbitrarily large, for sure, but how rapidly will it grow with triangle "size"? Each twin triangle within a cell has sides of length $\sim b^{-Hk}$, but a height of size $\sim b^{-k}$; it is very thin, and its area is $\sim b^{-(H+1)k}$. The total number of triangles is $b^{2k}=\alpha^{-2/(H+1)}$, and approximate area $(\alpha) \sim \alpha^{1-2/(H+1)}$. This is the counterpart to the relation $L(\eta) \sim \eta^{1-1/H}$ for curves, but the anomalous dimension is now $2/(H+1)$ instead of $1/H$.

The next grid to consider is itself self-affine and involves $(b'b'')^k$ rectangles b'^{-k} wide and b''^{-k} high, with $b'>b''$. Each triangle's area now becomes $\sim (b''^{-1}b'^{1-H})^k$. The anomalous dimension is now $\log(b'b'')/\log(b''b'^H)$. It can range between $2/(H+1)$ and $1/H$, a fractal form of the Schwarz area paradox.

FRACTALS IN PHYSICS
L. Pietronero, E. Tosatti (editors)
© Elsevier Science Publishers B.V., 1986

SELF-AFFINE FRACTAL SETS, III:
HAUSDORFF DIMENSION ANOMALIES AND THEIR IMPLICATIONS

Benoit B. MANDELBROT

Physics Department, IBM Research Center
Mathematics Department, Harvard University, Cambridge, MA 02138 USA

For certain self-affine fractals constructed recursively, the Hausdorff-Besicovitch dimension D_{HB} takes a doubly "anomalous" value: it is a fraction and it is less than D_{BL}. The question is raised, does this discrepancy point toward deep new developments, or does it cast doubt upon the special standing D_{HB} has held until now in fractal geometry? The paper also comments on section dimensions for certain self-similar fractals.

1. INTRODUCTION

Part I does not mention the value that the Hausdorff-Besicovitch dimension D_{HB} takes for recursively constructed self-affine fractals, because this D_{HB} raises interesting and partly wide open issues. Our point of departure will be a theorem due to McMullen[3,4], which shows that, for given b', b", and given generator, certain combinations of the signs of r_n' and r_n'' yield $D_{HB} < D_{BL}$. This positive "anomaly" $D_{BL} - D_{HB}$ came to me as a surprise. Additional results on these and related structures are given by Bedford.[5] We shall, however, either prove or argue that other combinations of signs — including some random combinations — yield $D_{HB} = D_{BL}$. The mathematical implications of this variability will be seen — unexpectedly — to involve certain fractal measures that I have introduced in 1974, in the study of turbulence[10], and that have since proven invaluable in the study of strange attractors[11].

Conceptual issues will also be discussed; they may mean that D_{HB} is not a physical notion, contrary to what I used to believe.

As background, we include some results, that seem novel, concerning cuts of certain self-similar fractals.

2. A THEOREM YIELDING D_{HB} AND COROLLARY

<u>Theorem</u> A[3,4]. Consider a recursive self-affine fractal generator in a lattice, with b' and b" the horizontal and vertical bases, and $H = \log b''/\log b'$. Assume $r_n' = \pm 1/b'$ and $r_n'' = 1/b''$ for all n (this means that all arrows point up). Denote by b'_j the number of cells contained in the j-th horizontal row of the generator. Then the limit fractal's D_{HB} is the solution of
$$b''^D = \sum b'^H_j.$$
<u>Corollary</u> B[3,4,5]. For the self-affine fractals covered in Theorem A, D_{BH} may be strictly less than the local mass-box dimension discussed in Part I.

Numerical example where $D_{HB} < D_{BL}$. Use the generator of Figure 2 in Part I. Here, $b'=4$, $b''=2$, $H=1/2$, $b'_1=1$ and $b'_2=3$, and the r_n' (i.e., the arrows) insure continuity. Theorem A yields $2^D = 1 + \sqrt{3}$, hence $D_{HB} = 1.44998$. But $D_{BL} = 2 - H = 1.5$.

Observation. In this example, the Y-projected measure is singular. Indeed, it is the well-known Besicovitch measure (my book, p.377) with $p_1 = 1/4$ and $p_2 = 3/4$.

3. EXPRESSION FOR THE VERTICAL ANOMALY $A'' = D_{BL} - D_{HB}$

The "anomaly" $A'' = D_{BL} - D_{HB}$ is an intrinsic measure of dispersion of the non-vanishing values of b'_j. For example, the stick generator of the record of a continuous function may either make a few large swings or many small ones; the discrepancy is larger in the second case. (The reader is encouraged to construct specific illustrations.)

First, we express the anomaly A'' in terms of $p_j = b'_j / N$, where the notation is obviously intended to bring to mind probabilities corresponding to the Y-projected measure. The dimensional anomaly A'' takes the form

$$A'' = (1-H)(\log_{b''} N'' - I''_H),$$

where $I''_H = (\log_{b''} \sum p_j^H)/(H-1)$.

The term $\log_{b''} N''$ is the dimension of the set that supports the Y-projected measure: the interval when $N'' = b''$, but a Cantor dust when $N'' < b''$. As to I''_H, it is familiar to the reader acquainted with the Besicovitch measure. It is the H-information dimension of the Y-projected measure.

Next, we note that the Besicovitch measure is a special case of a random fractal measure[10], often called M-measure (also called "multiplicative measure", or "Mandelbrot measure" in the terminology of Peyriere; discussed as "nonlacunar weighted curdling measure" in my book, p. 378). To prepare this generalization — and for its own sake — it is useful to adopt the same notation, $W_j = b'_j b'' / N = p_j b''$ and $\langle W^h \rangle = \sum b''^{-1} W_j^h$. This notation attributes equal probabilities to b'' possible values of a (random) "weight" W that satisfies $\langle W \rangle = 1$. Now,

$$A'' = -\log_{b''} \langle W^H \rangle = -\log_{b'} (\langle W^H \rangle^{1/H}) > 0.$$

Corollary C^4. When $b' = b''$, a necessary and sufficient condition for the anomaly to vanish, is that all the b'_j that do not vanish be identical.

Observation. The Y-projected measure is uniform on its support when $A'' = 0$ is singular on its support when $A'' > 0$.

4. HORIZONTAL CUTS' DIMENSION

4.1. Background: horizontal cuts of certain recursive self-similar fractals. Expressions for their anomaly.

To appreciate the next results, it is necessary to understand fully the corresponding results relative to the self-similar fractals that correspond to $b' = b''$. These results seem new, and are also interesting in themselves. Specifically, it is known that for the ordinary Sierpinski carpet, the horizontal cuts' dimensions are <u>not</u> obtained from the overall dimension $\log_3 8$ by subtracting 1. These cuts will be called "anomalous". To study their dimensions, let us place a uniform mass on the carpet, each construction stage spreading it into 8 pieces of density 9/8. Clearly, the X- and Y- projected measures are both Besicovitch measures, with $b = 3$, $p_1 = 3/8$, $p_2 = 2/8$, and $p_3 = 3/8$. Let me show that cut dimensions are related to these measures.

To carry out the argument, take general values of b, N and $p_j=b_j/N$, not excluding the possibility of some $p_j=0$. Write the intersecting horizontal line's ordinate in base b as $y=0.y_1y_2\ldots$ and let k_j be the number of repetitions of j in the first k digits. Define $\beta(y_h)$ to equal b'_j if $y_h=j$. In the k-th approximation, the horizontal line of ordinate y intersects a number of cells of side b^{-k} that is equal to the product of the $\beta(y_h)$ from h=1 to h=k. Thus, the k-th approximation to the horizontal cut has the finite dimensional exponent
$$\log N_k(y)/\log(b^k) = \sum(k_j/k)\log_b b_j.$$
When y is such that $k_j/k \to q_j$ for every j, this expression has a limit, the limit is a box dimension, and that dimension depends on y. Now, choose a point on our fractal with uniform measure, i.e., choose y with the Y-projected Besicovitch singular measure. In this case, $(k_j/k) \to p_j$. Subtracting and adding $\sum(k_j/k)\log_b N-1$ to both sides yields the result that almost surely the cut's dimension is
$$D = (\log_b N - 1) + (1-I),$$
where $I = -\sum p_j \log_b p_j$.

Since $\log_b N$ is the dimension of our original planar fractal, $\log_b N-1$ is the dimension given by the rule "subtract one". We see that the rule fails: there is an anomaly that depends on the 1-information dimension of the Y-projected measure, and that is either =0 or >0.

If (and only if) I=1, which requires $p_j \equiv N/b$, the anomaly vanishes.

If $p_j=1/N''$ for N'' values of j and $p_j=0$ for the other values, the anomaly is $1-\log_b N''$. In that case, the Y-projected measure is uniform over a Cantor dust. The horizontal cut is a.s. empty with respect to the uniform measure on [0,1], but with respect to the uniform measure on the Cantor dust the horizontal cut is a.s. of dimension
$$\log_b - \log_b N''.$$
This generalizes the standard rule in very interesting fashion to cuts that are conditioned to be nonempty.

4.2. Horizonal cuts of certain recursive self-affine fractals

With obvious changes, the results are parallel to those of Section 4.1. We shall sketch them, then comment on them in Section 4.3.

When the anomaly A'' vanishes, the horizontal cuts of the fractal in Theorem A are either empty or Cantor dusts of dimension $\log_{b'}(N/N'')$. When $N''=b''$, the cut is never empty and is of dimension
$$\log_{b'}(N/b'') = D_{BL} - 1,$$
which fulfills the standard rule about cuts' dimensions. When N''<b'' and the cut is conditioned to be non-empty, it is of dimension
$$\log_{b'}(N/N'') = (D_{BL} - 1) + (\log_{b''} N''),$$
which expresses the generalized standard rule of Section 4.1. in self-affine terms. But one can also write
$$\log_{b'}(N/N'') = D_{BL}^* - 1,$$
which fits the standard rule in a manner appropriate for self-affinity, by replacing D_{BL} by D_{BL}^*, the dimension the fractal takes after it has been squeezed vertically to eliminate its gaps.

When A''>0, almost surely the cut's dimension is
$$D = \log_{b'} N - \sum p_j \log_{b'} p_j.$$
It is best to restate the right hand side in terms of dimensions. Since
$$\log_{b'} N = D_{ML} - (1/H-1)\log_{b'} N''$$
$$= D_{ML} - (H-1)\log_{b''} N'',$$
$$D = D_{BL} - (1-H)\log_{b''} N'' - HI''_1,$$

where
$$I''_1 = -\sum p_j \log_{b''} p_j = 1 - \langle W \log W \rangle$$
is the 1-information dimension of the Y-projected measure. It follows that the cut's dimensional anomaly takes the form
$$D - (D_{BL} - \log_{b''} N'') = H[\log_{b''} N'' - I''_1].$$
This is >0, and is $> H[\log_{b''} N'' - I''_H]$, because $I''_1 < I''_H$.

4.3. Summary

In the self-similar case, the whole curve shows an anomaly (i.e., $D_{BH} = D_{BL}$), even where the horizontal cuts a.s. show an anomaly (i.e., D_{cut} is not $D_{whole} - 1$).

In the self-affine case, the whole shows an anomaly if and only if the horizontal cuts a.s. show an anomaly.

5. A GLOBAL COUNTERPART, D^*, FOR D_{HB}, AND THE HORIZONTAL ANOMALY $A' = D^* - D_{BG}$

It was observed in Part I that the formulas for $D_{BL} = D_{ML}$ and $D_{BG} = D_{MG}$ are symmetric of each other: obtained by exchanging the roles of b' and b'', and of N' and N''. Love of symmetry led me immediately to seek a global counterpart to D_{BH} in the solution D^* of the equation
$$b'^D = \sum b''^{1/H}_j.$$
The corresponding "horizontal anomaly" is taken with respect to D_{BG}, and is given by $A' = D_{BG} - D^* = -\log_{b''} \langle W^{1/H} \rangle < 0$.

The case of function records. here, $D^* = 1$, a value Parts I and II have made familiar for all the other global dimensions. The anomaly $A' = 0$, and the vertical cuts' dimension is 0, as it should be. The anomaly A' plays for vertical cuts the role that A'' plays for horizontal cuts but does D^* have anything else to recommend itself as dimension?

6. SELF-AFFINE CONTINUOUS RECORDS NOT COVERED BY THEOREM A

Example where $D_{HB} = D_{BL}$ and Question. The two generators shown on Figures 2 and 3 in Part I, only differ by the direction of the arrow placed on the second stick.

Figure 3, used recursively, yields a fractal that is the $X(t)$ coordinate record of a well-known Peano curve due to Cesaro, called "triangle sweep" in my book, p. 64. The $Y(t)$ coordinate record is $1 - X(1-t)$. In this case, it has been shown[4] that $D_{HB} = 1.5$.

Observe that the X-projected measure is the same for the record $X(t)$ and for the Peano trail in the XY plane. Since the Peano trail fills ("sweeps") uniformly a triangle, we find the projected measure to be differentiable, of density $2(1-x)$.

More generally, suppose that the Y-projected measure has right and left derivatives $f'(y+)$ and $f'(y-)$. A heuristic application of the formula in Theorem A to the generator after k stages yields this formula, with the sum from $p=1$ to $p=b''^k$.

$$b''^{kD} = \sum_{\angle y = pb''^{-k}} (\Delta f(y))^H b''^k$$
$$\sim \sum b''^{-k} (f'(y+))^H b''^{(2-H)k}$$
$$\sim b''^{(2-H)k} \int (f'(y+))^H dy.$$

When $k \gg 1$,
$$D = 2 - H + (1/k) \log_{b''} \int (f'(y+))^H dy,$$
Asymptotically, $D = 2 - H$.

The same result extends, obviously, to the case where the Y-projected measure is absolutely continuous on a Cantor dust of y's.

Reminder. For the process $B_H(t)$, $D_{BH} = 2 - H$, hence $A'' = 0$. Also, it is known that the Y-projected measure is differentiable. Observe that, for random processes, the abcissas of the horizontal level cuts through a level y form the set of recurrences to the point y (also called "local time").

Question: Is there a one-to-one relation between A being =0, and the H-information dimension of the Y-projected measure being the dimension of its support?

6.2. Random generators.

Randomly selected generators are manageable when they involve geometrically imbedded birth processes. (The study of these processes deserves to be pushed beyond the properties sketched here.)

The simplest function $M_H(t)$ in my Scripta paper[1]. My "pedagogical" function $M_H(t)$ is obtained by positioning all 4 arrows on Figure 2 or 3 at random, with equal probabilities for "up" and "down". A partly heuristic argument to be given momentarily suggests that the Y-projected measure of $M_H(t)$ is differentiable. Assuming that the question at the end of Section 6.1. should be answered by the affirmative, this measure's differentiability suggests that $D_{HB}=D_{ML}=2-H$ (as asserted without evidence in the Scripta paper, before I knew that the issue is a difficult one).

The first half of the argument is rigorous. The mass in a y interval of length b''^{-k} of the y-coordinate is contributed by rectangles b'^{-k} wide and b''^{-k} high of weight $N^{-k}=b'^{-k}$. Their number is given by a simple birth random process with an average of $N/b'' = b'/b''>1$ offspring per generation. Consider for each y the sequence of nested b-adic intervals of length b''^{-k} that defines y (if y is not b''-adic) or the sequences that define y+ and y− (if y is b''-adic). The average measures over the intervals in one of these sequences are of the form $(b'/b'')^{-k} \times$ (number of offspring at the k-th generation). A standard theorem on birth processes (13, p.12) tells us that these average measures almost surely converge to a limit, which is the value of a strictly positive random variable W, satisfying $\langle W\rangle=1$.

Therefore, $\mu(y)$, defined as the cumulative measure between 0 and y, has the property that over the nested intervals of length $(b'')^{-k}$ that define y, y+, or y−, the average slope of $\mu(y)$ converges to a limit.

The next step becomes much easier to state for $b'=4$, $b''=2$, and $N=4$. Select a finite $k\gg 1$ and consider the k-th approximate measure $\mu_k(y)$ over a dyadic interval (y', y'') of length 2^{-k}. In the left half of this interval, the number of offspring at stage $k=1$ is the sum of $2^k W$ independent random variables of expected value 2 and finite variance σ^2. This number can be written as $2^{k+1}W+(2.2^k W)^{1/2}\sigma G$, where G is a normalized Gaussian variable. In the right half of the same interval, the same holds with + replaced by −. Therefore, the slopes of $\mu_{k+1}(y)$ over the two halves of our binary interval are $W\pm\sqrt{W}G(\sqrt{2})^{-k-1}$. The approximate measure $\mu_{k+1}(y)$ is obtained from $\mu_k(y)$ by midpoint displacement (my book, Chapter 26). That is, the midpoint value $\mu_k((y'+y')/2)$ is to be displaced by the amount $2^{-k-1} \times \sigma\sqrt{W}G(2\sqrt{2})^{-k-1}$. When k is very large, one can neglect variations of W between construction stages, and one finds for $\mu(y)$ the series

$\mu(y)\sim\mu^*(y)=\mu_k(y)+\sigma\sqrt{W}\sum_{h>k}a^h\sigma(2^h y)$, where $a=(2\sqrt{2})^{-1}$ and $\sigma(y)$ is a random sawtooth function, namely a function that vanishes for integer y's, takes independent normalized Gaussian values at half integer y's, and is linear over dyadic intervals of length 2^{-k-1}. The resulting function $\mu^*(y)-\mu_k(y)$ is familiar. For $a>1/2$, it is continuous but nondifferentiable (and serves in several rough fractal algorithms meant to model mountains). For $a<1/2$, hence for our present value $a=(2\sqrt{2})^{-1}$, the function is continuous and right and left differentiable. (In fact, the derivative $\mu^{*'}(y+)$ is near Brownian; it is a variant of the Rademacher series close to the Fourrier series of the Brown-Wiener process).

I expect — though I had no time to check all details — that the heuristics in the last paragraph can be made rigorous, or there is a rigorous short argument.

Conclusion: $\mu(y)$ is differentiable: the above W is/are its derivative for non-b"-adic y's or its right or left derivatives for b"-adic y's. (At the b"-adic points, the right and left derivatives are negatively correlated.)

6.3. More general $M_H(t)$-like random functions.

The reason why the above discussion is written in terms of b', b", and N, instead of 4,2 and 4, is that the same argument holds more generally, as long as the stick generators that yield continuous records are assigned certain special probabilities. (When b">1 and b'≫1, randomly generated stick generators will do.)

For other probability assignments, however, the situation is more complex. An interval of length $b"^{-k}$ may be nested in either of b" "locations" within an interval of length $b"^{-k+1}$, and the expected number of offspring usually depends on the "location". The corresponding Y-projected measure is <u>not</u> expected to be differentiable.

Conjecture. Scattered examples make me expect that both $D_{HB}<D_{BL}$ a.s., or $D_{HB}=D_{BL}$ a.s. can be achieved by recursive self-affine continuous random records, and that $\Delta"=D_{BL}-D_{HB}$ is a continuous function of the probabilities allocated to the acceptable stick generators.

7. OTHER RANDOM SELF-AFFINE SETS NOT COVERED BY THEOREM A

7.1. Generators obtained by curdling with N/b">1.(resp., N/b">)

The idea of selecting the generator completely at random can take one of several forms. One can attribute equal probabilities to every way of drawing N among the b'b" cells. When all the choices are statistically independent, the ultimate fractal is obtained by the self-affine counterpart of the version of the process of "microcanonical" version of the curdling that I introduced in 1974 (my book, chapter 10). Alternatively, one can give the probability p to choosing each cell to be part of the generator. Each generator then includes pb'b" cells on the average. When all the choices are statistically independent, the ultimate fractal is obtained by the self-affine counterpart of the "canonical" version of the process of curdling that I also introduced in 1974 (my book, chapter 10).

In either case, the expected number of offspring per generation is N/b" (resp., pb'). The birth process becomes a birth and death process, but the argument of Section 3 remains generally valid if $N/b''>1$ (resp., when $pb'>1$). The novelty is that the derivative can now be zero with a probability that is >0 and <1. That is, in a given sample, the derivative may vanish over some intervals of y.

7.2. Generators obtained by curdling with $N/b''<1$ (resp., $pb'<1$).

In the combined generator after k stages, $b_j'=0$ for most values of j. In a first examination, let us disregard these values and consider only the j such that $b_j>0$. A standard theorem in the theory of birth processes (4, p. 12) is that the conditional distribution of b_j', knowing that $b_j'>0$, tends for $k\to\infty$ to the distribution of a limit random variable. If we denote this limit by W, like in section 3, the measure carried by a non-empty interval of length b''^{-k} is again the product of W by the measure N^{-k} of a cell $b''^{-k} \times b'^{-k}$. The average number of non-vanishing b'_j is, therefore, $N^k/\langle W\rangle$. In the limit $k\to\infty$, the y- projected measure is carried by a fractal dust of dimension $\log_{b''} N$ (resp., $\log_{b''}(pb'b'')$). This could have been guessed. What was less easy to guess is that, on this dust, the distribution of the measure is near-uniform, namely, uniform except for the factor W. This shows the situation to be parallel to that in Section 7.1., with the exception that $\langle W\rangle >1$ here.

Alternatively, one can study the measure over a fractal dust obtained as follows: At each stage, pick those cells of length b''^{-k} in which $b'_j>0$, plus any number of empty cells needed to add up to N. This amounts to "diluting"

W to allow $W=0$ and make $\langle W\rangle=1$. The resulting situation is parallel to that encountered earlier in this section.

Either way, the heuristic use of Theorem A suggests that D_{HB} is arbitrarily close to D_{BL}. That is, $D_{HB}=D_{BL}$.

Conjecture. I expect $D_{HB}=D_{BL}$ to hold widely for randomly generated self-affine sets that are not constrained to be records of functions.

Comment. Canonical curdling generates a special planar M-measure (see Section 2). It would be interesting to investigate the projections of more general planar M-measures, both in the self-similar and in the self-affine cases.

8. DISCUSSION

The value of D_{HB} yielded by Theorem A is usually "anomalous" because it is a fraction, but this has been exorcised by fractal geometry. The second anomaly is due to its being the wrong fraction. But Theorem A had not been contrived for this purpose. To the contrary, earlier known "second anomalies" had been specifically contrived, for example, there were highly non-uniform, like the Bouligand anomaly in Section 1 above. This is why, in every previous case of interest to physics, the fractal dimension could first be obtained by some rough and ready method, usually based on D_{BL} or D_{ML}, and later "confirmed" by more elaborate and technical calculations of D_{HB}.

Second observation: when dealing with records of functions such as B(t), it is natural to attach equal measures to records that correspond to time intervals of equal duration. This property is satisfied in the case of

B(t) by the Hausdorff measure relative to a suitable gauge function. Does such a gauge function exist for the recursive self-affine functions covered by Theorem A? If it does, the resulting Hausdorff time is not real time.

My first gut reaction was to view the assignment of arrows (signs to the r'_n and r''_n) in the self-affine fractal construction as being a "non-physical" fine detail, hence a quantity that depends on this assignment could not be physical. Next, history seemed to repeat itself, when this fine detail turned out to affect the (unique) a.s. dimension of the horizontal cuts, which is meaningful physically. But then the argument of Section 4.1. came to mind, so the anomaly in D_{HB} again looks non-physical. Even if it should eventually reveal some useful new physical intuition, I now fear that the Hausdorff Besicovich definition has lost its earlier "special standing."

ACKNOWLEDGMENTS FOR PARTS I, II, AND III

I: Sections 5.1. and 5.2. (from my Scripta paper) have greatly benefited from discussions with Richard F. Voss.

The main reason for undertaking this work arose when I was teaching fractal geometry (spring 1985). When discussing the usual continuous differentiable functions, which are self-affine, I saw that, while the subject was already defined in my 1977 book, little was known about it. Its investigation eventually grew to explain the strange results or theoretical difficulties encountered by self-affine models of surfaces (Part II) and of clouds (Section 7.3.).

II: Section 2.3., the first to be written, arose from puzzlement at the published estimates of various rough surfaces' fractal dimensions. This section was part of my Scripta paper. Penetrating comments by Michael V. Berry stimulated me to expand it.

III: Conversations with Curt McMullen have been extremely valuable.

REFERENCES FOR PARTS I, II, AND III

1. B. B. Mandelbrot, Physica Scripta
2. B. B. Mandelbrot, The Fractal Geometry of Nature (Freeman, New York, 1982)
3. C. McMullen, Nagoya Math. J. 96 (1984)1.
4. C. McMullen (to appear).
5. T.J. Bedford. Crinkly Curves, Markov Partitions and Dimension. Ph.D. Thesis (Warwick University, U.K. 1964).
6. S.A. Kline, J. London Math. Soc. 20(1945)79.
7. S. Lovejoy and B. B. Mandelbrot, Tellus A37 (1985) 209.
8. D. Schertzer and S. Lovejoy, in Turbulent Shear Flow 4 (ed. L.J.S. Bradbury et al.)
9. J. Peyrière, Bulletin Soc. Math. Fr. 114(1986).
10. B. B. Mandelbrot, J. Fluid Mechanics 62(1974) 331. Also C. R. Acad. Sc. (Paris) 278A (1974) 289 and 355.
11. H. G. E. Hentschel and I. Procaccia, Physica 8D (1983) 435.
12. T.H. Harris. The Theory of Branching Processes (Springer, Berlin, 1963).

FRACTALS IN PHYSICS
L. Pietronero, E. Tosatti (editors)
© Elsevier Science Publishers B.V., 1986

RANDOM FRACTALS, FLOW FRACTALS AND THE RENORMALISATION GROUP

John MELROSE

Department of Chemistry, Royal Holloway College, Egham, Surrey, TW20 OEX, U.K.

The renormalisation group is demonstrated to be a powerful tool for both the analysis and construction of ensembles containing self-similar fractals. The ensembles are hierarchical in that renormalisation can be carried out in a finite parameter space. At fixed points of the RG the self-similar constructions are found, away from fixed points non self-similar constructions are found and termed flow-fractals here. An example the Ising Snowflake is used to motivate discussion. Ensembles which are subsets of well known models are easily constructed, the construction of hierarchical SAW ensembles is described.

1. INTRODUCTION

Mandelbrot[1] introduced random fractal constructions to aid the modelling of natural fractals. Such constructions can generate an ensemble of configurations and it is useful to introduce a partition (generating) function[2] with weights conjugate to appropriately chosen basic shapes making up the possible configurations. In this work such partition functions are formed from a direct combination of the familiar iterative constructions with a simple renormalisation group scheme. Within this scheme fractal transfer matrices[3] arise naturally.

2. CONSTRUCTION

The ensembles and partition functions introduced here will be hierarchical:
let $Z_n(\bar{r})$ be the partition function for the ensemble, at the nth iteration, then the ensemble is said to be hierarchical, if $Z_n(\bar{r})$ obeys the recursion

$$Z_n(\bar{r}) = Z_{n-1}(R(\bar{r}))K(\bar{r}) \qquad (1)$$

where $\bar{r}$ is some finite set of q weights, $R(\bar{r})$ is a renormalisation recursion relation and q and $K(\bar{r})$ are independent of n (this a direct generalisation of the definition of hierarchical lattice).

An example the Ising Snowflake[4] will motivate discussion. This construction is an iterative decoration of boundaries separating black and white regions in the plane. Starting from a black triangle in a white background, at each step all black triangles are subdivided into nine sub-triangles. Then each boundary of a subdivided triangle is independently decorated according to some distribution of possible decorations.

Figure 1 shows the possible decorations on a triangle with one boundary; those for 2 and 3 bounded triangles are the obvious direct product of these[4]. Black regions are shaded.

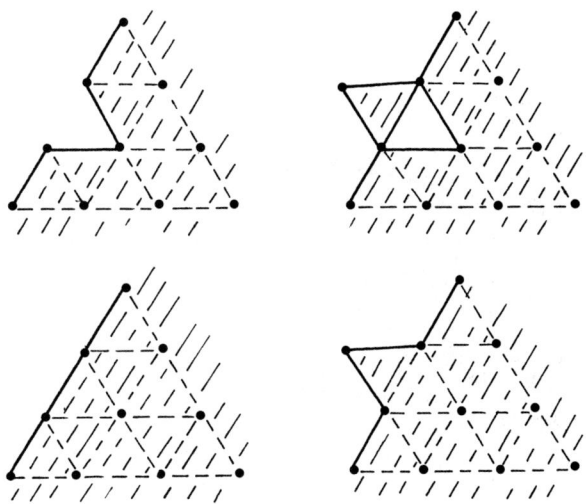

FIGURE 1

To select the decoration distribution weights r_0, r_1, r_2, r_3 are introduced conjugate to the basic shapes making up the configurations: black triangles with 0 to 3 boundaries, all white triangles are given weight unity.

The distribution functions for all possible decorations of each basic shape are the renormalisation recursion relations. Let a general multinominal recursion relation be:

$$r_i' = \Sigma_j \, b^i_{j_0\ldots j_{q-1}} \, r_0^{j_0} \ldots r_{q-1}^{j_{q-1}} \qquad (2)$$

where i denotes the weight in $\bar{r}$. Construction of nth level configurations at weight $\bar{r}$ proceeds as follows: the recursion (2) is iterated n times, $\bar{r}_1 = R(\bar{r})$, $\bar{r}_2 = R(R(\bar{r}))\ldots \bar{r}_N$; the notation r_{ip} will be used for the ith weight in r_p. Then starting from an initially chosen basic shape an iterative decoration is carried out with at the kth iteration each basic shape, say i^{th} type, being decorated with decorations, say j^{th}, chosen, afresh for each occurence of each shape with probability

$$(b^i_{j_0\ldots j_{q-1}} \, r_{0\,n-k}^{j_0} \ldots r_{q-1\,n-k}^{j_{q-1}})/r_{i\,n-k+1} \qquad (3)$$

The construction is simply the reverse of the renormalisation.

At fixed points of (2) the construction described above is self-similar (3 is constant) and *random fractals* are generated, away from fixed points the non-self-similar configurations constructed by following some renormalisation group trajectory are here termed *flow fractals*.

If all the b coefficients in (2) are unity all configurations with the same numbers of each basic shape are equiprobable, for general choices of the b's this is not so. For the snowflake with b's unity an nth level configuration is given a weight.

$$r_0^i r_1^j r_2^k r_3^l / Z n(r_0 \ldots r_3) \qquad (4)$$

if it contains respectively i to l black triangles with 0 to 3 boundaries. Recursion relations can be found by decorating in all ways each basic shape and weighting appropriately. With the b's unity one finds for the snowflake $r_0' = r_0^9$ and from figure 1

$$r_1' = r_0^6 r_1^3 + r_0^7 r_1^2 r_2 + r_0^4 r_1^4 + r_0^4 r_1^2 r_3 \qquad (5)$$

relations for r_2' and r_3' are given in Melrose[4].

A subtle point has passed unmentioned: the set of decorations in figure 1 are chosen such that any protrusion is not allowed within a distance from the end matrices of the decoration equal to its own height, hence under decoration of different boundaries multiple edges are not generated. However the intersection of vertices under different decorations is allowed, but in the weighting of the ensemble vertex sets are ignored. To this degree both geometric realisation and hierarchical construction are both satisfied.

3. EXPECTATIONS

Expectations formulated as derivatives of the partition function[2] can be found via familiar R.G. matrix algebra. Let some quantity of interest, Q_x, be associated with a conjugate field x and $r_i = f_i(x,\phi)$ with ϕ denoting some set of other fields also present in the problem, then

$$\langle Q_x \rangle_n = (x \, dZ_n(\bar{r}(x))/dx)/Z_n(\bar{r}(x)) \qquad (6)$$

Using the chain rule and the hierarchical property (1) one finds after rearrangement:

$$\langle Q_x \rangle_n = \bar{V}_n \prod_{m=1}^{n} \bar{T}_m \bar{V}_0 \qquad (7)$$

where the ith element of $\bar{V}_n$ obeys $(V_n)_i = \delta_{ik}$, with k the index of the basic shape used to initiate decoration, $(V_o)_i = x\, dr_{i,o}/dx$, and $\bar{T}_m$ is a set of q x q matrices.

$$(\bar{T}_m)_{ij} = \{r_{jm-1}\, (dR_i(\bar{r})/dr_j|_{\bar{r}=\bar{r}_{m-1}}\}/r_{im}$$

$\bar{T}_{mij}$ is the expected number of jth type basic shapes introduced on an ith basic shape at the n-m+1th decoration. $\bar{T}_m$ is the natural generalisation both to an ensemble and flowing fractals of the fractal-transfer-matrix introducted recently[3].

For the snowflake example a choice of fields

$$r_o = r,\ r_1 = rs,\ r_2 = rs^2,\ r_3 = rs^3 \qquad (8)$$

has r conjugate to the black area and s conjugate to boundary length.

Some expectations of interest not formulated as derivative of Z_n may be calculated as a sum of products of Tm's; expectations of vertex sets on the snowflake construction are an example of this.

4. SUBSETS OF KNOWN MODES

It is straightforward to construct ensembles as above with weighting (4) and (8) which are subsets of known models. The snowflake with weighting (4) and (8) is a subset of the low-temperature graphs of an Ising model on a hexagonal lattice[4]. However the snowflake subset is a poor approximation to the full model. The phase diagram has a fixed point on the invariant subspace r = 1 in (8). The black areas of the configurations have dimensionality $D_A = 2$ for all parameter values and the phase transition is between sinks at (r=1, s→o) and (r=1, s→∞) with boundary dimensions $D_b=1$ and $D_\infty=2$ respectively. Eigenvalues and eigenvectors of the snowflake and a more general construction without constant D_A are described in ref. 4. The fixed point has more than the 2 relevant fields expected in the full model. Critical singularities are modulated by periodic amplitudes[4]. In the snowflake construction $k(\bar{r})$ of (1) is unity.

SAW hierarchical ensembles are straight forward to construct. Construction is based on a square grid, decorations of independent edges are chosen which do not intersect at vertices. Considering the sufficient case of two edges at right angles, to be independent self avoiding walk decorations meeting at (A) can not pass through any of the diagonal bonds indicated by dashed lines in figure 2. Recursion relations can be found by enumerating all SAW's crossing clusters as shown in figure 3.

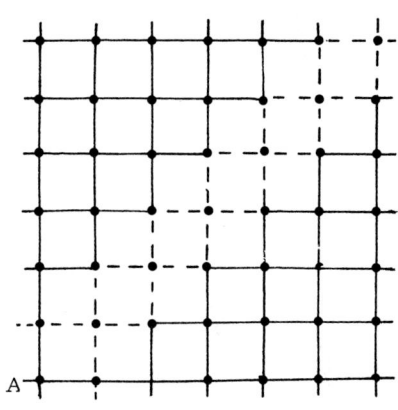

FIGURE 2

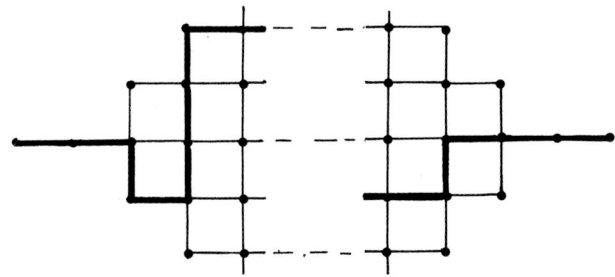

FIGURE 3

CONCLUSIONS AND OUTLOOK

The elementary steps required to unite random fractal constructions with traditional ensemble analysis via renormalisation have been described. It was noted that such hierarchical ensembles consisting of subsets of well known models can easily be found; this may be useful in constructing more controlled real space renormalisation approximation.

ACKNOWLEDGEMENT

The author thanks the S.E.R.C. for a postdoctorial research scholarship and Mrs. J. Evans for typing.

REFERENCES

1. B.B. Mandelbrot, The Fractal Geometry of Nature, (Freeman: San Francisco 1982).

2. W. Feller Introduction to Probability Theory and its Applications Vol (Wiley: New York 1950).

3. B.B. Mandelbrot, Y, Gegen, A. Aharony and J. Peyrière, J.Phys.A: Math Gen. 18 (1985), 335.

4. J.R. Melrose, submitted for publication 1985.

FRACTALS IN PHYSICS
L. Pietronero, E. Tosatti (editors)
© Elsevier Science Publishers B.V., 1986

ON FINITELY RAMIFIED FRACTALS AND THEIR EXTENSIONS

R. HILFER and A. BLUMEN

Lehrstuhl f. Theoretische Chemie, Techn. Universität München, Lichtenbergstr. 4, D-8046 Garching and Max-Planck-Institut f. Polymerforschung, Jakob-Welder-Weg 15, D-6500 Mainz, Germany (FRG)

We construct deterministic fractal lattices using generators with tetrahedral symmetry. From the corresponding master equation we determine the spectral dimension $\tilde{d}$ and prove that $\tilde{d}<2$. Furthermore we extend our set of fractals (with $\tilde{d}$ dense in [1,2]) by direct multiplication, thus obtaining fractals whose $\tilde{d}$ are dense in $[1,\infty[$.

It was suggested that many disordered media are fractal structures.[1] This motivates us to study deterministic fractal lattices by investigating random walks on them. We concentrate on lattices with finite order of ramification.[1,2] These are characterized by the property that the elimination of a preassigned finite number of lattice bonds is sufficient to isolate an arbitrarily large compact subset of the infinite structure.

We start from the class of finitely ramified fractals having tetrahedral symmetry whose best known representative is the Sierpinski-gasket. We briefly describe their construction and give their fractal (Hausdorff-) dimensions. Analysing the master equation we then turn to their spectral (fracton) dimension. Finally we consider also a class of infinitely ramified fractals obtained by direct multiplication.

Deterministic fractal lattices are completely described through a geometrical generator and an iteration prescription. Figure 1 shows a variety of two-dimensional (d=2) generators whose sidelength is called b. The Sierpinski-gasket corresponds to the special case b=2. In Figure 2 we exemplify one step of the iterative construction for the generator with b=5 of Figure 1. Infinite repetition yields the full fractal lattice. A finite number of iterations will be called a stage-n-structure, where n=1 corresponds to the generator.

b=5   b=7   b=9

FIGURE 1
Three generators for finitely ramified fractals

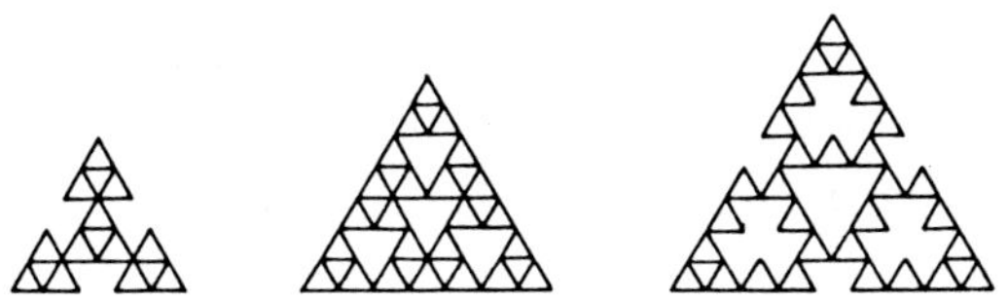

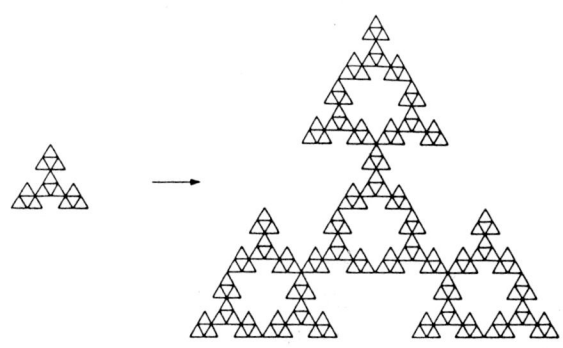

FIGURE 2
One step in the iterative construction of the fractal

We now turn to the Hausdorff-dimension $\bar{d}$ of the fractal[1] which may be expressed as $b^{\bar{d}} = \lim_{n\to\infty} N(n+1)/N(n)$ where N(n) is the number of lattice points in a stage-n-structure. From the relation N(n+1) = N(1) + N[N(n)-d-1] we have

$$\bar{d} = \ln N/\ln b \qquad (1)$$

where N denotes the number of upward pointing triangles (resp. hypertetrahedrons) inside a generator. Note that the number $E(n)$ of edges in a stage-n-structure obeys the recursion relation

$$E(n+1) = N\, E(n) \qquad (2)$$

which allows to determine $\bar{d}$ through $E(n)$.

To obtain the spectral dimension $\tilde{d}$, we use the master equation describing a nearest neighbour random walk on the lattice. Suppose the walker starts at site $\vec{r}_0$ at time $t=0$. The probability $P(\vec{r}_i,t)$ to find the walker at site $\vec{r}_i$ at time t obeys the master equation

$$\frac{d}{dt} P(\vec{r}_i,t) = \sum_{j(i)} [w_{ij} P(\vec{r}_j,t) - w_{ji} P(\vec{r}_i,t)] \qquad (3a)$$

where the sum runs over all $\vec{r}_j$ that are nearest neighbours to $\vec{r}_i$. We specify the transition rates w_{ij} from $\vec{r}_j$ to $\vec{r}_i$ through

$$z(\vec{r}_j) w_{ij} = w = \text{const} \qquad (3b)$$

where w is a constant rate and $z(\vec{r}_j)$ is the number of nearest neighbours of $\vec{r}_j$. Laplace-transforming Eq.(3) with initial condition $P(\vec{r}_i,0) = \delta_{\vec{r}_i,\vec{r}_0}$ gives

$$(1-\alpha)P(\vec{r}_i,u) - \sum_{j(i)} P(\vec{r}_j,u)/z(\vec{r}_j) = \delta_{\vec{r}_i,\vec{r}_0}/w \qquad (4)$$

where we have set $\alpha = -u/w$. Due to the self-similarity of our lattices Eqs.(4) may be solved via a decimation procedure which inverts the iterative construction. Only fractals with connected generators will be considered here. For these every point of the generator can be reached from any other point through a succession of bonds in which two consecutive bonds have one point in common.

To be specific we write Eqs.(4) restricted to a single generator whose corners $\vec{s}_0, \vec{s}_1, \ldots, \vec{s}_d$ survive the decimation step. The deleted interior sites are labelled $\vec{d}_1, \ldots, \vec{d}_M$ ($M=N(1)-d-1$) beginning with the d nearest neighbours of $\vec{s}_0$ such that $\vec{s}_0, \vec{d}_i$ and $\vec{s}_i$ are collinear ($1 \le i \le d$). Introducing the vectors $\vec{Q}_1 = (Q(\vec{d}_1,u), \ldots, Q(\vec{d}_M,u))$ and $\vec{Q}_2 = (Q(\vec{s}_0,u), \ldots, Q(\vec{s}_d,u))$ with $Q(\vec{r}_i,u) = P(\vec{r}_i,u)/z(\vec{r}_i)$ we get for the single generator

$$[(1-\alpha)D - A_1]\vec{Q}_1 = A_2 \vec{Q}_2 \qquad (5)$$

Here D is the diagonal $M \times M$-matrix given by $(D)_{ii} = z(\vec{d}_i)$. The matrices A_1 and A_2 are submatrices of the adjacency matrix A of the generator. The $M \times M$-matrix A_1 is obtained from A by eliminating the rows and columns corresponding to the corners, while the $M \times (d+1)$-matrix A_2 results from elimination of $d+1$ rows corresponding to the corners and M columns corresponding to interior sites.

Before proceeding with the decimation we have to analyse the invertibility of the matrix $(1-\alpha)D - A_1$. First we note that the matrix $D - A_1$ is diagonally dominant, i.e. that we have $|(D-A_1)_{ii}| \ge \sum_{j \ne i}^M |(D-A_1)_{ij}|$ for all $1 \le i \le M$. This can be seen as follows: The element $(A_1)_{ij}$ equals 1 or 0 depending on whether the point $\vec{d}_i$ is connected to $\vec{d}_j$ or not. Thus $|(D-A_1)_{ii}| = z(\vec{d}_i)$ and $\sum_{j \ne i}^M |(D-A_1)_{ij}| = \sum_{j \ne i}^M (A_1)_{ij}$. Moreover the row sums $\sum_{j \ne i} (A_1)_{ij}$ give the number of nearest neighbours of $\vec{d}_i$ which are also interior points of the generator and hence equal at most $z(\vec{d}_i)$. In addition $D-A_1$ is irreducible which means that there is no permutation matrix B such that $B(D-A_1)B^{-1}$ reduces to block form. This is equivalent to the connectedness of the generator.[3] Knowing that $(D-A_1)$ is an irreducibly diagonally dominant matrix with positive diagonal elements and nonpositive off-diagonal elements, we infer from the theory of nonnegative matrices that its inverse $(D-A_1)^{-1}$ exists and is elementwise positive.[3] Furthermore, since for $u \ge 0$ we have $\alpha \le 0$, it follows that $1-\alpha \ge 1$ and therefore also $[(1-\alpha)D - A_1]^{-1}$ exists and fulfills

$$([(1-\alpha)D - A_1]^{-1})_{ij} > 0 \qquad (6)$$

for all i,j and $\alpha \le 0$.

We can now solve Eq.(5) for one of the nearest neighbours of $\vec{s}_0$, say $\vec{d}_1$, to obtain

$$Q(\vec{d}_1,u) = g(\alpha)Q(\vec{s}_0,u) + \sum_{i=1}^{d} h_i(\alpha)Q(\vec{s}_i,u) \quad (7a)$$

with

$$g(\alpha) = ([(1-\alpha)D-A_1]^{-1}A_2)_{11} \quad (7b)$$

and

$$h_i(\alpha) = ([(1-\alpha)D-A_1]^{-1}A_2)_{1,i+1} \quad (7c)$$

Using the rotational symmetry of the tetrahedral generator the same result obtains for all nearest neighbours of $\vec{s}_0$. The calculation is then repeated for the l generators to which $\vec{s}_0$ belongs. We now write Eq.(4) for $P(\vec{s}_0,u)$ as

$$-\delta_{\vec{r}_0,\vec{s}_0}/w + ld(1-\alpha)Q(\vec{s}_0,u) = \sum_{k=1}^{l} \sum_{i=1}^{d} Q(\vec{d}_i^k,u) \quad (8)$$

where the different generators are distinguished by upper indices k, $1 \leq k \leq l$. For the $Q(\vec{d}_i^k,u)$ we insert the results from Eq.(7) and get

$$-\delta_{\vec{r}_0,\vec{s}_0}/[wh(\alpha)] + ld(1-\phi(\alpha))Q(\vec{s}_0,u) = \sum_{k=1}^{l} \sum_{i=1}^{d} Q(\vec{s}_i^k,u) \quad (9)$$

with $\phi(\alpha) = 1 - \{[1-\alpha-g(\alpha)]/h(\alpha)\}$ and $h(\alpha) = \sum_{i=1}^{d} h_i(\alpha)$. Eq.(9) involves only $\vec{s}_0$ and the corners of adjacent generators, thus completing the decimation.

We proceed to show that Eqs.(8) and (9) for $\vec{s}_0$ in the full resp. decimated lattice are indeed identical in the limit $\alpha \to 0$. Consider a stage-n-structure for large n. The stationary solution ($t \to \infty$) is then

$$Q(\vec{r}_i) = P(\vec{r}_i)/z(\vec{r}_i) = \text{const} \quad (10a)$$

as can be seen from Eq.(3). The constant follows from conservation of probability $\sum_{i=1}^{N(n)} P(\vec{r}_i,t) = 1$,

$$\text{const} = 1/\sum_{i=1}^{N(n)} z(\vec{r}_i). \quad (10b)$$

Inserting (10a) into Eq.(7) gives $g(0)+h(0) = 1$ which in turn means that $\phi(0)=0$ via Eq.(9). With $\sum_{i=1}^{N(n)} z(\vec{r}_i) = 2 E(n)$ where $E(n)$ is the number of edges in the stage-n-structure we insert Eq.(10) into Eq.(8) for $\vec{r}_0 = \vec{s}_0$ to obtain

$$-1/w + ld(1-\alpha)/2E(n) = ld/2E(n)$$

valid for $n \to \infty$ and $\alpha \to 0$. After one decimation step this becomes Eq.(9)

$$-1/[wh(\alpha)] + ld[1-\phi(\alpha)]/2E(n-1) = ld/2E(n-1)$$

Eliminating w from these equations we get $h(\alpha) = \alpha N/\phi(\alpha)$ and after taking the limit $\alpha \to 0$ we are left with

$$h(0) = N/\kappa \quad (11)$$

where $\kappa = \phi'(0)$. This equality has been called fractal Einstein relation[4]. Consider now Eq.(9) for α around 0 where we have $\phi(\alpha) \cong \kappa\alpha$. In this region Eq.(9) for the renormalized quantities $NQ(\vec{s}_i,u)$ and rates w/κ is identical to Eq.(8) for the original quantities Q and w.

We now determine the spectral (fracton) dimension $\tilde{d}$ which for random walks follows from the probability to be at the origin $P(\vec{r}_0,t)$. For longer times one has[5]

$$P(\vec{r}_0,t) \sim (1/wt)^{\tilde{d}/2}$$

which after decimation reads

$$N P(\vec{r}_0,t) \sim (\kappa/wt)^{\tilde{d}/2}$$

and thus

$$\tilde{d} = 2 \ln N/\ln \kappa = 2(1-\ln h(0)/\ln N)^{-1} \quad (12)$$

where Eq.(11) was used for the second equality. In previous works we have evaluated $\tilde{d}$ explicitly for many structures[6,7] and we have also shown how to construct fractals for prescribed $\tilde{d}$-values[7], $1 \leq \tilde{d} \leq 2$, densely filling the interval [1,2].

Here we note that for the above fractals $\tilde{d} < 2$. From inequality (6) plus the fact that A_2 is nonnegative we have, using Eq.(7), $g(\alpha) > 0$ and $h(\alpha) > 0$ for all α. Since $g(0)+h(0)=1$ it follows that $g(0),h(0)<1$. Thus $-\infty < \ln h(0) < 0$, while $\ln N > 0$ because of $N > 1$. This implies $(1-\ln h(0)/\ln N)^{-1} < 1$ and from Eq.(12) therefore $\tilde{d} < 2$.

Finally we extend the types of fractals by direct multiplication and thereby provide a dense set of $\tilde{d}$-values in $[1,\infty[$. As an example we show in Figure 3 the stage 2 result of multiplying a Sierpinski-gasket with a one-dimensional lattice. We call this the "Toblerone"-lattice[8]. Its spectral dimension is obtained from the low-frequency behaviour of its eigenmodes. In this model one envisages the lattice

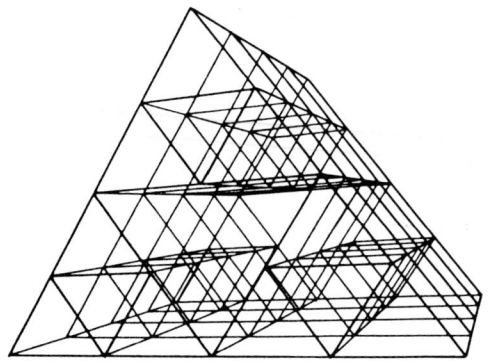

FIGURE 3
The Toblerone lattice: direct product of a Sierpinski-gasket with a linear chain

sites as occupied by masses $mz(\vec{r}_i)$ connected along the bonds through springs of strength f. For low frequencies ω the density of states obeys[5,9] $\rho(\omega) \sim \omega^{\tilde{d}-1}$. The equations of motion for the Fourier-transformed displacements $P(\vec{r}_i,\omega)$ are just Eqs.(4) without the inhomogeneity. Furthermore, α corresponds to $\alpha = m\omega^2/f$. Writing explicitly $\vec{r}_i = (x_i, y_i, z_i)$ we have

$$-\alpha P(x_i,y_i,z_i,\omega) = \sum_{j(i)} [P(x_j,y_j,z_j,\omega) - P(x_i,y_i,z_i,\omega)] \quad (13)$$

where we note that there are six nearest neighbours, $z(\vec{r}_i)=6$, on the Toblerone lattice. We can split the summation over nearest neighbours in Eq.(13) into a linear and a Sierpinski part as

$$\alpha P(x_i,y_i,z_i) = 2P(x_i,y_i,z_i) - P(x_i,y_i,z_i+1) - P(x_i,y_i,z_i-1) + \sum_{j(i)} [P(x_i,y_i,z_i) - P(x_j,y_j,z_i)] \quad (14)$$

where we assumed the lattice spacing to be one and have suppressed the ω-dependence for notational ease. Fourier-Transformation with respect to z, $P(x,y,k) = \Sigma_z e^{ikz} P(x,y,z)$, yields

$$(\alpha - 2 + 2\cos k) P(x_i,y_i,k) = \sum_{j(i)} [P(x_i,y_i,k) - P(x_j,y_j,k)] \quad (15)$$

For fixed k Eqs.(15) are exactly the equations for the Sierpinski-gasket if we take as spectral parameter $\alpha_k = \alpha - 2(1-\cos k)$. Since we are interested in the long wavelength limit $k \to 0$ we expand $\alpha_k \cong \alpha - k^2$. Thus from $\alpha = m\omega_T^2/f$, $\alpha_k = m\omega_S^2/f$ and $k^2 = m\omega_C^2/f$ we have $\omega_T^2 = \omega_S^2 + \omega_C^2$ where subscripts T, C or S refer to the Toblerone lattice, the chain or the Sierpinski gasket. To compute $\rho_T(\omega)$ we now count the number of modes with frequencies less than ω, $N_T(\omega) = \int_0^\omega \rho_T(\omega')d\omega'$. Since Eqs.(15) uncouple with respect to k we obtain $N_T(\omega)$ by summing $\rho_S(\omega')d\omega' \rho_C(\omega'')d\omega''$ subject to the condition $\omega'^2 + \omega''^2 \leq \omega^2$. Using $\rho_S(\omega) \sim \omega^{\tilde{d}_S-1}$ and $\rho_C(\omega) \sim$ const we get

$$N_T(\omega) = \iint_{\omega'^2+\omega''^2 \leq \omega^2} \rho_S(\omega')\rho_C(\omega'')d\omega'd\omega'' \sim \omega^{\tilde{d}_S+1}$$

Differentiation yields $\rho_T(\omega) \sim \omega^{\tilde{d}_S}$ and therefore $\tilde{d}_T = \tilde{d}_S + 1$ for the spectral dimension of the Toblerone lattice. Evidently one obtains higher dimensions $\tilde{d}$ by multiplying our fractals with higher dimensional regular lattices. The obtainable values are dense in $[1,\infty[$.

Summarizing we have concentrated here on general properties of finitely ramified fractals. For a general class we have shown that $\tilde{d} < 2$. In addition we have indicated how to build infinitely ramified fractals whose $\tilde{d}$-values are larger than 2 and form a dense set. -amdg1-

ACKNOWLEDGEMENT

We thank the Fonds der Chemischen Industrie and the Deutsche Forschungsgemeinschaft for support.

REFERENCES

1. B.B.Mandelbrot, The Fractal Geometry of Nature (Freeman, San Francisco, 1982)

2. P.Urysohn, Verhandelingen der Koninklijke Akademie te Amsterdam XIII No 4 (1927)

3. R.S.Varga, Matrix Iterative Analysis (Prentice Hall, Englewood Cliffs,1962)

4. J.A.Given and B.B.Mandelbrot, J.Phys.A 16 (1983) L565

5. S.Alexander and R.Orbach, J.Physique Lett. 43 (1982) L625

6. R.Hilfer and A.Blumen, J.Phys.A 17 (1984) L537

7. R.Hilfer and A.Blumen, J.Phys.A 17 (1984) L783

8. compare A.Maritan and A.L.Stella, Stat. Mech. of self-avoiding random surfaces, this volume

9. R. Rammal and G. Toulouse, J.Physique Lett. 44 (1983) L13

Part II
ANALYSIS OF FRACTAL PROPERTIES OF MATERIALS

STRUCTURE OF RANDOM SILICATES: POLYMERS, COLLOIDS, AND POROUS SOLIDS*

D.W. SCHAEFER and K.D. KEEFER

Sandia National Laboratories, P.O. Box 5800, Albuquerque, New Mexico 87185, USA

Small angle x-ray scattering and light scattering are used to characterize structures grown by random processes within the silica system. Dense colloids, rough colloids, and branched polymers are grown by polymerization in solution. Supermolecular structures are also studied including gels, colloidal liquids, and aggregates.

1. INTRODUCTION

With the exception of single crystals, almost all materials are, in some way, disordered. In spite of this fact, little is known about the structure of disordered materials or the relationship between structure and fabrication processes. This paucity of knowledge also extends to the relationship between structure and properties. Basically the problem is that both the techniques to study random structures and models relating structure to precursor growth processes have only recently been developed.

The purpose of this paper is to show that substantial information on random structures can be extracted from small-angle x-ray scattering (SAXS) curves. In addition, observed structures can often be explained in terms of random growth processes such as polymerization, colloid aggregation, and phase separation. We concentrate on the silica system which is very rich and displays many different random structures. These structures can be tailored through both chemical and physical growth processes.

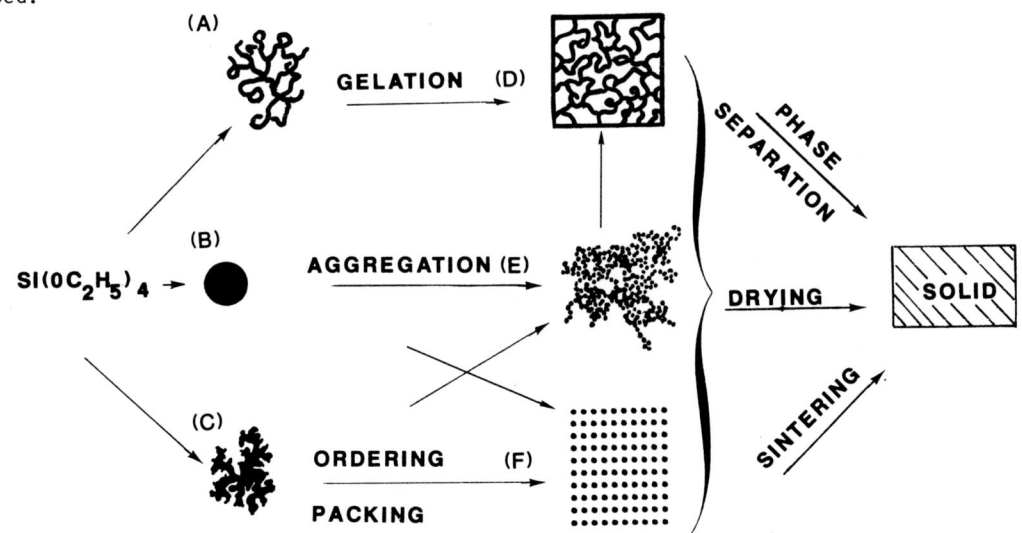

Fig. 1 Precursors to glass and ceramic materials

---

*This work performed at Sandia National Laboratories, Albuquerque, NM and supported by the U.S. Department of Energy under Contract No. DE-AC-04-76DP00789.

Figure 1 is a schematic diagram illustrating the variety of structures which have already been found in silicates. Generally the materials scientist is interested in the properties of a solid material such as illustrated on the right of Fig. 1. The solid might be anything from a dense glass to a highly porous silica gel. The goal is to control the growth processes on the left side of the figure to achieve any desired structure.

Two classes of growth processes can be distinguished: chemical and physical. Polymerization is a typical chemical growth process, as illustrated on the left of Fig. 1. The alcohol-soluble silicate monomer $Si(OC_2H_5)_4$ [tetraethylorthosilicate, TEOS] can be polymerized to create branched polymers(A), dense colloidal particles(B), and rough colloidal particles(C). Colloid aggregation(E) is an example of a physical growth process, whereas gelation(D) can occur by either a chemical or physical mechanism. Liquid crystalline structures(F), as shown in the center column of Fig. 1, can be induced by physical means such as an a.c. electric field.[1]

2. SCATTERING TECHNIQUES

This report concentrates on the structure of materials on dimensional scales of 5A - 1μm. Structures in this range can be determined using small-angle scattering of x-rays or neutrons as well as by light scattering. In many cases, the concept of fractal geometry is the key to understanding scattering curves, so substantial discussion is devoted to scattering from fractal objects.[2]

Fractal geometry[3] provides a quantitative measure of randomness and thus permits characterization of random systems such as polymers,[4] colloidal aggregates,[5] rough surfaces,[6,7] and porous materials.[6,7,8] Qualitatively, fractal objects show dilation symmetry,[3] meaning that the essential geometric features are invariant to scale changes (such as change of magnification in a microscope).

Here we exploit the technique of small-angle x-ray scattering (SAXS) to characterize fractal materials. Basically, all fractals show a power-law dependence of the scattered intensity,[2,8,9] I, on the momentum transfer, K ($K = 4\pi\lambda^{-1}\sin(\theta/2)$, θ = scattering angle)

$$I(K) \sim K^{-x}. \qquad (1)$$

Note that at a given K, one probes length scales (in the Fourier sense) of order K^{-1}. We call x the "Porod exponent" and refer to the power-law domain as the Porod region of the scattering curve.[2] Interpretation of the exponent, x, depends on the origin of the scattering. For so-called mass fractals (i.e., polymer-like structures) the exponent is simply D, the fractal dimension which relates the size R of the object to its mass N,

$$N \sim R^D. \qquad (2)$$

For a polymer-like fractal object with a 1-dimensional backbone, $1 \leq D \leq 3$, depending on the degree of branching and folding. For a sheet-like fractal object, $2 \leq D \leq 3$, where D is greater than two for branched and tortuous structures.

For scattering from 3-dimensional objects with fractal surfaces,[7]

$$x = 6 - D_s, \qquad (3)$$

where D_s is the fractal dimension of the surface ($2 \leq D_s \leq 3$). $D_s = 2$ represents a classical smooth surface. Finally, for fractally porous[6,7,10] materials, $x = 7 - \gamma$, where γ is the exponent describing the distribution P(r) of pores of radius r

$$P(r) \sim r^{-\gamma}. \quad (4)$$

The effect of power-law polydispersity on scattering curves has been studied in detail by Martin.[11]

3. SILICATE POLYMERS

Fig. 2 shows the SAXS profiles for silicates polymerized under a variety of conditions. The lower curve (E) is from a commercial colloidal silicate (Ludox^{TM}SM). This material is prepared by the Stöber process[12] in which the polymerizing species is orthosilicic acid, $Si(OH)_4$. The limiting slope of the Ludox data is -4 indicating that these particles are compact objects with smooth surfaces ($D_s = 2$ in eq. 3). Clearly the scattering curves are consistent with the common notion of colloidal structures.

Curve (D) of Fig. 2 represents scattering from rough colloidal particles prepared by the hydrolysis and polymerization of TEOS under base-catalyzed conditions.[13] In this case the polymerizing species are only partially hydrolyzed (e.g. $Si(OC_2H_5)_n(OH)_{4-n}$) and polymerization to dense structures is precluded because the alkoxide groups do not readily polymerize. The observed scattering curve for a base catalyzed system with $W = [H_2O]/[Si] = 2$ indicates that the structures are fractally rough with $D_s = 2.7$. We call these structures rough colloidal particles.[13] It should be noted, however, that eq. (4) provides an alternate interpretation of the data in terms of a power-law polydisperse collection of dense particles.

Polymer-like silicates can be synthesized in two ways. If TEOS is polymerized under base-catalyzed conditions with substoichiometric $[H_2O]/[Si]$ ratio, scattering curves like (C) result.[13] Here the slope of 2.8 indicates a mass-fractal object. Presumably this is a densely crosslinked polymer molecule. The structures cross over smoothly from mass fractals to surface fractals near $W = 2$ which is the stoichiometric water ratio.

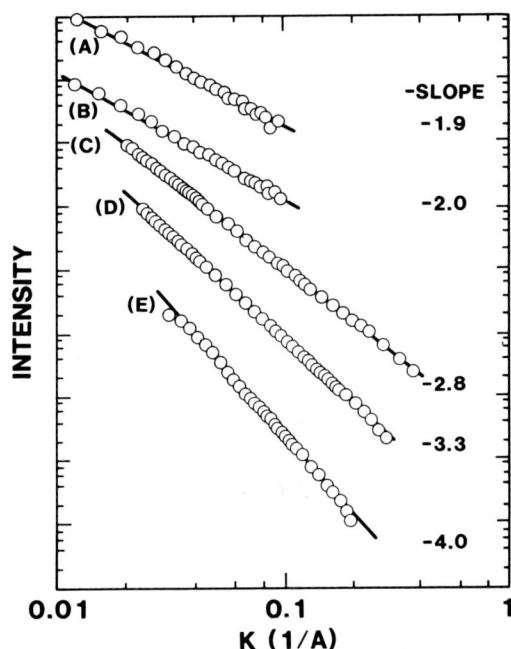

Fig. 2. Porod plots of scattering data for silicates polymerized under various conditions: (A) Two-step acid catalyzed;[2,14] (B) Two-step base-catalyzed;[2,14] (C) Single-step,[13] Base-catalyzed, W=1; (D) Single-step,[13] Base-catalyzed, W=2; (E) Ludox SM.

Polymers can also be synthesized if the polymerization is carried out in two stages.[4,14] In the first stage, small 5A prepolymers are grown under water-starved, acid-catalyzed conditions.[14] In a second stage, these prepolymers are linked under either base or acid-catalyzed conditions to yield the final polymers which are represented by curves (A) and (B) in Fig. 2. Regardless of the details of the second-stage polymerization, we always observe Porod slopes near -2 indicating polymer-like mass-fractal structures. By studying the evolution of the scattering curves with dilution, we have shown that the base-catalyzed systems are more high branched than their acid-catalyzed counterparts.[4] The fractal dimension, nevertheless, remains 2. This result is consistent with the renormalization calculations of Family[15] which show that, in the assymptotic limit of large structures, D is insensitive to the degree of branching.

4. SUPERMOLECULAR STRUCTURES

The polymeric species discussed in the previous section can form a variety of supermolecular structures via gelation, aggregation, phase separation, and ordering. These supermolecular structures are then dried or sintered or otherwise converted into the desired dry, solid material.

At least two distinct types of gels are possible: polymer gels prepared by chemically crosslinking branched polymer clusters and colloidal gels prepared by physical aggregation of colloidal particles. In both cases, the scattering curves are quite insensitive (for an undiluted system) to the gel point. Fig. 3, for example, shows the measured scattering curves for a polymeric silicate before and after the gel point. In spite of a drastic change in mechanical properties, at the gel point, there is no structural signature in the scattering profiles.

SAXS is insensitive to the gel point because the method is sensitive to electron density rather than connectivity. Near the gel threshold, the formation of a few crosslinks drastically modifies connectivity, with essentially no effect on the spatial distribution of atoms. Being sensitive only to the latter, SAXS profiles are unchanged.

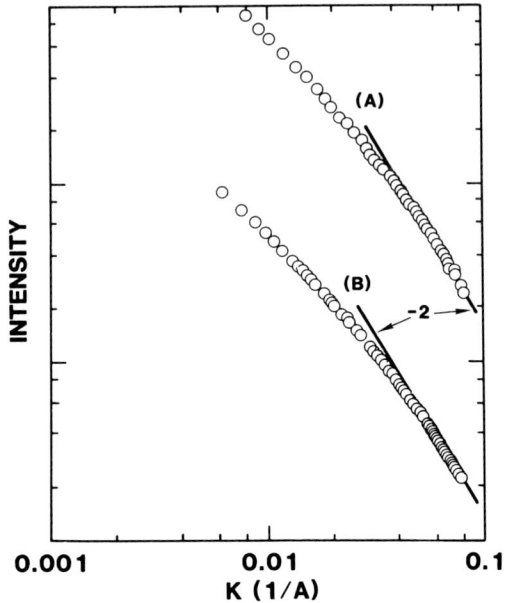

Fig. 3. Porod plots for a two-step, base catalyzed silicate: (A) Pre-gel; (B) post-gel.

Near the gel point, the observed scattering patterns can be understood by analogy to semidilute polymer solutions.[16] In the early stages of polymerization, far from gelation, branched clusters scatter independently. As polymerization proceeds, the Guinier radius, R_G (measured from the initial curvature of the scattering profiles) of the molecules increases, consistent with growth of clusters. If the concentration is such that the clusters begin to overlap and entangle, however, R_G will saturate or may even decrease as growth continues. In the overlapped (semidilute) regime, concentration fluctuations are correlated only over length scales comparable to the distance between interchain contacts.[17] Since this distance is unchanged by gelations, no divergence of R_G is observed. If the solution is diluted, however, then R_G reduces to the radius of the clusters and divergence is expected at gelation.

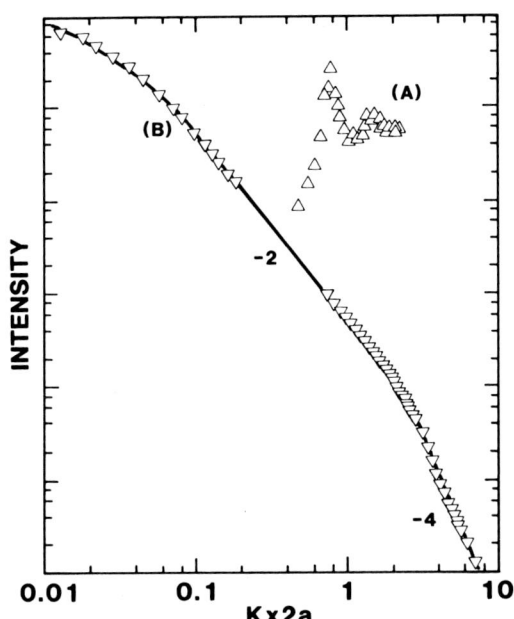

Fig. 4. Scattering from supermolecular structures prepared from colloidal precursors: (A) Deionized polytyrene Latex suspension;[19] (B) Colloidal aggregate of Ludox particles.[5]

In contrast to scattering in the Guinier regime, in the Porod regime scattering profiles are insensitive to long-range correlations and remain sensitive only to local chain topology. Since local structure is unchanged at gelation, however, Porod exponents are constant through the gel point.

Colloidal systems also produce novel supermolecular structures as illustrated by Fig. 4. Curve (B) is the scattering profile (light and x-ray) of a colloidal aggregate prepared by destabalizing to a charge-stabilized suspension of silica particles[5] (Ludox SM). This scattering curve shows two power-law regimes with Porod exponents of -2 and -4. The slope of -4 is consistent with scattering from a smooth surface on length scales smaller than the radius, a, of the primary colloidal particles. The slope of -2 on the other hand, is characteristic of a mass-fractal and is close to the value of 2 found in simulations of chemically-limited aggregation.[18]

If a colloidal suspension is deionized using a mixed-bed ion-exchange resin, then novel ordered arrays called colloidal crystals[19] can result. These structures show sharp Bragg scattering in the light scattering regime similar to those observed in the x-ray regime for atomic crystals. These ordered arrays result from strong repulsive coulomb forces between the colloidal particles. Depending on particle and salt concentrations, liquid-like order also exists[20] as illustrated in curve (A) of Fig. 4. The scattering profile is very similar to that of an atomic liquid, but the pattern is observed in the light-scattering region because the particles are separated by thousands of particle radii.

5. POROUS SOLIDS

As examples of solid silicate materials we consider porous silica aerogels prepared by the sol-gel process. Depending on the details of the preparation procedures, three distinct structures are produced: non-fractal, mass-fractal and surface-fractal.

First consider a porous solid prepared by air drying the fractally rough colloidal suspension studied in Fig. 2 curve (C). Scattering from this materials is shown[13] in Fig. 5 curve (C). In the power-law regime the profile is essentially identical to the solution precursor showing that the fractal surface is preserved on drying. The material can be considered to have uniform porosity with fractally rough surfaces. From the scattering curves alone, however, a power-law distribution of uniform pores cannot be excluded as a possible structure.

Curve (C) contrasts with that of porous solids made from uniform colloidal precursors with non-fractal surfaces. Two such materials are shown in curves (A) and (B). In both cases, these materials are prepared[21] from base-catalyzed hydrolysis and condensation of $Si(OCH_3)_4$. In this case, the polymerizing species $Si(OH)_4$ and dense colloidal particles form. Curve (B) is a relatively high density (.21 gm/cm^3) aerogel. The peak in the structure factor is reminiscent of the liquid-like structure factor in Fig. 5. We interpret this gel a collection of packed spheres and attribute the peak in the structure factor to the fact that the spheres cannot overlap (the correlation hole effect).[22]

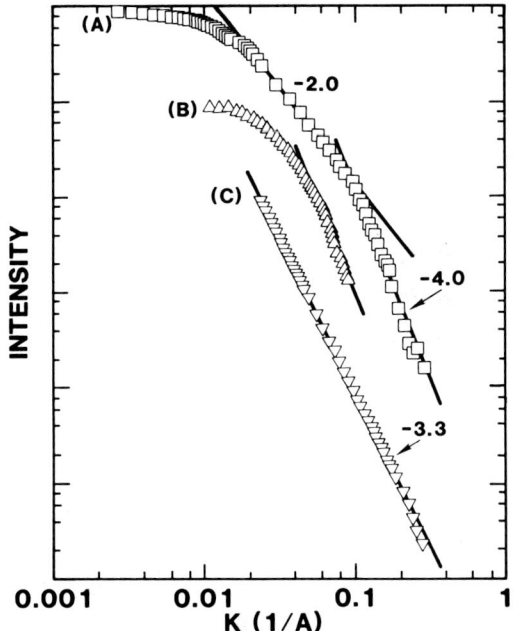

Fig. 5. Porod plots of porous silicates: (A) Aerogel prepared from colloidal precursor,[10] 0.09 gm/cm^3; (B) Aerogel prepared from colloidal precursor 0.21 gm/cm^3; (C) Aerogel prepared from fractally rough precursor.[13]

The low density (.09 gm/cm^3) aerogel shown in curve (A) gives a substantially different scattering curve.[10] Here the high-angle portion gives a slope of -4 consistent with a colloidal structure. At intermediate length scales (.01 ⩽ K ⩽ .08), however, a second power-law regime is seen with a slope near two indicating a mass-fractal object. The most reasonable interpretation of this behavior is in terms of a colloidal aggregate. Basically the porous solid looks like a random aggregate of primary colloidal particles.

Support for the aggregate-type structure for the low-density aerogel comes from comparison of the scattering curve to that of the solution aggregate (Fig. 4, curve B). Since the solution aggregate is prepared at very low SiO_2 concentration, the intermediate power-law regime is much more extensive. The similarity of the two curves confirms the interpretation of the porous solid in terms of an aggregate precursor.

6. CONCLUSION

In this review we exploit the concept of fractal geometry to interpret scattering curves of silicate materials. Depending on chemical conditions silicon-based polymers can be prepared with mass-fractal, surface-fractal, and uniform structures. Gels, both polymeric and colloidal, can be prepared from the polymeric precursors, and porous solids with different structures can be prepared from the gels. Finally, under proper conditions, liquid crystalline arrays and colloid liquids are formed. These numerous different structures demonstrate that the materials scientist can exploit chemical and physical growth phenomena in precursor solutions to control the properties of solid materials.

REFERENCES

1. A.J. Hurd, "A-C Field Induced Order in Dense Colloidal Systems" in <u>Ultrastructure Processing of Ceramics, Glasses and Composites</u>, Vol. 2, L.L. Hench and D.R. Ulrich Eds, Wiley-Interscience, N.Y., 1986.

2. D.W. Schaefer and K.D. Keefer, Mat. Res. Soc. Proc. <u>32</u>, 1 (1984).

3. B.B. Mandelbrot *Fractals, Form and Chance,* Freeman, San Francisco, 1977.

4. D.W. Schaefer and K.D. Keefer, Phys. Rev. Lett. **53**, 1383 (1984).

5. D.W. Schaefer, J.E. Martin, P. Wiltzius, and D.S. Cannell, Phys. Rev. Lett. **52**, 2371 (1984).

6. P. Pfeifer and D. Avnir, J. Chem. Phys. **79**, 3558, 3566 (1983).

7. H.D. Bale and P.W. Schmidt, Phys. Rev. Lett. **53**, 596 (1984).

8. D.W. Schaefer, J.E. Martin, A.J. Hurd, and K.D. Keefer in *Physics of Finely Divided Matter,* Springer-Verlag, N.Y. (1985).

9. S.K. Sinha, T. Freltoft, and J. Kjems, in *Kinetics of Aggregation and Gelation,* F. Family and D.P. Landau, Eds., North Holland, 1984, p. 87.

10. D.W. Schaefer and K.D. Keefer, "Structure of Porous Materials: Silica Aerogel," to be published.

11. J.E. Martin, J. Appl. Cryst., to be published.

12. W. Stöber, A. Fink and E. Bohn, J. Coll. and Interface Science, **26**, 62 (1968).

13. K.D. Keefer and D.W. Schaefer, "Growth of Fractally Rough Colloids," to be published.

14. C.J. Brinker, K.D. Keefer, D.W. Schaefer, R.A. Assink, B.P. Kay, and C.S. Ashley, J. Non Cryst. Solids **63**, 45 (1984).

15. F. Family, J. Phys. A **13**, L325 (1980).

16. M. Daoud, J.P. Cotton, B. Farnoux, G. Jannink, G. Sarma, H. Benoit, R. Duplessix, R. Picot and P.G. de Gennes, Macromolecules **8**, 804 (1975).

17. D.W. Schaefer, Polymer, **25**, 387 (1984).

18. M. Kolb and R. Jullien, J Physique Lett. (Paris), **45**, L-977 (1984).

19. D.W. Schaefer and B.J. Ackerson, Phys. Rev. Lett. **35**, 1448 (1975).

20. D.W. Schaefer, J. Chem. Phys. **66**, 3980 (1977).

21. S. Henning and L. Svensson, Physica Scripta **23**, 697 (1981).

22. T. Freltoft, J.K. Kjems, and S.K. Sinha, "Power-Law Correlations and Finite Size Effects in Aggregates," to be published, Phys. Rev.

FRACTALS IN PHYSICS
L. Pietronero, E. Tosatti (editors)
© Elsevier Science Publishers B.V., 1986

INTERACTION OF FRACTALS WITH FRACTALS: ADSORPTION OF POLYSTYRENE ON POROUS Al_2O_3

Peter PFEIFER

Fakultät für Chemie, Universität Bielefeld, D-4800 Bielefeld, West Germany

The problem is addressed to what extent a fractal interface (surface fractal or mass fractal, dimension D) can modify the fractal structure of flexible polymer chains as a result of adsorption. The chain conformation in solution and in the adsorbed state is described by the fractal dimension D_c^{sol} and D_c^{ads}, respectively. It is shown how from the number of adsorbed chains as function of their radius of gyration in solution, the product $D \cdot D_c^{sol}/D_c^{ads}$ is obtained; and how the pore-size distribution of the adsorbent separately yields D (the dimension D_c^{sol} is given by the solvent quality). For the system polystyrene/Al_2O_3 analyzed, there results $D \cong 2.90 \pm 0.10$ and $D_c^{ads}/D_c^{sol} = 1.04 \pm 0.04$. It implies that conformation is conserved and demonstrates the chains' inability to unfold in the pores of the solid to form a quasi monolayer of monomers. This differs drastically from the situation on low-D surfaces where unfolding does occur and leads to $D_c^{ads} = D$. This strong influence of D on the (effective) adsorption equilibrium is explained in terms of an increasing entropy barrier, as D increases, against true thermodynamic equilibration. Parallels to other cases of two unlike fractals interacting with one another are discussed.

1. INTRODUCTION, STATEMENT OF THE PROBLEM

The interface between a solid and its surroundings may be fractal, with fractal dimension D, in three different ways. It may be

(a) a surface fractal, in which case only the surface scales with exponent D while both the whole solid and the pore space scale like a volume in the range of scales at issue;

(b) a mass fractal, in which case the surface and the solid scale identically, i.e., both with exponent D, while the pore volume scales like a volume; or

(c) a pore fractal, for which the surface and pore space scale identically while the solid scales like a volume.

[To scale with exponent D or like a volume, respectively, means that the set in question obeys the mass-radius relation[1] with exponent D or 3. So, only for $D \to 3$ do the three classes merge. A more detailed discussion will be given elsewhere.] Most fractal interfaces discovered so far belong to the first two classes, and of these the ones with fractal properties from atomic length scales up[2-9] are of particular, surface-chemical interest. The fractal dimension D then controls the nature of adsorption complexes, distribution of active sites, surface diffusion, transport through pores, etc.[4,8] The most direct method of fractal surface analysis from Angstroms up, applicable equally to surface and mass fractals, is to verify[2,4,8,9] that the number of molecules of radius r per respective monolayer on the surface, n(r), as obtained from adsorption experiments, follows the power law

$$n(r) \propto r^{-D}. \qquad (1)$$

The only requirement for (1) is chemical homogeneity of the surface with respect to the employed yardstick molecules (most other methods[2-9] require additional conditions to be satisfied). With small and medium-sized molecules, essentially rigid and thus possessing a well-defined geometry, Eq. (1) may be established over a range of r values of roughly $10^0 - 10^1$ Å. For a considerable part of surface chemistry this is the entire range of interest (examples: prediction of adsorption capacities, orientation of adsorbed nonspherical molecules,[9] solvation properties of the surface,[10] microscopic contact angles between liquids and the surface, effective diffusion coefficients on some materials).

In order to observe (1) beyond that, one has to switch to polymers as probe molecules.

But for flexible chain molecules as used in most work on polymer adsorption, the geometry of a molecule in the adsorbed state need no longer be the same as in solution, so that the r dependence of n (r is used throughout to denote the radius in solution) may become different from (1). Indeed, on flat surfaces (D=2) it is well known that the chains, randomly coiled in solution, unfold to form in essence a monolayer of monomers (repulsion between monomers in the chain, attraction between monomers and the surface). The experimental manifestation is that the adsorbed mass is independent of the number of monomers per chain, N (see, e.g., Ref. 11 and work quoted there; the fact that measured thickness of the adsorbed layer does depend on N,[12] is not in contradiction with this picture but reflects that the layer is actually a diffuse one,[13,14] including loops and tails of unadsorbed segments).

So, the question is whether the same unfolding occurs on rough, porous surfaces (D>2) as has recently been suggested[15] (in which case the measured function n(r) would depend only on the relation between r and N in the given solvent, but not on the surface dimension D); or whether adsorption is irreversible[14] so that the chains, by confinement in pores of radius ≈ r, retain their original conformation[2,16] (in that case, Eq. (1) continues to hold). The two extremes are depicted in Fig. 1. More generally, the question is how much the adsorption process may alter the chains' fractal geometry. Thus, denoting the dimension of the chains in solution and on the surface by D_c^{sol} and D_c^{ads}, respectively (see also Sec. 2), we want to study D_c^{ads} as a function of D and D_c^{sol} (and possibly other factors).

This paper reports the first determination of D_c^{ads} on a fractal surface: Sec. 2 describes how earlier analyzed[2] data[11] for adsorption of polystyrene on porous Al_2O_3 yields the value of

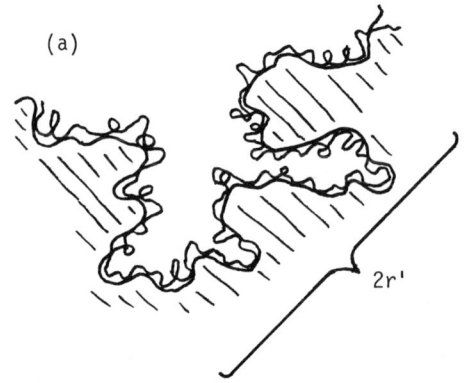

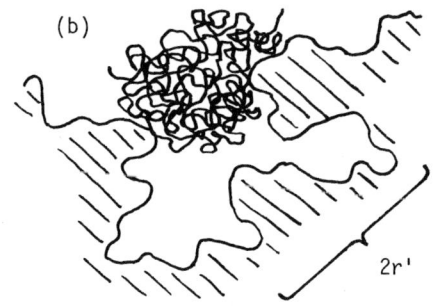

FIGURE 1
Two extreme modes of polymer adsorption on a porous surface. (a) Unfolding of chains similar to the situation on a flat surface. (b) Retention of the chain conformation present in solution, on the surface (steric hindrance). The figure also defines the on-surface coil diameter 2r' entering Eq. (4).

$D \cdot D_c^{sol}/D_c^{ads}$ (and more, if combined with additional data). The surface dimension D is obtained from the pore-size distribution of the same Al_2O_3 sample (Sec. 3), while D_c^{sol} is known from the solvent condition. The result, $D_c^{ads} \simeq D_c^{sol}$ on $D \simeq 2.9$, makes a strong case for previous conclusions[2,16] and future strategies that, on high-D surfaces, Eq. (1) holds for chain polymers just as if they were rigid structures. A detailed discussion and interpretation is given in Secs. 4 and 5.

2. THE FUNCTION n(r) IN TERMS OF D, D_C^{sol}, D_C^{ads}

We take a phenomenological viewpoint and treat the adsorption process formally as a chemical reaction,

$$\text{surf}(D) + n \text{ polym}(D_C^{sol}) \rightarrow \text{surf}(D)[\text{polym}(D_C^{ads})]_n. \quad (2)$$

The fractal dimensions in parentheses are considered as given quantities, and n(r) will be the "mass-geometry" balance for (2). Thus on the reactants' side, we have a solution of polymer coils, each of radius of gyration r and of degree of polymerization N, where

$$r \propto N^{1/D_C^{sol}} \quad (3)$$

and D_C^{sol} = 5/3 (self-avoiding random walk) for good solvents, and D_C^{sol} = 2 (simple random walk) for poor (θ) solvents. On the product side, a single adsorbed chain extends over a spatial region of radius r' (Fig. 1) where r' is assumed to scale with N according to

$$r' \propto N^{1/D_C^{ads}}. \quad (4)$$

This defines the fractal dimension of the adsorbed chain, D_C^{ads} ("length-volume relation"[1]). The adsorbed layer as a whole is taken to be a monolayer of such coils of radius r' (densely packed without overlap). By definition of a D-dimensional surface, then, the number of chains per monolayer, n, decreases with increasing r' as $(r')^{-D}$. Hence, using (4) and (3), we obtain

$$n \propto N^{-D/D_C^{ads}} \quad (5)$$

$$\propto r^{-D \cdot D_C^{sol}/D_C^{ads}}. \quad (6)$$

Eq. (6) is the desired relation allowing for structural relaxation of adsorbed molecules. Unlike Eq. (5), it applies also to "mixed yardstick series" where different yardstick molecules may consist of different monomers, provided that all rearrangement follows the same r' vs. r dependence (this will be exploited below). Conversely, Eq. (5) (but not (6)) extends to "mixed solvent series" where different pairs (n,N) may come from different solvents, provided that the structure of adsorbed chains does not depend on the solvent. Thus, (5) and (6) are geared to tests of two separate universality hypotheses, but are otherwise equivalent.

The two extremes in Fig. 1 correspond to

$$D_C^{ads} = D \quad \text{(unfolding of coils)}, \quad (7a)$$
$$D_C^{ads} = D_C^{sol} \quad \text{(coil structure is conserved)}, \quad (7b)$$

respectively, and substitution of (7a/b) into (5) and (6) yields all that was discussed in Sec. 1 for the two situations. Note that Eq. (7a) results from the fact that in this case the chains follow (isotropically) all surface details resolvable by a monomer. This shows that the structure of a 2-dimensional adsorbed chain is very different, depending on whether D_C^{ads}=2 is realized by (7a) on a flat surface or by (7b) in a poor solvent. It also shows that, in the case of (7a), the self similarity of adsorbed chains as described by D_C^{ads}, refers to the layer structure "parallel" to the surface; whereas the self similarity discussed in Refs. 13-15 (for the same case) refers to the layer structure "normal" to the surface. Finally, it is to be noted that since (7b) does not depend on prefactors in (3) and (4), it may well come from, say, a 50% reduction of all coil radii upon adsorption.

We now use Eq. (6) to reexamine the polystyrene/Al_2O_3 analysis in Ref. 2: There it was observed that the experimental data by Burns and Carpenter[11] for this system obey the power law $n(r) \propto r^{-2.79 \pm 0.03}$ over a range of radii r from 2Å to 380Å (the value r = 2Å refers to nitrogen and is the only yardstick molecule other than polystyrenes); and that the nitrogen data point is perfectly extrapolated by the power law for the polystyrenes alone. Since the polystyrenes were in a θ-solvent,[11] we additionally know that $D_c^{sol}=2$ (cf. also the data for r vs. N in Ref. 11). Combining all this with Eq. (6), one arrives at the following conclusions:

(i) The result, $D \cdot D_c^{sol}/D_c^{ads} = 2.8$ and $D_c^{sol} = 2$, rules out (7a). Instead we have $D/D_c^{ads} = 1.4$.

(ii) For nitrogen, there is trivially no conformational rearrangement, so that the data point for nitrogen should lie on the curve (1). But since it lies on the polystyrene curve, we must have $D_c^{ads} = D_c^{sol}$ for the polystyrenes as well. Whence D = 2.8 from (i).

This reinforces the argument in Ref. 2 that the Al_2O_3 surface under consideration has a fractal dimension of $\simeq 2.8$. The definitive, new piece of evidence to that effect, however, comes from experimental data in Ref. 11 that were ignored so far. It is presented in the next section.

3. THE PORE-SIZE DISTRIBUTION OF Al_2O_3

Besides the adsorption measurements, Burns and Carpenter[11] also determined the pore-size distribution of the Al_2O_3 sample: From the nitrogen desorption isotherm (hysteresis loop, capillary condensation), they determined A(ρ), the area inside pores of radius greater than ρ, from ρ=25Å to ρ=195Å. The surface area measured by nitrogen adsorption (monolayer) gives the additional value A(ρ≈2Å). Indeed, A(ρ) may be identified with the surface area as measured by (rigid) molecules of radius ρ. So, for surface and mass fractals, Eq. (1) predicts

$$A(\rho) \propto \rho^{2-D}. \quad (8)$$

A rigorous derivation (to be presented elsewhere) of (8) starts from the pore-size distribution in terms of the *volume* of pores of radius > ρ, V(ρ), defined for an arbitrary set K as the volume of the space around K that cannot be occupied by balls of radius ρ without intersecting K. The key is then to show that regardless of whether K is a fractal surface, curve, or dust, i.e.,

under conditions much more general than considered previously,[17] there holds the relation

$$-\frac{dV(\rho)}{d\rho} \propto \rho^{2-D}. \quad (9)$$

In particular, (9) holds for surface and mass fractals. Together with

$$-dV(\rho) = A(\rho)d\rho,$$

it implies (8).

Fig. 2 shows a log-log plot of the mentioned experimental values of A(ρ). A well-defined power law obtains indeed and leads to a fractal surface dimension of $\simeq 3.0$.

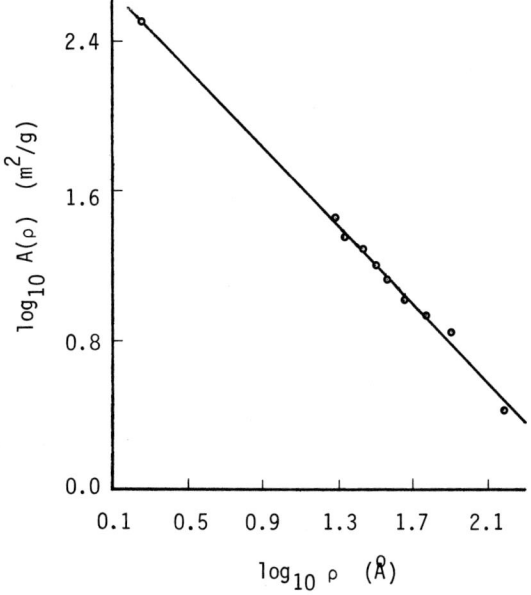

FIGURE 2
The pore-size distribution, A(ρ), of the alumina surface on which the polystyrenes were adsorbed. The ρ range covered is 2 - 195 Å (the leftmost point corresponds to the BET surface area). From the slope and Eq. (8) there results D = 3.04 ± 0.02.

4. DISCUSSION, THE ENTROPY BARRIER

We are now in the position to estimate D_c^{ads} independently of the considerations in Sec. 2, conclusion (ii). From the results $D \cdot D_c^{sol}/D_c^{ads} = 2.8$, $D_c^{sol} = 2$ (Sec. 2), and $D = 3.0$ (Sec. 3) we obtain $D_c^{ads} = 2.1$. But this is practically the same as conclusion (ii) in Sec. 2, and in substance also agrees with the conclusions by Burns and Carpenter, of course.

Conversely, we may identify the two surface dimensions, $D = 2.8$ from adsorption capacities and $D = 3.0$ from porosimetry, to within experimental error. In fact, the agreement of the two values is deemed remarkable in view of the very different underlying data sets, each carrying its own experimental uncertainties. [The error bars for D in Secs. 2 and 3 are standard deviations of the regression coefficients and presumably underestimate the actual uncertainties here. See also Refs. 1 and 18.] A conservative estimate is that $D = 2.90 \pm 0.10$ over an r range from 2 to 200 Å, for the present Al_2O_3. This compares well with recent results[19] from alcohol adsorption and small-angle X-ray scattering for a similar Al_2O_3 sample, and with the fact that structurally analogous silica-gel surfaces are equally close to 3-dimensional.[4,9,20-22] However, in none of those cases does the explored/verified r range come anywhere close to the range of interest here.

Altogether then, we see that the experimental data very persistently lead to $D_c^{ads} \simeq D_c^{sol}$ and $D \simeq 2.9$. This outcome is to be contrasted with the mentioned prediction[15] of $D_c^{ads} = D$ (in present terminology), for all $D \geq 2$. The argument[15] is that adsorption without significant distortion of coils (i.e., $D_c^{ads} = D_c^{sol}$) cannot correspond to thermodynamic equilibrium. Indeed, general arguments suggest that configurational entropy should increase with increasing D_c^{ads}, so $D_c^{ads} = D$ should minimize the free energy both on energy and entropy grounds if $D > D_c^{sol}$, i.e., if $D \geq 2$. (We do not consider the possibility of collapsed chains, $D_c^{sol} > 2$.) However, on a very porous surface as implied by a dimension D close to three, such equilibration, i.e., that the chain segments attached to the surface on first contact, partly detach again in favor of a configuration allowing more segments than previously to enter small pores, and so on, is virtually impossible. Under the geometric constraints generated by the pore network, a coil of radius r initially trapped in a pore of the same radius would have to undergo prohibitively concerted, hence improbable rearrangements in order to reach equilibrium, $D_c^{ads} = D$. In terms of transition-state theory of chemical kinetics, this amounts to an insuperable entropy barrier between reactants and products in the corresponding reaction (2): the transition state, overdrawn as stretched chain in front of the original pore of radius r, has zero entropy. For a very different example of states separated by an infinite entropy barrier, see Ref. 23.

Thus, while it is true that $D_c^{ads} = D_c^{sol}$ does not correspond to equilibrium as discussed in Ref. 15, it does correspond to an effectively stable state, or de facto equilibrium, on high-D surfaces - much as the supporting Al_2O_3 itself (or any other amorphous solid, glass, etc.) is a stable phase for all practical purposes. One expects therefore that the result $D_c^{ads} \simeq D_c^{sol}$ is not unique to the polystyrene/Al_2O_3 system, but that one has quite generally

$$D_c^{ads} = \begin{cases} D & \text{for } D \sim 2 \quad (10a) \\ D_c^{sol} & \text{for } D \sim 3 \quad (10b) \end{cases}$$

(recall Sec. 1 for (10a)). Adsorption of random copolymers (in a good solvent) on charcoal[24] is presumably a second example for (10b): Ref. 24 gives substantial evidence for $D_c^{ads} \simeq D_c^{sol}$; and Ref. 16 shows under this very assumption that $D \simeq 2.8$.

The fundamental question, then, to be answered by future investigations, concerns the behavior of D_c^{ads} at intermediate D values. Eq. (10) and $D_c^{sol} \le 2$ imply a local maximum of D_c^{ads} at some D value between 2 and 3. So of particular interest will be whether D_c^{ads} jumps at this D value, or whether it gradually decreases to D_c^{sol} as $D \to 3$; i.e., whether the transition from true equilibrium to de facto equilibrium is discontinuous in D_c^{ads} or not. If true (de facto) equilibrium corresponds to a finite (infinite) entropy barrier, one expects the barrier to increase with increasing D so as to become infinite at some critical value D_{crit}, $2 < D_{crit} < 3$, and to remain so beyond. This predicts a jump, namely $D_c^{ads} = D$ for $D < D_{crit}$ and $D_c^{ads} = D_c^{sol}$ for $D \ge D_{crit}$; and a critical slowing down of equilibration as D approaches D_{crit}.

5. CONCLUSIONS

It has been shown

- that the adsorption of flexible polymer chains on a porous surface may very naturally be described as interaction of two fractals ("guest" dimension D_c^{sol}, "host" dimension D);
- that the result of this interaction, the fractal dimension D_c^{ads} of the adsorbed guest, is well accessible to experimental determination;
- that D_c^{ads} is governed by the "parent" dimensions D_c^{sol} and D alone (at least in the cases $D \sim 2$ and $D \sim 3$ studied so far), the decisive variable being D;
- and that this dependence on D and D_c^{sol} is not simply the analytic continuation of the Euclidean case D = 2 but rather encompasses a drastic change, possibly discontinuous in D_c^{ads}, of the nature of the (effective) adsorption equilibrium as D increases from 2 to 3.

Such nonanalyticity with respect to the host dimension D, i.e., that the physics for nonintegral D may differ from what the interpolation/extrapolation from classical Euclidean situations suggests, is known for other fractal-fractal interactions: For phase transitions on fractals,[25] the fractal dimension of the critical Ising clusters ("guest") depends not only on the dimension D of the support lattice, but also on order of ramification, lacunarity, etc. of the lattice. For diffusion on fractals,[26] the fractal dimension of random walks ("guest") is given by $D \cdot \min\{1, 2/\tilde{D}\}$ (not counting multiplicities) where the additional parameter $\tilde{D}$ is the spectral dimension of the lattice.

ACKNOWLEDGMENTS

Helpful discussions with D. Avnir and S. Alexander, and two grants by the University of Bielefeld are gratefully acknowledged.

REFERENCES

1. B.B. Mandelbrot, The Fractal Geometry of Nature (Freeman, New York, 1982).
2. D. Avnir, D. Farin, and P. Pfeifer, Nature 308 (1984) 261.
3. P. Pfeifer, Applic. Surf. Sci. 18 (1984) 146.
4. P. Pfeifer, D. Avnir, and D. Farin, J. Stat. Phys. 36 (1984) 699; 39 (1985) 263.
5. H.D. Bale and P.W. Schmidt, Phys. Rev. Lett. 53 (1984) 596.
6. D. Avnir, D. Farin, and P. Pfeifer, J. Colloid Interface Sci. 103 (1985) 112.
7. P. Pfeifer, U. Welz, and H. Wippermann, Chem. Phys. Lett. 113 (1985) 535.
8. P. Pfeifer, Chimia 39 (1985) 120.
9. D. Farin, A. Volpert, and D. Avnir, J. Am. Chem. Soc. 107 (1985) 3368; 107 (1985) in press.
10. A. Levy, D. Avnir, and M. Ottolenghi, Chem. Phys. Lett. (1985) in press.
11. H. Burns and D.K. Carpenter, Macromolecules 1 (1968) 384.
12. M. Kawaguchi and A. Takahashi, Macromolecules 16 (1983) 1465.
13. P.G. deGennes, Macromolecules 14 (1981) 1637.
14. J. Klein and P. Pincus, Macromolecules 15 (1982) 1129.
15. P.G. deGennes, C.R. Acad. Sc. Ser. II 299

(1984) 913.

16. D. Avnir, D. Farin, and P. Pfeifer, J. Chem. Phys. 79 (1983) 3566.

17. P. Pfeifer and D. Avnir, J. Chem. Phys. 79 (1983) 3558; 80 (1984) 4573.

18. D.A. Weitz and J.S. Huang, in: Kinetics of Aggregation and Gelation, eds. F. Family and D.P. Landau (Elsevier, Amsterdam, 1984) pp. 19-28.

19. D. Avnir and P.W. Schmidt, private communication.

20. D. Avnir and P. Pfeifer, Nouv. J. Chim. 7 (1983) 71.

21. P.W. Schmidt and H.D. Bale, in: Fractal Aspects of Materials, eds. B.B. Mandelbrot and D.E. Passoja (Materials Res. Soc., Pittsburgh, 1984) pp. 14-16.

22. M. Drake and S.K. Sinha, private communication.

23. P. Pfeifer, Phys. Rev. A 26 (1982) 701.

24. A. Hopkins and G.J. Howard, J. Polym. Sci. A 9 (1971) 841.

25. Y. Gefen, A. Aharony, and B.B. Mandelbrot, J. Phys. A 17 (1984) 1277; and references therein.

26. R. Rammal, J. Stat. Phys. 36 (1984) 547; and references therein.

FRACTALS IN PHYSICS
L. Pietronero, E. Tosatti (editors)
Elsevier Science Publishers B.V., 1986

SCATTERING BY FRACTALS

Eric JAKEMAN

Royal Signals and Radar Establishment, St Andrews Road, Malvern, Worcestershire WR14 3PS, UK

The statistical characteristics of intensity fluctuations generated when waves are scattered by Gaussian random fractal phase screens are reviewed.

1. INTRODUCTION

Although the concepts of fractal geometry introduced by Mandelbrot[1] have found wide application in many forefront areas of Physics they also have interesting implications of a more practical nature for the older classical areas of science. One of these areas is the scattering of waves - electromagnetic or scalar - by random variations in refractive index whose largest scale sizes are larger than the incident wavelength so that Physical Optics or Kirchhoff diffraction theory applies. Such scattering is of interest in both the remote sensing context - the remote, or contactless measurement of roughness of surfaces, or layers or extended regions of turbulence for example - and in the context of noise - the modelling of noise and clutter limiting the performance of optical, microwave and acoustic systems.

The simplest physical optics scattering system is the random phase changing screen, which merely introduces spatially random phase distortions into an incident wave. The subsequent propagation of the "scattered" wave results in the development of amplitude fluctuations which, in optical experiments, are revealed as a more or less complicated pattern of bright and dark regions when the scattered light is intercepted by a screen. The random phase screen can be used as a model for rough surfaces, thin diffusing layers and, in some situations, more extended regions of varying refractive index.

In this brief review we discuss the properties of waves which have been scattered by Gaussian random phase screens characterised by power law spectra - so called "Gaussian random fractals"[2]. These self-affine models lead to amplitude fluctuations with statistical properties which are very different from those generated by smoothly varying phase screens. We shall assume throughout that path differences greater than a wavelength are introduced by the screen. In the next section we define the scattering models and geometries usually investigated theoretically and in experiments. In sections 3 and 4 we examine the predictions of these models and make comparisons with experimental data, whilst in section 5 the short wave limit of one scattering model is examined in more detail. A brief summary and conclusions are presented in section 6.

2. SCATTERING MODEL

The geometry of a simple random phase screen scattering experiment is illustrated in Figure 1. The Physical Optics solution to the problem is given by the Huyghens-Fresnel diffraction integral[3]

$$\mathcal{E}(\underline{r}) \propto \int_{-\infty}^{\infty} d^2\underline{r}' \exp[ikr'^2/2z - i\underline{k}\cdot\underline{r}'\sin\theta + i\phi(\underline{r}')] \times A(r') \quad (1)$$

where z and $\sin\theta$ are defined in Figure 1, ϕ is the random phase distortion, $k = 2\pi/\lambda$ is the

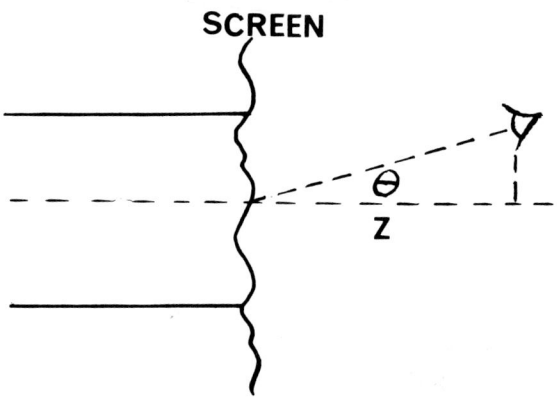

FIGURE 1
Scattering Geometry

wavevector and $A(r')$ is an aperture function. If the phase screen is placed at the waist of a laser beam then

$$A(r') = \exp(-r'^2/W^2) \qquad (2)$$

Assuming relation (2) often simplifies the calculations. Two scattering geometries are normally examined (1) the far field or Fraunhofer limit where $kW^2/2z \ll 1$ and (2) the Fresnel limit $kW^2/2z \gg 1$. In the far field it is well known that [5], when $W \gg \xi$, where ξ is the largest scale size of the phase fluctuations then the scattered field is a complex Gaussian process so the intensity $I = |\mathcal{E}|^2$ forms a speckle pattern with distribution

$$p(I) = \frac{1}{\langle I \rangle} \exp - \frac{I}{\langle I \rangle} \qquad (3)$$

and moments

$$\frac{\langle I^n \rangle}{\langle I \rangle^n} = n! \qquad (4)$$

When W is comparable to or smaller than ξ the statistics are non-Gaussian and the normalised moments exceed the values given by equation (4). It is simplest to evaluate the Fresnel limit by setting $A(r') \equiv 1$. For sufficiently large propagation distances z the field is again a complex Gaussian process, but close to the phase screen geometrical and diffraction effects limit the area contributing to the field at the detector and again non-Gaussian statistics are observed. It is these non-Gaussian regimes with which we are concerned. They are complex, containing more information but also often presenting more severe limitations on system performance than the Gaussian speckle regimes[5].

A measure of the information content of the intensity fluctuations is provided by their statistical properties. In the Fraunhofer region these depend on illuminated area and scattering angle θ. In the near field they are a function of propagation distance. In order to calculate the statistical properties of the scattered field a model has to be adopted for the properties of ϕ. For simplicity and for want of a better model it is generally assumed that ϕ is a stationary Gaussian Process. It then remains to choose the spectrum or autocorrelation function $\langle \phi(o)\phi(\underline{r}) \rangle$. In the case of fractal models, the structure function is more appropriate. It is useful to write the phase distortion in terms of a notional height $\phi(\underline{r}) = k\,h(\underline{r})$ with structure function

$$D(\underline{r}) = \langle (h(o) - h(\underline{r}))^2 \rangle \qquad (5)$$

For smoothly varying heights which are differentiable to all orders $D(\underline{r})$ can be expanded as an even powered series in $\underline{r}$ about the origin. For a simple corrugated fractal surface, however

$$D(x) = |x|^\nu L^{2-\nu} \text{ with } 0 < \nu < 2 \qquad (6)$$

The index ν is related to the fractal dimension D by $\nu = 2(2-D)$[6] whilst the length scale L is usually referred to as the "Topothesy"[7]. A height function with the property (6) is continuous but not differentiable. Another corrugated hierarchical model is the sub-fractal surface whose <u>slope</u> structure function is given by[8,9]

$$S(x) = |x|^{\nu-2}/L^{\nu-2} \quad \text{with } 2 < \nu < 4 \quad (7)$$

A surface with the property (7) is once differentiable and has a continuous slope.

3. FRACTAL HEIGHT MODEL

The spectral features associated with scattering by this type of model are a consequence of the fact that the surface is not differentiable. The concepts of rays and geometrical optics effects are thus not applicable. Since it is commonly assumed that the angular distribution of scattered intensity in the far field of a rough surface is determining by the proportion of surface facing the appropriate direction, ie by the surface slope distribution, the fractal assumption[6] is expected to have important consequences even for this simplest of statistical properties. Indeed it is found that $\langle I(\theta) \rangle$ is a stable distribution of $\sin\theta$

$$\langle I(\theta) \rangle \propto p_\nu([kL]^{1-\frac{2}{\nu}} \sin\theta) \quad (8)$$

where

$$\int_{-\infty}^{\infty} p_\nu(x) \exp(i\lambda x)\, dx = \exp(-A\lambda^\nu)$$

For a smoothly varying surface

$$\langle I(\theta) \rangle \propto p_m(\sin\theta) \quad (9)$$

where p_m is the surface slope distribution. The tail of the distribution (8) falls off like $\sin^{\nu+1}\theta$ beyond $(kL)^{1-2/\nu} \sin\theta \sim 1$. In principle this behaviour can be used to deduce both model parameters ν and L from measurements of $\langle I(\theta) \rangle$ in a way which is independent of the absolute value of this quantity. This may be contrasted with the implication of result (9) for the smoothly varying surface, that only properties of the slope distribution can be obtained from experimental measurements.

Both Fraunhofer and Fresnel region behaviour of the second normalised intensity moment $\langle I^2 \rangle / \langle I \rangle^2$ reflect the absence of geometrical optics effects. In neither region is significant enhancement above the Gaussian value of 2 (equation (4)) predicted[2,3,6]. Only single curve functions of products of k, L and W or z are obtained unlike the smoothly varying case where families of curves are obtained depending on the mean square deviation of height. For large values of the latter quantity very large fluctuations are predicted and observed in the smoothly varying case corresponding to the occurrence of caustics or geometrical singularities in the intensity pattern[10]. These are not predicted by the fractal model[6] and indeed are not observed in infrared scattering experiments using an artificially constructed Brownian fractal surface ($\nu = 1$, $D = 1.5$)[11,12].

4. FRACTAL SLOPE MODEL

Unlike the fractal height model, the subfractal or fractal slope model does lead to elementary geometrical optics or ray effects. As in the case of smooth surface models, the angular distribution of intensity versus angle is determined by the surface slope distribution, and it is the higher order statistics which show interesting features characteristic of the multiscale nature of the surface. For example in the far field a geometrical optics dominated regime exists when the illuminated area is smaller than the outer scale size but larger than the aperture size required to generate a diffraction spread greater than the spread of rays produced by the allowed range of tilts of the surface[13]. In this geometrical optics regime an incident beam is refracted through an angle Ω_1 whose rms slope is determined by the slope structure function at the outer scale size ie

$\Omega_1 = \sqrt{S(\xi)}$ and spreads out through an angle Ω_2 determined by the slope structure function corresponding to the aperture W ie $\Omega_2 = \sqrt{S(W)}$. In a crude approximation in which it is assumed that the beam has a rectangular profile, so that the intensity seen by the detector is a square wave, it is easy to show that[2,14]

$$\frac{\langle I^2 \rangle}{\langle I \rangle^2} \propto \left(\frac{\xi^2}{W^2}\right)^{(\nu-2)/2} \qquad (10)$$

and similar results can be derived for the higher moments. The power law dependence on W is characteristic of the model and can be deduced from more exact calculation. It differs from the behaviour predicted for smoothly varying surfaces which give a deviation from Gaussian statistics inversely proportional to the illuminated area (ie $\propto W^{-2}$).

In the Fresnel region a very remarkable result is obtained: if no outer scale is included in the model then the amplitude fluctuations saturated at large distances at a value greater than for Gaussian speckle[15]. This is due to the fact that ray density fluctuations fail to average out even at large propagation distances where a large area of the surface contributes to the intensity pattern because the surface slope is correlated over an "infinite" range. If an outer scale is introduced then the intensity fluctuations eventually do subside to the Gaussian speckle value as uncorrelated regions of the surface begin to contribute to the scattered field. It can be shown that the intensity is approximately K-distributed over a substantial range of propagation distances for this model - a feature which has been observed in many experimental investigations - and that in the absence of an outer scale the intensity fluctuations for the Brownian case ($\nu = 3$) are described <u>exactly</u> by[15]

$$p(I) = 2K_0(2\sqrt{I}) \qquad (11)$$

Perhaps the most important feature of this model is the fact that although the surface slope is well defined, its curvature is not, so that ray density fluctuations are generated but not focussing or caustics. It follows that the statistics do not diverge in the short wave limit and diffraction smoothing need not be included in the calculations. Considerable simplification of the analysis of this limiting case ensues, in particular the ray density fluctuations are finite and may be studied as a simple geometrical property of random fractal functions.

5. RAY DENSITY FLUCTUATIONS

These may be studied through the functional (corrugated case)[15]

$$R(y,z) = \frac{1}{z}\int_{-\infty}^{\infty} \delta(m(x) - \frac{x-y}{z}) \, dx \qquad (12)$$

which leads to the results

$$\langle R \rangle = 1 \; ; \quad \langle R^2 \rangle = 2/(4-\nu) \qquad (13)$$

Higher moments can be calculated subject to the assumption that $m(x)$ is a Gaussian process and they are found to be close to those of the class of Gamma distributions with the same mean and variance. In the special case $\nu = 3$ (Brownian sub-fractal) the entire statistical problem can be solved exactly

$$\langle \prod_{j=1}^{N} R(y_j) \rangle = \sum_n \prod_{j=1}^{N} \exp\left[-\frac{L}{x^2}|y_j - g_j^{(n)}|\right] \qquad (14)$$

where $\{g_j^{(n)}\}$ are permutations of $\{y_j\}$. This shows that R is the intensity of a complex Gaussian-Markov process[16].

Because of the relative simplicity of the mathematics, more complicated ray problems can be analysed for the sub-fractal model. One problem of interest is scattering by multiple phase screens. The ray density fluctuations generated by two sub-fractal screens in series, for example, are characterised in a small angle approximation by the second moment

$$\frac{\langle R^2 \rangle}{\langle R \rangle^2} = \frac{2r}{\pi} \frac{1}{(4-\nu)} \int_{-\infty}^{\infty} \frac{|x|^{(\nu-2)/2} dx}{[1-(1+r)x]^2 + |x|^{\nu-2}(1-x)^2} \quad (15)$$

where $r = \ell_1/\ell_2$, ℓ_1 being the screen separation and ℓ_2 the second screen to detector distance. Note that whereas the results (13) for the single screen are independent of propagation distance (although the spatial structure of the pattern grows, equation (14)) result (15) depends on the ratio r. Equation (15) is plotted in Figure 2 for various values of ν. It reduces to the single screen result when r is large (detector at second screen) and when r is small (two screens close together) as expected, but shows significantly enhanced fluctuations when r lies between 1 and 10.

Finally it is perhaps worth noting that computer simulation of ray density patterns is relatively straightforward and can be used to confirm the theory. An example of ray propagation from a Brownian sub-fractal with finite outer scale is shown in Figure 3.

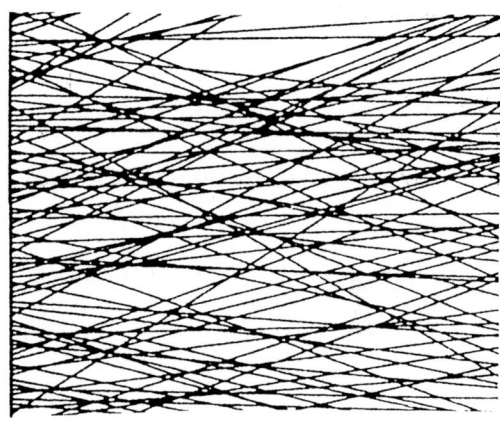

FIGURE 3
Ray propagation beyond a diffuser with Brownian fractal slope

6. CONCLUSIONS

It has only been possible in this brief review to mention a few of the more significant features which characterise refringement scattering from Gaussian random fractals. These serve to illustrate the qualitative differences with the more familiar scattering from smoothly varying objects. In practice it is found that most solid surfaces, for example, have a range of scales which may extend down through the wavelength level and exhibit fractal-like scattering properties at least in some geometries. Fluids, however, tend to behave in a more smoothly varying manner, although there is some evidence that sub-fractal models might be appropriate for turbulent systems. It is clear, however, that the commonly assumed Kolmogrov model ($\nu = 5/3$ in equation (6)) cannot predict the geometrical optics effects which are often generated when waves propagate through turbulent media: inner scale smoothing must be included in the scattering calculations.

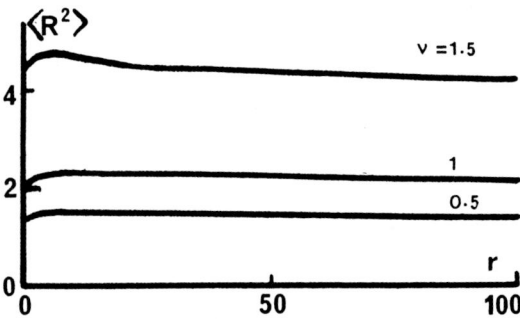

FIGURE 2
Scattering by two screens (equation (15))

REFERENCES

1. B.B. Mandelbrot, The fractal geometry of nature (Freeman, San Francisco 1982).

2. E. Jakeman, Coherence and Quantum Optics V (Plenum, New York, 1984) 1039.

3. E. Jakeman and J.G. McWhirter, J. Phys. A 10 (1977) 1599.

4. J.C. Dainty, ed., Laser Speckle and Related Phenomena, Vol. 9 of Topics in Applied Physics (Springer-Verlag, Berlin, 1975).

5. E. Jakeman, Optical Engineering 23 (1984) 453.

6. M.V. Berry, J. Phys. A 12 (1979) 781.

7. R.S. Sayles and T.R. Thomas, Nature 271 (1978) 431.

8. C.L. Rino, Radio Science 14 (1979) 1135.

9. E. Jakeman, J. Phys. A 15 (1982) L55.

10. J.G. Walker and E. Jakeman, Optica Acta 29 (1982) 313.

11. D.L. Jordan, R.C. Hollins and E. Jakeman, Appl. Phys. B 31 (1983) 179.

12. D.L. Jordan, R.C. Hollins and E. Jakeman, Opt. Commun. 49 (1984) 1.

13. E. Jakeman, Optica Acta 30 (1983) 1207.

14. J.G. Walker and E. Jakeman, Optica Acta 31 (1984) 1185.

15. E. Jakeman, J. Opt. Soc. Am. 72 (1982) 1034.

16. R.J. Glauber, Phys. Rev. 131 (1963) 2766.

OPTICAL FOURIER TRANSFORMS OF FRACTALS

Catherine ALLAIN, Michel CLOITRE

Laboratoire d'Hydrodynamique et Mécanique Physique UA CNRS/857, ESPCI, 10 rue Vauquelin, Paris 75005, France

Diffraction experiments are performed on one and two dimensional deterministic fractals. We show that this method allows a direct determination of several geometrical characteristics of fractals among these, the Hausdorff's dimension D. Applications to experimentally obtained objects are discussed.

1. INTRODUCTION

Most of the fractals encountered in experimental situations exhibit self-similarity, ie their geometric caracteristics are invariant over dilatations. In general, one of the purposes of an experiment is to determine the Hausdorff's dimension D and the limits within which the structure exhibits fractal properties, ε and L. For instance, in the case of an aggregate, these are the sizes of an individual particle, ε, and the largest scale of the cluster, L. Determination of D can be acheived by studying the way in which the mass M embedded in a sphere of radius R increases : $M(R) \sim R^D$ [1]. Another procedure uses the density-density correlation function g(R) which conforms to a power law variation : $g(R) \sim R^{D-d}$ [2]. Scattering techniques are well-suited for laboratory investigations and have been used in experiments on silica gels[3] and colloids[4].

In this paper, we present Optical Fourier Transforms (OFT) performed on fractal gratings. OFT provides a powerful method to analyze and manipulate the spatial frequencies of an object. In particular, OFT allows a direct determination of D. First, we describe briefly an experimental arrangement which performs OFT on two dimensional objects. Applications to some deterministic fractal gratings are discussed in details and compared with exact calculations of the Fourier transforms. Extensions of this analogic method to experimentally obtained aggregates are discussed.

2. EXPERIMENTAL PROCEDURES

The fractals studied in this paper are one and two dimensional self-similar objects which can be constructed recursively, for instance Cantor bars[1] and Vicsek fractals[5] represented in figure 1.

 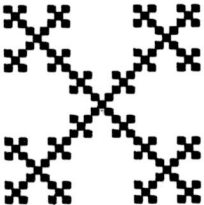

FIGURE 1
A Cantorian triadic bar and a Vicsek fractal

They are calculated on a microcomputer and drawn on a graphics plotter. Then, transparencies are realized on a 24×35 mm high resolution film. Because of the width of the lines drawn by our graphics plotter, we are limited to fractals obtained from about 7 iterations; then the ratio between the greatest scale L of the fractal and the smallest one is about $L/\varepsilon \sim 1000$.

This limitation is released if we are interested in the analysis of transparencies prepared from a real experimental fractal. In this case, the essential limitation arises from the finite size of an elementary grain on the film, which gives the smallest scale ϵ which can be recorded on the film ($\epsilon_{min} \simeq 2\mu m$) and from its largest scale L which determines the resolution of $I(\vec{Q})$ measurements; the resulting ratio L/ϵ, which is the maximum number of scales which may be recorded on a film, is about 10000. In practice, the magnification ratio is adjusted in order to make the picture of the object to be studied fit within these limits.

The optical arrangement on which the diffraction experiments are performed is represented in figure 2 :

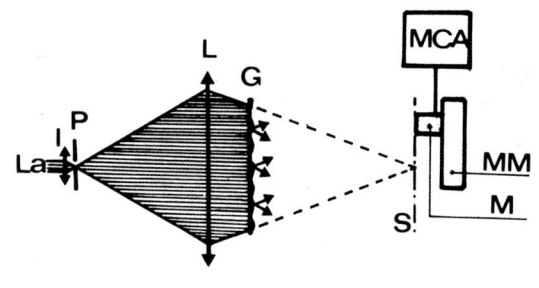

FIGURE 2
Schematic diagram of the experimental arrangement

The beam La of an argon laser (488 nm, 600 mW) is expanded by a microscope lens 1; at its focal point, a small pinhole P (25 μm in diameter) acts as a spatial filter selecting properly the transverse mode TE_{00}. L is a converging lens (Nikkor 690 mm, f/11). In the following, we shall neglect any diffraction limitation which might be introduced by lens L. The fractal grating to be studied is disposed between lens L and the screen S. The diffraction pattern $I(\vec{Q})$ observed on S is the Fourier Transform of the grating G^6 ($\vec{Q}$ is the spatial frequency of coordinates p,q). $I(\vec{Q})$ is recorded by means of a photomultiplier, M. Its displacement in the Fourier plane is controlled by a high precision micrometer, MM. Finally, the data are stored in a multichannel analyzer (MCA).

3. OFT OF CANTOR BARS.

We show in reference 6 that the Fourier transform of any self-similar fractal can be calculated analytically. For instance, the diffraction pattern of Cantor bars generated at iteration n (D=Log2/Log3) is one dimensional and is given by:

$$I_n(q) = 2^{2n}\epsilon^2 [\prod_{i=0}^{n-1} \cos(2\pi q 3^i \epsilon)]^2 [\frac{\sin \pi q \epsilon}{\pi q \epsilon}]^2$$

The coefficient $2^{2n}\epsilon^2$ is the intensity transmitted at q=0. It can be shown easily that it varies as :

$$\sqrt{I_n(0)} \sim L^D \epsilon^{1-D}$$

Experimentally, $\sqrt{I_n(0)}$ simply measures the mass M embedded in a fractal of linear size L. Using stops of different apertures to limit the part of G effectively illuminated, one can vary L and consequently determine D. This method is analogous to the determinations of D based on mass measurements which have been described in the introduction. In the following, we have normalized $I_n(0)$ to 1.

As usual in diffraction experiments, $I_n(q)$ is composed of a form factor F(q) and of a structure factor S(q). F(q) corresponds to the intensity diffracted by an elementary unit (here a bar of width ϵ) :

$$F(q) = [\frac{\sin \pi q \epsilon}{\pi q \epsilon}]^2$$

F(q) goes to zero at $q_{max} = 1/\epsilon$. In the following, we take $q_{max} = 1$.
The structure factor S(q) describes the way in which the elementary units are distributed in the fractal :

$$S(q) = [\prod_{i=0}^{n-1} \cos(2\pi q 3^i \epsilon)]^2$$

We have studied in detail this structure factor

in a previous paper[6]. Let us simply recall that the self-similarity of the object in real space is reflected in the OFT by the existence of n-1 frequency bands which are scale invariant over dilatations of factor 3 (B_j). The mean energy $<S(q)>$ scattered over the frequency bands B_j varies as :

$$<S(q)> \sim q^{-D} \quad (1)$$

This relation obtained after an averaging operation in order to break up any regularity due to the deterministic nature of Cantor bars is the same as the one observed for random structures[7].

Figures 3a and 3b show the experimental and calculated spectra of Cantor bars (n=7) :

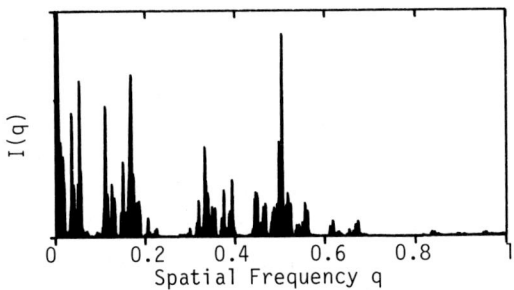

FIGURE 3a
Experimental diffraction spectrum of a Cantorian triadic bar (n=7).

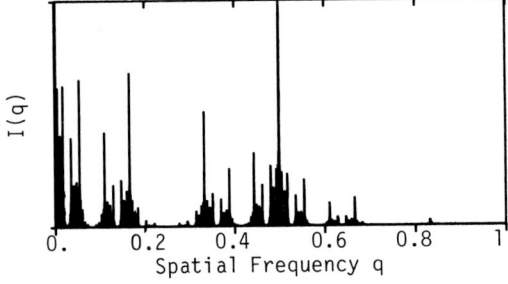

FIGURE 3b
Representation of the computed diffraction spectrum of a Cantorian triadic bar (n=7).

Since the first zero of F(q) occurs for q=1, at high spatial frequencies, F(q) smears out the structure of the grating; ε is determined from the measure of q_{max}. On the contrary, at low angles (q<0.25), F(q)~1 and F(q) is neglected. The frequency of the oscillations of the diffracted intensity on spectra 3a and 3b is related to 1/L : this provides a direct determination of L. Finally, calculations of $<S(q)>$ for the experimental spectrum lead to the power law variation (1); the value of D fits with the expected one within an uncertainty of 10%.

4. GENERALIZATION TO 2d FRACTALS.

The preceding method can be generalized to any deterministic or random fractal in 2d. The experimental procedure described above is applied identically with the proviso that a two dimensional scanning of the Fourier plane is necessary to record I(Q) completely. It is noteworthy that the intensity diffracted on one axis, q for instance, is:

$$I(p) = \int_{-\infty}^{+\infty} e^{2\pi i q y} \left(\int_{-\infty}^{+\infty} T(x,y) \, dx \right) dy$$

where T(x,y) is the optical transmittance of the fractal being studied. This formula can be interpreted as the OFT of the optical projection of the fractal on axis y. Note that this projection may not be a fractal. If it is, the value of the fractal dimension determined from I(q) measurements does not necessarily coincide with the fractal dimension of the initial object. Weitz et al have shown[8] that in general, the two values are identical only if :

D<d-1

d being the dimension of the euclidian space embedding the fractal.

Figure 4 is the OFT of a Vicsek fractal (n=5) It clearly exhibits invariance over dilatations of factor 3. The determination of D can be acheived through relation (1), averaging the diffracted intensity over each of the frequency bands which are scale invariant.

Now let us be interested in random fractals. Figure 5 is the OFT of a 1024 particles aggregate[9] generated out of lattice according to the Witten-Sander model[2].

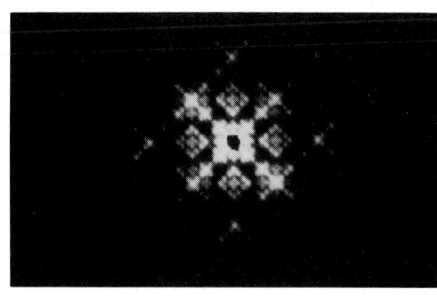

FIGURE 4
OFT of a Vicsek fractal (n=5).

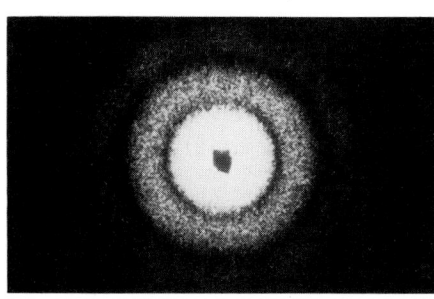

FIGURE 5
OFT of a 1024 particles aggregate generated out of lattice according to the Witten-Sander model.

The bright rings surrounding the central peak are due to the form factor, ie the diffraction pattern of a single spherical particle, which modulates the structure factor. As above, the form factor goes to zero at wavevectors $\vec{Q}$ such as $Q=1/\epsilon$. ϵ is the diameter of an individual particle. Here, because of the randomness of the initial fractal, no discrete structure is observed on the diffraction pattern. The determination of D is acheived through the application of formula (1), $<S(Q)>$ being the diffracted intensity averaged over circles of radius Q.

5. CONCLUSIONS

Summarizing, we have presented an analogic method for performing OFT of 2d fractal objects It allows a direct determination of L, ϵ and D. Applications have been given in detail for a Cantorian triadic bar; the experimental spectra have been compared to the exact one obtained from analytical calculations. Extensions to more complicated cases, 2d deterministic or random fractals have been discussed. Our purpose is now to apply OFT to the analysis of experimentally obtained aggregates, for instance the 2d clusters which are formed during the aggregation of small particles floating on a liquid[10].

REFERENCES

1. B.B. Mandelbrot, The fractal geometry of Nature (W.H. Freeman and Company, New York, 1982)

2. T.A. Witten and L.M. Sander, Phys. Rev. Lett. 47(1981)1400.

3. D.W. Shaefer and K.D. Keefer, Phys. Rev. Lett. 53(1984)1383.

4. D.W. Shaefer, J.E. Martin, D. Cannel and P. Wiltzius, Phys. Rev. Lett. 52(1984)2371.

5. T. Vicsek, J. Phys. A 16(1983)L647.

6. C. Allain and M. Cloitre, preprint submitted to Phys. Rev. B.

7. J.K. Kjems and T. Freltoft, to appear in the Proceedings of the NATO ASI (Geilo, 1985).

8. D.A. Weitz and M. Oliveira, Phys. Rev. Lett. 52(1984)1433.

9. This aggregate was computed by R. Jullien and M. Kolb (Orsay).

10. C. Allain and M. Cloitre, this volume.

FRACTALS IN PHYSICS
L. Pietronero, E. Tosatti (editors)
© Elsevier Science Publishers B.V., 1986

ON THE MEASURE OF FRACTAL DIMENSIONALITIES THROUGH PHYSICAL PROPERTIES

Constantino TSALLIS

Centro Brasileiro de Pesquisas Físicas/CNPq, Rua Xavier Sigaud, 150, 22290 - Rio de Janeiro, Brazil[*] and Centre de Recherches sur les Très Basses Températures/CNRS, B.P. 166, Centre de Tri, 38042 Grenoble Cedex, France

The measure of the *fractal dimensionalities* of real substances through direct use of the mathematical definitions can be extremely cumbersome. We claim that in the case of arbitrarily rough and irregular conductors, the task can be considerably simplified by using the traditional skin effect. Indeed both *electrical resistance* and *electromagnetic power dissipation* should present anomalous power-law dependences on the applied frequency, the corresponding exponents being directly related to relevant fractal dimensionalities.

1. INTRODUCTION

An enormous amount of real physical systems present, within appropriate length ranges, a fractal nature which is commonly characterized by the values of one or more relevant fractal dimensionalities. The usefulness of such approach in the case of porous coals[1], proteins[2,3], sintered metallic powders[4], cement gels[5] and many other substances[6] needs no more to be proved. However the practical determination of the fractal dimensionalities is quite frequently cumbersome, which in turn implies in results whose degree of confidence is not always satisfactory. We claim[7] that the traditional skin effect can be an excellent tool for overcoming the above difficulties whenever we are dealing with substances presenting a non neglectable electrical conductivity.

There are physical properties (e.g., spin relaxation[2], specific heat[8]) which are closely related to both *static* (purely geometric) and *dynamic* fractal dimensionalities. For example, the low temperature (T) spin-lattice relaxation time t_1 of some hemoproteins is given[2] by $1/t_1 \propto T^n$, where the non-integer exponent n can be related to a non-integer dimensionality $\bar{d}$ through[2,3] $n = 3 + 2\bar{d}$. Under certain circumstances, the dimensionality $\bar{d}$ can be identified[3] with the *fracton dimensionality*[9] $d_{fr} = 2d_f/d_w$, where d_f is the structural fractal dimensionality of the hemoprotein, and d_w is the fractal dimensionality of a random walk (not self-avoiding) constrained to the fractal (d_w is defined through $<r^2>^{d_w} \sim t^2$, where $<r^2>$ is the mean square end-to-end distance of the walk after t steps). We then see that the experimental data (say n) yields information on the purely geometric dimensionality d_f, but the relationship is rather indirect, and mixed with the "dynamic" dimensionality d_w (reflecting diffusion and vibrational aspects of the problem). Consequently we would hardly consider spin relaxation measurements as a practical tool for determining d_f.

On the other hand, other physical phenomena (e.g., small-angle X-ray scattering[1,6]) exist which provide experimental data *directly* related to *purely geometric* fractal dimensionalities of the system, and are therefore convenient for determining those quantities. For example, in the just quoted X-ray experiments, the scattering intensity I is proportional[1] to

---

[*]Permanent address

q^{d_S-6} where q is the scattering angle (radians). Consequently a $\ln I$ vs. $\ln q$ representation of the experimental data immediately provides the fractal surface dimensionality d_S (which equals 2 in the Euclidean case, thus providing the well known $I \propto 1/q^{-4}$ law). We argue herein that the standard skin effect belongs to this same category of methods, and should therefore constitute a convenient way for determining fractal dimensionalities of arbitrarily rough and irregular conductors, the use of the mathematical definitions of those quantities being replaced by relatively simple physical measurements.

2. SKIN EFFECT

Let us consider a roughly cylinder-like conductor (see Fig. 1). We denote ℓ_L and ℓ_T its fractal longitudinal and transverse (perimeter) lengths respectively, and S the fractal area they support. For arbitrary fractal surfaces it will in general be $S \neq \ell_L \ell_T$. We assume that an alternate electric potential with voltage V and frequency ω is applied along the cylinder. The corresponding skin depth is given[10] by

$$\delta \propto \frac{1}{\sqrt{\omega}}$$

where the proportionality factor depends on the electromagnetic properties (such as the electrical conductivity) of the (homogeneous) substance.

The fractal quantities ℓ_L, ℓ_T and S respectively yield the fractal dimensionalities d_L, d_T and d_S through the following relations:

$$d_L = \frac{\ln(\ell_L/\delta)}{\ln(1/\delta)} \tag{2}$$

hence

$$\ell_L = \frac{1}{\delta^{d_L-1}}, \tag{2'}$$

$$d_T = \frac{\ln(\ell_T/\delta)}{\ln(1/\delta)} \tag{3}$$

hence

$$\ell_T = \frac{1}{\delta^{d_T-1}}, \tag{3'}$$

and

$$d_S = \frac{\ln(S/\delta^2)}{\ln(1/\delta)} \tag{4}$$

hence

$$S = \frac{1}{\delta^{d_S-2}} \tag{4'}$$

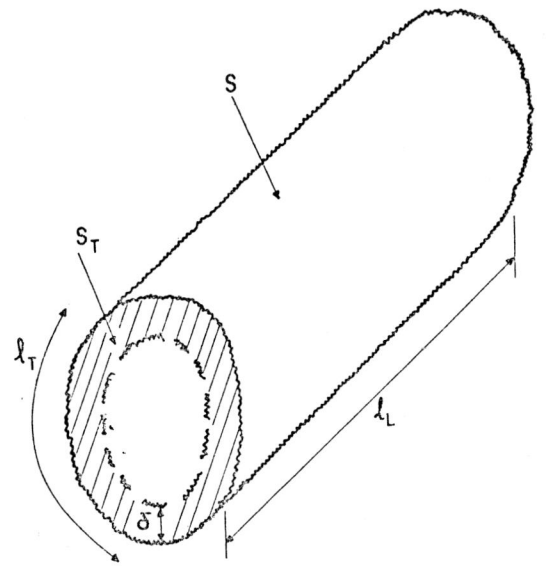

FIGURE 1
Cylinder-like conducting sample with fractal external surface; ℓ_T refers to the perimeter. The alternate voltage is applied between the two (equipotential) bases. The indicated skin depth δ is out of scale (too large) if assumed to be appropriate for probing the "fractality" of the illustrated "rough" surface.

It will in general be $1 \leq d_L$, $d_T < 2$ (the equality holds for smooth differentiable curves), and $2 \leq d_S < 3$ (the equality holds for smooth differentiable surfaces). An interesting particular situation is that in which $S \simeq \ell_L \ell_T$, which implies $d_S = d_L + d_T \geq 2$. Another interesting particular situation is that of a (statistically) homogeneous and isotropic fractal surface; in that case it will in general be $d_L = d_T = d_S - 1 \geq 1$.

The power P dissipated by the substance is given by

$$P = \int \vec{j}(\vec{r}) \cdot \vec{E}(\vec{r}) \, d^3r \quad (5)$$

where $\vec{j}(\vec{r})$ and $\vec{E}(\vec{r})$ respectively are the current density and the electric field at the point $\vec{r}$. By using Ohm's law $\vec{j}(\vec{r}) = \sigma \vec{E}(\vec{r})$ ($\sigma \equiv$ electrical conductivity), Eq. (5) becomes

$$P = \sigma \int E^2 d^3r$$
$$= \sigma <E^2>_{skin} \int_{skin} d^3r$$
$$\simeq \sigma <E^2>_{skin} S\delta \quad (6)$$

where "skin" refers to the volume $v \equiv \int_{skin} d^3r \simeq S\delta$ of the substance where the electric field is sensibly different from zero (we recall that the electric field vanishes exponentially while entering into the material).

If we perform a *fixed electromagnetic field density experiment* (in the interior of an appropriate cavity), Eq. (6) yields

$$P \propto S\delta \simeq \delta^{3-d_S} \quad (7)$$

where we have used Eq. (4'). Therefore, using Eq. (1), we obtain

$$P \propto \frac{1}{\omega^{\frac{3-d_S}{2}}} \quad (8)$$

This relation reproduces the standard one[10] ($P \propto 1/\sqrt{\omega}$) when $d_S=2$. Note also that Eq. (7) implies $v \simeq \delta^{3-d_S}$, which through appropriate change of the characteristic length, recovers Eq. (4) of Ref. 1.

If we perform instead an *electric current flow experiment* we will have $<E^2>_{skin} \simeq V^2/\ell_L^2$ hence, using Eq. (6),

$$P \simeq \frac{\sigma V^2 S\delta}{\ell_L^2} \quad (9)$$

which, if identified with $P = V^2/R$ (R being the electrical resistance), provides

$$R \simeq \frac{\sigma \ell_L^2}{S\delta} \quad (10)$$

Finally, using Eqs. (2') and (4'), we obtain

$$R \propto \delta^{d_S - 2d_L - 1} \quad (11)$$

hence, through Eq. (1),

$$R \propto \omega^{\frac{1 + 2d_L - d_S}{2}} \quad (12)$$

This relation reproduces the standard one[10] ($R \propto \sqrt{\omega}$) when $d_S = 2$ and $d_L = 1$.

Summarizing, the representation of the experimental data in $\ln P$ vs. $\ln \omega$ and $\ln R$ vs. $\ln \omega$ graphs would provide straight lines whose respective slopes would be $(d_S-3)/2$ and $(1+2d_L-d_S)/2$. The determination of d_S and d_L would then be straightforward.

In the particular case mentioned before, namely $d_S = d_L + d_T$, Eqs. (8) and (12) provide

$$P \propto \frac{1}{\omega^{\frac{3-d_L-d_T}{2}}} \quad (13)$$

and

$$R \propto \omega^{\frac{1+d_L-d_T}{2}} \quad (14)$$

In the other particular case (quite frequent in nature), namely $d_L = d_T = d_S - 1$, Eqs. (8) and (12) provide

$$P \propto \frac{1}{\omega^{\frac{2-d_L}{2}}} \quad (15)$$

and

$$R \propto \omega^{d_L/2} \quad (16)$$

In this case a single experience, say R vs. ω, would determine d_L.

In both types of experiment (electric current flow and electromagnetic cavity) the spatial distribution of the electromagnetic field presents an evolution with ω: it is through this evolution that the field probes the fractality of the system. We have qualitatively represented in Fig. 2.a (Fig. 2.b) the spatial distribution of the electric field outside (inside) of the substance in a electromagnetic cavity (electric current flow) experiment.

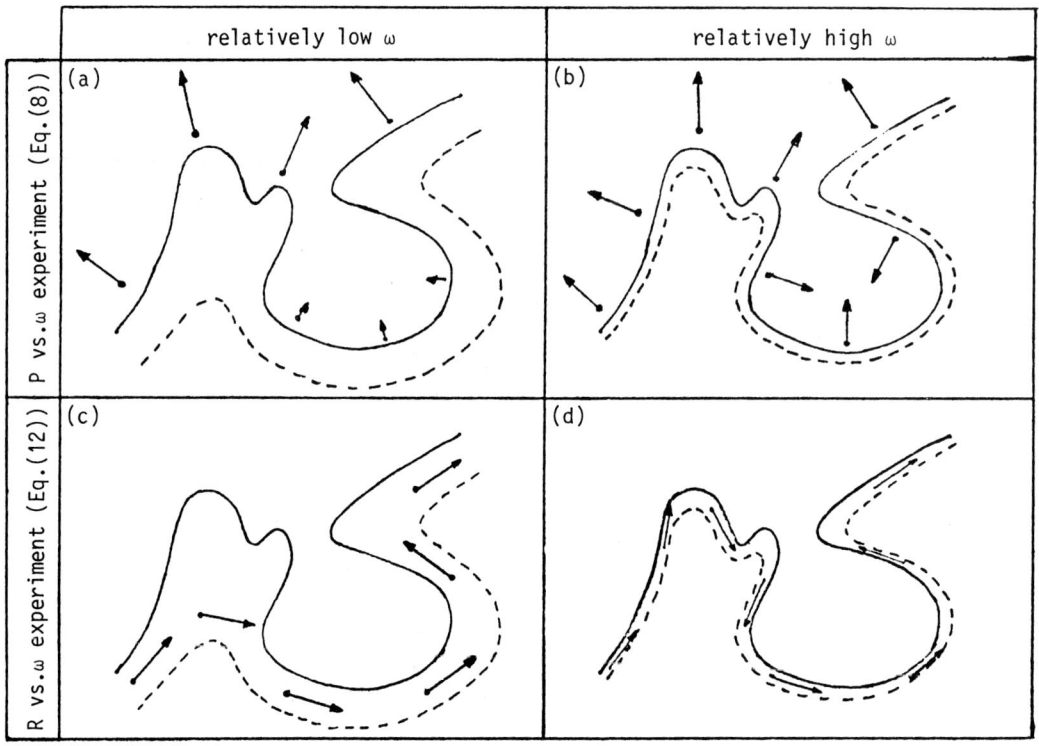

FIGURE 2
Magnified view of the rough surface of the conductor: the dashed line qualitatively indicates the skin of the substance at a particular frequency. For the cavity experiment we have indicated ((a) and (b)) the electric field *outside* of the conductor; for the current flow experiment we have indicated ((c) and (d)) electric field *inside* of it.

3. CONCLUSION

As a consequence of the scaling arguments presented above, the skin effect appears to be a promising tool for measuring fractal dimensionalities of electrically conducting materials. Real substances are naturally not expected to exhibit this type of anomalous skin effect for all frequencies. Therefore an actual experiment, say a resistance measurement, should present various regimes, making crossovers from one into the other at appropriate frequencies. A *low frequency regime* will always be present in which the resistance is independent from ω (δ larger than the transverse linear size of the sample); this regime ends when δ becomes comparable with this size. A *high frequency regime* will also exist always; it corresponds to the break-down of the condition $\omega \ll \sigma/\varepsilon$ ($\varepsilon \equiv$ dielectric constant), which is necessary for the validity of Eq. (1) (if $\omega > \sigma/\varepsilon$, the polarizability itself starts depending on ω, and non trivial effects appear). Between these low and high frequency regimes, one or more fractal and/or euclidean skin regimes can appear, corresponding to scales of δ within which the surface of the conductor is seen as "rough" or "smooth". In Fig. 3 we have illustrated these concepts by assuming, in the intermediate, frequency region, a fractal regime (slope $(1 + 2d_L - d_S)/2$ different from $1/2$) followed by a standard one (slope $1/2$).

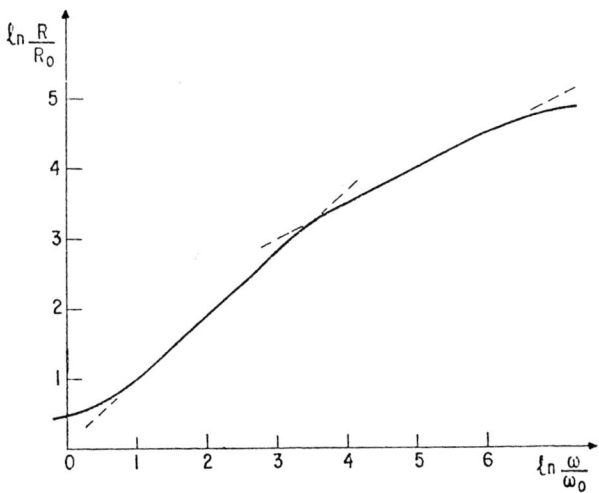

FIGURE 3
Possible result for a resistance measurement of a conducting rough sample, exhibiting a fractal regime (slope $\neq 1/2$) which crosses over to a standard one (slope $1/2$). $\omega_0(R_0)$ is a reference frequency (resistance) adapted to the particular sample.

Let us conclude by saying that experiments testing Eqs. (8) and (12) would be extremely welcome. Systems like copper, gold, silver porals[4], porous coals, metallic sponges could be good candidates.

ACKNOWLEDGEMENTS

I acknowledge interesting discussions with A.O. Caride, C.G. Bollini, J.J. Giambiagi, J.S. Helman, A.F. Craievich and H.J. Herrmann; fruitful remarks from L. Pietronero, L.M. Sander and P. Pfeifer are acknowledged as well.

REFERENCES
1. H.D. Bale and P.W. Schmidt, Phys. Rev. Lett. 53 (1984) 596.
2. H.J. Stapleton, J.P. Allen, C.P. Flynn, D.G. Stinson and S.R. Kurtz, Phys. Rev. Lett. 45 (1980) 1456; J.P. Allen, J.T. Colvin, D.G. Stinson, C.P. Flynn and H.J. Stapleton, Biophys. J. 38 (1982) 299.
3. J.S. Helman, A. Coniglio and C. Tsallis, Phys. Rev. Lett. 53 (1984) 1195 and 54 (1985) 1735; M.E. Cates, Phys. Rev. Lett. 54 (1985) 1733; H.J. Stapleton, Phys. Rev. Lett. 54 (1985) 1734.
4. H. Franco, J. Bossy and H. Godfrin, Cryogenics (September 1984) 477.
5. A.J. Allen and P. Schofield, preprint (1985), to be published in the Proceedings of the NATO meeting on "Scaling Phenomenon in Disordered System", Geilo, April 1985 (Plenum).
6. P. Pfeifer and D. Avnir, J. Chem. Phys. 79 (1983) 3558; D. Avnir, D. Farin and P. Pfeifer, J. Chem. Phys. 79 (1983) 3566; D. Avnir, D. Farin and P. Pfeifer, Nature (London) 308 (1984) 261.
7. C. Tsallis, A.O. Caride, C.G. Bollini, J.J. Giambiagi and J.S. Helman, to be published.
8. L.D. Dillon, R.E. Rapp and O.E. Vilches, J. Low Temp. Phys. 59 (1985) 35; R.E. Rapp, L.D. Dillon and H. Godfrin, Cryogenics 25 (1985) 152.
9. S. Alexander and R. Orbach, J. Phys. (Paris) Lett. 43 (1982) L 625.
10. J.D. Jackson, Classical Electrodynamics, 2nd edition (J. Wiley and Sons, 1975).

Part III
POLYMER STATISTICS AND
SELF-AVOIDING-WALKS

FRACTALS IN PHYSICS
L. Pietronero, E. Tosatti (editors)
© Elsevier Science Publishers B.V., 1986

RANDOM WALKS WITH MEMORY

L. PELITI

Dipartimento di Fisica, Università "La Sapienza", Piazzale Aldo Moro, 2, I-00185 ROMA (Italy) and GNSM-CNR, Unità di Roma.

Random walks with memory are a popular arena for methods aimed at describing irreversible aggregation phenomena. They produce topologically trivial aggregates whose embedding in ambient space can be described in terms of fractal concepts. The analogy with scaling properties of critical phenomena is well established and has led to a more complete understanding of their properties than for general aggregation processes.

1. INTRODUCTION

One of the simplest fractals generated by a physical process is the track of a Brownian particle: it appears as the sole illustration taken from experiment in B. Mandelbrodt's recent book [1]. One may model the unfolding of the track as a simple aggregation process: the Random Walk (RW) . It is a simple exercise to interpret the properties of the RW from the point of view of the theory of fractals. One has therefore a model which exhibits simple but nontrivial fractal properties and is both exactly tractable and physically realized.

Building on this success, one has introduced variations on the theme of the RW in the hope of obtaining tractable models with different fractal properties. This program is all the more exciting since de Gennes's proof [2] of the strict analogy between the phase transition in a general ferromagnetic model and the statistical properties of self-avoiding configurations of long chains, which are easily modelled in turn as tracks of a kind of random walk with memory: the self-avoiding walk (SAW). One has witnessed therefore a blossoming of models of random walks with memory, in which different "rules of the game" lead to different geometrical properties of the aggregates produced. These models do not as a rule aim at describing an actual physical process: one hopes instead to understand by their study the physical processes which lie at the roots of fractality. The analogy with critical phenomena is strict and powerful, allowing for the success of renormalization group methods in most cases. On the other hand this same success may be at the origin of an error in perspective, leading to considering as essential those aspects of aggregation processes which are most easily interpreted in terms of the current views of scaling in critical phenomena. Random walks with memory provide a counterweight to this tendency, with the striking (and still quite unexplained) success of self-consistent (Flory [3], Pietronero [4]) arguments. They, contrary to the Landau arguments in critical phenomena, provide us with nontrivial, dimensionality-dependent exponents, which are in general in close agreement with experimental or simulation data. These arguments are peculiar to the theory of linear aggregates, and the explanation of their efficacity is still a challenge to the theoreticians.

I shall only treat in this paper <u>walks</u>, considered as kinetic processes producing topologically linear aggregates. I do not discuss therefore equilibrium properties of linear polymers subject to different constraints, except insofar as they correctly describe the asymptotic properties of walks. The simplest walk, RW, is introduced in sec. 2, where the effects of a finitely lived memory are briefly discurred. The self-avoiding walk (SAW) - an aggregation process which reproduces the statistics of chains with excluded volume - is introduced and discussed in sec. 3. The so-called "true" self-avoiding walk (TSAW) is described and discussed in sec. 4. Further variations and developments are sketched in sec. 5. A few conclusions are drawn in sec. 6.

Most of the contents of this paper and its general structure are borrowed from a review article [5] which has been written in collaboration with Prof. L. Pietronero. I refer the interested reader to this article for more details.

2. RANDOM WALKS WITH SHORT MEMORY

The random walk (RW) on a lattice models a diffusing (Brownian) particle. Let us consider discrete epochs t_o, t_1, t_2, ... At each epoch t_n the walker is at a site $\vec{r}$ of a d-dimensional (simple cubic) lattice. At epoch t_{n+1} the walker may be at any of the nearest neighbors $\vec{r}' = \vec{r} + \vec{\delta}$ of the site $\vec{r}$: $\vec{\delta}$ is a vector leading from $\vec{r}$ to one of its nearest neighbors (a nearest neighbor vector). For a random walk without memory the conditional probability $p(\vec{r}', t_{n+1}|\vec{r}, t_n)$ that the walker is at site $\vec{r}'$ at epoch t_{n+1} given that it was at $\vec{r}$ at epoch t_n fully describes the process, and only depends on $\vec{r}$, $\vec{r}'$ (and possibly on t_n). In the simplest case it is a constant, which equals the inverse of the coordination number q of the lattice. It is straightforward to derive in this case the asymptotic probability distribution $P(\vec{r},t)$ of the location $\vec{r}$ of a walker which left off the origin at epoch $t_o = 0$. It is a Gaussian of variance $\xi^2(t) = Dt$ where D is the diffusion constant, expressed in terms of the lattice constant $a = |\vec{\delta}|$ and of $\Delta t = t_{n+1} - t_n$ by

$$D = a^2/\Delta t. \qquad (1)$$

The fractal we are interested in is the track $(0 = \vec{r}(t_o), \vec{r}(t_1), ... \vec{r}(t))$ left by the walker. It has a linear extension of the order of $\xi(t)$ and for lengths ℓ satisfying $a \ll \ell \ll \xi(t)$ exhibits a self-similarity which may be characterized by a fractal dimension d_f which equals 2 in all dimensions. We may consider indeed stepping sites at epochs separated by a time interval $\Delta t'$ which is a large integer multiple of Δt, but still much smaller than t. These sites are each separated from the next by a distance ℓ of order $(D\Delta t')^{\frac{1}{2}}$, since the walk which takes place in each time interval of duration $\Delta t'$ is itself a random walk with the same diffusion constant. As a consequence the whole track can be covered by a number M of boxes of radius ℓ, where

$$M = t/\Delta t' = Dt/\ell^2, \qquad (2)$$

which corresponds to the usual "box counting" definition of the fractal dimension. This reasoning allows one to relate the fractal dimension d_f to the exponent which characterizes the increase in the end-to-end distance of the walk with increasing duration, and which is denoted by ν by analogy with the theory of critical phenomena:

$$\xi^2(t) = \langle|\vec{r}(t)|^2\rangle \propto t^{2\nu}. \qquad (3)$$

We have:

$$d_f = 1/\nu. \qquad (4)$$

To the extent that self-similarity is possessed at all by the tracks of the walks we are going to discuss, this relation carries over to more general cases. Another exponent, γ, is related to the total number $\mathcal{N}$ of walks of duration t which may be generated. Denoting by $N = t/\Delta t$ the total number of steps we have in general:

$$\mathcal{N} \propto (q^*)^N N^{\gamma-1} \qquad (5)$$

For the RW we have $q^* = q$ and $\gamma = 1$.

It is easy to convince oneself that these properties still hold for walks with finitely lived memory. The process is described in this case by the conditional probabilities $p(\vec{r}_{n+1}, t_{n+1}|\vec{r}_n,t_n; \vec{r}_{n-1}, t_{n-1}; \ldots; \vec{r}_{n-k}, t_{n-k})$ that the walker is at site $\vec{r}_{n+1}$ at epoch t_{n+1} if its track between epochs t_{n-k} and t_n is given by $(\vec{r}_{n-k}, \vec{r}_{n-k+1}, \ldots, \vec{r}_n)$. In this way memory of the recent history of the walk (for a duration $\Delta t' = k \Delta t$) modifies the probability of taking the next step. But one only has to consider the track as being defined by the stopping sites at epochs separated by time intervals larger than $\Delta t'$ to recover the previous model. The conclusion is that finitely lived memory effects do not modify the asymptotic properties of the RW.

3. THE SELF-AVOIDING WALK (SAW)

The self-avoiding walk (SAW) is a kinetic process which asymptotically reproduces the equilibrium statistics of linear polymers in a good solvent. In the process a walker steps at random on a lattice, but is suppressed (and the corresponding track is removed from the statistics) whenever it happens to step on a site which has already been visited in the past. The properties of this process can be analyzed by a simple self-consistency argument [4] which reproduces the results obtained by Flory [3] for the equilibrium properties of polymers. Let us denote by $G(\vec{r},t)$ the probability distribution function of the location at epoch t of a walker which leaves off the origin at epoch 0.

This is given by the product of the distribution $G_o(\vec{r},t)$, calculated for a standard random walk with the same stepping probabilities, with a survival probability $S(\vec{r},t)$. This function gives the probability that a walk of duration t which lands at $\vec{r}$ has no self-intersections. It is clear that $S(\vec{r},t)$ is a growing function of $|\vec{r}|$, because for more elongated walks the probability of self-intersections is correspondingly smaller. For long durations t and not too large values of $|\vec{r}|$ $G_o(\vec{r},t)$ can be approximated by a Gaussian:

$$G_o(\vec{r},t) \simeq (2\pi Dt)^{-d/2} \exp(-|\vec{r}|^2/2Dt). \quad (6)$$

Let us now evaluate $S(\vec{r},t)$. The walk extends over a volume of order $|\vec{r}|^d$ and has therefore an average density $\rho \propto t|\vec{r}|^{-d}$. Let us look at the fully developed walk of duration t. At each step, the probability to overlap with another position of the walk is of order ρ. The probability to survive after each step is therefore $(1-\rho)$ and the probability to survive after $N = t/\Delta t$ steps is given by:

$$S(\vec{r},t) \sim (1-\rho)^N \sim \exp(-N\rho), \quad (7)$$

which we may write

$$S(\vec{r},t) \sim \exp(-gt^2|\vec{r}|^{-d}). \quad (8)$$

where g is some (positive) constant. Let us now compute the mean square end-to-end distance $\xi^2(t)$. We have:

$$\xi^2(t) = <|\vec{r}|^2> = \int d^d\vec{r} \, |\vec{r}|^2 \, G(\vec{r},t)$$
$$= \int d^d\vec{r} \, |\vec{r}|^2 \, G_o(\vec{r},t) \, S(\vec{r},t) \quad (9)$$
$$= \int d^d\vec{r} \, |\vec{r}|^2 \, \exp[-F(\vec{r},t)],$$

where

$$F(\vec{r},t) = \frac{|\vec{r}|^2}{2Dt} + g\frac{t^2}{|\vec{r}|^d}. \quad (10)$$

If we compute the integral in eq. (9) by the saddle point approximation we obtain

$$\xi(t) \sim (gDt)^{\nu_F}, \quad (11)$$

where the exponent ν_F is given by the Flory expression:

$$\nu_F = \frac{3}{d+2} \quad (12)$$

and corresponds to a fractal dimension $d_f = (d+2)/3$. Let us remark that when the ambient dimension, d, exceeds four, and eq. (12) would give an exponent ν smaller than the value 1/2 corresponding to the RW, the above reasoning does not apply. In this case the saddle point value of F decreases with increasing t and the approximation is no more valid. It is indeed the Gaussian behavior of the first term which dominates, and the asymptotic values of the SAW are the same in this case as those of the RW. Dimension four thus appears as the upper critical dimension of the SAW. Eq. (12) is in striking agreement with the results of both experiment or simulation: while correctly identifying the upper critical dimension with four, and giving for d = 1 the obviously correct prediction $\nu = 1$, it is off by no more than 1% for $d = 3$ and yields for $d = 2$ the probably exact value $\nu = .75$. There is at present no convincing explanation of the striking success of this approximation [6]. A similar argument can be developed [7] to obtain the following prediction for the exponent γ defined in eq. (5). One obtains:

$$\gamma = 6/(d + 2). \qquad (13)$$

This result is not so good as that for ν, being off by 3% at $d = 3$ and by 10% at $d = 2$.

The SAW can also be considered from a field theoretical point of view [8,9]. Such an approach yields the proof that the asymptotic behavior of the SAW does reproduce the equilibrium statistics of polymers in a good solvent (and hence, by on Gennes's argument [2], can be studied as the $n \to o$ limit of a generalized Heisenberg model of ferromagnetism, where the order parameter has dimensionality n). It allows therefore the use of the powerful renormalization group (RG) methods, which provide us with an expansion of the exponents γ and ν in powers of $\varepsilon = 4 - d$. These methods can be derived by applying a formalism introduced by Martin, Siggia and Rose [9] to describe classical evolution equations by means of path integrals. This formalism has been applied by Doi [10] to the treatment of chemical reactions, in a form more suitable to the present purposes (see also Grassberger and Scheunert [11]; a pedagogical introduction can be found in ref. [12]). One can thus formally derive the equivalence between the asymptotic properties of the SAW and the equilibrium properties of a polymer chain in a good solvent.

4. "TRUE" SELF-AVOIDING WALKS (TSAW)

Suppose we devise the following model [13] to grow a chain on a d-dimensional lattice in self-avoiding configurations:

(i) a walker leaves from the origin at epoch zero on a d-dimensional simple cubic lattice;

(ii) at each time step the walker can go from the site $\vec{r}$ it is at to one of its nearest neighbors $\vec{r} + \vec{\delta}$;

(iii) the probability of stepping on site $\vec{r} + \vec{\delta}$ depends on the number of times, $n_{\vec{r}+\vec{\delta}}$, the site has been visited in the past, in such a way that the walker is discouraged from stepping on sites it has already visited. We take in particular

$$p_{\vec{r} \to \vec{r}+\vec{\delta}} \propto \exp(-g n_{\vec{r}+\vec{\delta}}). \qquad (14)$$

The probability is normalized in much a way that the walker performs a step at each time. The factor g, assumed to be positive, represents the intensity of repulsion from areas which have already been visited. Now, these rules do not prevent the walker from visiting twice (or more) the same lattice site, even in the limit of g going to infinity. Indeed, in such a case the walker will choose with equal probability among the sites which have been visited least often. Quite surprisingly, the walk which we have defined has a different asymptotic behavior than the SAW. It has been unfortunately called the "True" Self-Avoiding

Walk (TSAW), although probably the name "Houdini" Walk, suggested by M.F. Shlesinger at this Conference, is more appropriate, since the walker succeeds in escaping from all traps in which it may inadvertently fall. (If one were to introduce the rule that, if no previously unvisited neighbors to the actual location are available, the walker is suppressed [14, 15] one would obtain a model whose asymptotic behavior is the same as the SAW).

A heuristic argument, due to Pietronero [16] yields the upper critical dimension and the exponent ν (and therefore the fractal dimension $d_f = 1/\nu$) for this walk. If we assume that the root mean square end-to-end distance ξ (defined in eq. (3) above) is the only relevant length measuring the size of the walk, we expect the density $\rho(\vec{r},t)$ of points visited in a walk of duration t to have the form:

$$\rho(\vec{r},t) \sim t\,\xi^{-d}\,f(|\vec{r}|/\xi) \quad (15)$$

We make the usual power law assumption about the t dependence of ξ ($\xi \sim t^\nu$). The increase in ξ due to a prolongation of the walk for a duration Δt should be given by:

$$\Delta\xi \sim t^{\nu-1}\,\Delta t. \quad (16)$$

On the other hand this increase is due to the outward thrust due to repulsion effects, which may be estimated to be proportional to the gradient of ρ calculated at distances of order ξ from the origin. We have therefore from eq. (15):

$$\left.\frac{d\rho}{dr}\right|_{|\vec{r}|\sim\xi} \sim t\,\xi^{-(d+1)}\,f(1) \sim t^{1-(d+1)\nu} \quad (17)$$

Comparison of eqs. (16) and (17) yields:

$$\nu = 2/(d+2) \quad (18)$$

This expression cannot be valid if d is larger than two, since in this case the repulsion effects are negligible compared with the outward thrust due to simple diffusion. Two appears therefore as the upper critical dimensionality, where one should expect logarithmic deviations from RW behavior [13,16]. Such deviations have been calculated by RG methods and can be checked by careful analysis of series expansion [17,18]. The situation is still controversial. Let us remark that from the fractal point of view eq. (18) gives a fractal dimensionality which is larger than the ambient dimensionality d for d between two and one. This is not contradictory, since d_f in this case characterizes a mass distribution (a measure) and not a set.

One can consider two interesting nontrivial cases of the TSAW: the one-dimensional (1D) case [19-21] and the TSAW on a fractal [22]. The 1D turns out to be less trivial than one would first guess, yielding a value of ν indiscernably close to 2/3 (in accordance with eq. (18)). Although heuristic arguments have been presented to justify this value [21] one is still in presence of the same striking and quite unexplained success of Flory-like arguments as for the SAW. TSAW on a Sierpiński gasket have been simulated and a generalization of the argument presented above has been introduced. The condition for the critical dimensionality turns out to be:

$$2 - d_s/2 = d_s/d_f \qquad (19)$$

where d_s is the spectral dimension [23]. When the rhs of this equation is smaller than the lhs the exponent ν is estimated to be given by:

$$\nu = \frac{d_s}{d_f} \cdot \frac{2}{d_s + 2} \quad . \qquad (20)$$

This expression is suggested on the one hand by the requirement that the quantity $d_f \cdot \nu$ should be intrinsic, i.e. independent of the embedding of the fractal in the ambient (Euclidean) space; on the other by the observation that d_s is the simplest intrinsic dimension. Substitution of d_s to d in Pietronero's expression (18) yields the result. It agrees surprisingly well with simulation data on the 2D Sierpiński gasket.

5. VARIATIONS

The TSAW met with some success since it is a simple model, easily implementable on a computer, which shows nontrivial but still tractable behavior at low dimensionality. This prompted the birth of a few other models of random walk with memory which we sketchily and incompletely mention in this section. We have mentioned in passing [14, 15] a variation on the TSAW which was shown to be in the same universality class as the SAW. Another modification, which allows the walker on a 2D lattice to recognize traps before falling into them was proposed by Kremer and Lyklema [24], who called it Indefinitely Growing Self-Avoiding Walk (IGSAW). (It can be obtained as a suitable limit of the Laplacian Walks (LW) discussed by Lyklema [25] at this conference). It does not obviously fit into one of the universality classes which we have discussed so far. This is all the more puzzling since its version on a honeycomb lattice appears to correspond to the equilibrium statistics of self-avoiding chains with suitabel nearest-neighbor and next-nearest-neighbor interactions [26]. There are actually some configurations in the equilibrium problem which do not arise in the kinetic one, so that nothing prevents IGSAW to be in an universality class of its own instead, has it has been conjectured, in the universality class of polymers in a θ-solvent.

TSAW with long-range repulsion have also been considered [27,28] both on the basis of a Flory-Pietronero argument and of a field theoretical approach. It may well be the case that heuristic arguments heat here in accuracy the more sophisticated ones, just as it happens in the case of linear polymers with long-range interactions [29,30]. The Flory results lie quite close to simulation data, whereas ε-expansion results appear to be quite off to mark, except very near the trivial cases.

One should use some care with Flory arguments, as shown in the equilibrium case by the fact that the so-called k-tolerant walks [31] where configurations are allowed provided they entail no more than k intersections at nodes, which obviously belong to the same universality class as polymers in a good solvent, would seem to have a different asymptotic behavior on the basis of a Flory argument. By the same token, it is not obvious to what extent one should trust the predictions made by Ottinger [32] about the asymptotic behavior of a variation on the TSAW, defined by the transition probability

$$P_{\vec{r} \to \vec{r}+\vec{\delta}} \propto \exp(-gn^{\alpha}_{\vec{r}+\vec{\delta}}) \qquad (21)$$

One would not expect a different asymptotic behavior for $\alpha = 1$ (ordinary TSAW) and, say, $\alpha = 2$, just as in the case of k-tolerant walks. On the other hand simulations do corroborate the predictions of Flory-Pietronero arguments. My conclusion is that a deepening of our understanding of these arguments is urgent.

6. CONCLUSIONS

One of the main challenges facing the theory of fractals is to understand which caracteristics of a physical process give rise to fractal structures. In a sense, the theory of random walks with memory avoids the question, since it gives the problem too simple a solution. On the one hand, tracks left by random walks with memory are obviously lacunar, on the other their self-similarity can usually be explained on the same basis as that shown by critical fluctuations, i.e. as a consequence of the fixed point of a suitable renormalization group. By the same token one should not expect in these objects such interesting phenomena as those which appear in DLA clusters on in percolating networks, namely the coexistence of several populations, each characterized by a different behavior under scaling, which are selected by looking at different moments of the mass distribution. In my opinion the very success of our understanding of the behavior of random walks with memory hides the fact that while we might have a good handle on the kinematics of aggregation (and their study has undoubtedly improved our understanding in this domain) we do not yet have a clue to the most relevant dynamics. And to get such a clue, we can only look at the nasty world of real aggregation processes.

I am grateful to L. Pietronero to have shared with me many of his insights on this subject. I also acknowledge fruitful discussions with D.J. Amit, L. de Arcangelis, A. Coniglio, P. Grassberger, S.P. Obukhov, H.C. Ottinger, G. Parisi and Zhang Y.C.

REFERENCES

[1] B.B. Mandelbrot, The Fractal Geometry of Nature, (Freeman, San Francisco, 1982).

[2] P.G. de Gennes, Scaling Concepts in Polymer Physics, (Cornell University Press, Ithaca, N.Y., 1979), p. 272 ff.

[3] P. Flory, Principles of Polymer Chemistry, (Cornell University Press, Ithaca, N.Y., 1971). Chap. XII.

[4] L. Pietronero, Phys. Rev. B$\underline{27}$ (1983) 5887 Phys. Rev. Lett. $\underline{55}$, 2025 (1985)

[5] L. Peliti and L. Pietronero, Random Walks with Memory, La Rivista del Nuovo Cimento, in press.

[6] S.P. Obukhov J. Phys. A. Math. Gen. $\underline{17}$ (1984) L-965.

[7] L. Pietronero and L. Peliti, Phys. Rev. Lett. 55, 1479 (1985).

[8] L. Peliti, J. Physique Lettres 45 (1984) L-925.

[9] P.C. Martin, E.D. Siggia and H.A. Rose, Phys. Rev. A7 (1973) 423.

[10] M. Doi, J. Phys. A. Math. Gen. 9 (1976) 1465; 9 (1976) 1479.

[11] P. Grassberger and M. Scheunert, Fortschritte der Physik, 28 (1980) 547.

[12] L. Peliti, Path Integral Approach to Birth Death Processes on a Lattice, J. Physique (in press).

[13] D.J. Amit, G. Parisi and L. Peliti, Phys. Rev. B27 (1983) 1635.

[14] I. Majid, N. Jan, A. Coniglio, H.E. Stanley Phys. Rev. Letts. 52 (1984) 1257.

[15] J.W. Lyklema and K. Kremer, J. Phys. A. Math. Gen. 17 (1984) L-691 and in press.

[16] S.P. Obukhov and L. Peliti, J. Phys. A. Math. Gen. 16 (1983) L-147.

[17] C. Byrnes and A.J. Guttman, J. Phys. A. Math. Gen. 17 (1984) 3335.

[18] J. Adler, private communication. The series were generated by R. Dekeyser, unpublished.

[19] J. Bernasconi and L. Pietronero, Phys. Rev. B29 (1984) 5196.

[20] R. Rammal, J.C. Anglès d'Auriac and A. Benoît, J. Phys. A. Math. Gen. 17 (1984) L-9.

[21] S.P. Obukhov, J. Phys. A. Math. Gen. 17 (1984) L-7.

[22] J.C. Anglès d'Auriac and R. Rammal, J. Phys. A. Math. Gen. 17 (1984) L-15.

[23] R. Rammal and G. Toulouse, J. Physique Letters, 44 (1983) L-13.

[24] K. Kremer and J.W. Lyklema, Phys. Rev. Letts. 54 (1985) 267.

[25] J.W. Lyklema, Laplacian Random Walks, this volume.

[26] A. Coniglio, N. Jan, I. Majid and H.E. Stanley, to be published.

[27] Zhang Y.C., Brookhaven preprint.

[28] L. Peliti and Zhang Y.C., J. Phys. A. Math. Gen. 18 (1985) L-709.

[29] P. Grassberger, J. Phys. A. Math. Gen. 18 (1985) L-463.

[30] J.W. Halley and H. Nakanishi, Phys. Rev. Letts. 55 (1985) 551.

[31] Cf. A.L. Stella, R. Dekeyser and A. Maritan, Self-similarity of mutual and Self-Intersections of Random Fractals, this volume.

[32] H.C. Ottinger, J. Phys. A. Math. Gen. 18 (1985) L-363.

SURVIVAL PROBABILITY AND ENHANCEMENT FACTOR IN POLYMER STATISTICS

L. PIETRONERO[+] and L. PELITI[++]

+ University of Groningen, Melkweg 1, 9718 EP Groningen, The Netherlands
++Dipartimento di Fisica, Università "La Sapienza", I-00185 Rome, Italy
  and GNSM-CNR, Unità di Roma

The method of survival probability provides a new perspective for Flory-type approaches and elucidates the mathematical basis of the approximations involved. It also allows to compute an explicit expression for the exponent γ: $\gamma = 3-d\nu$, bringing to completion the mean field theory of polymer statistics.

1. INTRODUCTION

In the preceding paper[1] we have seen how a number of kinetic problems with memory can be treated within selfconsistent scaling of theories. We have also seen that the standard equilibrium problem of the self-repelling chain (SRC)[2] can be formulated from a kinetic point of view with use of the survival probability concept[3,4]. In the first part of this short note we discuss how approximations involved in the Flory approach can be better defined from this new point of view. In the second part we show how to generalize the Flory theory to compute the second independent exponent of polymer statistics γ.

2. THE APPROXIMATIONS OF THE FLORY APPROACH

Let us consider the Flory approach as rederived in sect. 3 of the preceding paper with the survival probability approach[1]. In the present paper we use N for the length of a chain instead of t (time) used for the kinetic walks. The basic idea is to try to give a theoretical description to the method used in numerical simulations to produce the statistics of self-repelling chains. Many standard random walks are started and as soon as one steps on a previously visited site the entire walk is eliminated from the statistics. The crucial question is therefore how to compute the probability that a given walk, with end to end distance r, will never cross itself and therefore will be part of the final statistics. This clarifies that, contrary to the usual point of view[2], there is no approximation involved in the use of a gaussian in the "entropy" term of the standard Flory argument. The approximation is made in the "energy" term that in our approach corresponds to the probability of self-intersection. This probability per step is assumed to be proportional to the average coil density

$$\rho(N) \simeq \frac{N}{R^d} \simeq N^{(1-\nu d)} \qquad (1)$$

where $R \simeq N^\nu$. This is correct for the return probability corresponding to a gaussian distribution. On the other hand this probability of self-intersection should actually be computed for those walks whose end to end distance is $R \simeq N^\nu$ with ν arbitrary to be specified at the end from the selfconsistency relation. Since we know that the final ν is appreciably different from 1/2 this implies that the main contribution to our final distribution comes from a very particular subset of the initial configurations. Since $\nu > 1/2$ these configurations are "stretched" with respect to the average configurations[5] corresponding to the gaussian distribution. In view of this fact the return probability cannot be computed simply by using the expression corresponding to a gaussian distribution (Eq. 1) even if we allow the exponent ν to differ from 1/2. A more general form of the return probability for these "stretched" configurations could then be of type

$$P_N(1) \simeq \frac{1}{R^d} \left(\frac{1}{R}\right)^{\tilde{g}(\nu)}, \qquad (2)$$

in analogy to the return probability corresponding to the scaling form of generalized random walks[2,4]. Here the exponent $\tilde{g}(\nu)$ is reminiscent of the exponent g[2,4] but in the present case it should depend on the preselected value of ν corresponding to the subset of configurations we consider. Equation (1) is then generalized to

$$\tilde{\rho}(N) \simeq \frac{N}{R^{[d+\tilde{g}(\nu)]}} \simeq N^{1-\nu[d+\tilde{g}(\nu)]}. \qquad (3)$$

The selfconsistency condition for ν is then

$$\nu = \frac{3}{2+d+\tilde{g}(\nu)} \qquad (4)$$

and contains a nonlinear feed back term for the evaluation of the final ν. Clearly $\tilde{g}(1/2) = 0$; the problem of computing corrections to the Flory theory is now reduced to the calculation of $\tilde{g}$ for the subset of "stretched" configurations corresponding to a given value of ν. At the moment we are not able to compute this correction term but at least the problem is now related to a mathematically well defined question.

3. THE ENHANCEMENT FACTOR EXPONENT γ

Considering the SRC as a critical problem there are actually two independent exponents, while the Flory approach[2] only gives ν. Here we sketch how, using the approach of the survival probability, the second independent exponent γ can also be computed. A more complete derivation can be found in Ref. (4).

The total number of SRC of N steps has the asymptotic form (at large N)

$$Z_N \simeq \tilde{z}^N N^{\gamma-1} \qquad (5)$$

where the term $\tilde{z}$ represents the "effective" number of available neighbours at each step[2]. The total number of random walks of N steps is (in a lattice with coordination z) z^N. The survival probability for a walk of N steps is then

$$S_N \simeq \left(\frac{1}{z^N}\right) Z_N \simeq \left(\frac{\tilde{z}}{z}\right)^N N^{\gamma-1}. \qquad (6)$$

Consider now a walk that has survived N steps. The probability to encounter another portion of the walk that is distant exactly $\tilde{N}$ steps ($1 < \tilde{N} < N$) from the tip is related to the probability to return to the origin for a walk of $\tilde{N}$ steps

$$P_{\tilde{N}}(1) \simeq \tilde{N}^{-\nu(d+g)}. \qquad (7)$$

The probability $p(N)$ to encounter some portion of the walk no matter at what distance $\tilde{N}$ is then

$$p(N) = \sum_{\tilde{N}=1}^{N} P_{\tilde{N}}(1) \simeq \int_1^N d\tilde{N}\, P_{\tilde{N}}(1) =$$
$$= p(\infty) - \Delta p(N) \qquad (8)$$

where

$$p(\infty) = \int_1^\infty d\tilde{N}\, P_{\tilde{N}}(1) \qquad (9)$$

is an asymptotic encounter probability of order of unity and

$$\Delta p(N) = \int_N^\infty d\tilde{N}\, P_{\tilde{N}}(1) \simeq N^{-\nu(d+g)+1} \qquad (10)$$

is a correction term due to the fact that the length of the walk is actually N and not infinite.

Let us now follow a walk from its start n=1 until it reaches n=N and compute its total survival probability. At a given length n this is given by 1-p(n). The total survival probability is then

$$S_N = \prod_{n=1}^{N}[1 - p(n)] = \prod_{n=1}^{N}[1 - p(\infty) + \Delta p(n)] =$$
$$[1 - p(\infty)]^N \prod_{n=1}^{N} x(n) \simeq e^{-p(\infty)N} \prod_{n=1}^{N} x(n) \quad (11)$$

where we have introduced

$$x(n) = \frac{1 + \Delta p(n)}{1 - p(\infty)} \quad . \quad (12)$$

By comparing Eq.(11) with Eq.(6) we can make the identification

$$p(\infty) = \ln(\tilde{z}/z) \quad (13)$$

which enlightens the asymptotic meaning of $\tilde{z}$. The term

$$f_N = \prod_{n=1}^{N} x(n) \simeq N^{\gamma-1} \quad (14)$$

corresponds to the enhancement of the survival probability due to the fact that the chain is actually finite at each step. The requirement that f_N behaves asymptotically as a power law implies that

$$\frac{df_N}{dN} \simeq (\frac{1}{N})f_N \simeq f_{N+1} - f_N = f_N[x(N+1)-1] =$$
$$= f_N \frac{\Delta p(N+1)}{[1-p(\infty)]} \quad . \quad (15)$$

This gives rise to the scaling condition

$$\Delta p(N) \sim N^{-1} \quad (16)$$

which leads to the relation

$$\nu(d + g) = 2 \quad . \quad (17)$$

By using the des Cloiseaux relation[2] $g=(\gamma-1)/\nu$ Eq.(17) becomes

$$\gamma = 3 - d\nu \quad (18)$$

which is our final result. In a similar way one can derive for the θ point[4]

$$\gamma^\theta = 2(2 - \nu^\theta d) \quad . \quad (19)$$

These results bring to completion the mean field theory of polymer statistics. They are discussed in some detail in Ref. (4).

Finally we would like to mention an interesting approach by Alexandrowicz[6] that also produces closed form expressions for ν and γ: $\nu \simeq (4+d)/4d$ and $\gamma \simeq 8/(4+d)$. His method is constructed more in the spirit of an expansion from the upper critical dimension than as a standard mean field theory. The expressions derived result very close to the exact values and they agree with the ε expansion to first order.

In summary we have shown that the concept of survival probability leads to a deeper understanding of the approximations involved in Flory-type approaches. In addition it produces for the first time a mean field expression for the second independent exponent of polymer statistics (γ).

REFERENCES

1. L. Peliti, this volume.

2. P.G. de Gennes, Scaling Concepts in Polymer Physics (Cornell Univ. Press, 1979).

3. L. Pietronero, Phys. Rev. Lett.,55, 2025 (1985)

4. L. Pietronero and L. Peliti, Phys. Rev. Lett. 55 (1985) 1479.

5. S.P. Obukhov, in print. This author has made similar remarks with respect to the importance of "stretched" configurations.

6. Z. Alexandrowicz, Phys. Rev. Lett. 53 (1984) 1088.

THE LAPLACIAN RANDOM WALK

J.W. LYKLEMA

Institut für Festkörperforschung der Kernforschungsanlage Jülich, D-5170 Jülich, West Germany

Carl EVERTSZ

Laboratorium voor Vaste Stof Fysica, Rijks Universiteit Groningen, Melkweg 1, 9718 EP Groningen, The Netherlands

We introduce a one parameter (η) family of random walks which are truly kinetic and strictly self-avoiding. They provide an example of linear aggregation phenomena and are closely connected to the dielectric breakdown problem. For $\eta = 0$ we recover the recently introduced indefinitely growing self-avoiding walk in any dimension. The asymptotic behaviour of these walks is studied by a series analysis of exact enumeration data. We find that the critical exponent ν varies continuously between 1/2 and 1 with the parameter η.

1. INTRODUCTION

Aggregation phenomena have attracted widespread attention[1] recently. A particular simple example is the so-called indefinitely growing self-avoiding walk[2] (IGSAW). This walk is strictly self-avoiding (i.e. every site can be visited only once) and it is truly kinetic (i.e. all trajectories are of infinite length). The IGSAW therefore constitutes an example of linear aggregation. For a review on other kinetically growing walks see ref. 3.

Apart from its possible physical applications, this walk is important from a theoretical point of view. Although aggregation has been studied numerically very extensively, even a mean field theory of the Flory type is lacking. Since a self-averaging effect may obscure the consequences of the approximations, we expect that to develop a theory for these phenomena, it is easier to study a non-branching problem like the IGSAW. Unfortunately an algorithm which applies for the IGSAW in three dimension has not yet been developed.

In this paper we introduce a linear version of the dielectric breakdown problem[4] and we discuss the connection with the IGSAW[11]. We show that the properties of the IGSAW can be recovered if the jump probability of the walk is related to the solution of the Laplace equation with appropriate boundary conditions. This new walk, the Laplacian Random Walk (LRW), can be defined in any dimension. As in the full dielectric breakdown problem, the jump probability of the LRW depends on the field strength through a power η. For $\eta = 0$ we recover the IGSAW, also in three and higher dimensions. In the following we study the LRW on the square lattice by means of an exact enumeration procedure. From these numerical results we analyze the asymptotic behaviour of the mean square end-to-end distance ($<R^2(N)> \propto N^{2\nu}$). In particular we study the effect of the parameter η on the critical exponent ν. In section 2 we define the LRW and we describe its construction. In section 3 we analyze the numerical data and discuss the results.

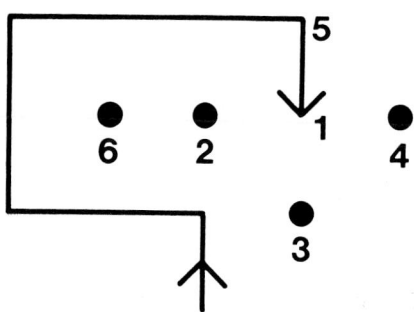

FIGURE 1: see text

2. DEFINITION AND CONSTRUCTION OF THE LAPLACIAN RANDOM WALK

The connection between the IGSAW and the dielectric breakdown problem is illustrated in Fig. 1, which is a possible configuration of the IGSAW on the square lattice. A jump to site 2 is not allowed, because afterwards the walk gets trapped. Thus the probabilities for the next steps are $P(1 \to 5) = 0$ (self-avoiding condition), $P(1 \to 2) = 0$ (truly kinetic condition) and $P(1 \to 3) = P(1 \to 4) = 1/2$, because the jump probability is defined as 1/(number of jump sites). In this definition jump sites are empty sites which do not lead into a cage. We now consider the same figure centered on the middle of a large circle with radius R_c. This radius is much larger than the typical size of the walk ($R_c \gg R$). For this configuration we solve the Laplace equation $\nabla^2 \Phi = 0$. The walk and the circle at R_c are perfect conductors with $\Phi(\text{walk}) = 0$ and $\Phi(R_c) = 1$. The potential at the other sites is found from the solution of the discretized Laplace equation

$$\Phi_i = \frac{1}{4} \sum_{nn} \Phi_j \qquad (1)$$

Here the sum extends over all nearest neighbours (nn). From this we find $\Phi_2 = 1/4\, \Phi_6$ and $\Phi_6 = 1/4\, \Phi_2$ with the solution $\Phi_2 = \Phi_6 = 0$. This is a consequence of the so-called Faraday screening and it is clear that it can be extended to cages of all sizes. As soon as a cage can be closed by a nearest neighbour or a next nearest neighbour connection in the forward direction, the potential inside the cage becomes zero. If we define the jump probability of the walker to be proportional to the potential, we obtain properties similar to those of the IGSAW. The walker will be self-avoiding and truly kinetic because $\Phi = 0$ on the walk and in the cages. We generalize this, as in the full dielectric breakdown process, to

$$P_i = \Phi_i^\eta / \sum_{nn} \Phi_j^\eta \qquad (2)$$

The difference between our walk and the full dielectric breakdown problem is that we allow only for growth at the tip of the structure. In this way we create a linear fractal instead of a branched one. From the dielectric breakdown point of view, our definition of P_i has the "unphysical" feature that only the field strength at the tip determines the new discharge. From the statistical point of view however, the model is a well-defined random walk with non-trivial long-range interactions.

The parameter η governs the asymmetry between the allowed directions. For $\eta = 0$ the probability P_i is independent of the field strength and we recover the IGSAW. For $\eta > 0$ we have a repulsive walk. The direction with the largest field has the largest probability and therefore the directions which point outward will be favoured over the other ones. The extreme case is the $\eta \to \infty$ limit. Then one expects a straight line, which results in an exponent $\nu = 1$. For $\eta < 0$ we have an attractive chain. Now the directions which point inward will be favoured and we get a denser object. In order to prevent a collapse of the chain, we exclude the points with $\Phi_i = 0$ from the definition (eq. 2) of the probability. The walk is then strictly self-avoiding and truly kinetic also for $\eta < 0$. The limit $\eta \to -\infty$ will be a perfect spiral because the direction which is closest to the object will always be chosen. For the asymptotic behaviour this gives $\langle R^2(N) \rangle \propto N$ or $\nu = 1/2$. It is known already that for $\eta = 0$ (IGSAW) the exponent $\nu = 0.567$[1]. So the question to be studied is whether the asymptotic behaviour varies smoothly from $\nu = 1/2$ ($\eta \to -\infty$) to $\nu = 1$ ($\eta \to \infty$) or if it is equal to $\nu = 0.567$ for all η values except for the two limits. As the trajectories for all η values are exactly the same the question can also be reformulated. Namely, does the critical behaviour of the walk change if we shift the weight of the trajectories from dense to stretched ones. A similar question has been studied for the usual SAW[5] and for the true SAW[6]. In both cases it has been shown that a local interaction does not change the asymptotic properties. For the LRW on the other hand, the calculation of the jump probabilities is based on the long range interaction present in the Laplacian and one can expect that a variation in the parameter η changes the critical behaviour. This prediction would be in agreement with the results[4] for the full dielectric breakdown. However it is not clear in advance if this result still holds if one studies a linear fractal instead of a branched one.

3. ANALYSIS AND RESULTS

As mentioned above, we want to study the asymptotic scaling behaviour of the mean square end-to-end distance $\langle R^2(N) \rangle = AN^{2\nu}$. To this end we have performed an exact enumeration for $N \leq 18$ on the square lattice. The alternative would have been a Monte Carlo calculation. We have taken the former approach because it has the enormous advantage that we can use the solution of the Laplace equation of a particular configuration for the calculation of the jump rates for a whole series of η values. In this study we have taken 49 different η values. Clearly this is impossible for a Monte Carlo calculation where one has to sample the walks for one value at a time. The disadvantage is that we have only relatively short series. For the $\eta = 0$ case[2] (IGSAW), however, we know that the enumeration result agrees exactly with the value from the Monte Carlo calculation ($N \leq 100$) and we expect this to hold also for non-zero η values.

The solution of the Laplace equation eq. (1) is obtained by iteration. The new value for Φ_i is calculated from the old value of its nearest neighbours. If we introduce a sublattice structure[7], it is then possible to vectorize the algorithm. Only because of this feature it is possible to perform an exact enumeration of this length. The actual CPU time needed for this project was 75 hours on the Cray X-MP at Jülich. To estimate the exponent ν from an accurate Monte Carlo simulation of chains of length $N \leq 100$[2,5] one certainly needs much more time, even if one is satisfied with a much smaller number of η values. Clearly it is desirable to have exact results for longer chains as for the $\eta = 0$ case ($N \leq 22$). However for every new generation the total amount of CPU time increases a factor 2.7, and we therefore stopped the enumeration at the 18-th generation. This required the solution of the Laplace equation for $9 \cdot 10^6$ different trajectories. We expect that a maximum length of $N = 18$ is sufficient to study our main objective, the η dependence of the critical exponent ν.

For the analysis of the mean square end-to-end distance we use the well known methods[2,5,8,9] from series analysis. For the asymptotic scaling behaviour we assume

$$\langle R^2(N) \rangle = AN^{2\nu}(1 + BN^{-\Delta} + CN^{-1} + \ldots) \quad (3)$$

and we define an effective exponent $\nu(N)$

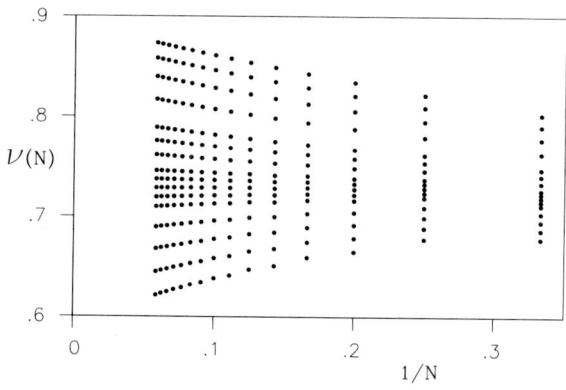
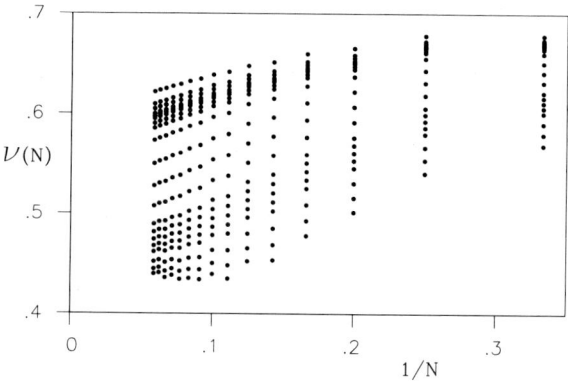

FIGURE 2
Plot of the effective exponent (N) versus 1/N on the square lattice. The value from top to bottom are: 2.0, 1.75, 1.5, 1.25, 1.0, 0.9, 0.8, 0.7, 0.65, 0.6, 0.55, 0.5, 0.4, 0.3, 0.2, 0.1.

$$\nu(N) = \frac{1}{2} \frac{\ln\{<R^2(N+1)>/<R^2(N-1)>\}}{\ln\{N+1/N-1\}}$$

$$= \nu - \frac{1}{2} \Delta B N^{-\Delta} - \frac{1}{2} C N^{-1} + \ldots \quad (4)$$

We then plot $\nu(N)$ versus 1/N. From the extrapolation of $1/N \to 0$ we obtain an estimate for ν. In Fig. 2 we show the results for the repulsive case for 16 different values of η. The results from this plot strongly suggest a continuously varying exponent ν, thereby confirming the earlier result for the branching dielectric breakdown problem[4].

We now discuss a few special results. For η = 7.5 (not shown) we find ν = 0.997, practically the limiting value ν = 1 for the straight line. The linear "equivalent" of DLA (η = 1) has an exponent ν = 0.8 or in terms of the fractal dimension d_f = 1.25. This has to be compared with $d_f \simeq 1.75$ for branched DLA. Here one should not take the "equivalence" literally, because the boundary conditions are not identical. To create a DLA fractal[10] one starts random walkers at the circle R_c and lets them move until they reach the fractal. There they stick to the fractal at the first nearest neighbour site they encounter. In the LRW the sticking only occurs

FIGURE 3
Plot of the effective exponent (N) versus 1/N on the square lattice. The values from top to bottom are: 0.1, 0.05, 0.03, 0.01, 0.0, -0.01, -0.03, -0.05, -0.1, -0.2, -0.3, -0.4, -0.5, -0.55, -0.6, -0.65, -0.7, -0.8, -0.9, -1.0

at the tip, at all the other points the walker disappears without leaving a trace i.e., at these points the walker loses its sticking capability. A different situation is obtained when the tail of the walk acts like a hard wall. The random walker bounces back and continues its path. These boundary conditions are more in the spirit of the original DLA problem, but have not been realized in the context of the LRW.

The usual SAW exponent ν = 3/4 is obtained with η = ln 2. To appreciate this, one has to realize that the trajectories of the LRW are a vanishinly small subset of the trajectories of the SAW, but it is the subset of infinite length and these determine the asymptotic behaviour. At last, a somewhat unusual value is η = 0.6, where the walk behaves asymptotically direct from the beginning (ν = 0.73).

In Fig. 3 we show a second set of values. The first 5 are repulsive the other 15 are attractive. For η = 0 we recover the IGSAW value ν = 0.567. For -0.1 < η < 0.5, ν appears to be a linear function of η. Thus one expects that for small negative η values, a linear extrapolation of ν(N) versus 1/N is still valid. For larger negative values of η, a linear extrapolation is certainly not correct as can be seen from Fig. 3. The smallest possible value for ν in two dimensions is 1/2 but the ν(N) series for η = -0.3 seems to extrapolate to a value smaller than 1/2. This clearly shows that a linear extrapolation is not valid in this region and that one has to generate chains for larger N values. Thus one must conclude that ν(N) goes through a minimum and will increase for larger N values to an asymptotic value ν ≥ 1/2. This behaviour can be observed for η = -1. For this value the minimum is located at N ≃ 12. This form of the series suggests the possibility that the correction to scaling exponent Δ (see eq. 3) is smaller than one. A similar behaviour has been observed for the usual SAW[9]. The assumption, Δ > 1, has been made implicitly in the linear extrapolation of the repulsive LRW. This is justified by the η = 0 (IGSAW) result from Monte Carlo calculation of chains with $N_{max} = 100^2$. Because we find a continuously varying leading exponent, we also expect to have a continuously varying value for Δ, the correction the scaling exponent. Thus it is tempting to conclude from our data that for the repulsive case Δ > 1 and that it becomes smaller than one in the attractive regime. Of course these details cannot be extracted quantitatively from our series. For this one possibly needs accurate Monte Carlo calculations. An additional complication for large negative values of η is the oscillating behaviour of ν(N). This is due to the lattice structure. That it shows up only for these values is because the spiralling trajectories are now weighted heavily.

In Fig. 4 we show the η dependence of the critical exponent ν for the repulsive LRW. The asymptotic values for ν(= $1/d_f$, d_f: fractal dimension) are linear extrapolations from the series ν(N) of Fig. 2 and 3. We find a smooth

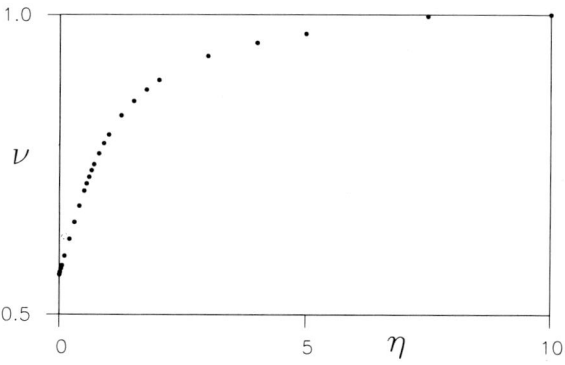

FIGURE 4
Plot of the estimated value for ν versus η.

curve running from ν = 0.567 for η = 0 to ν = 1.0 for η = 10. Unfortunately we are not yet able to explain this result, because even a mean field theory for this type of phenomena has not been developed.

To obtain an idea of the functional behaviour of ν(η), we have fitted the estimated values of ν to a rational function. We find that our results are reproduced to within an accuracy of 2 % by ν(η) = $(1+η+η^2)/(d_f^c+η+η^2)$. From this formula we find ν(-0.4) = 0.5. This suggests that the LRW with η = -0.4 has an upper critical dimension d_c = 2. This prediction cannot be verified from our data for reasons described above. It cannot, however, be excluded. Because we find a continuously varying exponent ν, we also expect a continously varying d_c. Thus there is no a priori reason to expect that ν reaches its limiting value 1/2 only for η → -∞.

In conclusion, we have shown that the introduction of a long range interaction as defined for the LRW, is relevant for the asymptotic properties. In addition, we find that a redistribution of the probabilities over all the available trajectories does change the critical exponent ν. A more detailed analysis of the results and those for the 3 dimensional LRW will be published elsewhere.

ACKNOWLEDGEMENTS

The authors thank L. Pietronero for suggesting this problem. One of us (C. E.) acknowledges the support of the "Stichting voor Fundamenteel Onderzoek der Materie" (FOM), which is financially supported by the "Nederlandse organisatie voor Zuiver Wetenschappelijk Onderzoek" (ZWO).

REFERENCES

1. See, for instance, Kinetics of Aggregation and Gelation, eds. F. Family and D.P. Landau (North Holland, Amsterdam, 1984)
   CECAM-Workshop on Kinetic Models for Cluster Formation, R. Jullien, M. Kolb, H. Herrmann and J. Vannimenus, J. Stat. Phys. 39 (1985) 241

2. K. Kremer and J.W. Lyklema,
   Phys. Rev. Lett. 54 (1985) 267
   K. Kremer and J.W. Lyklema,
   J. Phys. A 18 (1985) 1515

3. L. Peliti and L. Pietronero, La Rivista del Nuovo Cimento, in print.
   J.W. Lyklema, Kinetically Growing Self-Avoiding Walks, this volume.

4. L. Niemeyer, L. Pietronero and H.J. Wiesmann, Phys. Rev. Lett. 52 (1984) 1033
   L. Pietronero and H.J. Wiesmann,
   J. Stat. Phys. 36 (1984) 881

5. J.W. Lyklema and K. Kremer,
   J. Phys. A 17 (1984) L691
   K. Kremer and J.W. Lyklema,
   Phys. Rev. Lett, in print
   J.W. Lyklema and K. Kremer,
   J. Phys. A, in print

6. J. Bernasconi and L. Pietronero,
   Phys. Rev. B 29 (1984) 5196

7. W. Oed, Angewandte Informatik 7 (1982) 358

8. Z.V. Djordjevic, I. Majid, H.E. Stanley and R.J. dos Santos, J. Phys. A 16 (1983) L519

9. J.W. Lyklema and K. Kremer,
   Phys. Rev. B 31 (1985) 3182

10. T.A. Witten and L.M. Sander,
    Phys. Rev. Lett, 47 (1981) 1400; Phys. Rev. B 27 (1983) 5686

11. L. Pietronero, to be published

FRACTALS IN PHYSICS
L. Pietronero, E. Tosatti (editors)
© Elsevier Science Publishers B.V., 1986

KINETICALLY GROWING SELF-AVOIDING WALKS

J.W. LYKLEMA

Institut für Festkörperforschung der Kernforschungsanlage Jülich, Postfach 1913, D-5170 Jülich, Federal Republic of Germany

I discuss self-avoiding random walks with one-step jump probabilities which depend on the local environment. Because of linearity, these walks constitute the simplest examples of aggregation phenomena. Their asymptotic behaviours are studied using series analysis of Monte Carlo generated enumeration data. The results suggest that there are at least three different universality classes. (i) The SAW class with $\nu=0.75$ for $d = 2$, $\nu=0.59$ for $d = 3$ and $d_c = 4$. (ii) The growing trail with $\nu=0.535$ for $d = 2$, $\nu=0.5$ for $d = 3$ and $d_c = 3$ (iii) The indefinitely growing self-avoiding walk with $\nu=0.567$ for $d = 2$.

1. INTRODUCTION

In this contribution we discuss the asymptotic scaling behaviour of kinetically growing self-avoiding walks. These walks differ from the usual self-avoiding walk (SAW) in that their one-step transition probability depends on the environment. For the usual SAW this probability equals $1/q-1$ (for the first step $1/q$), where q is the coordination number of the lattice. This decribes for example, the equilibrium ensemble for a polymer where all bonds carry the same weight. Several modifications of this walk have been studied. These include the k-tuple SAW[1] (a SAW where the walker can visit a site at most k times), the k-tuple self avoiding trail (SAT)[2] (the self avoiding condition for the sites has been replaced by a similar condition for the bonds) and a ring closing version of the self-avoiding walk (SAR)[3]. All these modifications resulted in the same asymptotic behaviour as the usual SAW, $\nu=3/4$ for $d = 2$, $\nu= 0.588$ for $d = 3$ and an upper critical dimension $d_4 = 4$.

The first authors who obtained a different critical behaviour were Amit et al[4] who introduced the so called true self-avoiding walk (TSAW). In this walk the jump probabilities depend on how often the nearest neighbour sites have been visited previously. This results in an upper critical dimension $d_c = 2$ and $\nu=2/3$ in one dimension[5]. This walk is truly kinetic in the sense that it never stops, and it could model a growth phenomenon. Because it visits sites more than once, however, it is not strictly self-avoiding and therefore unsuitable for describing realistic growth processes.

2. THE GROWING SELF-AVOIDING WALK

In an attempt to correct this drawback a growing version of the SAW was introduced[6] (GSAW). For this model we modify the SAW transition probability to $p_i = 1/$(number of free sites). In this way we try to avoid trapping. However, if all nearest neighbours have been visited before the walk is terminated. The usual SAW is terminated already if we try to visit a previously visited site. Initially it was thought that this model belonged to a new universality class. However it was argued[7] that this could not be true, because the trap density is proportional to the density of visited points, as in the usual SAW. Consequently the GSAW should belong to the same universality class as the SAW. That this is indeed the case is shown by Kremer and Lyklema[8], using very extensive Monte Carlo calculations. From this work it became clear that the asymptotic regime can not be studied by series enumeration because the series are much too short, typically of the order of 20 steps for the square lattice. However as shown in ref. 8 one needs for the GSAW on the square lattice at least 200 steps, so posing a problem for the analysis. If we assume for the mean square end-to-end distance the usual scaling form

$$\langle R^2(N) \rangle = AN^{2\nu}(1 + BN^{-\Delta} + CN^{-1} + \ldots) \quad (1)$$

one can calculate from the slope of a lg-lg plot of R^2 versus N the exponent ν. However from such a plot it is very difficult to decide if the asymptotic region is reached. A slightly curved line can easily be interpreted as being linear, suggesting an asymptotic behaviour. This problem has been solved by applying and extending techniques from series analysis to high accuracy Monte Carlo data[9]. To this end we define an effective exponent $\nu(N)$

$$\nu(N) = \frac{1}{2} \frac{\ln\{\langle R^2(N+i)\rangle/\langle R^2(N-i)\rangle\}}{\ln\{N+i/N-i\}} =$$

$$= \nu - \frac{1}{2}\Delta BN^{-\Delta} - \frac{1}{2}CN^{-1} + \ldots \quad (2)$$

With the assumption that the correction to scaling exponent $\Delta \geq 1$, one can estimate ν from an extrapolation versus $1/N$. This not only gives ν, but the behaviour of $\nu(N)$ as a function of N gives a clear indication of whether the analysis covers the asymptotic regime.

3. THE GROWING SELF-AVOIDING TRAIL

We have found that every modification of the SAW which still has the property of being strictly self-avoiding, is in the SAW universality class. The next model which one can study is the growing self-avoiding trail (GSAT)[10]. This model is a growing version of the SAT. We have modified the one-step probabilities of the SAT analogous to the GSAW to $p_i = 1/$(number of free bonds). This walk allows for self intersection, however, it should not be confused with the k-tuple SAW. This walk can never be terminated except at the origin, because when the walker enters a site there are always three bonds or one bond available to leave. The origin is a special point. After a second visit there is only one empty bond left so that the walk is terminated after a third return. This walk can only be defined on lattices with an even number of bonds. From this construction, one may anticipate a different asymptotic behaviour from the SAW because this walk can be terminated only at one point whereas all the other walks can get trapped at every site of the lattice.

I[10] have performed extensive Monte Carlo calculations for this model on the square lattice and on the simple cubic lattice. These confirm the expectation of a different critical behaviour from the SAW universality class. In two dimensions we find $\nu=0.535$ and in three dimensions $\nu=0.5$. This latter result strongly suggests that the critical dimension is equal to three. If this were the case one expects a logarithmic correction of the following form

$$\langle R^{2n}(N) \rangle = A_n N^n (\ln N)^{n\alpha} \quad (3)$$

With $\alpha=0.025$, $A_1=1.5$ and $A_2=3.75$, this scaling behaviour was found to give a perfect fit to the Monte Carlo data. At the critical dimension one also expects to find a gaussian behaviour which means that the ratio $\langle R^4(N)\rangle/\langle R^2(N)\rangle^2 = A_2/A_1^2$ is equal to 5/3; the value I find from the Monte Carlo simulation. Thus I have found a self-avoiding walk with different asymptotic properties from the usual SAW universality class to which the other walks belong, with the exception of the not strictly self-avoiding TSAW.

4. THE INDEFINITELY GROWING SELF-AVOIDING WALK

The last walk we discuss is the indefinitely growing self-avoiding walk[11] (IGSAW). This walk is a self-avoiding walk with a one-step probability $p_i = 1/$(number of jump sites). Jump sites are nearest neighbour sites which do not lead into a cage, thus this walk will never terminate. This walk therefore comibines two important properties which are necessary for a growth model. It is truly kinetic and it is strictly self-avoiding. The trajectories of the IGSAW are a vanishingly small subset of those of the conventional SAW. One interesting question is whether the critical behaviour is changed if we use a different definition for the jump probability p_i [12,13]. We have found that this indeed happens for the Laplacian Random Walks[12] (LRW), which is a generalization of the IGSAW. For the IGSAW we find $\nu=0.567$, different from both the SAW value and the GSAT value. The question of the upper critical dimension also arises here. Unfortunately this cannot be answered yet because a construction algorithm for this model in three dimensions has not been found (see however ref. 12).

A second point which deserves attention is the number of different trajectories of length N, $\Gamma(N)$. For the SAW this has the asymptotic behaviour[14]

$$\Gamma(N) \propto q_{eff}^N N^{\gamma-1} \qquad (4)$$

The exponent γ has the meaning of a susceptibility exponent, as can be established through the use of the n vector model in the limit $n \to 0$[14]. One obtains

$$\chi T_{magn} \propto \sum_N \Gamma(N) q_0^{-N} \qquad (5)$$

The expression $\Gamma(N) q_0^{-N}$ equals $Z(N)=\sum_{C_N} \prod_{i=1}^N p_i$, the probability of having of walk of length N. Here $\Gamma(N) = \sum_{C_N} 1$ the sum over all possible configurations of length N and p_i is a constant $1/q_0$ ($q_0=q-1$, q=coordination number of the lattice). Clearly $Z(N)$ and the number of walks $\Gamma(N)$ have the same asymptotic behaviour because the jump probability p_i is constant. A natural generalization for a non-constant p_i is the assumption that $Z(N)$ asymptotically behaves like $q_{eff}^N N^{\gamma-1}$. The exponent γ defined in this way again has the interpretation of a susceptibility exponent (see Lyklema and Kremer, ref. 6).

For the IGSAW we have $Z(N)=1$ because of conservation of probability and consequently the susceptibility exponent γ for the IGSAW equals one.

We now return to the asymptotic behaviour of the number of walks $\Gamma(N)$. We have already remarked that these form a subset of the trajectories of the SAW. In fact it is the subset of asymptotic length, because finite walks which become trapped are removed by the construction procedure. One therefore may expect that the number of IGSAW's $\Gamma(N)$ behaves asymptotically as $\Gamma(N) \propto q_{eff}^N N^{\gamma-1}$, where q_{eff} and γ have the SAW values $\gamma=43/32$ and $q_{eff}=2.64$. This has been confirmed by Monte Carlo calculations[13]. The exponent γ can however not be interpreted as a susceptibility exponent. This result strongly suggests that the IGSAW can be used to study the critical behaviour of the SAW if we perform a biased sampling. Work along these lines is in progress[13].

REFERENCES

1. A. Malakis, J. Phys. A8 (1975) 1885
   A. Malakis, J. Phys. A9 (1976) 1283
   A.J. Guttmann, C. Byrnes and N.E. Frankel, J. Phys. A17 (1984) L457
   R. Dekeyser, A. Maritan and A.L. Stella, Phys. Rev. B31 (1985) 4659

2. A. Malakis, J. Phys. A17 (1984) L837
   A.J. Guttmann, J. Phys. A18 (1985) 567
   A.J. Guttmann, J. Phys. A18 (1985) 575
   D.C. Rapaport, J. Phys. A18 (1985) L475

3. P.D. Gujrati, Phys. Rev. B27 (1983) 4507
   A.J. Guttmann, J. Phys. A17 (1984) 455

4. D.J. Amit, G. Parisi and L. Peliti, Phys. Rev. B27 (1983) 1635

5. L. Pietronero, Phys. Rev. B27 (1983) 5887

6. I. Majid, N. Jan, A. Coniglio and H.E. Stanley, in: Kinetics of Aggregation and Gelation, eds. F. Family and D.P. Landau (North Holland, Amsterdam, 1984) pg 51.
   J.W. Lyklema and K. Kremer in: Kinetics of Aggregation and Gelation, eds. F. Family and D.P. Landau (North Holland, Amsterdam, 1984) pg 241
   I. Majid, N. Jan, A. Coniglio and H.E. Stanley, Phys. Rev. Lett. 52 (1984) 1257

J.W. Lyklema and K. Kremer, J. Phys. A17 (1984) L691
S. Hemmer and P.C. Hemmer, J. Chem. Phys. 81 (1984) 584

7. L. Peliti, J. Physique Lett. 45 (1984) 45
L. Pietronero, Phys.Rev.Lett., 55, 2025 (1985)

8. K. Kremer and J.W. Lyklema, Phys. Rev. Lett., 55, 2091 (1985).
J.W. Lyklema and K. Kremer, J. Phys. A, in print

9. J.W. Lyklema and K. Kremer, Phys. Rev. B31 (1985) 3182

10. J.W. Lyklema, J. Phys. A18 (1985) L617

11. K. Kremer and J.W. Lyklema, Phys. Rev. Lett. 54 (1985) 267
K. Kremer and J.W. Lyklema, J. Phys. A18 (1985) 1515

12. J.W. Lyklema and C. Evertsz, Laplacian Random Walks, this volume.

13. J.W. Lyklema and K. Kremer, in preparation

14. P.G. de Gennes, Scaling Concepts in Polymer Physics (Cornell University Press, Ithaca, 1979)

THE COIL-GLOBULE TRANSITION IN 2-DIMENSIONS

Naeem JAN[*], Antonio CONIGLIO, Imtiaz MAJID and H. Eugene STANLEY

Center for Polymer Studies, Boston University, Boston, MA 02215, USA

We show that the Interacting Self-Avoiding Walk/polymer chain at the theta temperature may be mapped onto the Indefinitely Growing Self-Avoiding Walk/Smart Kinetic Walk. Extremely accurate statistics for these walks exist which together with our independent numerical results support $\nu = 0.57$ (~4/7) where ν is the exponent describing the dependence of the polymerization index N on the end-to-end distance R_N.

The critical properties of the self-avoiding walk (SAW) are known exactly in 1- and 2-dimensions (d),[1] at its upper critical dimensionality, 4[2] and known to a high degree of accuracy at d equal to 3.[3,4,5] The Flory theory[2] gives precise values for the exponent ν which describes the dependence of the polymerization index N on the radius of gyrator R_N

$$R_N^2 \sim AN^{2\nu} \quad (1)$$

where $\nu = 3/(2+d)$ Flory. (2)

The interacting SAW shows distinct critical properties at the θ-temperature which heralds the onset of the collapsed phase. The upper critical dimensionality of the polymer chain at the θ-temperature is 3 and here mean-field exponents with logarithmic corrections are exact.[2] The 2-dimensional problem is of special interest as here we expect non-classical exponents. This problem - the determination of the critical properties at the θ-temperature has presented a severe challenge to the theorist. For example de Gennes[6] using the ε-expansion of Stephen and McCauley[7] reported $\nu = 0.505$ which is in agreement with the direct polymer approach of Duplantier.[8] The Monte Carlo data obtained with the reptation method was found to be in good agreement with the ε-expansion results referred to above. Kholodenko and Freed[10] carried out a different perturbation expansion and obtained $\nu = 0.55$. The recent preprint of Derrida and Saleur[11] using the transfer matrix technique and finite size scaling report a value of $\nu = 0.55$. A Flory type agreement[12] appropriate for the θ-point leads to $\nu = 2/3$ whereas the experimental work of Villanove and Rondelez[13] on a 2-d polymer monolayers indicate a ν of 0.56.

Our starting point is the well known relationship between the number of self avoiding polygons ¶ of N + 1 edges and the end-to-end distance R_N

$$¶ (N + 1) \sim z_{eff} R_N^{-d} \quad (a)$$
$$\sim N^{-\nu d} \quad (b) \quad (4)$$

z_{eff} is the effective coordination number.

If the polygons are appropriately weighted i.e. the nearest-neighbor monomer pairs, A_{NN} contribute an additional weighting factor to that polygon of $\exp(-A_{NN} \varepsilon/\tau)$ then

$$¶_W (N + 1) \sim R^{-\nu(\tau)d}$$

where from general considerations we expect

---

[*] Physics Department, St. Francis Xavier University, Antigonish, Nova Scotia B2G 1C0 CANADA

$$\nu(\tau) = \nu_{SAW} \text{ for } \tau > \theta$$
$$= \nu_\theta \text{ for } \tau = \theta$$
$$= \nu_{collapsed} \equiv \frac{1}{d} \text{ for } \tau < \theta$$

Consider the Smart Kinetic Walk on the honey comb lattice. This walk may be constructed as follows: At the start of the walk two hexagons are arbitrarily selected and a 'o' placed in one while an 'x' is inserted in the center of the other.

manner. These steps are made with probability 1/2, the same as a SAW but in instances where the walker is in the vicinity of a previously visited region the walker may have no option but to step in a particular direction i.e. with probability 1. This occurs when an alternative step would lead either to self intersection or entrapment. The walk may be considered as one on the perimeter of a cluster on the triangular lattice at P_c. The steps which avoid self

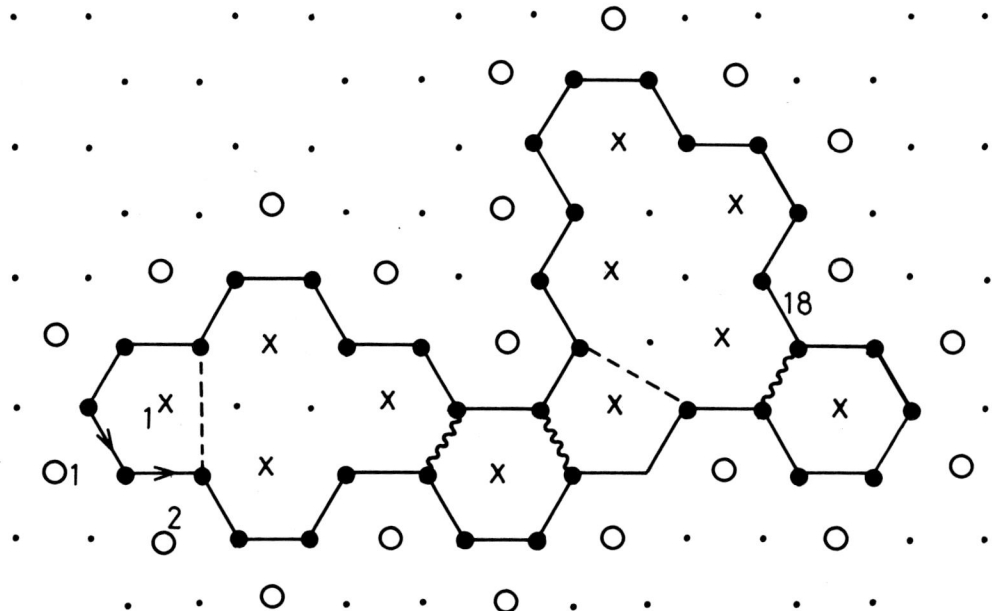

FIGURE 1
SKW of 36 steps. The hexagons labelled 1 indicate the first step. The status of hexagon 2 which is determined by the flip of a coin constrains the second step, which in the example is a 'o' and the walk continues along the perimeter of the cluster of 'x'. Note that the 'o' and 'x' are part of the triangular lattice. The nearest-neighbour interactions are denoted by ~~~~ and next-nearest neighbor interactions are denoted by ----. For example the eighteenth step is determined without having to consider the status of any hexagon and has a weight of 1, as distinct from the previous steps which are identical to a SAW.

A step is made along the bond separating the two hexagons. The second step is determined by the flip of a coin. In the example it is a 'o' and the second step is made along the direction shown. Subsequent steps are made in the same

intersection contribute $\exp(+\varepsilon)$, ($\varepsilon = \ln 2$) an attractive monomer-monomer nearest neighbour interaction whilst these that avoid entrapment contribute a next-nearest neighbour attraction of $\exp(\varepsilon)$. Thus the weight of a particular polygon, $W(N+1)$

$$W(N \pm 1) = W_{SAW} e^{\Sigma_{nn} \varepsilon + \Sigma_{nnn}^1 \varepsilon}$$

Hence the number of weighted polygons, $\P_w(N+1)$

$$\P_w(N+1) \sim R_N^{-d}$$
$$\sim N^{-\nu(\tau)d}$$

where we expect from general renormalization arguments $\nu(\tau)$ to belong to one of the three fixed points implied in equation 6. The exponent ν_{skw} has been determined to a high degree of accuracy and is $\sim 4/7$. Since this walk maps onto an interacting SAW and this exponent is distinct from the SAW and the collapsed phase then it follows that the SKW has the same exponent ν as the θ-point.

To check this result we have used the recently introduced Kinetic Growth Walk (KGW) which avoids wherever possible nearest-neighbor intersection but may be trapped in a cul-de-sac. One may generate very long SAW's in an efficient manner with only nearest neighbour attractive energy. These walks show the characteristics of the SAW[15] but if the attraction is increased we find $\nu = 0.57$ in good agreement with the analysis given above.

ACKNOWLEDGEMENTS

We wish to thank B. Derrida, J.W. Lyklema, L. Pelito and H. Saleur for useful discussions.

REFERENCES

1. B. Nienhuis, Phys. Rev. Lett., 49, (1982) 1062.

2. P.G. de Gennes, Scaling Concepts in Polymer Physics (Cornell U P, Ithaca, 1979)

3. J.C. Le Guillou and J. Zinn-Justin, J. Physique Lett., 46, (1985) L137.

4. I. Majid, Z. Djordjevic and H.E. Stanley, Phys. Rev. Lett., 51, (1983) 1282.

5. B. MacDonald, N. Jan, D.L. Hunter and M.O. Steinitz, J. Phys. A. 18, (1985) 2627.

6. P.G. de Gennes, J. Physique Lett., 36, (1975) L55.

7. M.J. Stephen and J.L. McCauley, Phys. Lett. 44A (1973) 89.

8. B. Duplantier, 'in print'.

9. A. Baumgartner, J. Physique, 43, (1982) 1407.

10. A.L. Kholodenko and K.F. Freed, J. Chem. Phys., 80, (1984) 900.

11. B. Derrida and H. Saleur, 'in print'.

12. I. Majid, N. Jan, A. Congilio and H.E. Stanley, Phys. Rev. Lett., 52, (1984) 1257.

13. R. Villanove and F. Rondelez, Phys. Rev. Lett., 45, (1980) 1502.

14. K. Kremmer and J.W. Lyklema, Phys. Rev. Lett. 54 (1985) 267.
    A. Weinrib and S.A. Trugman, Phys. Rev. B31 (1985) 2993.

15. J.W. Lyklema and K. Kremmer, J. Phys., A17 (1984) L691.

SELF-SIMILARITY OF MUTUAL AND SELF-INTERSECTIONS OF RANDOM FRACTALS

A.L. STELLA[+], R. DEKEYSER[§] and A. MARITAN[¤]

Dipartimento di Fisica della Università di Padova, I-35100 Padova, Italy

The fractal dimensions of various types of intersection sets of random fractals are discussed. This helps a better understanding of issues concerning upper critical dimensionality, Flory approximations or universality of critical behaviour with respect to different excluded volume mechanisms. Diffusion on self-intersecting fractals, and on fractals with local bridges, is also considered in the light of the above discussion, and accurate results for the relative exponents, obtained from analysis of Monte-Carlo generated enumerations, are presented.

1. INTRODUCTION

For many problems of critical behaviour in lattice statistics, like self-avoiding walks (SAW) or related models for linear and branched polymers, mutual or self-intersection properties of random fractals turn out to be relevant in several respects. It is e.g. well known that the upper critical dimension of a problem must be related to the mutual intersection properties of two replicas of the random fractal under consideration. On the other hand, the concept of self-intersections of a fractal enters as a very basic ingredient in the widely applied Flory approximation for the ν-exponent of polymer problems.[1] Indeed, the repulsive energy due to the excluded volume is in this approximation simply estimated as being proportional to the average number of self-contacts of a purely random linear (R.W.) or branched chain.

In the second section of this report, we will review the basic results of the (as far as we know) first systematic investigation of the scaling laws governing the statistical geometry of various kinds of intersections of random fractals on a lattice.[2] In spite of its relevance, which will be illustrated here, up to now only scarce attention has been paid to this subject, both in the mathematical and physical literature.[3] Besides throwing new light on important issues like the above mentioned connection between intersections and critical dimensionality, or the universality of the ν-exponent with respect to different excluded-volume mechanisms, the study of the laws determining the fractal dimensions of self- and mutual intersections will naturally lead us to a deeper understanding and a nontrivial reinterpretation of the Flory approximation itself, as will be illustrated in the third section.

Random fractals allowing for self-intersections, like k-tolerant walks (i.e. walks that can visit each lattice site at most k times), are not only interesting for the discussion of the above mentioned static properties. An important question is whether such self-intersections are able to allow a nontrivial value ($\neq 1$.) for the spectral dimension, which is related to diffusion on the fractal.[4,5] Such a type of dynamical question, for which new aspects of the geometry of self-intersections are coming into the game, is of particular interest also in con-

---

+ Unità G.N.S.M. del C.N.R. and C.I.S.M., Padova, and I.S.A.S., Trieste, Italy.
§ Instituut voor Theoretische Fysica, Katholieke Universiteit Leuven, B-3030 Leuven, Belgium.
¤ I.N.F.N., Sezione di Padova, and I.S.A.S., Trieste, Italy.

nection with recent debates about the possible mechanisms leading to the noninteger values for the spectral dimension of ferrodoxin and hemoproteins, as measured by spin-lattice relaxation experiments.[6]

2. FRACTAL DIMENSION OF MUTUAL AND SELF-INTERSECTION SETS

The fractal dimension $D(A \cap B)$ of the set of mutual intersection points of two random fractals A and B with fractal dimensions $D(A)$ and $D(B)$ respectively, is expected to be given by the law of codimension additivity:[3]

$$D(A \cap B) = D(A) + D(B) - d \quad (1)$$

if the r.h.s. is nonnegative, and $D(A \cap B) = 0$ otherwise. In (1), d is the dimension of the embedding Euclidean space. This law can be derived on the basis of simple scaling considerations.[2]

To be specific, let us consider the case of walks (SAW, R.W. or k-tolerant, etc.) on a d-dimensional lattice. Indicating by W a generic walk, with an extremum at a given origin $\vec{0}$ and with $|W|$ steps, the grandcanonical probability for a walk to visit a site $\vec{r}$ is given by

$$P(\vec{r}) = \sum_W K^{|W|} \chi_{\vec{r}}(W) \cdot \left[\sum_W K^{|W|} \right]^{-1} \quad (2)$$

where K is a step fugacity, and $\chi_{\vec{r}}(W)$ is 1 if W visits site $\vec{r}$, but 0 otherwise. If $\xi(K)$ is the grandcanonical root mean square end-to-end distance of the walks, then the natural scaling assumption for $P(\vec{r})$ is

$$P(\vec{r}) \simeq r^{-x} f(\frac{r}{\xi}) \quad (3)$$

for $r \to \infty$ and $K \to K_c^-$, such that $\xi(K) \to (K_c-K)^{-\nu} \to \infty$, with ν the end-to-end distance exponent. The function $f(y)$ is expected to approach a constant value for $y \to 0$. The exponent x can easily be derived in terms of d and ν. Indeed, it is straightforward that $\sum_r P(\vec{r})$ is equal to $<\|W\|>$, the average number of *distinct* points visited by the walk. By definition[3] we have that $<\|W\|> \sim \xi^{1/\nu} \equiv \xi^D$, where D is the fractal dimension of the walk; thus, on the basis of the above mentioned properties, we may conclude that $\xi^D \sim \xi^{d-x}$, or $x = d-D$.

Consider now two independent walks W_1 and W_2 with the same fractal dimension and origin, for simplicity. In this case one finds

$$<\|W_1 \cap W_2\|> = \sum_r P_1(\vec{r})P_2(\vec{r}) \underset{K \to K_c^-}{\sim} \xi^{2D-d}, \quad (4)$$

which is consistent with eq. (1).

The law of codimension additivity has been verified directly by Monte-Carlo simulations on k-tolerant walks with k up to 5, in d=2 and d=3.[2] In the case of purely random walks, $P(\vec{r})$ can be computed exactly, and turns out to satisfy (3) with $x = d-2$ for $d > 2$ and $x = 0$ for $d \leq 2$, as it should since the fractal dimension is $D = 2$ for $d > 2$ and $D = d$ for $d \leq 2$.[2] According to eq. (4), $d = 4$ is the dimensionality below which $<\|W \cap W'\|> \to \infty$, and above which $<\|W \cap W'\|> \to$ constant for $\xi \to \infty$, if W and W' are two independent random walks with the same origin. It is remarkable that the quantity $<\|W \cap W'\|>$ turns out to be proportional to the integral giving the first order perturbative correction to the 4-point 1-particle irreducible vertex function of a ϕ^4-theory with ultraviolet cutoff. In this way the appearance of infrared divergences in the ϕ^4-theory is seen to be directly related to the divergence of $<\|W \cap W'\|>$ for two random walks or, equivalently, to their intersection being a fractal with positive Hausdorff dimension.[2]

The situation is more difficult when we consider the fractal dimension of the self-intersections (if any) of a given fractal. It has been conjectured that the fractal dimension of multiple self-intersection points of m-th order of a fractal (in what follows we will call such points m-ple points) should be determined as if these points would be the mutual intersections of m independent replicas of the fractal.[3,7]

According to (1), this fractal dimension should then be $mD-(m-1)d$, if D is the dimension of the self-intersecting fractal. This replica idea, however, is contradicted by exact results for random walks and by extensive numerical investigations of k-tolerant walks in $d=2$ and $d=3$.[2] In the former case, one knows that the average number of m-ple points $(m=2, 3,...)$ for all random walks with $|W|=N$ grows as a finite fraction of N for $N\to\infty$.[8] On the other hand, when $K\to K_c$, one can exactly calculate the divergence of the correlation length for the probability that two given sites are both visited e.g. twice by a random walk.[2] This divergence turns out to be proportional to $(K_c - K)^{-1/2}$, which together with the above results implies that the fractal dimension of double points is equal to the fractal dimension of the random walk itself.

k-tolerant walks are an interesting case of nontrivial fractals, in which it is possible to test numerically the self-similarity of m-ple points $(m \leq k)$. Such an investigation is a very difficult task, because, already in order to obtain the asymptotic scaling properties of the end-to-end distance of these walks, one must estimate (e.g. by Monte-Carlo sampling) statistical properties of extremely long walks, especially at high k. Such difficulties are becoming much more severe if, like in the present case, one tries to discuss the scaling of the average radius of gyration with respect to the center of mass of e.g. double points. A succesful strategy[2] (a similar strategy was developed independently in ref. 9) consisted in evaluating up to a high degree of accuracy the average radius of gyration of m-ple points for walks of low and intermediate length (typically up to $N \sim 50$) by a Monte-Carlo biased sampling.[10] Such lengths are of course not accessible to exact enumeration, but they still do not require a too large sampling in order to reach a reasonably accuracy. To these approximate numbers we then applied some standard methods of asymptotic analysis of series expansions. While some techniques (like e.g. the ratio method) are too sensitive to uncertainties in successive data to be of much use in our case, other methods that make use of all the data simultaneously (like Padé approximant methods) are much less sensitive to noise, and they yield rather consistent results. With these methods, very clear evidence could be obtained for the fact that the replica trick does not apply to m-ple points. For each m, like in the random walk case, the average number of m-ple points grows as a finite fraction of N, and their average radius of gyration grows like the end-to-end distance of the walk. This clearly indicates that the fractal dimension of m-ple points is just the same as the one of the whole walk.

3. FLORY APPROXIMATION REVISITED

As mentioned in the introduction, in the Flory approximation for e.g. a linear, self-avoiding chain (polymer) of N steps (monomers), the repulsive energy is estimated as being proportional to the average number of double points of a random walk.[1] This average number is given as $\sim N^2 R^{-d}$, with R the characteristic size of the chain (end-to-end distance). Subtracting from this energy the elastic entropy term $\sim R^2 N^{-1}$, appropriate for a Gaussian chain of elongation R, and extremalizing this free energy with respect to R, leads to $N \sim R^D$ (for $N\to\infty$) with $D=(2+d)/3$.[1,11]

A first important remark is that the estimate of the average number of double points as being $\sim N^2 R^{-d}$ is not consistent with what we know about self-intersections. We know that, more correctly, this number should be proportional to N.[2] On the other hand, the term $N^2 R^{-d}$ can also be read as R^{2D-d} (since $N \sim R^D$). According to the law of codimension additivity, this should be interpreted as the number of mutual intersections of two *independent* replicas of the chain, which seems to contrast with the intuitive idea at the basis of the approximation.

In order to understand how these two apparently contradictory aspects (mutual vs. self-intersections) can be reconciled, one must arrive at a reinterpretation of the whole approximation, as explained below. The necessity of such a reinterpretation can best be seen when considering the case of the k-tolerant walks, for which a naive application of standard Flory arguments leads to incorrect results. Along the line of the usual derivation for the k=1 case, the Flory repulsive energy of a k-tolerant walk, being due to k+1 monomer encounters, would seem to be $\sim N^{k+1}R^{-kd}$, leading to a k-dependent fractal dimension $D = (2+kd)/(k+2)$.[7] This turned out, however, to be a wrong prediction since, for all k, D can be seen to remain equal to its value for k = 1 (SAW).[12,13] We clearly need a more reliable criterion for writing the repulsive energy term in a Flory free energy. As we will show, this criterion cannot ignore the correct geometry of both mutual and self-intersections of random fractals.

In order to estimate the free energy of a chain of N monomers, let us first imagine to partition it into N/ℓ segments in such a way that each segment contains a large number ℓ of monomers, and can be considered as a separate fractal. The free energy can then be estimated as the sum over the free energies of N/ℓ segments, plus the free energy due to segment-segment correlations. The first part can simply be written as $\frac{N}{\ell} \cdot F_0(\ell)$, where $F_0(\ell)$ is the free energy of a segment of size ℓ, which is clearly independent of the end-to-end elongation R of the whole chain. In the spirit of the Flory approach, we will treat the various segments as independent random chains (R.W.) of length ℓ. Their mutual repulsive energy is given by segment-segment repulsion effects. In a mean-field spirit, the number of segment-segment encounters will be proportional to $(N/\ell)^2 (R/\ell^{1/2})^{-d}$, taking into account that the elongation of each segment should be $\sim \ell^{1/2}$. Being fractals, two segments will generally have an average number of mutual intersection points proportional to $\ell^{1/2}$ (their elongation) elevated to the power $(2\cdot 2 - d)$, the fractal dimension of the mutual intersections of two random walks. The entropy of the segments can easily be estimated as the one pertaining to a random chain of N/ℓ ($\gg 1$) segments, with total elongation $R/\ell^{1/2}$. We can finally write for the total free energy:

$$F = \frac{N}{\ell}F_0(\ell) + a(\frac{N}{\ell})^2 (R/\ell^{1/2})^{-d}(\ell^{1/2})^{4-d} + b(R/\ell^{1/2})^2/(N/\ell) \quad , \tag{5}$$

with a and b suitable, weakly ℓ-dependent, dimensional factors. A remarkable feature of eq. (5) is that the basic dependence of the free energy on N and R is the same as obtained through the standard Flory derivation of the free energy, in the case of self-avoiding chains. Equation (5), with ℓ playing the role of a rescaling parameter, naturally evokes the inhomogeneous scaling equation for the free energy of a spin model at a fixed point of a renormalization transformation,[14] and shows that there is a form of scaling invariance hidden in the usual expression of the Flory free energy.

The above derivation makes use of the same basic approximations as the standard one,[1] plus the ingredient of a correct description of fractal intersections. An immediate bonus is that we can now understand why the free energy (5), leading to the SAW-exponent, should apply also to k-tolerant walks with k > 1. Indeed, the second term should be left invariant, regardless of k, because we may conclude from the considerations of the previous section that (k+1)-ple mutual intersection points of two random walks of ℓ steps should grow, on the average, like $(\ell^{1/2})^{4-d}$ for all k. These points can indeed be seen as simple intersections of n_1-ple points of the first walk with n_2-ple points of the latter, with $n_1+n_2 = k+1$. For each n, the n-ple self-intersections of a fractal form themselves a fractal with the same dimension.

Considerations similar to those presented in the present section can also be formulated for Flory appriximations for other problems, like e.g. branched polymers.[2]

4. SPECTRAL DIMENSION OF k-TOLERANT AND RANDOM WALKS

As mentioned in the introduction, self-intersecting fractals (like e.g. k-tolerant walks) are of interest also from a dynamical point of view. Well known experimental results on the low-frequency behaviour of the density of vibrational states in ferrodoxin and some hemo-proteins[6] were a very important stimulus for the general study of dynamics on fractals.[4,5] A specific problem connected with the interpretation of such experiments consists in understanding up to which extent the presence of crosslinks (e.g. H-bonds) between different segments of a protein backbone can alter its essentially one-dimensional character as far as dynamics is concerned, leading to a spectral dimension greater than 1.[15]

At the level of lattice models for diffusion on random fractals, one can ask the equivalent question whether the exponent characterizing e.g. a random walk on a self-avoiding chain is altered or not, when we allow for extra hoppings between non-adjacent sites of the chain. Extra hopping is generally considered only across short-range bridges, i.e. between neighbouring sites that are both visited by the chain with a certain number of steps in between. An alternative way of realizing a deviation from the law of diffusion on a self-avoiding chain, is to perform a random walk on e.g. a 2-tolerant walk. The fractal dimension of this 2-tolerant walk is known to be the same as if it would be self-avoiding;[12,13] the diffusion on it, however, can profit from the many extra hoppings available at the double points.

The problem of diffusion on self-avoiding walks with short-range bridges has very recently been investigated by several authors.[16,2] A sufficiently accurate determination of the exponent governing such diffusion in two dimensions could be obtained by sampling about 10^7 random walks (for each length from 1 to 50 steps), starting in the middle of self-avoiding walks of 200 steps (to avoid boundary effects).[2] A different self-avoiding walk configuration was used for every group of 50 random walks. When we apply the series analysis methods mentioned in the previous section to the obtained data, we obtain $D_W = 2.65 \pm 0.03$. This fractal dimension D_W for the diffusion is defined such that the average displacement R, after a large number t of diffusion steps, grows like t^{1/D_W}. The value obtained for this exponent is the same as one would find for simple diffusion on a self-avoiding chain; this implies a spectral dimension equal to 1.

A similar study has been performed for the diffusion on 2-tolerant walks in two dimensions. Also in this case we obtain a rather clear, even if less sharp, evidence for the absence of deviation from the self-avoiding walk case: $D_W = 2.65 \pm 0.15$.

A typical picture of a 2-tolerant walk consistent with this result and with those of the previous section, is that of a configuration in which most of the loops formed by the walk between two visits of its double points are of small scale. After some coarse graining of these small scale loops, the walk looks like a proper self-avoiding structure. Big loops are prevented by a kind of screening effect, due to the presence of a thick and uniform sequence of small loops along the structure. According to such a picture, the fractal dimension of the double points is clearly identical to the one of the whole walk, as it should be, and diffusion must asymptotically behave like on a simple self-avoiding walk.

We must expect a different situation for diffusion on a pure random walk. In this case, there is no k-tolerance to provide screening, and we can expect loops to be important at all length scales, thus affecting substantially the dynamical exponent. Indeed, extensive enumerations in $d = 2$ for this case produced $D_W = 2.99 \pm 0.04$. This result clearly supports the value $D_W = 3$ for this case, as indirectly, and less sharply suggested by random chain resistance calculations,[17] and in contrast with the results of reference 18. The spectral dimension[4,5] of a random walk in two dimensions is thus equal to $2D/D_W = 4/3$, where $D = 2$ is the fractal dimension of the random walk.

5. CONCLUDING REMARKS

The above results show that short-range bridges or self-contacts of a basically self-avoiding structure do not affect the diffusion properties on this structure, as was hoped by some authors, in order to give a simple explanation of experimental facts concerning proteins.[15] In order to obtain a spectral dimension greater than one, we must remove completely the self-avoiding, or k-tolerant constraint.

An alternative possibility, while preserving the self-avoiding structure, consists in allowing for long-range hopping rates among different sites of the chain. This was shown on the basis of exact renormalization group calculations for a model of diffusion on a deterministic fractal in two dimensions, which yielded non-universal spectral dimensionalities in the whole range between 1 and $4/3$.[19] Long-range hopping bridges in this model essentially play the role of the loops at all length scales, affecting diffusion on a purely random unconstrained chain.

REFERENCES

1. P.G. de Gennes, Scaling Concepts in Polymer Physics (Cornell University Press, Ithaca, 1979).
2. R. Dekeyser, A. Maritan and A.L. Stella, paper in preparation.
3. B.B. Mandelbrot, The fractal Geometry of Nature (Freeman, San Francisco, 1982).
4. S. Alexander and R. Orbach, J. Physique Lett. 43 (1982) L 625.
5. R. Rammal and G. Toulouse, J. Physique Lett. 44 (1983) L 13.
6. H.J. Stapleton, J.P. Allen, C.P. Flynn, D.G. Stinson and S.R. Kurtz, Phys. Rev. Lett. 45 (1980) 1456.
7. L. Turban, J. Phys. A 16 (1983) L 643.
8. E.W. Montroll and G.H. Weiss, J. Math. Phys. 6 (1965) 167.
9. J.W. Lyklema and K. Kremer, Phys. Rev. B 31 (1985) 3182.
10. M.N. Rosenbluth and A.W. Rosenbluth, J. Chem. Phys. 23 (1955) 356.
11. P.J. Flory, Principles of Polymer Chemistry (Cornell University Press, Ithaca, 1969).
12. A.J. Guttmann, C.Byrnes and N.E. Frankel, J. Phys. A 17 (1984) L 457.
13. R. Dekeyser, A. Maritan and A.L. Stella, Phys. Rev. B 31 (1985) 4659.
14. Th. Niemeijer and J.M.J. van Leeuwen, Physica 71 (1974) 17.
15. J.S. Helman, A. Coniglio and C. Tsallis, Phys. Rev. Lett. 53 (1984) 1195.
16. Y.S. Yang, Y. Liu and P.M. Lam, Z. Physik B 59 (1985) 445; D. Chowdhury and B.K. Chakrabarti, J. Phys. A 18 (1985) L 377.
17. J.R. Banavar, A.B. Harris and J. Koplik, Phys. Rev. Lett. 51 (1983) 1115.
18. S. Havlin, G.H. Weiss, D. Ben-Avraham and D. Movshovitz, J. Phys. A 17 (1984) L 849.
19. A. Maritan and A. Stella, University of Padua preprint DFPD 17/85 (1985).

FRACTALS IN PHYSICS
L. Pietronero, E. Tosatti (editors)
© Elsevier Science Publishers B.V., 1986

STATISTICAL MECHANICS OF SELF-AVOIDING RANDOM SURFACES

Amos MARITAN[*,†] and Attilio L. STELLA[+,†]

Dipartimento di Fisica dell'Università di Padova, Padova, Italy

Self-avoiding random surfaces (SAS's) are obtained as the $n \to 0$ limit of suitable lattice gauge models with n-component bond variables. The exponent ν of SAS's on a lattice with fractal dimension $1 + \ln 3/\ln 2$ is calculated exactly.

1. INTRODUCTION

Models of random surfaces and their fractal properties are of interest in several respects. From an experimental point of view, e.g., there is now clear evidence that the surfaces of many materials present irregularities which are self-similar over several length scales in the molecular range [1-3]. Random surfaces are also natural generalizations of the more familiar random walk models of lattice statistics. The former are expected to play, in the context of gauge theories, a role as important as the one played by walks in connection with spin problems.

SAS's, by analogy with self-avoiding walks, are defined as connected sets of plaquettes of a hypercubic, d-dimensional lattice. At most two plaquettes can meet along a common edge, and each plaquette of the lattice can enter in the set only once (excluded volume). For SAS's, a scaling theory, based on proper extensions of the concepts and methods of polymer statistics, was recently proposed [4,5]. Here a preliminary account will be given of new results concerning the connection of SAS's with lattice gauge theories (Section 2), and an exact study of a nontrivial model of SAS's on a fractal lattice (Section 3) [6].

In a grand canonical formulation of the SAS problem typically one would consider the one-loop correlation function

$$W_\Gamma(K) = \sum_{S: \partial S = \Gamma} K^{|S|} \qquad (1)$$

where K is a plaquette fugacity, and each S is a SAS, with $|S|$ plaquettes, and contour ∂S (edges belonging to a single plaquette) coinciding with the self-avoiding closed path Γ [5]. Often Γ can be chosen as the contour ∂p, of an elementary plaquette p, and the definition (1) can be obviously generalized to the case of more than one loop, e.g. to give the (connected) plaquette-plaquette correlation function, whose range, $\xi(K)$, gives a typical length characterizing e.g. the surfaces contributing to (1).

Both $W_{\partial p}(K)$ and $\xi(K)$ are expected to be singular when K approaches a critical value K_c from below: the leading singular terms are $W_{\partial p}(K) \sim (K_c - K)^{-\gamma + 1}$, and $\xi(K) \sim (K_c - K)^{-\nu}$.

The exponent ν is the reciprocal fractal dimension, whereas γ has an entropic meaning: indeed, consistently with (1), the number of surfaces with $|S|$ plaquettes and with contour

---

[*] I.N.F.N., Sezione di Padova, Italy.
[+] G.N.S.M. e C.I.S.M., Unità di Padova, Italy.
[†] I.S.A.S., Trieste, Italy.

∂p should grow like $|S|^{\gamma-2} K_c^{-|S|}$, for large $|S|$ [4,5].

For surfaces without handles an approximation of the Flory type yields $\nu = 3/(4+d)$ for $d \leq 8$ and $\nu = 1/4$ for $d > 8$ [4]. This formula is in fair qualitative agreement with results of real space renormalization group calculations [4]. The value $\nu = 1/4$ for SAS's above the upper critical dimension, $d_c = 8$, is also supported by calculations in the $d \to \infty$ limit, which yield $\nu = 1/4$ and $\gamma = 1/2$ [5,7]. The value $1/4$ for the ν exponent is also known to describe random surfaces without an excluded volume constraint, on the basis of exact model calculations [8].

Using arguments of a type already applied to percolation and lattice animals [9], one can conjecture a pattern of hyperscaling violation for SAS's, which would imply $\gamma = 2 - \nu(d-2)$ for $d \leq 8$ [6].

2. SAS's AND LATTICE GAUGE THEORIES

The connection between gauge theories and random surfaces can be made precise if the latter are self-avoiding. Indeed SAS's can be obtained as $n \to 0$ limits of suitable gauge models with n-component vector variables associated with the bonds [6,10]. Vector, rather than matrix variables must be used, in order to avoid problems which would prevent the existence of the $n \to 0$ limit [10,11].

To illustrate the main result, let us consider a gauge model on a lattice. To each link connecting a pair of nearest-neighbour sites (i,j), we associate variables S_{ij}^α, $\alpha = 1,2,...,n$, which are the conponents of a vector with $2n$ possible orientations: $\vec{S} = (0,...,0,\pm\sqrt{n},0,...,0)$.

Along lines similar to those in Ref. 12, it is possible to show that, for an Hamiltonian

$$H = K \sum_p \sum_{\alpha=1}^{n} \prod_{(i,j) \in \partial p} S_{ij}^\alpha \qquad (2)$$

where the sum runs over all plaquettes, the following holds

$$\lim_{n \to 0} \langle \prod_{(k,\ell) \in \Gamma} S_{k\ell}^\beta \rangle = W_\Gamma(K) \qquad (3)$$

where β is an arbitrary label, and Γ is a self-avoiding closed path. The function on the right hand side is defined by (1) and contains contributions from surfaces with and without handles.

The Hamiltonian (2) has a global symmetry with respect to permutations of the components. Its local, gauge invariance is with respect to the Z_2 group, i.e. under transformations like $\vec{S}_{ij} \to \varepsilon_i \vec{S}_{ij} \varepsilon_j$, with $\varepsilon_\ell = \pm 1$. The above results can be easily generalized to other correlations, or to the case of vector variables, which take all possible orientation in n-dimensional space.

3. SAS's ON A FRACTAL LATTICE

Consider the fractal lattice in $d = 3$ sketched in Fig. 1

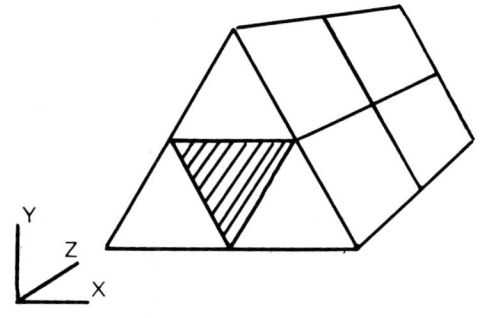

FIGURE 1
Toblerone Lattice

It consists of a sequence of Sierpinski gaskets parallel to the XY plane. Each site on a given gasket is also connected by bonds to the corresponding nearest neighbour sites on the adjacent gaskets. The same lattice, which we will call Toblerone lattice, has been considered independently by Hilfer and Blumem in a different context [13]. For the construction of SAS's one can use all the square plaquettes which contain bonds in the z direction. Only elementary up triangles parallel to the XY plane can be used as plaquettes for the same construction (unshaded).

By a recursive method, which essentially amounts to a renormalization group calculation, it is possible to calculate exactly the ν exponent appropriate to SAS's on this lattice. One finds $\nu = \ln 2/\ln(7-\sqrt{5}) \simeq 0,44$. It should be remarked that this value of ν corresponds to a fractal dimension, which is just equal to the one of a self-avoiding walk on a Sierpinski gasket in two dimensions [14], increased by 1. The planar Sierpinski gasket and the Toblerone lattice have fractal dimensions which differ by 1 ($\ln 3/\ln 2$ and $1 + \ln 3/\ln 2$, respectively). Our Flory approximation for SAS's [4] and the well known one for self-avoiding walks [15] show a similar property. The fractal dimension predicted for surfaces in $d + 1$ dimensions is just equal to that of walks in d dimensions, increased by 1 ($d \leq 4$).

Substituting into our Flory expression for ν the fractal dimension of the Toblerone lattice, in place of d, we find $\nu \simeq .45$, in reasonable qualitative agreement with the exact result.

4. CONCLUDING REMARKS

The development of Section 2 is a non-trivial generalization to surfaces of the well known result concerning the $n \to 0$ limit of $O(n)$ spin models [16].

Those reported in Section 3 are, to our knowledge, the first exact results on a non-trivial model of SAS's. Moreover they give further support to the validity of the renormalization approach and Flory approximation introduced for the general study of SAS's [4,5].

REFERENCES

1. D. Avnir, D. Farin and P. Pfeifer, Nature (London) 308 (1984) 261.

2. P. Pfeifer, D. Avnir, D. Farin and H. Wipperman, Fractal surfaces and pore size distribution of disordered materials, this volume.

3. D.W. Schaefer, Structures of random materials from scattering techniques, this volume.

4. A. Maritan and A.L. Stella, Phys. Rev. Lett. 53 (1984) 123.

5. A. Maritan and A.L. Stella, Scaling approach to self-avoiding walks and surfaces, in : Applications of field theory to statistical mechanics, ed. L. Garrido (Springer, Berlin 1985) pp. 316-327.

6. A. Maritan and A.L. Stella, paper in preparation.

7. J.M. Drouffe, G. Parisi and N. Sourlas, Nucl. Phys. B161 (1980) 397.

8. T. Eguchi and T. Filk, Phys. Lett. 149B (1984) 381.

9. F. Family, J. Phys. A15 (1982) L583.

10. A. Maritan, Topics on static and dynamic properties of fractals, Ph.D. thesis, International School for Advanced Studies (Trieste 1985).

11. B. Durhuus, J. Fröhlich and J. Jonsson, Nucl. Phys. B225 (1983) 185.

12. H.J. Hilhorst, Phys. Rev. B16 (1977) 1253.

13. R. Hilfer and A. Blumen, On finitely ramified fractals, this volume.

14. D.J. Klein and W.A. Seitz, J. Physique Lett. 45 (1984) L241.

15. P.G. de Gennes, Scaling concepts in polymer physics (Cornell Univ. Press, Itacha 1979).

16. P.G. de Gennes, Phys. Lett. 38A (1972) 339.

BETHE - LIKE APPROXIMATION FOR SELF-AVOIDING RANDOM WALKS AND SURFACES (AND FRUSTRATIONS)

Andrea CAPPELLI[§], Roberto LIVI[§], Amos MARITAN[+], Stefano RUFFO[§]

[§] Dipartimento di Fisica and INFN Sezione di Firenze, Largo E. Fermi 2 I-50125 FIRENZE

[+] Dipartimento di Fisica and INFN Sezione di Padova, Via Marzolo 8 I-35131 PADOVA

1. INTRODUCTION

The standard application of the Mean Field (MF) method to lattice statistical models amounts i) to reduce a many body problem to a "one body" (or few bodies) problem in an effective field and ii) to determine this field by a self-consistency equation on the local order parameter (the magnetization).

A large variety of statistical models, like random walks and frustrated models, are difficult to describe within the MF because the suitable order parameter is unknown or possibly does not even exist.

Here we shall discuss some Bethe-like methods which avoid these difficulties. The main idea is to use a consistency equation between correlation functions evaluated on clusters with many sites. The symmetry breaking term is replaced by an effective coupling on the boundary of the cluster.

For self-avoiding walks and surfaces and for a fully frustrated 2d spin model the order parameter remains to be identified, while for lattice gauge theories a more precise identification of the order parameter, the "frustration", is possible through the use of the duality transformation: a relevant feature of this method is, in the last case, its manifest gauge invariance.

2. WALKS AND SURFACES

Let us consider the self-avoiding random walks model on a square lattice. The generating function of the walks is[1]:

$$G(x,y;k) = \sum_{w:\, \delta w=\{x,y\}} k^{|w|} \qquad (2.1)$$

where the sum runs over all the self-avoiding walks from x to y, made of $|w|$ steps between nearest neighbour sites, and k is the monomer fugacity.

The Bethe-like approximation is defined on the cluster in Figure 1 through the introduction of a surface effective fugacity k'[2] and by imposing

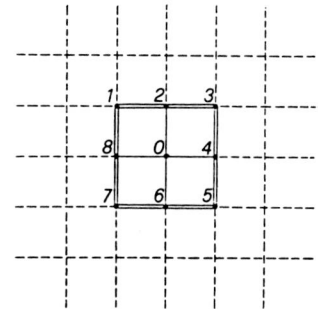

FIGURE 1
The cluster considered for the two dimensional self-avoiding walks problem. The effective fugacity k' is associated to the double-line links

the following consistency equation between n.n. correlations [2]

$$G(1,2;k,k') = G(0,8;k,k') \qquad (2.2)$$

The solution of such equation $k'(k)$ disappears at $k_c = .406$. This can be interpreted as the transition point above which the sum in eq.(2.1) does not converge(Hammersley theorem[3]). The accepted value is $k_c = .38$. At the transition the "internal energy"

$$U(k) = k \sum_{\hat{\mu}=1}^{d} G(x, x+\hat{\mu};k) \qquad (2.3)$$

(where $\hat{\mu}$ are unit vectors of the lattice) behaves as expected:

$$U(k) \underset{k \to k_c}{=} \text{const.} \cdot (k_c - k)^{1-\alpha} \qquad (2.4)$$

with $\alpha = \frac{1}{2}$.

The self-avoiding random surfaces represent a generalization of the previous model. The surfaces are built up by elementary squares (plaquettes) P on a hypercubic lattice. The constraint is that each bond b of a surface S is shared by two plaquettes, unless $b \in \delta S$; in this case b belongs to only one plaquette. A fugacity k is associated to each plaquette on S. The "contour function", i.e. the Green's function associated to a closed non intersecting path γ is:

$$G(\gamma) = \sum_{S: \delta S = \gamma} k^{|S|} \qquad (2.5)$$

where the sum runs over all the self-avoiding surfaces of area $|S|$ and boundary $\delta S = \gamma$.

The method can be applied in the simplest case for d=3 taking into account the cluster in Figure 2. k and k' are the fugacities of the bulk and of the boundary respectively. Also in this

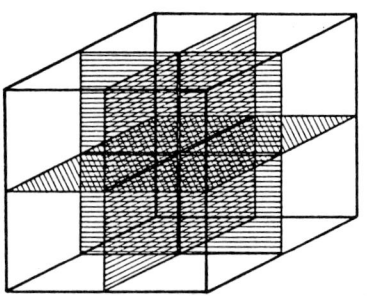

FIGURE 2
The cluster considered for the self-avoiding surfaces problem. Bulk plaquettes are shaded.

case we impose the consistency equation:

$$G(\delta P; k, k') = G(\delta P'; k, k') \qquad (2.6)$$

where P' and P are one of the boundary and internal plaquettes respectively. The solution of eq. (2.6) is lost at $k_c = .537$. According to the generalization of Hammersley theorem [4] this can be interpreted as the signal of a phase transition. The MC estimate is $k_c = .588$ [5] and a renormalization group calculation [6] gives $.62 \div .64$. The critical exponent of the "internal energy" is again $\alpha = \frac{1}{2}$.

In both the cases that we have discussed, improvements can be obtained by enlarging the size of the cluster and/or by introducing further consistency equations.

3. TRIANGULAR FRUSTRATED MODEL

The antiferromagnetic Ising model on a triangular lattice has not an ordered phase at zero temperature [7]. As a consequence a standard MF for the magnetization is not defined.

Also in this case we can apply our method on the simple cluster in Figure 3. The reduced action of this cluster reads:

$$A_\Lambda = -\beta \sigma_1 \sigma_3 - \beta' \sum_{i=1}^{4} \sigma_i \sigma_{i+1}, \quad \sigma_5 = \sigma_1$$

We impose a consistency equation between the "internal" and the "external" two-point Green's functions

$$<\sigma_1 \sigma_3>_\Lambda = <\sigma_1 \sigma_2>_\Lambda \qquad (3.1)$$

The solution $\beta'(\beta)$ of eq. (3.1) is always pre-

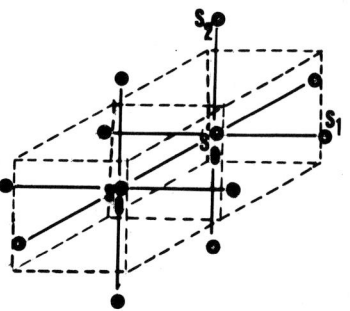

FIGURE 4
The cluster considered for the Z_2 gauge theory in d=3 and its dual, the spin cluster (full links).

$$A_\Lambda = \beta(S_0 S'_0 + \sum_{i=1}^{5} S_0 S_i + S'_0 S'_i)$$
$$+ b \sum_{i=1}^{5} S_i + S'_i \qquad (4.1)$$

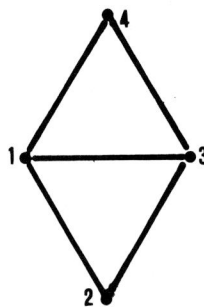

FIGURE 3
The cluster considered for the triangular frustrated model.

sent and gives no signal of a phase transition (except at zero temperature) in agreement with the exact result[7]. (In the ferromagnetic case the transition is found at $\beta_c = .346$). We obtain a reasonable value for the entropy at zero temperature $S_0 = .2877$ (exact, $S_0 = .3231$[7,8]). Improvements and extensions of our method to other frustrated lattice models are under analysis.

4. Z_2 LATTICE GAUGE THEORY

Let us restrict ourselves to describe the extension of our method to the Z_2 lattice gauge model in d=3 (for a more general analysis see Ref. (9)).

On the simple Ising cluster Λ (Figure 4) we define the reduced action

where a symmetry breaking term b is introduced from the beginning. By a dual transformation A_Λ is mapped in the gauge action

$$A_{\tilde{\Lambda}} = \tilde{\beta} \sum_{p \in \tilde{\Lambda}} \sigma_p + \tilde{b} \sum_{p \in \delta \tilde{\Lambda}} \sigma_p \qquad (4.2)$$

defined on the dual cluster $\tilde{\Lambda}$, where $th(\tilde{\beta})=e^{-2\beta}$, $th(\tilde{b})=e^{-2b}$ and $\sigma_p = \sigma\sigma\sigma\sigma$ is the plaquette variable.

The Bethe-Peierls consistency equation for the magnetization on the Ising cluster transforms under duality into the following equation

$$< e^{-2(\tilde{\beta}+\tilde{b})\sigma_{\bar{p}}} >_{A_{\tilde{\Lambda}}} = < e^{-2\tilde{b}\sigma_{\bar{p}}} >_{A_{\tilde{\Lambda}}} \qquad (4.3)$$

The (l.h.s.) is the "frustration" ($\pi \text{sign } \tilde{\beta} = \pi \text{sign } \tilde{b} = -1$) on an internal cube of $\tilde{\Lambda}$, while the (r.h.s.) corresponds to the "frustration" on a cube external to $\tilde{\Lambda}$ sharing a plaquette $\bar{p} \in \delta\tilde{\Lambda}$.

This equation shows a second order phase transition at $\tilde{\beta}_c = .805$ (MC, $\tilde{\beta}_c = .761$) between an ordered phase with vanishing expectation values in eq. (4.3) and a disordered one where isolated frustrations on cubes condensate.

The broken symmetry can be identified as a dual parity symmetry[9].

Equation (4.3) provides a manifestly gauge invariant MF approach. Improvements of the method can be obtained by considering larger clusters (see Figure 5).

The extension to the d=4 case gives a first order phase transition and the possibility to reinterpret in a more physical way the "mean link" approximation[10] for lattice gauge theories, which is not manifestly gauge invariant.

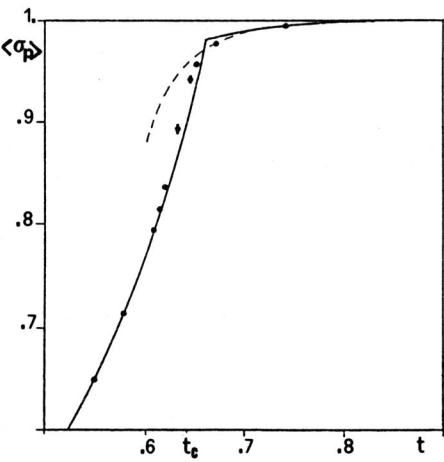

FIGURE 5
The mean value of the plaquette as a function of $t=\text{th}(\tilde{\beta})$ for a 3×3×2 cluster, compared with MC results (dots) and the weak coupling series (dashed line)

REFERENCES

1. P.G. De Gennes, Scaling Concepts in Polymer Physics (Cornell University Press, 1979).
2. R. Livi, A. Maritan, S. Ruffo, J. Phys. A, in press.
3. J.M. Hammersley, Proc. Camb. Phyl. Soc. 57 (1961) 516.
4. B. Durhuus, J. Fröhlich, T. Jonsson, Nucl. Phys. B225 (1983) 185.
5. J. Greensite, T. Sterling, Phys. Lett. 121B (1983) 345.
6. A. Maritan, A. Stella, Phys. Rev. Lett. 53 (1984) 123.
7. G.H. Wannier, Phys. Rev. 79 (1950) 357.
8. S. Caracciolo, G. Parisi, N. Sourlas, Nucl. Phys. B205 (1982) 345.
9. A. Cappelli, R. Livi, A. Maritan, S. Ruffo, Mean field for topological excitations of Z_2 spin and gauge models, University of Florence, preprint (1985).
10. J.M. Drouffe, J.B. Zuber, Phys. Rep. 102 (1983) 1.

ON THE SELF-AVOIDING WALKS ON DISORDERED LATTICES

S. MILOŠEVIĆ and A. CHERNOUTSAN*

Department of Physics and Meteorology, University of Belgrade, P.O.Box 550, 11001 Belgrade, Yugoslavia

We study self-avoiding walks (SAWS) on the random Manhattan square lattice (RM), which is an archetype of disordered lattices. Within the position-space renormalization group theory, we show that SAWS on RM and SAWS on the ordinary square lattice are in the same universality class. This result is discussed in connection with recent controversial predictions for SAWS on fractals.

1. INTRODUCTION

The self-avoiding walk (SAW) model represents a random walk that contains no self-intersections. Its statistical properties comprise a set of well-established research problems[1,2]. Recently, the problem of SAWS on infinite percolation clusters at the percolation threshold has attracted a particular attention, as a number of controversial predictions have been put forth (see, for example, the comprehensive introduction of Nadal and Vannimenus[3]). The infinite percolation clusters provide a realization of fractal structures in physics, and the question with controversial answers concerns critical properties of SAWS on such structures compared with criticality of SAWS on homogeneous lattices.

In this work we study the SAW problem on the square random Manhattan lattice (RM). A prototype of this lattice (see fig.1) could be created if all the one-way signs in downtown Manhattan were randomized at every intersection[4]. Conceived as a traffic barrier, such a lattice is at a percolation threshold. Indeed, the position-space renormalization group (PSRG) analysis[5] shows that changing signs of an arbitrary small part of, for example, eastward (E)

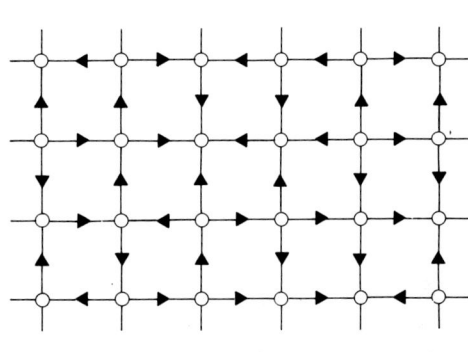

FIGURE 1
A part of the random Manhattan lattice (RM).

and northward (N) oriented bonds (streets) would induce a directed percolation in a south-west (SW) direction, whereas depriving RM of an equally small parts that contribute separately to the NE and to the SW percolations would cause no percolation at all (the randomized downtown would be impassable). Besides, introducing non-oriented bonds instead of an equally small parts of the bonds that contribute to the NE and SW percolations would promote the isotropic percolation. Thus, RM is at a multiple percolation threshold, and, for this reason, it should be a suitable testing ground for the SAW problem. With this object in view we apply the PSRG analysis of the type

---

*Present address: Department of Physics, Gubkin's Institute of Petrochemical and Gas Industry, Moscow 117296, USSR.

recently introduced by Prentis[6]. We find that SAWS on RM and SAWS on the ordinary square lattice are in the same universality class. This finding is discussed in connection with the controversial predictions mentioned in the preceding paragraph.

2. RENORMALIZATION GROUP ANALYSIS

The statistics of SAWS is determined by the total number C_N of N-step SAWS and by the corresponding mean square end-to-end distance $\langle R_N^2 \rangle$. The criticality of SAWS is manifested in the assumed power-laws, valid for large N,

$$C_N \sim \mu^N N^{\gamma-1}, \quad \langle R_N^2 \rangle \sim N^{2\nu}, \quad (1)$$

where μ is the connectivity constant, while γ and ν are the associated critical exponents (it should be noted that $1/\nu$ is a fractal dimension of SAWS). Adopting the fugacity parameter K as the fundamental scaling field the generating functions for C_N and $\langle R_N^2 \rangle$ appear to be the grand canonical ensemble formulae[7]

$$Z(K) = \sum_{N=0}^{\infty} K^N C_N \quad (2)$$

and

$$\xi^2(K) = \sum_{N=0}^{\infty} K^N \langle R_N^2 \rangle C_N / Z(K), \quad (3)$$

which should exhibit the following critical behaviour

$$Z(K) \sim (K_c - K)^{-\gamma}, \quad \xi^2(K) \sim (K_c - K)^{-2\nu}, \quad (4)$$

for K less than, but very close to, the critical fugacity

$$K_c = 1/\mu. \quad (5)$$

Thereby, a group of SAWS (e.g. SAWS on different two-dimensional lattices) constitute one universality class if all members of the group have same critical exponents γ and ν.

We shall first compare values of ν for SAWS on RM and on the ordinary square lattice, calculated by the PSRG method. To construct the PSRG for SAWS on RM we should first rescale the RM lattice to a new coarse-grained RM lattice. For this purpose we use the cell-to-bond mapping with the scale factor b=3, and the *majority rule*[6] illustrated in fig.2. Next, we

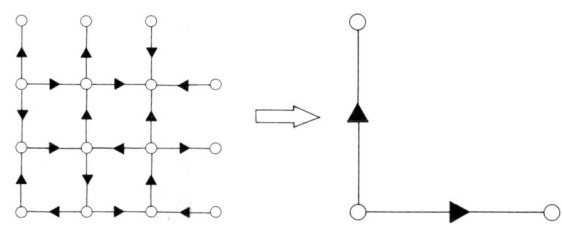

FIGURE 2
Rescaling of a b=3 cell of RM.

have to construct RG transformation, $K'=F(K)$, for the fugacity K' which is the weight of a step on the rescaled RM. The transformation should conserve the total weight of all SAWS, that is to say $Z(K)$. This can be approximately achieved by the *connectivity rule*[8], which requires that F(K) is calculated as the total weight of all feasible SAWS that traverse a renormalized cell in, for instance, northward direction, so that each n-step SAW brings on a K^n term in F(K). We use the "equal-averaging" variant of the connectivity rule, which means that we allow three possible entries of SAWS, on the bottom of the b=3 cell, and count all SAWS with the equal weight 1/3. Furthermore we have to average F(K) over the 2^{14} different b=3 cells that correspond to the northward oriented renormalized bonds. This arduous programme can be facilitated by performing the averaging separately for each term in F(K), i.e. for each SAW that traverses the cells.

Within the one-parameter (fugacity) PSRG scheme we permit each SAW step to take place only in the direction determined by the underlying bond of RM. Then the programme described in the preceding paragraph leads to the following RG transformation

$$K' = \frac{1}{3}[(10944\, K^3 + 21888\, K^4 + 21888\, K^5 + 9120\, K^6 + 4608\, K^7 + 936\, K^8 + 482\, K^9)/2^{14}] \quad (6)$$

whereby we locate the critical fugacity $K_c = 0.9078$ as the corresponding fixed point value $K_c = K^*$ of (6). The critical exponent $\nu = 0.7234$ follows from the formula[9] $\nu = \ln b / \ln \lambda$, where λ is the eigenvalue $\lambda = dK'/dK$ of the linearised transformation (6). A comparison of the foregoing value of ν with the corresponding value for SAWS on the ordinary square lattice[6] $\nu = 0.7283$ reveals a small (but finite) difference. Such a difference does not permit any positive conclusion concerning the universality classes of the two SAWS.

To resolve the universality class query we are going to enlarge the parameter space in such a way that models in question are encompassed with one PSRG scheme. To this end we introduce, following Prentis[6], the probability w that a SAW's step is made in the direction of the underlying bond, and, consequently, the probability $1-w$ that a step is made against the bond direction. In the former case weight ascribed to the step is wK, whereas in the latter case step's weight is $(1-w)K$. Hence, by changing w from 1 to 0.5 we induce an evolvement of SAWS on RM (with fugacity K) to SAWS on the ordinary square lattice (with the effective fugacity $K/2$). The corresponding RG transformations have the schematic form

$$w'K' = F_1(K, w, 1-w),$$
$$(1-w')K' = F_2(K, w, 1-w), \quad (7)$$

where w' and K' are the renormalized parameters, while function F_1 (function F_2) corresponds to the total weight of SAWS averaged over all cells that give the northward (southward) oriented renormalized bonds. We have found the following property

$$F_2(K, x, y) = F_1(K, y, x), \quad (8)$$

and the following explicit expression

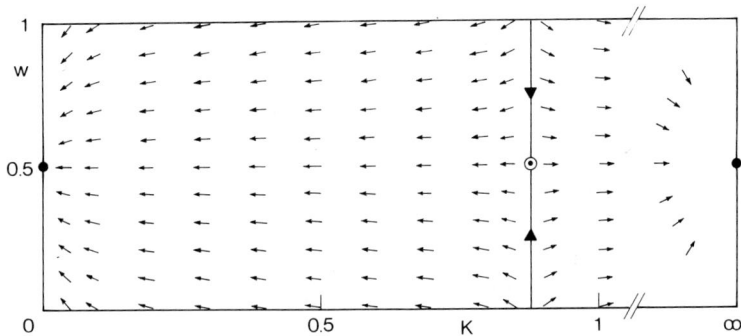

FIGURE 3
Flow diagram generated by the RG transformations (7). The points on the critical line flow into the nontrivial fixed point (⊙) at $w=0.5$. The trivial fixed points (•) are also shown.

$$F_1(K,x,y) = \frac{1}{3} \frac{1}{2^{14}} \sum_{i=3}^{9} \sum_{j=0}^{i} a_{ij} x^{i-j} y^j b_i K^i, \quad (9)$$

with the corresponding sets of exact coefficients

$\{b_i\} = \{192, 384, 384, 160, 64, 8, 2\}$,
$\{a_{3j}\} = \{57, 126, 66, 7\}$,
$\{a_{4j}\} = \{57, 183, 192, 73, 7\}$,
$\{a_{5j}\} = \{57, 240, 375, 265, 80, 7\}$, (10)
$\{a_{6j}\} = \{57, 297, 615, 640, 345, 87, 7\}$,
$\{a_{7j}\} = \{72, 444, 1137, 1565, 1235, 543, 116, 8\}$,
$\{a_{8j}\} = \{117, 831, 2526, 4307, 4480, 2861, 1058, 193, 11\}$,
$\{a_{9j}\} = \{241, 1924, 6701, 13609, 17551, 14705, 7895, 2515, 380, 15\}$.

The flow diagram that corresponds to the above transformations is depicted in fig.3. It reveals only one nontrivial fixed point, located at $w^* = 0.5$ and $K^* = 0.8788$, with the relevant eigenvalue $\lambda_k = 4.5198$ and the related critical exponent $\nu = 0.7283$. Besides, the critical line definitely intersects the lines $w=1$ and $w=0$ that correspond to SAWS on the RM lattice and SAWS on the inversed RM lattice, respectively. This fact and the RG flow on the critical line make one conclude that SAWS on RM are in the same universality class with SAWS on the ordinary square lattice.

3. DISCUSSION

The main result of the preceding section is the conclusion that SAWS on RM and SAWS on the ordinary square lattice are, as far as the critical exponent ν is concerned, in the same universality class. In order to challenge this conclusion we pose two questions. First, could the character of the obtained flow diagram be a consequence of the used parametric representation of SAWS, and, the second question, is this diagram in agreement with similar results obtained for other (simpler) models? In answering the first question we note that one may wish to introduce a more physical parameter u instead of the parameter w. The new parameter u may be, for instance, the probability that the SAW walker makes a step against the bond direction, whereas with the probability $1-u$ he restrains from violating the bond directions. However, it can be verified that the corresponding flow diagram, in the (K,u) plane, vindicates the above-mentioned conclusion. The only difference would be a new critical line, which turns out to be neither vertical nor straigt, and which bears no point that corresponds to SAWS on the inversed RM lattice.

Concerning our second question we would like to mention the recent work of Sahimi[10], who applied the PSRG method to study SAWS on a square lattice with a fraction $(1-p)$ of sites eliminated. Sahimi found that for all fractions above the percolation threshold $(p > p_c)$ SAWS are in the universality class of SAWS on the ordinary square lattice, while only SAWS on the infinite percolation cluster $(p = p_c)$ are described by a different (larger) exponent ν_1. We may argue that our results support the work of Sahimi, on the ground that by applying the majority rule mapping, instead of the percolation rule as Sahimi[10] did, we forced the underlying lattice to behave as a lattice above p_c. On the other hand, we could not have applied the percolation rule mapping, for it would lead[4] to a lattice that consists of oriented bonds, nonoriented bonds and vacancies. Such a lattice would evidently furnish an inconvenient substratum for studying the SAW problem.

REFERENCES

1. C. Domb, Adv.Chem.Phys. 15 (1969) 229.
2. D.S. McKenzie, Phys.Rep. 27C (1976) 35.
3. J.P. Nadal and J. Vannimenus, J.Physique 46 (1985) 17.
4. S. Redner, in: Percolation Structures and Processes, eds. G.Deutscher, R.Zallen and J.Adler (Adam Hilger, Bristol, 1983) pp. 447-474.
5. S. Redner, Phys.Rev. B 25 (1982) 3242.
6. J.J. Prentis, J.Phys. A 17 (1984) L21.
7. S. Redner and P.J. Reynolds, J.Phys. A 14 (1981) L55.
8. S. Redner and P.J. Reynolds, J.Phys. A 14 (1981) 2679.
9. H.E. Stanley, P.J. Reynolds, S. Redner and F. Family, in: Real Space Renormalization, eds T.W. Burkhardt and J.M.J. van Leeuwen (Springer, Berlin, 1982) pp. 171-208.
10. M. Sahimi, J. Phys. A 17 (1984) L 379.

PROTEINS IN THE EXPERIMENT

Y.S. YANG

International Centre for Theoretical Physics, Trieste, Italy.[+]

The backbone of ferredoxin and hemoproteins are described by SAWs in two and three dimensions. But the spin-lattice relaxation process of Fe^{3+} ions cannot be described by pure fractal model. The spectral dimensions observed in experiment is defined through $d_s = d_f/a$, a is given by the scaling form of the low frequency mode $\omega(bL) = b^a \omega(L)$ of the whole system consisting of proteins and the solvent upon a change of the length scale.

For temperature $T \gg g \mu_B H/k$, the spin-lattice relaxation of Fe^{3+} ions in low-spin ferredoxin and hemoproteins is dominated by a two-phonon process (Raman). The temperature dependence of the relaxation time τ is given by[1,2]

$$1/\tau \propto T^{3+2d_s} F(T/\Theta, d_s) \quad (1)$$

where $F(x,y)$ is a smooth function of x, and Θ is the Debye temperature. The fracton or spectral dimension d_s[3,4] enter Eq.(1) through

$$\rho(\omega) \propto \omega^{d_s - 1} \quad (2)$$

where $\rho(\omega)$ is the phonon density of states at low frequencies and ω is the frequency. For an ordinary d-dimensional object, d_s coincides with the Euclidean dimension d, and Eq.(1) is in good agreement with experiment.

However, experiment[2,5-7] shows that d_s and the fractal dimension d_f are 4/3 for ferredoxin and 5/3 for hemoproteins. Some authors[2,6] identified d_s with the fractal dimension d_f and postulated the validity of Eq.(1) also for noninteger value of d_s. As these values coincide with the fractal dimensions for two and three dimensional self-avoiding walks (SAWs), they also argued that the protein backbones can be modelled by SAWs. However, the spectral dimension d_s is in general different from the fractal dimension d_f and the space dimension d[8]. For SAWs, $d_s = 1$. The substitution $d_s = 1$ into Eq.(1) gives a result which disagrees with experiment.

To solve this conflict, Helman et al[9] proposed a fractal model which incorporated massless bridges. They argued that for a high enough density of the bridges, d_s equals d_f. But theoretical analysis and computer simulations[12,13] show that the inclusion of finite length bridges do not change the spectral dimension of the system.

I think that the disagreement between theory and experiment is because the previous theory was based on "pure" fractal models, but the experiment was done in the solvent (water). The phonon density of states is different for these two cases. Physically, the relaxation process involves a modulation of the orbital

---

[+]Permanent address: Institute of Physics, Chinese Academy of Sciences, Beijing, China.

electronic wave function by structural vibrations. The spins are coupled to vibrations via spin-orbit interactions. Therefore, only those phonon modes that affect the orbital wave function of Fe^{3+} ions contribute to the relaxation process. In this letter, I propose a fractal model that incorporates the solvent effect. The protein molecules, and therefore Fe^{3+} ions form self-avoiding walk configurations on the lattice, the other sites are occupied by solvent molecules, Fig.(1).

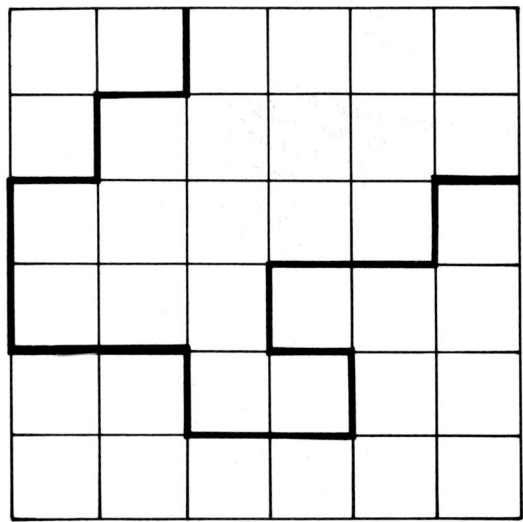

FIGURE 1

Lattice model for proteins in the solvent. Sites linked by the heavy lines are protein molecules and they form self-avoiding walk configurations. Other lattice sites are occupied by the solvent molecules.

Since only when the distance between Fe^{3+} ions changes can the spins flip, only the vibrational modes in the Fe^{3+} ions contribute to the relaxation process. Following the scaling argument given by Rammal and Toulouse[4], under a change of the length scale $L \to bL$, the number of Fe^{3+} ions in an unit volume changes as

$$N(L) = b^{-d_f} N(bL) \qquad (3)$$

where d_f is the fractal dimension of the protein. Assume the protein backbones to be SAWs, $d_f = (d+2)/3$. If under this scale transformation, the mode frequency has also a scaling behavior

$$\omega_L = b^{-a} \omega_{bL} \quad , \qquad (4)$$

as the number of phonon modes related to Fe^{3+} ions is proportional to the number of the ions, denoting $\rho(\omega)$ as the density of states of phonon which are related to the Fe^{3+} ions, it follows

$$\rho_L(\omega_L) = b^{-d_f} \rho_{bL}(\omega_{bL}) \frac{d\omega_{bL}}{d\omega_L} = b^{a-d_f} \rho_{bL}(\omega_{bL}). \qquad (5)$$

Taking $b = \omega^{-1/a}$, it is easily seen that the spectral dimension is given by

$$d_s = d_f/a \quad . \qquad (6)$$

From the presentation given above, one sees that the solvent effect is reflected in the scaling form of the mode frequency, Eq.(4). If protein-protein, protein-solvent, solvent-solvent interactions are the same, the vibrational spectrum is just the same as that of a homogeneous system, i.e. a=1. It is conjectured that if all the three interactions are finite, one also has a=1. This is just the case with the experiment, and the agreement between theory and experiment is restored. If the interactions between molecules are isotropic, then the equations of motion of the system are

$$m_i \frac{\partial^2}{\partial t^2} u_i = \sum_\delta K_{i,i+\delta}(u_{i+\delta} - u_i) \qquad (7)$$

where I denotes the type and the position of

the atoms. $K_{i,j}$ is the force constant between atom i and j. m_i is the mass of atom i. δ is the nearest neighbour vector. Fourier transform of Eq.(7) is

$$-\omega^2 m_i u_i = \sum_\delta K_{i,i+\delta}(u_{i+\varepsilon} - u_i) \quad . \tag{8}$$

Recall the Laplace transform of the master equation for the random walk problem

$$-\varepsilon p_i = \sum_\delta W_{i,i+\delta}(p_{i+\delta} - p_i) \quad , \tag{9}$$

where $p_i(t)$ is the probability that the walker is at position i at time t, $w_{i,j}$ the jump probability from i to j. If the mean displacement of the random walker at time t has the scaling form

$$R(t) \propto t^{1/d_w} \quad . \tag{10}$$

From the similarity of Eq.(8) and Eq.(9), the following relation exists

$$a = d_w/2 \tag{11}$$

Transformed from vibrational problem to random walk problem, the jump probability $W_{i,j}$ are dependent on i and j, but all are finite. Therefore, a finite lower bound W_{min} and a finite upper bound W_{max} exist. Denote $R_{min}(t)$ as the mean displacement of the random walker at time t with all jump probabilities equal W_{min}, and $R_{max}(t)$ with all jump probabilities equal W_{max}, it follows

$$R_{min}(t) \leq R(t) \leq R_{max}(t) \quad . \tag{12}$$

since

$$R_{min}(t) \propto t^{1/2}$$
$$R_{max}(t) \propto t^{1/2} \tag{13}$$

it follows

$$R(t) \propto t^{1/2} \quad . \tag{14}$$

This proves the conjecture given above for the isotropic interactions.

In conclusion one cannot explain the experimental results for ferredoxin and hemoproteins by pure fractal model. The spectral dimension observed in experiment is in general different from that of the pure fractal model. One should be careful while applying the results from pure fractal model to real physical systems.

ACKNOWLEDGEMENTS

The author wishes to thank Professors P.M. Lam, A. Coniglio, A.L. Stella, B.B. Mandelbrot, S.R.A. Salinas, D. Kumar and K.A. Chao for useful discussions. He is also grateful to Professor Abdsu Salam, the International Atomic Energy Agency and UNESCO for hospitality at the International Centre for Theoretical Physics, Trieste.

REFERENCES

1. R. Orbach, Proc. Roy. Soc. London 264 (1961) 458.
2. J.P. Allen, J.T. Colvin, D.G. Stinson, C.P. Flynn, H.J. Stapleton, Biophys. J. 38 (1982) 299.
3. S. Alexander and R.J. Orbach, J. Physique Lett. 43 (1982) L625.
4. R. Rammal and G. Toulouse, J. Physique Lett. 44 (1983) L13.
5. C. Mailer and C.P.S. Taylor, Biochimica et Biophysica Acta 322 (1973) 195.
6. H.J. Stapleton, J.P. Allen, C.P. Flynn, D.G. Stinson, S.R. Kurtz, Phys. Rev. Lett. 45 (1980) 1456.
7. J.T. Colvin and H.J. Stapleton, J. Chem. Phys. 82 (1985) 4699.

8. H.E. Stanley, in: N. B. S. Conference on Fractals, J. Sta. Phys. 35 (Sept. 1984).

9. J.S. Helman, A. Coniglio, C. Tsallis, Phys. Rev. Lett. 53 (1984) 1195 and 54 (1985) 1735.

10. M.E. Cates, Phys. Rev. Lett. 54 (1985) 1733.

11. H.J. Stapleton, Phys. Rev. Lett. 54 (1985) 1734.

12. Y.S. Yang, P.M. Lam, Commun. Theor. Phys. in press.
    Y.S. Yang, Y. Liu, P.M. Lam, Z. Phys. B-Condensed Matter 59 (1985) 445.

Part IV
BRANCHED POLYMERS, GELATION AND PERCOLATION

FRACTAL DIMENSION AND THE SYNTHESIS OF BRANCHED POLYMERS

Z. ALEXANDROWICZ

Department of Polymer Research, Weizmann Institute of Science, Rehovot 76 100, Israel.

1. INTRODUCTION

The statistical behaviour of branched polymers is commonly identified with that of "animals", viz. an unweighted ensemble of distinct graphs drawn from M bonds on a lattice. The identification is based on that polymers and animals alike, consist of randomly oriented, non-overlapping bonds. It seems to me however that that a dogmatic identification of branched polymers *in toto*, with the animals, is open to the following criticism (see Ref. 1). Branched polymers may be formed with the help of a reversible or irreversible linking of individual bonds (monomers). In the former case we deal with an ensemble of N mutually connected or disconnected monomers and the polymers (clusters) constitute a sort of *transient* topological feature. Hence, as de Gennes has pointed out long ago[2], this reversible linking or "gelation", corresponds to the phenomena of percolation. Clearly therefore the similarity of polymers to animals is meant to apply to the second case of an irreversible linking of monomers. But here we are faced with the following difficulty: The construction of an ensemble of animals assumes an equilibrium with respect to the bonds' rearrangement. The irreversible synthesis of branched polymers proceeds generally in a stepwise fashion. Bifunctional or polyfunctional monomers become irreversibly attached to current "growing tips" of the chain. The formation of various topological structures is kinetic, it is not clear whether these structures are weighted equally, like in the equilibrium ensemble of animals. It seems therefore worthwhile to study the effect of definite kinetic models on the distribution of topological structures and hence on the statistics of shape. One possible kinetics is that of unlimited growth. During a given interval of time, all growing tips propagate, either linearly or with branching (corresponding respectively to the addition of a bifunctional or polyfunctional monomer), which leads to an exponential increase of the mass with the time. Such growth would overfill the space and exhaust the monomer supply. Hence sustained growth necessarily, involves some limiting mechanism.

2. MODELS OF POLYMERS' SYNTHESIS.

Two very simplified kinetic models will be discussed in this connection. In the first model, S, growth is limited by a fixed rate of supply of a homogeneously distributed monomer. In a given interval of time *not all* but only a fixed number of tips (say one) adds a bifunctional or trifunctional monomer. This number is assumed to be randomly distributed among the current set of G tips. As a result of branching, G increases with the mass M, forming an ever growing waiting list (with randomly ordered service!). In the absence of interactions, for dimension $D \geq D_c$, $G \sim M$. Model S corresponds to that studied by Render[3]. In the second kinetic model, T, growth is limited through *termination*. A growing tip may either add a bifunctional or trifunctional group, with probabilities P_1 and P_2, respectively, *or* become terminated with probability $1 - (P_1+P_2)$. (The termination may be due to addition of a monofunctional monomer or to loss of a radical.) Sustained growth, for which branching precisely compensates termination[4], requires the critical value $P_1^c + 2P_2^c = 1$. In the absence of fluctuations this value would lead to a linear growth, viz., to G = const. Because of fluctuation, however, G will nonetheless increase with M. In the absence of correlation between consecutive stochastic steps, G increases as $M^{1/2}$. Since $G/M \to 0$, the limitation to a fixed rate of monomer supply becomes immaterial. The fractal dimension d_f, or the mean radius (R) versus mass (M) dependence,

$$R \sim M^{\nu_M} \qquad (1)$$

(where clearly $\nu_M = 1/d_f$), has been studied for branched polymers synthesized with the help of the S or T kinetics.

3. THE TIME VARIABLE (t).

The foregoing discussion repeatedly refers to a "time" of growth of the branched chain. This may be defined as follows. The chain starts to grow on a lattice from an origin, at $t = 0$. Nearest neighbors to the origin may join the chain; those that do constitute the shell $t = 1$. In turn, nearest neighbors to the sites constituting the shell $t = 1$ may join the chain; those that do constitute the shell $t = 2$, etc. (In actual kinetics, like models S and T, entire iso-t shells need not be filled one after another in succession; some branches may grow faster in real time than others, viz., take a **larger number** t of steps from origin). The time t provides a *length* variable, which enables one to study the mean radius versus length dependence, like in a linear chain,

$$R \sim t^{\nu_t}. \qquad (2)$$

R and t may be measured between a point on the chain and the origin, or, indeed, between any pair of points, one of which may be taken as the "origin." In the absence of loops the definition of a length t between a pair of points is unique (in the presence of loops an average of different pathways may be used instead). The increase of M with t may be studied as well,

$$\langle M \rangle \sim t^{\gamma_t} \qquad (3)$$

From Eqs. (1)-(3) it follows that $\nu_t = \nu_M \gamma_t$. If, as here, the growth is not in iso-t shells, one should measure γ_t^{-1} from $\langle t \rangle \sim M^{1/\gamma_t}$. The description of branched chains in terms of t, ν_t, and γ_t has been first introduced by the author, in connection with a construction of percolating clusters as "critically branched chains."[5] It has been adopted in several subsequent studies of percolation and of animals [6(a-e)]. (Some of these call t the "chemical" or "topological" distance.) In what follows the comparison between models S and T relies to a large extent on the measurement of the t dependence.

4. RESULTS

Branched polymers corresponding to models S and T have been constructed with the help of Monte Carlo simulation. Three lotteries are made at each step. The first picks one out of a current set of growing tips. The second, with P_1 and P_2 determines whether one or two bonds grow from that tip. With model S, $P_1+P_2 = 1$, with model T, termination, with probability $1 - P_1 - P_2$, constitutes a third possibility. A third lottery determines the bond's direction in D-dimensional space. The ends of the bonds constitute the new tips. If a new tip falls into an unoccupied lattice site, the growth continues. If, however, it falls into a site which is already occupied by a preceding chain segment, violating excluded volume, the entire construction is discarded and the process restarted from the origin.

Samples of 2,000-10,000 chains, of maximum mass ranging from 150 to 2,000 as dimension D varies from 2 to 8, have been constructed with the help of the "enrichment" method[7]. In order to discern an inhomogeneity of the segments' density with respect to the origin, the radius-mass exponent is measured in three different ways: (i) $\nu_{M,g}$ has been determined from the average radius of gyration of M segments; (ii) $\nu_{M,0}$ has been determined from the average square distance of M segments *from the origin*; and (iii) a fractal dimension, d_f, has been determined with the help of a recently proposed technique[8]. A branching exponent g, defined through $G \sim M^g$, has been determined as well.

The results are displayed in Table I; superscripts S and T denote the two models. The ideal chain, without the excluded-volume restriction (constructed in D = 8), is denoted by I. The following observations can be made.

(a) *Branching*. - In the absence of an effect of excluded volume, we expect that $g^S = 1$, while g^T should be approximately equal to $\frac{1}{2}$ (Ref. 1). This is borne out by the results. As D decreases, excluded volume favors more linear chains, hence g decreases.

TABLE I

D	g^S	g^T	$1/\gamma_t^S$	$1/\gamma_t^T$	$\nu_{M,0}^S$	$\nu_{M,g}^S$	$1/d_f^S$	$\nu_{M,0}^T$	$\nu_{M,g}^T$	$1/d_f^T$	ν_M^{animal}
2	0.83	0.25	0.57	0.74	0.44	0.57	0.57	0.62	0.62	0.63	0.64
3	0.91	0.34	0.46	0.66	0.32	0.42	0.42	0.47	0.47	0.50	0.50
4	0.96	0.38	0.38	0.63	0.24	0.32	0.34	0.40	0.40	0.43	0.42
5	0.97	0.41	0.33	0.60	0.20	0.26	0.30	0.35	0.35	0.38	0.36
6	0.98	0.44	0.30	0.58	0.17	0.23	0.27	0.31	0.31	0.35	0.32
7	0.99	0.46	0.26	0.54	0.14	0.19	0.24	0.29	0.29	0.32	0.28
8	0.99	0.46	0.24	0.54	0.13	0.17	0.23	0.28	0.28	0.30	0.25
I	1.0	0.46	0.20	0.52	0.10	0.13	0.22	0.26	0.26	0.28	0.25

(b) <t> vs M. - Conversely, γ_t^{-1} is minimum in the absence of excluded volume; as D decreases, γ_t^{-1} increases towards 1, corresponding to linear topology. Model T obeys very well the equation $\gamma_t^{-1} = 1 - g$, which has been derived recently for branched chains growing at criticality[9]. Very clearly, $1/\gamma_t^T > 1/\gamma_t^S$ at all D, which indicates that model T is always more linear than model S. This disparity may be traced to the kinetics of formation. With model T there is no overproduction of tips ($G/M \rightarrow 0$). With model S, however, there is a vast overproduction, and growth proceeds through random access to a fast-expanding waiting list of tips. This results in the formation of a large number of relatively short branches, viz., a bulky, as opposed to linear, topology. In addition, because of repeated random access to the waiting list, monomers are more likely to join tips created at early stages of the kinetics. This should give rise to a relatively denser distribution of segments around the origin.

(c) *Radius mass*. - The bulky topology of model S is further revealed by the values of the radius of gyration; thus $\nu_{M,g}^T > \nu_{M,g}^S$ at all D. Furthermore, with model S the values of $\nu_{M,0}^S$ are significantly lower than the corresponding $\nu_{M,g}^S$, which indicates that indeed, as M increases, the packing of segments around the origin becomes increasingly dense. With model T, $\nu_{M,0}^T \simeq \nu_{M,g}^T$, which indicates the absence of an appreciable inhomogeneity at the origin. This observation is further supported by the results for the fractal dimension d_f, measured over sets of neighbors to arbitrarily chosen segments (see Ref. 8), without reference to the actual stepwide growth of M. This effaces the singular role of the origin in model S. Thus the values of $1/d_f^S$ are comparable with $\nu_{M,g}^S$, or higher than that. (d_f^{-1} listed in Table I have been measured with respect to segments picked at random; d_f^{-1} measured with respect to the origin, though expected to be smaller, turn out to be equal to within experimental accuracy). The values of $1/d_f^T$ for model T are approximately equal, as they should be, to $\nu_{M,0}^T$ and $\nu_{M,g}^T$. (Still $1/d_f^T$ seem to be slightly but consistently higher, see Ref. 1 footnote 12). Altogether, the radius-mass exponents for model T are similar (possibly a little smaller at low D) to ν_M^{animal} calculated for lattice animals[10a-d].

5. CONCLUSIONS

It appears that the kinetics of formation indeed affects the distribution of topological structures and consequently the mean shape of branched polymers. The applicability of animal exponents should be therefore limited to certain branched polymers only, notably to those not synthesized in an irreversibly stepwise fashion and also *perhaps* to those that are so synthesized but, like model T, grow without overproduction of tips. True , models S and T represent two extremes, of a great overproduction of *equally accessible* tips and of a complete compensation of growth by termination. Actual kinetics presumably exhibits intermediate types of behavior. Thus, with model S, the tips created at early times are buried within the chain's bulk and less accessible, viz. become to some extent terminated. A deviation from the equal accessibility of tips, due to a sort of "diffusion limited aggregation" effect , is also to be expected.

REFERENCES.

1. Z. Alexandrowicz, Phys.Rev.Lett. 54 (1985) 1420.
2. P.G. de Gennes, J. Physique Lett. 37 (1976) L1.
3. S. Redner, J. Phys. A 12 (1979) L239.
4. P. Flory, *Principles of Polymer Chemistry* (Cornell Univ., Ithaca, 1953), Chap. IX, Sect. 1.
5. Z. Alexandrowicz, Phys.Lett. 80A (1980) 284.
6. [a]R. Pike and H.E. Stanley, J.Phys. A14 (1981) L169; [b]J. Chalupa and F.G. Krausz, Phys. Lett. (1984) 115; [c]S. Havlin and R. Nossal, J. Phys. A17 (1984); L427; [d]P. Grassberger, Math.Biosci. 63 (1983) 157; [e]S. Havlin, Z.V. Djordjevic, I.Majid, H.E. Stanley and G.H. Weiss, Phys.Rev.Lett. 53 (1984) 178.
7. F.T. Wall and J.J. Erpenbeck, J.Chem.Phys. 30 (1959) 634.
8. Y. Termonia and Z. Alexandrowicz, Phys. Rev.Lett. 14 (1983) 1265.
9. Z. Alexandrowicz, Phys.Lett. 109A (1985)169.
10. [a]T.C. Lubenski and J. Issacson, Phys.Rev. Lett. 41 (1978) 829; and Phys.Rev.A20 (1979) 2130 ; [b]Parisi and N.Sourlas, Phys.Rev/ Lett. 46 (1981) 871; [c]D. Dhar,Phys.Rev.Lett. 51 (1983) 853; [d]B. Derrida and L. De Seze, J.Phys. (Paris) 43 (1982) 475.

FRACTAL DIMENSIONALITIES OF BACKBONES AND CLUSTERS IN A KINETIC GELATION MODEL

Ashvin CHHABRA[a,b], H.J. HERRMANN[c] and D.P. LANDAU[a]

[a]Department of Physics, University of Georgia, Athens, GA 30602, U.S.A.
[b]Mason Laboratory, Yale University, New Haven, CT 06520, U.S.A.
[c]Service de Physique Theorique, CEN Saclay, Gif-sur-Yvette Cedex, France

We present results of a computer simulation study of the fractal dimensionality of the largest cluster, backbone and the elastic backbone of a radical initiated irreversible kinetic gelation model in three dimensions. This work was motivated by earlier observations, that although the bulk exponents of this model are compatible with those of percolation, the cluster size distribution is vastly different (damped oscillatory) and obeys different scaling forms. On contrasting these dimensionalities with those from percolation models we find that while the fractal dimensionalities of the elastic backbone and the largest cluster are similar (to percolation) the dimensionality of the backbone is significantly different.

1. Introduction

Recently considerable attention has been focussed on the utility of fractals as models of random media[1]. Clusters generated by various growth models (e.g. percolation, kinetic gelation, DLA) are examples of such fractals. A description of the geometrical structure of these clusters is closely related to the problem of describing the propagation of order at a critical point. Questions such as a description of the flow of fluids through a porous media, the electrical conductivity of a random resistor network, the elasticity of a polymer are all offshoots of the above problem.

In this paper we calculate the fractal dimensionalities of the largest incipient cluster, backbone and elastic backbone of the clusters generated by an irreversible kinetic gelation model. These dimensionalities are contrasted with similar values that have been previously obtained for random bond percolation.

2. Description of the Growth Model[2].

The model consists of tetra-functional monomers placed on the sites of a simple cubic lattice with periodic boundary conditions. A fixed number of initiators are randomly sprinkled on the lattice sites.

The initiator concentration C_i is defined as the total number of initiators normalised by the total number of possible bonds in the lattice, here $3L^3$. Each initiator breaks one of the monomers double bonds and bonds with it. This creates a monomer with a single unsaturated bond defined as an active center. An active center is then randomly chosen and forms a bond with a randomly chosen neighboring site (provided that site is not already saturated). The number of active centres decreases with time due to annihilation and trapping. A combination of cluster growth and cluster merging leads to the formation of an infinite cluster that marks the onset of a sol-gel phase transition The reader is referred to [2,4] for further details of the model.

3. Definitions and procedure for computing the fractal dimensionalities.

Once the infinite cluster is selected, we proceed to cut it i.e. remove the periodic boundary conditions. Doing so causes the cluster to break into multiple pieces. We search for the end points of the largest remaining piece. Clusters that break into many small pieces on cutting or those that have their end points too close to each other

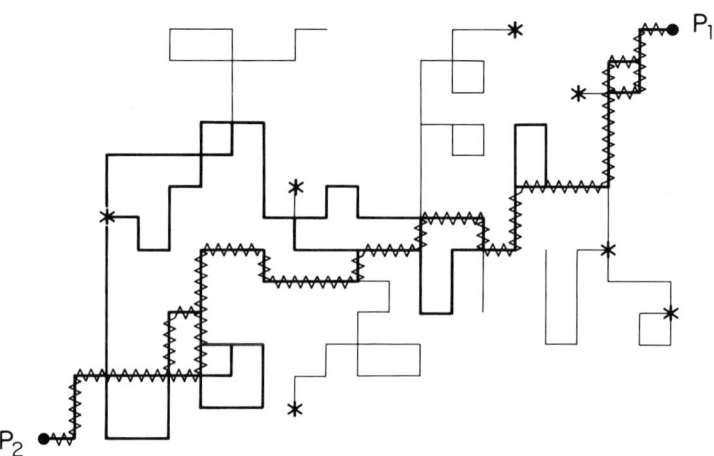

FIGURE 1
Schematic diagram of the largest cluster (—), backbone (▬) and elastic backbone (∧∧∧).

are rejected to ensure certain uniformity in the samples. We then proceed to construct the backbone and elastic backbone.

To define the backbone consider two endpoints (P_1 and P_2) which are separated by a distance comparable to the correlation length of the system. If a fluid is made to flow from P_1 to P_2, then the set of paths (bonds) through which the fluid flows is called the backbone. Related to the backbone is another fractal object called the elastic backbone. The motivation for introducing this comes from an attempt to describe the elasticity of disordered systems. If we consider the bonds between sites to be small springs, then it is reasonable to assume that only the shortest paths connecting the points P_1 and P_2 would offer significant resistance to any deformation of the system. It is the union of these paths (bonds) that make up the elastic backbone.

The number of sites, loops etc in the largest cluster and the backbones are calculated using the method of burning of clusters. The reader is referred to Herrmann et al[3] for the details of the algorithm. The mean density of sites ρ_s goes as

$$\rho_s(p) \sim (p - p_c)^\beta$$

Using finite size scaling theory we know that the number of sites in the cluster goes as

$$S(p_c) = L^d \rho_s(p_c) \sim L^{d-\beta/\nu} = L^D$$

where d is the spatial dimensionality and D the fractal dimensionality. (Note that the exponents β and ν now refer to the exponents describing the critical behavior of the relevent quantity i.e. the backbone or elastic backbone etc.) By plotting log-log plots of the relevent quantities with lattice size we obtain the various fractal dimensionalities.

4. Motivation for comparison of results with percolation.

Although the bulk exponents of the kinetic gelation model are compatible with those of percolation[2] there are several interesting differences between the two models. While the ratio of the susceptibility amplitudes below and above the critical point are universal for percolation, this ratio varies with the concentration of initiators in the kinetic gelation model. Another difference is that the cluster size distribution of the kinetic gelation model shows a damped oscillatory behavior as oppsed to the smooth monotonic decay exhibited in percolation models[4]. This leads them to obey different scaling forms. It is therefore of interest to see if the critical behavior and the fractal dimensionalities of the clusters and backbones of the kinetic gelation model are different from those of percolation.

5. Results and Conclusion.

The lattice sizes studied and the number of samples over which the various quantities were averaged over are summarised in the following table.

Table 1
Statistics for $C_i = 3 \times 10^{-4}$.

Lattice Size	Growth Samples	(Accepted) Backbones
20	4000	653
30	1000	164
40	500	50
60	200	15

Figure 2 shows log-log plots of the data with respect to the lattice size. Our values for the fractal dimensionalities thus computed are summarised and contrasted with those of percolation.

Table 2
Comparision of Fractal Dimensionalities

Fractal Dimension	Kinetic Gelation	Percolation
D_{lcl}	2.34±0.14	2.49±0.01
D_{bb}	2.22±0.10	1.74±0.04
$D_{cl.bb}$	1.47±0.08	1.37±0.07

We note that while the fractal dimensionality of the elastic backbone and the largest cluster are within error bars similar to those of percolation, the fractal dimensionality of the backbone is significantly different.

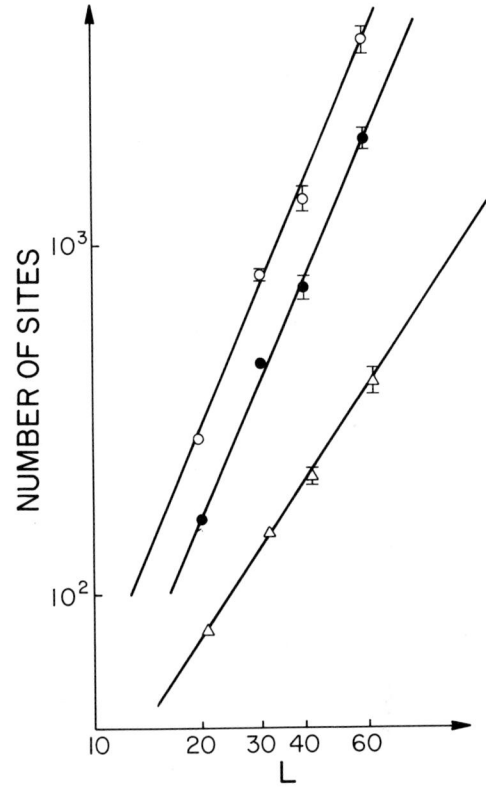

FIGURE 2

Log-log plot of the average number of sites in the largest cluster (○), backbone (●) and elastic backbone (△) against the lattice size.

References

[1] B.B. Mandelbrot, The Fractal Geometry of Nature (W.H. Freeman and Company 1983)

[2] H.J. Herrmann, D. Stauffer and D.P. Landau, J. Phys. A16, 1221 (1983).

[3] H.J. Herrmann, D.C. Hong and H.E. Stanley, J. Phys. A17, L261 (1984).

[4] Ashvin Chhabra, D. Matthews-Morgan, D.P. Landau and H.J. Herrmann, J. Phys. A18, L575 (1985).

FRACTALS IN PHYSICS
L. Pietronero, E. Tosatti (editors)
© Elsevier Science Publishers B.V., 1986

A LATTICE MAGNETIC MODEL FOR BRANCHED POLYMERS AND THE SOL-GEL TRANSITION

Agustín E. GONZALEZ

Instituto de Física UNAM, Apdo. Postal 20-364, Delegación Alvaro Obregón, 01000 México, D.F., México

It is shown that the limit n=0 of an nq-component spin model on a lattice gives us a correspondence with a system of branched polymers in a good solvent on a related lattice. The regimes below, at and above the sol-gel transition can be described by the model.

1. INTRODUCTION

The sol-gel transition undergone by a system of branched polymers under certain conditions has been studied for a long time. In this transition, the branched polymers react between them to produce larger molecules up to a point, called the sol-gel transition point, where an "infinite" molecule is produced, with an extent of the size of the container in which the reaction is taking place. In the 1940's, Flory[1] and Stockmayer[2] proposed a sol-gel transition theory which worked well for the transition point and the different average molecular weights of these systems. There was however a shortcoming in the theory, the neglect of cycles and loops, which led workers[3,4] in the 1970's to try to model the transition using the random percolation problem, in which the cycles or loops appear naturally. The random site (or random bond) percolation problem on a lattice has, however, only one degree of freedom: The occupation probability of the sites (or bonds). The clusters (branched polymers) that are formed have a well defined fractal dimensionality or compactness below, at and above the percolation transition point[5]. On the other hand, the branched polymers that are formed in an actual gelation transition can have a whole variety of fractal dimensionalities, depending on the mechanism that produces the larger molecules from the smaller ones. For example, in some cases this mechanism can be thought to be cluster-cluster aggregation[6], a topic which is of considerable study nowadays. In this case there are at least two fractal dimensionalities that are obtained for the aggregates below the transition point and do not coincide with the fractal dimensionality of the percolation clusters below the percolation threshold. In the magnetic model[7] I want to consider now, this deficiency is surmounted: We will see that we are going to have a fugacity for the number of monomers (occupied sites), another one for the number of polymers (clusters) and still one more for the number of loops (compactness of the clusters). This will allow us to change the compactness of the clusters at our will. Although a similar theory, for the continuum case and using field theoretical methods, was introduced before by Lubensky and Isaacson[8], the motivation for developing this lattice magnetic correspondence is twofold: First, because it is more easily understandable, at least by persons who are not so familiar with the field theoretical methods, and secondly, because of the possibility of doing a real space renormalization group for the physically interesting cases of 2 and 3 dimensions, which are far away from the upper critical dimensionality for these systems (8 in the case of dilute branched polymers[8]).

2. THE MODEL

Let us consider for simplicity the two-dimensional case, and take a regular lattice of N sites in which an nq-component spin $\sigma_{l\alpha}$, such that $\Sigma_l \Sigma_\alpha \sigma_{l\alpha}^2 = nq$ with $l = 1,\ldots,q$ and $\alpha = 1,\ldots,n$, is assigned to each site of the lattice. We choose the following hamiltonian[7]:

$$-\beta H = k \Sigma_I \Sigma_l \Sigma_\alpha \sigma_{I_1 l\alpha} \cdots \sigma_{I_f l\alpha} + h \Sigma_i \Sigma_l \sigma_{i l 1} \quad (1)$$

where the first subindex in the spin variables indicates the spin of the lattice we are considering, and the other two label the components of that spin. The first sum in the interaction term runs over alternate cells of the lattice, chosen in such a way that no two such cells have more than one spin in common and every spin belongs to two cells; the interaction term is a f-spin interaction between the f spins belonging to each cell. In figure 1 an example is given for $f = 4$ (square lattice). Note that the magnetic field is coupled to q components of the spins.

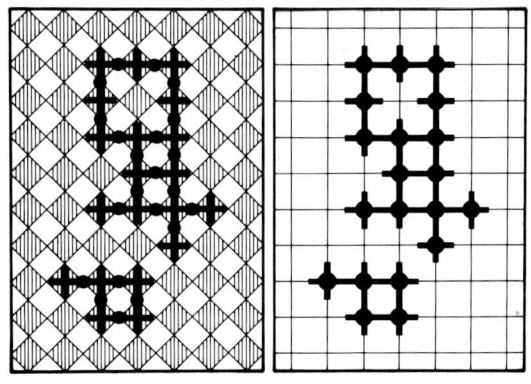

FIGURE 1
Left: A section of the original lattice, showing the alternate cells and two clusters. In this section, the number of monomers S is 20 (crosses), the number of vertices N_v is 36 (unreacted functionalities), the number of reactions N_R is 22 (black dots), the number of loops N_l is 4, and the number of polymers N_p is 2. Right: The same clusters in the new lattice.

The partition function for this system Z is given by $Z/\Omega = \langle e^{-\beta H} \rangle_0$, where $\langle A \rangle_0 \equiv (\P_i \int d\Omega_i \, A)/(\P_i \int d\Omega_i)$ indicates an average over the spin variables, $d\Omega_i$ is the differential of solid angle in nq dimensions, and $\Omega \equiv \P_i \int d\Omega_i$. In the limit $n = 0$, this partition function can be shown[7] to be

$$Z/\Omega = [2V/(2V-q)]^N \, \Xi_p \,, \quad (2)$$

where

$$\Xi_p = \Sigma_{N_p} \Sigma_S \Sigma_{N_R} \exp\left[(\ln q) N_p + (\ln U) S + (\ln V) N_R\right] Q(N_p, S, N_R) \quad (3)$$

is a grand partition function for a branched polymer solution in good solvents, defined on a new lattice whose M $(=2N/f)$ sites correspond to the alternate cells chosen in the old lattice (Cf. fig. 1). $Q(N_p, S, N_R)$ is the number of ways of inscribing N_p polymers, with a total of S monomers and N_R reactions, in the new lattice of M sites. Finally, U and V are defined by $U = k h^f / (1 + q h^2/2)^f$ and $V = (1 + q h^2/2)/h^2$.

3. THE CORRESPONDENCE

By defining a free energy $g \equiv \lim_{N\to\infty}(1/N)\ln(Z/\Omega)$, the magnetization $m \equiv (\partial g/\partial h)_{q,k}$ and the internal energy $u \equiv -k_B T \, k \, (\partial g/\partial k)_{q,h}$ of the magnetic system, we can get the monomer concentration ϕ, polymer concentration ϕ_p and reaction concentration ϕ_R of the branched polymer solution as[7]

$$\phi = -\frac{f}{2} \frac{u}{k_B T} \quad (4)$$

$$\phi_p = -\frac{f}{2}\left[\frac{qh^2}{2} + \frac{fqh^2}{4} \frac{u}{k_B T} - q\left(\frac{\partial g}{\partial q}\right)_{k,h} - \frac{qh^2}{4} hm\right] \quad (5)$$

$$\phi_R = -\frac{f}{4}\left[f\left(1 - \frac{qh^2}{2}\right)\frac{u}{k_B T} + \left(1 + \frac{qh^2}{2}\right)hm - qh^2\right]. \quad (6)$$

The conversion factor, ρ, can be obtained from (4) and (6), using the relation $\rho = 2\phi_R/f\phi$. Finally, the osmotic pressure, π, is given by $\pi V_o/k_B T = (f/2)g - (f/2)\ln(1 + qh^2/2)$, where V_o is the volume of a unit cell in the new lattice (an area in two dimensions).

By using the relation $N_R = S - N_p + N_l$—which indicates that, for a fixed number of monomers and polymers, an increase in the number of reactions gives an equal increase in the number of loops N_l—it is possible to eliminate the number of reactions in favor of the number of loops in the grand partition function[7] and to have fugacities for the number of monomers, polymers and loops, as mentioned before. Therefore, N_R is also a measure of the compactness of the clusters.

It is clear that there is going to be a critical surface in the space of the basic fugacities (q,k,h), separating the non-percolated phase (where only finite clusters are present) from the percolated phase (where, besides the finite clusters, an infinite cluster spreads the whole lattice). Two points on this critical surface deserve mentioning[7]:
a) $q = 1$, $h = \sqrt{2}$ and $k = 2^{f/2} p_c (1-p_c)^{-1}$ is a random site percolation point. Here p_c is the percolation threshold of the new lattice.
b) $q = 0$, $h = 1$ and $k = k_c$ is a random site animal point, that is, the point where a single cluster in the lattice becomes infinite in extent. Also, a tree percolation point and a random animal point without loops can be located[7].

Vertex-vertex correlation functions between vertices belonging to the same or different polymers and between vertices belonging to the same polymer can also be obtained in terms of magnetic quantities[9]. In particular, the correlation function P_{ij} between vertices belonging to the same polymer (pair connectedness) takes the simple form[9]

$$P_{ij} = \lim_{n \to 0} q\, h^2 \frac{<\sigma_{i\perp} \sigma_{j\perp} e^{-\beta H}>_o}{<e^{-\beta H}>_o}, \quad (7)$$

where $\sigma_{i\perp}$ is a component of the spin on site i, perpendicular to the direction of the field. As a generalization, consider the hamiltonian

$$-\beta H = K_1 \sum_I \sum_l \sum_\alpha \sigma_{I_1 l \alpha} \cdots \sigma_{I_f l \alpha} +$$
$$+ K_2 \sum_{<ij>} \sum_l \sum_\alpha \sigma_{il\alpha} \sigma_{jl\alpha} + h \sum_i \sum_l \sigma_{il1} \quad (8)$$

where the first sum is as before and <ij> denotes nearest neighbor sites in the lattice. In this case we are going to obtain branched polymers made of bifunctional and f-functional units[9]. By letting $K_2 \to 0$, we get the case considered before, while by making $K_1 \to 0$, we should recover the linear polymer solution[10], allowing us to study this crossover.

REFERENCES

1. P. J. Flory, J. Am. Chem. Soc. 63 (1941) 3083, 3091, 3096.
2. W. H. Stockmayer, J.Chem. Phys. 11 (1943) 45; 12 (1944) 125.
3. P. G. de Gennes, J. Phys. (Paris) 37 (1976) L1.
4. D. Stauffer, J. Chem. Soc. Faraday Trans. II 72 (1976) 1354.
5. D. Stauffer et al, Adv. Polym. Sci. 44 (1982) 103.
6. M. Kolb, R. Botet and R. Jullien, Phys. Rev. Lett. 51 (1983) 1123 and this volume.
7. A. E. González, J. Phys. (Paris) in print.
8. T. C. Lubensky and J. Isaacson, Phys. Rev. Lett. 41 (1978) 829; Phys. Rev. 20 A (1979) 2130; J. Phys. (Paris) 42 (1981) 175.
9. A. E. González, to be published.
10. P. G. de Gennes, Scaling Concepts in Polymer Physics (Cornell University Press, Ithaca, 1979).

FRACTALS IN PHYSICS
L. Pietronero, E. Tosatti (editors)
© Elsevier Science Publishers B.V., 1986

PERCOLATION IN A CONCENTRATION GRADIENT

Jean-François GOUYET, Michel ROSSO and Bernard SAPOVAL

Laboratoire de Physique de la Matière Condensée, Ecole Polytechnique
91128 Palaiseau, France

The problem of site percolation on lattices with a concentration gradient appears to bring a new promising approach in studying diffusion fronts, and in determining percolation parameters such as percolation thresholds and critical exponents. We recall here the main results in two dimensions and present some 3D preliminary results.

1. INTRODUCTION

The study of site percolation on lattices on which a concentration gradient has been introduced is of great interest to the understanding of the diffusion fronts,[1] but it also brings a drastic simplification to determine quantities which are specific to percolation problems.[2]

In a constant concentration gradient along an x-direction with a concentration decreasing from one at $x = 0$ to zero, an infinite cluster appears in the high occupation domain. This cluster has an external frontier which is located in a region of concentration close to the critical percolation value p_c.[1]

Using numerical simulations, we studied the general structure of this frontier. First, on a 2D lattice, it has a fractal dimension in a range which increases as the inverse of the concentration gradient ∇p. The fractal dimension is found[1,3] to be $d_H = 1.75$. Second, the number of particles on the frontier and its width follow power laws of ∇p. The associated exponents and the fractal dimension can be expressed as simple functions of the critical exponents of the percolation problem in two dimensions.[1]

On the basis of an adequate definition of the frontiers we are able to determine the critical percolation probability p_c to a very high precision with small computation time.

The results are given in section 3.

Some preliminary results for 3D systems are also developped in section 4.

However before discussing these results, it is necessary to recall some definitions used in percolation problems or that we previously introduced in the particular case of the concentration gradient.

2. DEFINITIONS

In the following we use as an example the square lattice case.

A cluster of occupied sites (A-sites) linked via nearest-neighbor connections (four possible connections) will be called an A-cluster. The percolation probability is the minimum concentration probability for the onset of an infinite A-cluster. Considering now the remaining empty sites (B-sites) we define in a similar manner a B-cluster as being composed of empty sites connected via first and second-nearest-neighbor connections (eight possible connections). A- and B-clusters are defined on two lattices (L_A and L_B) with different coordination numbers (four or eight connections for each site) : These lattices are said to form a "matching pair".[4]

We deal with a monotonously decreasing concentration probability of A-sites, starting from a concentration $p = 1$ at $x = 0$ (Fig. 1) and reaching $p = 0$ at enough large x. In opposition to the usual percolation problem we remark that there exist both infinite A- and B-clusters (In the limit of an infinitely large system). The infinite A-cluster is located in the high concentration region, whereas the infinite B-cluster is in the low concentration region. They are in contact with each other (Fig. 1).

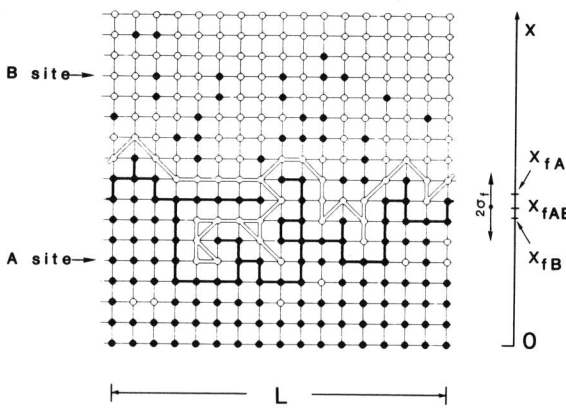

FIGURE 1
Frontiers f_A (black circles connected by black line) and f_B (white circles connected by white line), on a square lattice with a concentration gradient

The sites of the infinite A-cluster which are connected with sites of the infinite B-cluster via first- or second-nearest-neighbor connections, constitute the frontier as defined in reference 1. We call this frontier f_A; similarly we define the frontier f_B as composed with the sites of the infinite B-cluster which are connected via first-nearest-neighbor connection with the sites of the infinite A-cluster (Fig.1).

For each frontier f_A or f_B we consider the following quantities: $p_{fA}(x)$ (resp. $p_{fB}(x)$) is the probability of finding one site of f_A (resp. f_B) at distance x.

x_{fA} (resp. x_{fB}) is the mean position:

$$x_f = \int_0^\infty x\, p_f(x)\, dx / \int_0^\infty p_f(x)\, dx$$

σ_{fA} (resp. σ_{fB}) is the width:

$$\sigma_f^2 = \int_0^\infty (x-x_f)^2 p_f(x)\,dx / \int_0^\infty p_f(x)\,dx$$

and N_{fA} (resp. N_{fB}) is the number of sites of the frontier f_A (resp. f_B). The width L of the lattice being always taken much larger than the width σ_f, N_f is proportional to L.

$$N_f = L \int_0^\infty p_f(x)\, dx$$

As shown in ref. 2 it is appropriate to consider the double frontier f_{AB}, union of frontiers f_A and f_B; its mean position x_{fAB} is simply the mean position of all the lattice points belonging to the frontiers f_A and f_B. Similarly we also introduce

$$p_{fAB}(x) = p_{fA}(x) + p_{fB}(x)\ ;\ \sigma_{fAB};\ N_{fAB}$$

Gradients of concentration may appear in different physical contexts: In the diffusion problem considered in reference 1 the concentration of particles varies as a complementary error function of x due to the boundary condition used. However it is easy to show[5] that the asymptotic behavior of the frontiers is completely determined by the local gradient on this frontier. We present here the study of percolation in a constant gradient of concentration.

3. CONSTANT GRADIENT IN TWO DIMENSIONS

The concentration $p(x)$ varies from $p = 1$ at $x = 0$ to $p = 0$ at $x = (\nabla p)^{-1}$. The gradient ∇p is constant and

$$p(x) = 1 - x\,|\nabla p|$$

The frontiers are shown[1,2,5] to be located around p_c the critical percolation threshold and the parameters x_f, σ_f, N_f are found to follow power laws as functions of ∇p. The concentrations at the mean positions $p(x_{fA})$, $p(x_{fB})$, $p(x_{fAB})$ tend very rapidly to p_c.

i) The particularly rapid convergence of $p(x_{fAB})$ allows for a precise determination of p_c. From numerical results in ref. 2 we obtained

$$x_{fAB} \simeq (1 - p_c)(\nabla p)^{-1} + \delta x_{fAB}$$

with the values

$$p_c \text{ (square lattice)} = 0.59280_2 \pm 10^{-5}$$

and

$$\delta x_{fAB} \text{ (sq. lattice)} \simeq 0.0162.$$

Using a similar procedure Cassereau and Couture[6] obtained for the honeycomb lattice :

$$p_c = 0.69703_5 \pm 10^{-5}.$$

ii) x_{fA} and x_{fB} contain[5] more important corrections to scaling :

$$x_{fA} \simeq (1-p_c)(\nabla p)^{-1} - 0.250(\nabla p)^{(1-\nu)/(1+\nu)} + \delta x_{fAB}$$

$$x_{fB} \simeq (1-p_c)(\nabla p)^{-1} + 0.250(\nabla p)^{(1-\nu)/(1+\nu)} + \delta x_{fAB}$$

where $\nu = 4/3$ is the critical exponent of the coherence length of the percolation problem in 2D.

iii) Simple considerations show that[1,5]

$$\sigma_{fA} \simeq \sigma_{fB} \simeq 0.500 |\nabla p|^{-\alpha_\sigma}$$

while $N_{fA}/p_c \simeq N_{fB}/(1-p_c) \simeq N_{fAB} \propto |\nabla p|^{-\alpha_N}$

with $\alpha_\sigma = \dfrac{\nu}{1+\nu}$ and $\alpha_N = \dfrac{\nu}{1+\nu}(d_H - 1)$

where d_H is the fractal dimension of the frontier when $\nabla p \to 0$.

With the support of numerical[1] and scaling[3] arguments it has been proposed that

$$d_H = 1 + \nu^{-1} = 1.75.$$

This fractal dimension, the same for f_A, f_B and f_{AB} is obtained in a range starting from some intersites distances to the width σ_f of the frontier. For larger distances a crossover to dimension one appears. The dimension d_H is also very close to that found for the hull[7] or external frontier of the infinite percolation cluster. More precisely the frontier f_A appears to be an exact representation of this hull in the $\nabla p = 0$ limit.[1]

A "Dressed Self Avoiding Walk" can be defined[5,8] which allows to systematically generate these frontiers and to obtain a better understanding of their structure.

4. CONSTANT GRADIENT IN THREE DIMENSIONS

The 3D case is more complicated than the above 2D systems. The reason lies in the existence of two different critical percolation thresholds. In the 2D case, and for a given constant concentration there can only exist one infinite cluster either A or B. So that if p_{cA} and p_{cB} are the percolation thresholds for the matching lattices L_A and L_B one has $p_{cA} + p_{cB} = 1$. On the contrary in the 3D case, there exists a concentration range $1 - p_{cB} > p > p_{cA}$ where the two infinite A- and B-clusters coexist. As a consequence in the case of a concentration gradient the frontier is no more located at a given p, but extends between $p_1 = p_{cA}$ and $p_2 = 1 - p_{cB}$.

It is illuminating to represent the probability distributions in reduced coordinates : $p_{fA}(x)/p(x)$ and $p_{fB}(x)/(1-p(x))$ then reduce to a single curve (Fig. 2). In other words, the reduced distributions for the <u>infinite</u> A-clusters (resp. B) and that of the <u>finite</u> B-clusters (resp. A) are identical.

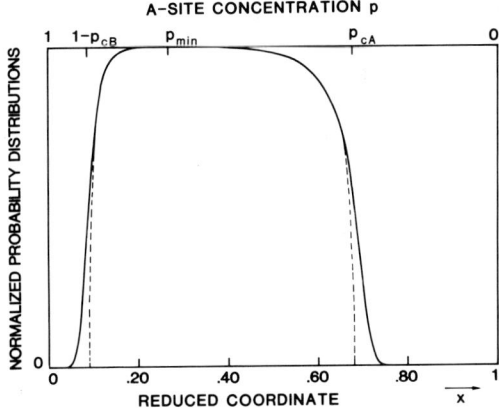

FIGURE 2
Normalized probability distributions $p_{fA}(x)/p(x)$ or $p_{fB}(x)/(1-p(x))$ for $\nabla p = 1/256$ and sample size 64 x 64 x 256. The dashed line indicates the limit $\nabla p = 0$.

We have then obtained the following striking result : Almost all A-sites (resp. B) in the region (p_1, p_2) belongs to the frontier f_A (resp. B).

This is clearly shown on figure 2. At a concentration p_{min} only $2.5 \; 10^{-4}$ sites do not belong to any frontier. This value may be easily determined using duality consideration : Around p_{min} the sites which are not on the frontiers are (essentially) isolated A or B sites completely surrounded either by f_A sites or by f_B sites. The reduced probability of such sites satisfy

$$p_i = (1 - p_{min})^6 = p_{min}{}^{26}$$

One finds $p_{min} = 0.73256$ and $p_i \simeq 2.5 \; 10^{-4}$.

Due to the existence of two percolation thresholds, the concentration at x_f is neither close to p_{cA} nor to $1 - p_{cB}$. However considering the behavior at $\nabla p \to 0$, one can convince oneself that the reduced probability density curve on the frontier tends towards the dashed curve shown on figure 2. This suggests a determination of p_c based on the derivative of the distribution $p_f(x)$. The derivative presents two extrema which must tend asymptotically towards p_{cA} and $1 - p_{cB}$ (see figure 3). These two peaks are asymmetrical.

With $\nabla p = 1/256$ and averaging on 100 samples of $64 \times 64 \times 256$, we obtain a first promising estimation

$$p_{cB} \simeq 0.091 \text{ and } p_{cA} \simeq 0.31.$$

The present accepted[9] values being respectively $p_{cB} \simeq 0.098$ and $p_{cA} \simeq 0.3117$. Using short range coherence length behavior apart from the two critical regions we can moreover use limited samples around these regions as in the 2D case.[3]

REFERENCES

1. B. Sapoval, M. Rosso and J.-F. Gouyet, J. Physique Lett. 46 (1985) L149.

2. M. Rosso, J.-F. Gouyet and B. Sapoval, Phys. Rev. B., to appear September 85.

3. A. Bunde and J.-F. Gouyet, J. Phys. A.18 (1985) L285.

4. M.F. Sykes and J.W. Essam, J. Math. Phys. 5 (1964) 1117 ; V.K.S. Shante and S. Kirkpatrick, Adv. Phys. 20 (1975) 325.

5. J.-F. Gouyet, M. Rosso and B. Sapoval, to be published.

6. S. Cassereau and B. Couture, private communication.

7. R.F. Voss, J. Phys. A17 (1984) L373.

8. J.-F. Gouyet, A. Bunde and H. Harder, in preparation.

9. D.W. Heermann and D. Stauffer, Z. Phys. B 44 (1981) 339 ; H. Saleur and B. Derrida, J. Physique 46 (1985) 1043.

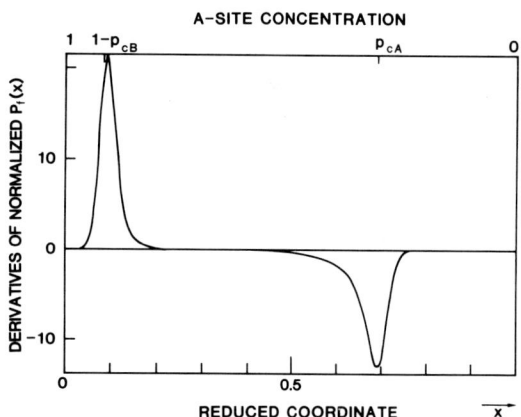

FIGURE 3
Derivative of the normalized distributions given on figure 2.

FRACTALS IN PHYSICS
L. Pietronero, E. Tosatti (editors)
© Elsevier Science Publishers B.V., 1986

SURFACE TENSION IN POTTS MODELS AND PERCOLATION

C. K. HARRIS

Department of Physics, University of Edinburgh, King's Buildings, Mayfield Road, Edinburgh EH9 3JZ, UK.[†]

The Kasteleyn-Fortuin relation is used to give a simple proof for the s-state Potts model of the relation $\beta\sigma\xi = 1$ between the correlation length in a particular direction on a planar lattice and the surface tension of an interface with corresponding orientation on the dual, thus generalising a well known result for the Ising model. The relation is also used to give an interpretation of the surface tension in the percolation problem in general dimension.

1. INTRODUCTION

The partition function and correlation function of s-state Potts models with various interactions can be expressed as weighted sums over graphs of occupied or unoccupied bonds[1] or sites[2,3] on the lattice leading to a correspondence between the one-state limit of these models and the bond or site percolation problems.

This procedure is here generalized to Potts partition functions with fixed state boundary conditions, in terms of which the Potts interfacial free energy per unit area or 'surface tension' may be defined[4], and leads to a direct physical interpretation of the latter quantity in the percolation problem.

2. GENERAL THEORY

We consider the following Potts hamiltonia:

$$-\beta H_B = K_S \sum_{(ij)} [\delta(\sigma_i, \sigma_j) - 1]$$

$$-\beta H_S = K_S \sum_i [\delta\{\sigma_\alpha\}_i - 1]$$

i, j are sites, and (ij) nearest neighbour bonds, of a lattice L, and α are sites of the covering lattice L_C, which lie at the midpoints of the bonds of L. The notation $\{\sigma_\alpha\}_i$ is a shorthand for the set of Potts states on sites of L_C which are adjacent to i.

$\sigma_i, \sigma_\alpha = 1,2,...s$

and

$\delta(\sigma_i, \sigma_j) = 1$ if $\sigma_i = \sigma_j$, 0 otherwise.

$\delta\{\sigma_\alpha\}_i = 1$ if all states in $\{\sigma_\alpha\}_i$ are equal
= 0 otherwise.

In order to define a surface tension, we consider finite portions $L^{(f)}$ of L, and $L_C^{(f)}$ of L_C, as shown in figures 1a and 1b respectively. A region R (convex, but otherwise of arbitrary shape), and a plane surface S, perpendicular to $\hat{n}$, are superimposed on the lattices, as illustrated in figure (1). The area of intersection of the plane and the region is A. Referring to figure (1), let $Z^{\mu\nu} \equiv$ the Potts partition function on $L^{(f)}$ or $L_C^{(f)}$ with C_1, C_2 fixed in states μ, ν respectively. Then the Potts surface tension of an interface normal to $\hat{n}$ may be defined as

$$\beta\sigma(\hat{n},s) \equiv \lim_{A \to \infty} -A^{-1}\ln(Z^{12}/Z^{11})$$

Let us carry out the mapping of the rhs. onto weighted sums of graphs on the lattice for H_B, H_S:

2.1 Bond percolation.

$$Z^{12} = \text{Tr}\left(\Pi'\delta(\sigma_i,1)\delta(\sigma_j,2)\right)\exp{-\beta H_B}$$

$$= \text{Tr}\left(\Pi'\delta(\sigma_i,1)\delta(\sigma_j,2)\right)\Pi_{(ij)}(1 - p + p\delta(\sigma_i,\sigma_j))$$

where $p = (1 - \exp{-sK})$
and Tr denotes a sum over all states of the system and Π' a product over sites $i \in C_1$ and $j \in C_2$.

Each term in the expansion of the second product on the rhs. corresponds to a graph of occupied and unoccupied bonds on the lattice L, where (ij) is occupied in G if $\delta(\sigma_i,\sigma_j)$ is present in the term, otherwise

---

[†] Work supported in part by SERC grant no. NG96322.

unoccupied. On performing the trace over states, each cluster of occupied sites in G picks up a factor s, 1 or 0 according to whether it is not connected to any sites of C_1 or C_2, connected to at least one site of C_1 or C_2 but not both, or connected to at least one site of both C_1 and C_2 respectively.

$$Z^{12} = \langle \delta[N(G,C_1,C_2),0] s^{n_I(G)} \rangle_G \quad (1)$$

$$Z^{11} = \langle s^{n_I(G)} \rangle_G \quad (2)$$

Here,

$N(G,C_1,C_2)$ = no. of clusters in G connected to at least one site of each of C_1 and C_2.

$n_I(G)$ = no. of clusters in G connected to no boundary sites.

$$\langle ... \rangle_G = \sum_G p^{N_B(G)} (1-p)^{N_B - N_B(G)}$$

$N_B(G)$ = no. of occupied bonds in G.

2.2 Site percolation.

$$Z^{12} = \mathrm{Tr}\left(\Pi'\delta(\sigma_\alpha,1)\delta(\sigma_\beta,2)\right)\exp-\beta H_S$$

$$= \mathrm{Tr}\left(\Pi'\delta(\sigma_\alpha,1)\delta(\sigma_\beta,2)\right)\prod_i\left(1-p+p\delta\{\sigma_\alpha\}_i\right)$$

Here Π' denotes a product over sites $\alpha \in C_1$, $\beta \in C_2$ on $L_C^{(f)}$.

In this case, each term in the expansion of the second product on the rhs. corresponds to a graph of occupied and unoccupied sites on the lattice L, such that i is occupied in G if $\delta\{\sigma_\alpha\}_i$ is present in the term, otherwise unoccupied. On performing the trace over states, each cluster of occupied sites and each bond between a pair of unoccupied sites picks up a factor s, except that perimeter bonds and the clusters to which they are attached pick up a factor unity, and clusters attached to at least one bond of each of the perimeter sets C_1 and C_2 contribute a factor zero.

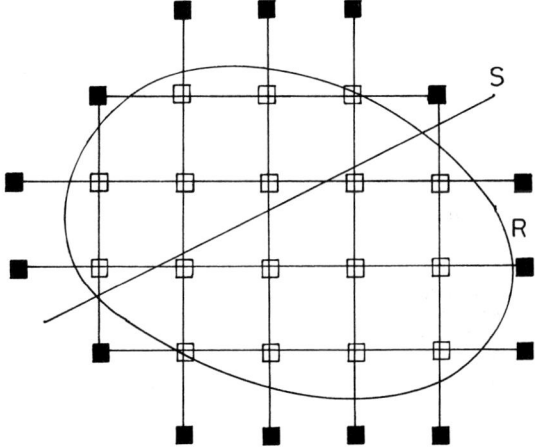

FIGURE 1a

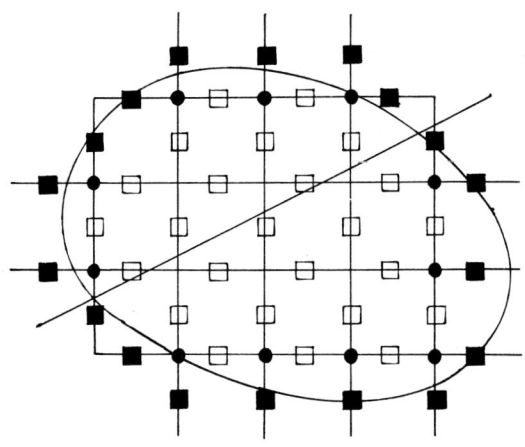

FIGURE 1b

FIGURE 1
Showing construction defining Potts surface tension for H_B (1a) and H_S (1b). Key is as follows:

▫ Potts sites inside the region — i.e. located at vertices (1a) or bonds (1b) of cells of L inside the region.

■ boundary sites — C_1, C_2 according to which side of S they are on.

● perimeter sites P_1, P_2 of L for H_S. For H_B we define $P_i = C_i$, $i = 1,2$.

Then,

$Z^{12} = \langle \delta[N(G,C_1,C_2),0] s^{n_I(G) + n_U(G)} \rangle$

$Z^{11} = \langle s^{n_I(G) + n_U(G)} \rangle_G$

$n_U(G)$ = no. of pairs of adjacent unoccupied sites in G.

$\langle ... \rangle_G = \sum_G p^{N_S(G)} (1-p)^{N - N_S(G)}$

$N_S(G)$ = no. of unoccupied sites in G.

3. THE ONE STATE LIMIT AND PERCOLATION

On taking the one-state limit in either case, we have

$Z^{12}/Z^{11} = \langle \delta[N(G,P_1,P_2),0] \rangle$

$= Q(A,p),$

where $Q(A,p)$ is the probability that no site of P_1 is connected to any site of P_2 in the bond or site percolation problem, and

$\beta\sigma(\hat{n},1) = - \underset{A \to \infty}{Lt} A^{-1} \ln Q(A,p)$ (3)

This is the desired physical interpretation of the surface tension in the percolation problem.

We see immediately that the surface tension given by (3) has the following properties:

$\sigma(\hat{n},1) > 0$ always
$\sigma(\hat{n},1) = 0$ for $p < p_C$

Further, from finite size scaling arguments we expect

$Q(A,p) = A^v f(A/\xi^{d-1})$

and hence

$\sigma(\hat{n},1) \sim (p - p_C)^{(d-1)\nu}$

at criticality.

As an application of (3), let $P(S)$ be the probability that there is a large rent in the percolating cluster for $p > p_C$ along a surface S, of area A, as shown in figure (2). Then,

$\underset{A \to \infty}{Lt} A^{-1} \ln P(S) = - \int dS \beta\sigma(\hat{n},1) / \int dS$

Such an argument can be applied to the distribution of large dead ends and to show that it is asymptotically the same as that of large finite clusters.

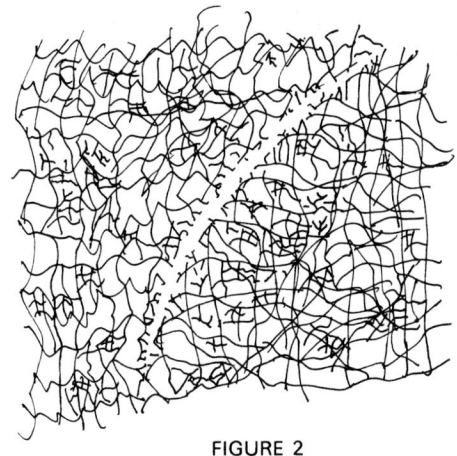

FIGURE 2
Percolating cluster with large 'tear'.

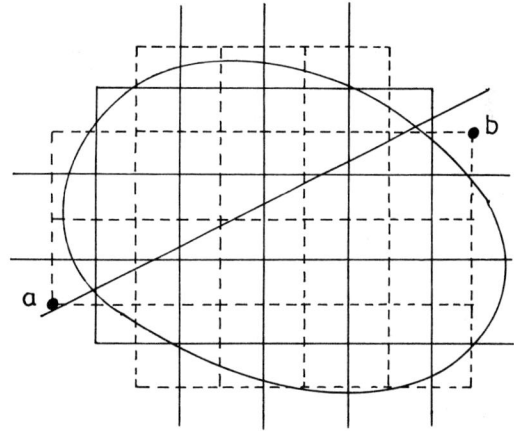

FIGURE 3
Construction of figure 1 with a finite portion $(L^*)^{(f)}$ of the dual lattice L^* (shown dashed).

4. DUALITY RELATION FOR PERCOLATION ON PLANAR LATTICES

4.1 Bond percolation

We consider the finite portion of the dual lattice L^*, $(L^*)^{(f)}$, in which every bond crosses a bond of $L^{(f)}$, as illustrated in figure (3). The two sites a and b on $(L^*)^{(f)}$ are the ones in those faces of L which have Potts sites from both C_1 and C_2 in the boundary of the face.

With each graph G on $L^{(f)}$ may be associated a unique graph G^* on $(L^*)^{(f)}$ such that each occupied bond of G crosses an unoccupied bond of G^* and vice-versa. It is easy to see that G for which P_1 and P_2 are connected corresponds to G^* for which a and b are not connected and viceversa. Then,

$$\delta[N(G,P_1,P_2),0] = 1 - \delta[N(G^*,a,b),0] \quad (4)$$

and since

$$N_B(G) = N_B - N_B(G^*) \quad (5)$$

we deduce that

$$<\delta[N(G,P_1,P_2),0]> = <1 - \delta[N(G^*,a,b),0]> \quad (6)$$

provided that we identify

$$p^* = 1 - p$$

4.2 Site percolation

In this case G is a graph of occupied and occupied sites on $L^{(f)}$, and a bond of G is considered occupied if it connects a pair of occupied sites, otherwise unoccupied. $(L^*)^{(f)}$ forms a lattice of interaction vertices through which occupied sites of G^* on L interact with other occupied sites in the same face of $L^{(f)}$. The occupied sites of G^* are those which are unoccupied in G, and the interaction vertices, and the bonds of G^* are occupied if they join a pair of occupied sites, otherwise unoccupied. Consider the face formed by a pair of neighbouring sites of $(L^*)^{(f)}$ and the pairs of sites on $L^{(f)}$ which are neighbours of both. If G connects the latter across the face, G^* does not connect the former around the face, and vice-versa. Then the relations (4), (5) and (6) above will hold as before (with N_B replaced by N_S). $<...>_{G^*}$ is a site percolation problem on $L^{(f)}$, in which occupied sites in the same face are connected.

Thus for the bond or site case, $<1 - \delta[N(G^*,a,b),0]>_{G^*}$ is the probability that a and b are connected and decays as $\exp-A/\xi^*$ where ξ^* is the correlation length for the dual bond or site percolation problem. Then we deduce that

$$\beta\sigma\xi^* = 1$$

5. THE DUALITY RELATION FOR GENERAL s

This is easily established for the Potts model with nearest neighbour interactions. The bonds of $L^{(f)}$ can be put into a one-to-one correspondence with those of the lattice dual to $(L^*)^{(f)}$, in which P_1 and P_2 are coalesced into a single site, the exterior site e. The quantities on the rhs. of (1) and (2) are then related to corresponding quantities on this lattice, using the invariance of $n_l(G)$. Graphs on $L^{(f)}$ for which P_1 and P_2 are not connected correspond to graphs for which no circuit passing through e crosses the line S an odd number of times in the interior of the region R. Standard duality arguments[5] and (3) may then be used to show that Z^{12}/Z^{11} is equal to the correlation of the Potts states at a and b on $(L^*)^{(f)}$ and the result follows. A detailed derivation and discussion is given in reference 6.

REFERENCES

1. Kasteleyn P.W. and Fortuin C.M., J. Phys. Soc. Japan Suppl. 16 (1969) 11
   Fortuin C.M. and Kasteleyn P.W., Physica 57 (1972) 536

2. Giri M.R., Stephen M.J. and Grest G.S., Phys. Rev. B 16 (1977) 4971.

3. Essam J.W., J. Math. Phys. 20 (1979) 1769.

4. Earlier work on surface tension in general spin systems, including Potts models, has been carried out by mathematicians concerned with using it to establish a rigorous criterion for the existence of a phase transition in such systems:
   Fontaine J.R. and Gruber Ch., Commun. Math. Phys. 70 (1979) 243.

5. Wu. F.Y., Rev. Mod. Phys. 54 (1982) 235 and refs. therein.

6. Harris C.K., J. Phys. A:Math Gen. 18 (1985) 2259.

PERCOLATION ON THE DAP

R.DEWAR and C.K.HARRIS

Department of Physics, University of Edinburgh, Mayfield Road, Edinburgh EH9 3JZ U.K.

We discuss various algorithms for the enumeration of cluster properties in the percolation problem on a parallel computer, the ICL Distributed Array Processor. A recent conjecture of G.Jug concerning percolation critical behaviour in two dimensions is critically examined in the light of our results.

1. INTRODUCTION

The ICL Distributed Array Processor (DAP) is a parallel processing machine upon which many interesting calculations in physics can be carried out efficiently. Although not widely available, it belongs to a growing class of machines with novel architecture designed to tackle large-scale problems in computational physics and other disciplines.

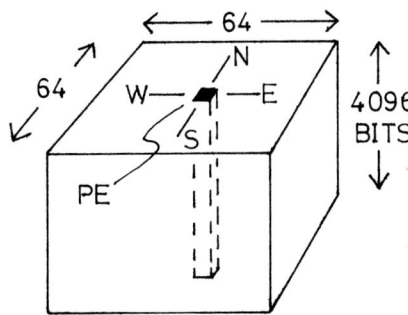

FIGURE 1
Schematic diagram of the DAP

The DAP (Fig.1) is a 64 × 64 array of processing elements (PE's) each of which accesses 4096 bits of storage and has interconnections to its 4 nearest neighbours (north,south,east,west). Connections at the array edges can be chosen to implement planar or cyclic boundary conditions in either direction. The processors operate synchronously.

The software used is DAPFORTRAN, a version of familiar FORTRAN that expresses the parallelism of the DAP. For example, addition of 64 × 64 matrices is accomplished by

C = A + B

where the addition and assignment are performed simultaneously on all elements. The processors operate on one 64 × 64 bit plane at a time so the fastest operations are Boolean operations involving logical matrices, whose elements are TRUE or FALSE. This is why the DAP is particularly efficient for "two-state" problems such as Ising simulations[1] and percolation.

Logical arrays are extremely useful as conditional masks, to incapacitate selected processors during an instruction. For instance, we may write

A(LMASK) = B

which assigns the elements of B to the corresponding elements of A only where LMASK is TRUE. Elsewhere the value of A is unchanged.

Processors communicate with each other via global shift operations. For example, using a cyclic shift "east" with logical arrays L,M,N,

L = M.AND.SHEC(N,3)

simultaneously assigns to every element of L the logical .AND. of the corresponding element of M with the element of N that lies 3 sites to the "west", with cyclic boundary conditions imposed in the east-west direction.

The classic percolation problem considers clusters of occupied sites or bonds on a regular lattice. An inherent difficulty lies in the non-local nature of the cluster connectivity. In order to extract interesting information, such as the cluster size distribution, from percolation configurations, the sites must be labelled to keep track of those sites belonging to the same cluster and these labels must be updated when two candidate clusters are found to be connected.

The parallelism of the DAP cannot be exploited fully in such a procedure (although we remark as an aside that for directed percolation, because there is no "back-tracking" of directed paths connecting a pair of sites, it has been possible to write an efficient parallel algorithm for cluster connectedness properties)[2]. However, the nearest-neighbour local environment of single sites can be efficiently analysed. This is exploited in the following algorithm for counting the number of clusters in a two-dimensional percolation configuration.

2. A PARALLEL ALGORITHM FOR COUNTING CLUSTERS

The algorithm is most easily illustrated for site percolation on a square lattice with planar boundary conditions. A configuration of occupied sites (up to linear size L = 64) with concentration p is generated as a random logical matrix in the DAP. Iteration of the central algorithm reduces clusters of occupied sites to single-site clusters, each of which augments the number of clusters by one and is then removed. The reduction is depicted in Figure 2 for a typical 4 × 4 configuration containing 3 clusters.

The steps that achieve this reduction, implemented by the parallel Boolean logic and shift operations of DAPFORTRAN, are as follows:
(1) Identify all occupied sites having no occupied nearest-neighbours to the north and east, i.e. the 4 cases shown (a)-(d):

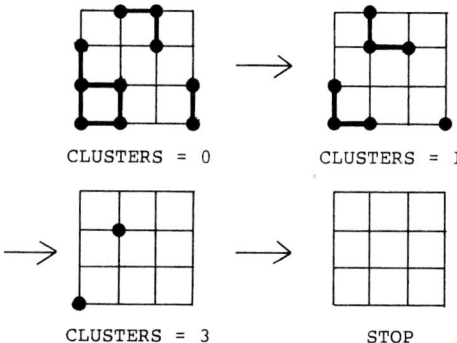

FIGURE 2
Illustrating the cluster algorithm

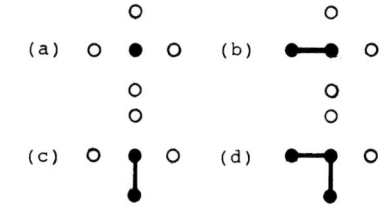

(2) For each (a)-type site, augment the cluster total by one.
(3) Remove all the sites found at step (1), replacing the (d)-type sites with new occupied sites lying to the south-west (if not already occupied) thus:

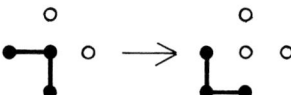

Note that since operations are done in parallel, no two clusters are joined inadvertantly by this step.

Repeat steps (1)-(3) until no occupied sites remain.

When cyclic boundary conditions apply, some clusters are never reduced, looping endlessly around the lattice. In this case, the algorithm is terminated after a number of iterations sufficient to ensure the complete reduction of all reducible clusters while any remaining sites are assumed to constitute a single spanning cluster that contributes one to the cluster total.

This algorithm, and a similar one for bond

percolation, can be used on various types of lattice, up to size L = 64, and is particularly efficient for L = 8,16 or 32 when many such lattices can be stored and analysed simultaneously. For these sizes, the number of configurations analysed per hour falls roughly as L^{-2} and for L = 32 is approximately 2 million and 1.5 million for bond and site percolation respectively. This is to be compared with 70000 for bond percolation with L = 30, using the powerful serial machine CRAY1[3].

3. CRITICAL BEHAVIOUR IN 2D PERCOLATION

3.1. Mean number of clusters

In the bond or site percolation problem, the mean number of clusters per site $K(p)$ at concentration p is analogous to the free energy per site in a thermal problem. Near the percolation threshold p_c, it has a singular part of the form

$$K_{sing}(p) = C|p-p_c|^{2-\alpha} \quad (1)$$

where in two dimensions the value of the exponent α is currently accepted to be

$$\alpha = -2/3 \quad (2)$$

Recently, however, Jug[4] has challenged this value, and hence either the validity of hyperscaling or the values of other exponents in 2D percolation. Starting from the 2D bond-diluted Ising model and using a Grassmann path integral (GPI) method exact for small dilution, he calculates the free energy $f(T,p)$ and finds that it exhibits dilute Ising critical behaviour everywhere along the critical curve $T_c(p)$, including the percolation threshold $T_c = 0$, $p = p_c$, near which it is found to have a singular part of the form

$$K_{sing}(p) = C(p-p_c)^2 \ln|\ln|p-p_c|| \quad (3)$$

Hence the GPI theory predicts $\alpha = 0$, in contrast to (2), and calls into question the existence[5] of a separate universality class for 2D percolation processes.

Jug[3] has produced evidence in support of this controversial result from numerical work and reanalysed series expansions. We shall discuss the former, and results of our own using the DAP, for the remainder of this paper.

3.2. Numerical investigation

The singular behaviour in (1) or (3) appears as a divergence in the third derivative of $K(p)$ at p_c. A finite-size scaling analysis then leads to the expectation

$$K'''(p_c,L) = \begin{cases} A + BL^{1/4} & (4) \\ \text{(conventional theory)} \\ \\ A + B(L)L^{1/\nu} & (5) \\ \text{(GPI theory)} \end{cases}$$

as the system size $L \to \infty$, where $K''' = d^3K/dp^3$ and $B(L)$ is a logarithmic correction. In contrast to ref.3, whose numerical work supports form (5) with dilute Ising value $\nu = 1$, our results favour the conventional form (4).

Following ref.3, we randomly generate site and bond percolation configurations on lattices of various sizes, and evaluate $K'''(p_c,L)$ numerically via the fluctuation formula

$$K'''(p,L) = C_1\{<N_0n_c> - <N_0><n_c>\} + C_2\{<N_0^2n_c> \\ - 2<N_0><N_0n_c> + 2<N_0>^2<n_c> - <N_0^2><n_c>\} \\ + C_3\{<N_0^3n_c> - 3<N_0><N_0^2n_c> + 6<N_0>^2<N_0n_c> \\ - 3<N_0^2><N_0n_c> - 6<N_0>^3<n_c> + 6<N_0><N_0^2><n_c> \\ - <N_0^3><n_c>\} \quad (6)$$

where $<.....>$ is an average over a large number of configurations. N_0 and n_c are the number of occupied bonds (or sites) and the number of clusters per site in each configuration, respectively, and the constants C_1,C_2,C_3 are given by

$$C_1 = 2(p^{-3} + q^{-3})$$
$$C_2 = -3(p^{-3} + p^{-2}q^{-1} - p^{-1}q^{-2} - q^{-3})$$
$$C_3 = (p^{-1} + q^{-1})^3 \quad (7)$$

where $q = 1 - p$.

Our results for bond percolation on square lattices (□) up to size L = 32 are shown below (Fig.3), in a direct comparison with the corresponding results reported in ref.3 (O), and similar results for site percolation on triangular lattices (△) are also shown. We

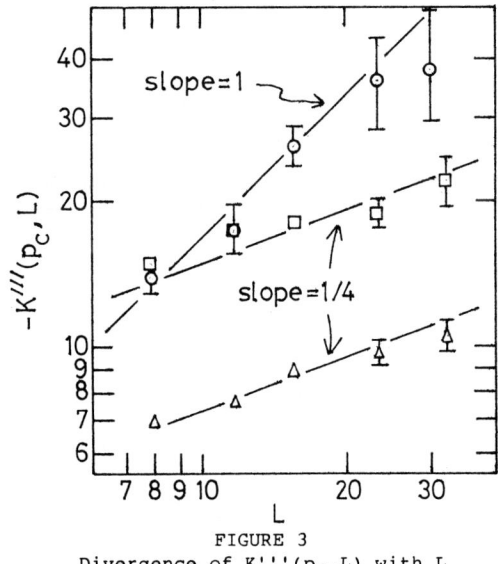

FIGURE 3
Divergence of $K'''(p_c,L)$ with L

believe the discrepancy evident in Fig.3 is due to an underestimate in ref.3 of the noise generated by equation (6). In order to reduce the statistical fluctuations, we adopt the following approach.

$K(p)$ contains an analytic contribution
$$K_{an}(p) = a + b(p-p_c) + c(p-p_c)^2 + d(p-p_c)^3 + \ldots \quad (8)$$
where constants a,b,c,d are either known analytically or may be estimated from series expansions[6]. The terms in b and c do not contribute to $K'''(p_c)$ but they do contribute significantly to the noise when the averages in (6) are taken over a finite sample of configurations. This noise may be subtracted off by replacing n_c on the r.h.s. of (6) by
$$n_c^* = n_c - b(N_0-Np_c)/N - c(N_0-Np_c)^2/N^2 \quad (9)$$
where N is the total number of nodes(edges) in the lattice for the site(bond) problem. The d term in (8) contributes to the constant A in (4) or (5) and similarly may be subtracted off to facilitate the interpretation of the data.

The effect of subtracting off this noise is shown graphically in Figure 4 from which we conclude that the error estimates reported in ref.3 are too low. A straight line fit to our

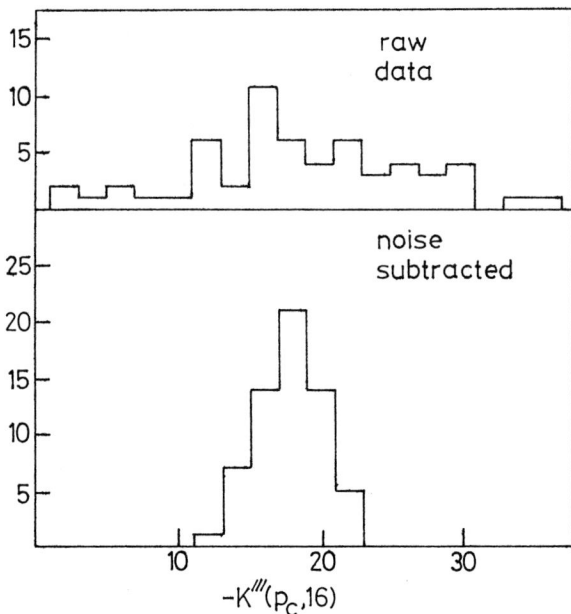

FIGURE 4
Histogram for $K'''(p_c,16)$, bond percolation. 1 box = average over 1638400 configurations. Total histogram = 62 boxes.

site and bond percolation data yields slope 0.25 ± 0.02 and 0.28 ± 0.03 respectively, consistent with (4).

ACKNOWLEDGEMENTS

We would like to thank G.Jug for drawing our attention to his work and to ref.6, ERCC for DAP facilities and SERC for financial support.

REFERENCES

1. See e.g. G.S.Pawley et al, Phys. Rev. B29 (1984) 4030; J.K.Williams, J. Phys. A18 (1985) 49

2. J.K.Williams and N.D.McKenzie, J. Phys. A17 (1984) 3343

3. G.Jug, (1985) in print

4. G.Jug, Phys. Rev. Lett. 53 (1984) 9

5. D.Stauffer, Z. Phys. B22 (1975) 161; R.J.Birgeneau et al, Phys. Rev. Lett. 37 (1976) 940

6. C.Domb and C.J.Pearce, J. Phys. A9 (1976) L137

Part V (a)
IRREVERSIBLE GROWTH MODELS: LAPLACIAN FRACTALS, DIELECTRIC BREAKDOWN, FRACTURE PROPAGATION AND VISCOUS FINGERS IN LIQUIDS

PROPERTIES OF LAPLACIAN FRACTALS FOR DIELECTRIC BREAKDOWN IN 2 AND 3 DIMENSIONS

H.J. WIESMANN* and L. PIETRONERO**

* Brown Boveri Research Center, CH-5405 Baden, Switzerland
** University of Groningen, Melkweg 1, 9718 EP Groningen, The Netherlands

We investigate properties of random fractal structures S in 1,2 and 3 dimensions. The growth process is governed by a probability measure which is based on the solution of the boundary value problem of potential theory (*Laplacian Fractals*). The relation between growth probability and local electric field is of the form $p \propto E^\eta$. Fractal dimension D, average radius and growing zone are characterized by the first and second moment of the charge distribution on S. Particular attention is focused on the nontrivial dependence of D on η.

1. INTRODUCTION

Many different forms of dielectric breakdown are known. They range from atmospheric lightnings to electrical treeing in polymers. Depending on the dielectric (gas, liquid, solid or polymer) the microscopic process leading to breakdown might be different. Nevertheless we observe very similar global or structural properties of the discharge pattern, and our main intentions tend towards an understanding of these aspects of the phenomenon.

Dielectric breakdown is an example of a random growth process. If an insulating material is exposed to an electric field which exceeds a certain critical value E_c, then a conducting phase is created within the material. This phase grows with time in a stochastic way and forms typical discharge patterns. Different geometries are observed ranging from diffuse quasihomogeneous to filamentary with weak or strong ramification.

Questions which are as well of technical as of scientific relevance are the following ones:
(i) What is the value of the critical field E_c? Whereas in gases the situation is rather well understood[1] only a few recent results are known for the critical space charge injection field in polymers[2].

(ii) Does the growth process extend to global (catastrophic) breakdown or will it be limited to a bounded volume in space? Of course this question depends on the geometrical configuration of the electrodes but, more interesting, it has also an intrinsic aspect: maximum extension and critical field are closely related.
(iii) Can we characterize the geometry of the breakdown pattern and can its structure be explained by a growth law? To answer this question at least partially, we have introduced a mathematical description of the discharge process which leads to fractal structures[3]. Consequently it becomes possible to characterize patterns by numbers: fractal dimension and critical exponents. These numbers turn out to depend in a nontrivial way on the specific form of the growth law and therefore are not universal. In this paper we shall present extended studies of the problem in euclidean dimension d = 2 and new results for d = 3.

2. THE REAL AND THE IDEALIZED PROCESS

In order to understand the limitations (from the physical point of view) of the model and to perceive possible generalizations, we need an outline of a "realistic" phenomenological description of the discharge process. During breakdown the material is divided into two different phases P_d (dielectric or insulating) and P_c (conducting). They are separated by an interface, which is a dynamical object of more or less complicated stochastic form. In each of the phases we need a set of equations to determine the physical quantities of interest: electric field E, potential ϕ, charge density q and possibly others. Excluding photoionization processes we have

$$-\Delta\phi = q \quad , \quad E = -\mathrm{grad}\phi \qquad (1)$$

$$q = e \sum_{\sigma,k} \sigma n_k^\sigma \qquad (2)$$

$$n_k^\sigma = 0 \qquad \text{in } P_d \qquad (3)$$

$$\dot{n}_k^\sigma = D_k^\sigma n_k^\sigma \qquad \text{in } P_c \qquad (4)$$

where n_k^σ denotes the density of charge carriers with charge $\sigma \cdot e$. k distinguishes different species if necessary. D_k^σ is a local but nonlinear operator depending on E and the n_k^σ. The right hand side of (4) represents charge creation, annihilation and transport processes (field and diffusion currents). The dynamics of the interface is of course related to the particle velocities in its neighborhood.

Because in P_d the charge density is zero, (1) reduces to the Laplace equation. For ϕ we have to solve the boundary value problem of potential theory with boundary conditions on the interface.

The solution of the coupled system of Poisson-(1) and transport-(4) equations in P_c is a very ambitious program to understand dielectric breakdown. Much of the theoretical work deals with these equations within a great variety of different, more or less artificially restricted boundary conditions. Unfortunately it turns out to be much too complicated in order to get an insight into the origin of the typical discharge structures. In addition the theory as described above does not yet contain an element leading to the apparent stochastic nature of the breakdown process.

Our idealized model leads to an understanding of these important aspects at the price, that we have to ignore all details described by equations (4). They are replaced by the assumption that $\phi = \phi^o = $ cst in P_c which, by the way, is a trivial solution of the Laplace equation. We can say that P_c is assumed to be ideally conducting. This assumption is in rather good agreement with observation for example for leader discharges in gases. As a consequence the charge density q will be different from zero only on the interface (induced charge on the surface of the conductor P_c). Furthermore q is proportional to the field strength in P_d at the interface, it is therefore determined by the solution of the boundary value problem in P_d. We interpret the surface field in a stochastic sense as the driving force for the growth process: interface points with high surface field move fast on average, points with low field move slowly or are stationary.

3. LATTICE MODEL FOR LAPLACIAN FRACTALS

We realize the ideas outlined above in a lattice model, in order to simplify computer simulations and to reduce the manifold of discharge configurations. The object of investigation is a lattice structure S of the d-dimensional (cubic) lattice, which represents the conducting phase P_c, whereas P_d corresponds to its complement $Z^d \backslash S$. It is useful to distinguish between site and bond structure. Let S_n denote the connected site configuration, containing (n+1) sites (black dots in figure 1) and S_n' the corres-

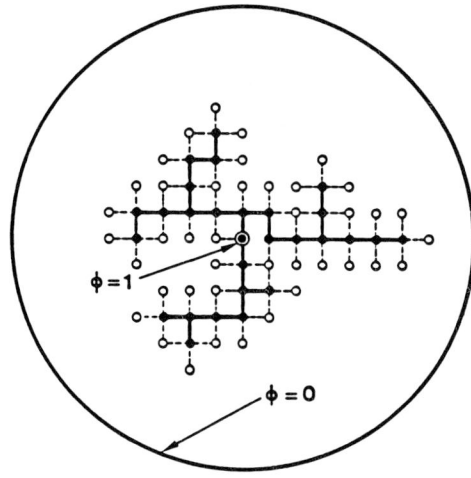

FIGURE 1
Lattice discharge structure: S_n (sites, black dots) and S_n' (bonds, solid lines). The probability associated with each bond of $\partial S_n'$ (surface bonds, dashed lines) depends on the corresponding field.

ponding bond structure with n bonds (solid lines).

$$S_n = \{k_m | k_m \in Z^d, \quad m = 0,\ldots,n\}$$
$$S_n' = \{(k,k') \,|\, |k-k'| = 1, \quad k,k' \in S_n\} \quad (5)$$

The irreversible growth process takes place in elementary steps $S_n \to S_{n+1}$, $S_n \subset S_{n+1}$. In order to maintain the connectedness of S_n after each step we have to impose the restriction

$$S_{n+1} = S_n \cup \{h\} \quad \text{with } \text{dist}(h, S_n) = 1$$
$$S_{n+1}' = S_n' \cup \{(k,h)\} \quad \text{with } (k,h) \in \partial S_n' \quad (6)$$

where

$$\partial S_n' = \{(k,h) \,|\, |k-h| = 1, \; k \in S_n, \; h \notin S_n\} \quad (7)$$

is the set of surface bonds of S_n' (dashed lines in figure 1). In each step we choose therefore one bond out of the candidate-set $\partial S_n'$ at random. The associated probability is given by a measure μ_{S_n} which is based on the solution of the boundary value problem for ϕ on the d-dimensional lattice:

$$\Delta \phi_h = \sum_{|e|=1} (\phi_{h+e} - 2d\, \phi_h) = 0, \; h \in Z^d \setminus S \quad (8)$$

with Dirichlet boundary conditions

$$\phi_k = \phi^o \quad , \; k \in S_n \quad (8a)$$
$$\phi_k \to 0 \quad , \; |k| \to \infty \text{ in } d \geq 3$$
$$\phi_k \sim -\phi^o \cdot \ln|k| \quad , \; |k| \to \infty \text{ in } d = 2 \quad (8b)$$

We define

$$\mu_{S_n}(k,h) = \begin{cases} Z_{S_n}^{-1} \cdot f(\phi_k - \phi_h) & \text{if } (k,h) \in \partial S_n' \\ 0 & \text{otherwise} \end{cases} \quad (9)$$

with a normalization factor

$$Z_{S_n} = \sum_{(k,h) \in \partial S_n'} f(\phi_k - \phi_h) \quad (9')$$

i.e. the probability to choose the surface bond (k,h) in the (n+1)-st step is given by a function f of the local field attributed to this bond in the configuration S_n. In the choice of f we still have an open degree of freedom. There are two different possibilities to determine it. Either, on the basis of microscopic arguments, one might calculate at least a threshold value E_c, or one models different structures on the basis of various functions f, in order to enable a comparison with experimental patterns. We follow the second way and restrict f to a one parameter set of functions

$$f(y) = \text{cst } y^\eta \quad (10)$$

μ is then given by

$$\mu_{S_n}(k,h) = Z_{S_n}^{-1} \cdot (\phi_k - \phi_h)^\eta, \; (k,h) \in \partial S_n' \quad (11)$$

This choice actually neglects the existence of a critical field E_c. μ is independent of the applied voltage ϕ^o, as the potential ϕ is linear in ϕ^o. There is no natural *physical* (like field or voltage) unit in the problem and μ depends only on the *geometrical* configuration of S_n. In fact (10) is the most general choice for f which is consistent with a purely geometrically defined measure. Why? μ is independent of ϕ^0 if and only if f fulfils a relation of the type

$$f(\lambda y) = g(\lambda) f(y) \qquad (12)$$

for some function g, which is equivalent to

$$F(\lambda y) = F(\lambda) F(y) \qquad (12')$$

with $F(y) = f(y)/f(1)$.

The only measurable solution of (12') is a power law and leads to (10).

With (8) and (11) the growth process is perfectly defined if it starts with a given initial configuration S_{n_0}, which in the following always will be the lattice origin $S_0 = \{0\}$. The resulting infinitely extended structures turn out to have fractal properties. We call them *Laplacian Fractals* because the dynamical process which creates them is intimately related to the Laplace equation.

4. PROBABILITY MEASURE AND INDUCED CHARGE

The structure S_n and its dynamical behavior are characterized by two different density functions related to each other through the boundary value problem (8). These are the occupation density ρ

$$\rho_{S_n}(k) = \begin{cases} 1 & \text{if } k \in S_n \\ 0 & \text{otherwise} \end{cases} \qquad (13)$$

and, assuming $\eta = 1$ for the present, the charge density q

$$q_{S_n}(k) = -\Delta\phi_k = -\sum_{h,|h-k|=1}(\phi_h - \phi_k) \qquad (14)$$

Of course (14) is zero for $k \notin S_n$ according to (8), but in general not on S_n. As the potential on S_n is constant, terms in the sum of (14) with $h \in S_n$ can be dropped such that from (11) we get

$$q_{S_n}(k) = Z_{S_n} \cdot \sum_{h,(k,h) \in \partial S_n'} \mu(k,h) \qquad (15)$$

Eventually, summation over all $k \in S_n$ yields

$$Q_{S_n} = Z_{S_n} \qquad (16)$$

i.e. the normalization factor of μ is nothing but the total charge supported by S_n. Due to equations (15) and (16) the normalized charge density q_{S_n}/Q_{S_n} can be interpreted as the probability distribution for the (n+1)-st growth step to start in point k of S_n. We define the following moments of this distribution

$$r_n^m = Q_{S_n}^{-1} \sum_k q_{S_n}(k) \cdot |k|^m, \quad m = 1, 2 .. \qquad (17)$$

r_n^m as defined in (17) represents an average value with respect to a fixed realization S_n. Replacing q_{S_n} by its average q_n over all possible realizations with n elements leads to $\langle r_n^m \rangle$. The moments m=1 and 2 are of practical importance, as they allow numerical estimates. We use the abbreviation $\langle r_n \rangle$ for $\langle r_n^1 \rangle$ and $\langle \delta r_n \rangle$ for the fluctuations

$$\langle \delta r_n \rangle = \langle (r_n - \langle r_n \rangle)^2 \rangle^{1/2} \qquad (18)$$

Whereas $\langle r_n \rangle$ can be interpreted as the radius of the average structure with n elements, $\langle \delta r_n \rangle$ is a measure for the radial thickness of the growing zone. We find the following scaling behavior for these quantities with respect to n

$$\langle r_n \rangle \sim a_1 n^{\nu_1}, \quad \langle \delta r_n \rangle \sim a_2 n^{\nu_2} \qquad (19)$$

Before presenting results we have to mention that in the case $\eta \neq 1$ the charge density q has to be replaced by a "generalized" charge density

#○ : $\rho(r)$
#◇ : $\rho_+(r)$, #□ : $\rho_-(r)$

FIGURE 2
Characterization of a discharge pattern by counting the number of branches at a given radius, $\rho(r)$ and the density of bifurcation points $\rho_+(r)$ and endpoints $\rho_-(r)$.

$$q_{S_n}^{(\eta)}(k) = - \sum_{h,|h-k|=1} (\phi_h - \phi_k)^\eta \quad (20)$$

which allows an analogous procedure as for $\eta = 1$.

Finally figure 2 illustrates the significance of the exponent ν_1 for discharge figures in a spherically symmetric electrode configuration. The asymptotic structure $S = \lim_{n \to \infty} S_n$ is characterized by its fractal dimension D which is related to the exponent ν_1 by

$$D = \nu_1^{-1} \quad (21)$$

as was pointed out already by Plischke and Racz[4] under the assumption of radially Gaussian distributed charge. If $\rho(r)$ denotes the number of discharge lines crossing r, $\rho\pm(r)$ the radial density of branching points (+) and dead ends (-) and $L(r)$ the total length of discharge lines within r, then the following scaling relation holds:

$$L(r) \propto r^D, \quad \rho(r) \propto r^{D-1}, \quad \rho\pm(r) \propto r^{D-2} \quad (22)$$

RESULTS

We are far from a complete solution of the dynamical growth problem for Laplacian fractals. Exact solutions are known only in dimension $d = 1$. For $d = 2$ and 3 we can present numerical results. They are restricted to radial densities.

The problem in $d = 1$ is equivalent to the ordinary random walk in one dimension. The essential structural complexity of higher dimensions is absent. The (connected) structure S_n is simply characterized by its endpoints, carrying each the normalized charge $\frac{1}{2}$. If r_n denotes the coordinate of the (right) endpoint, we have, independent of η

$$r_n = \frac{1}{2} \sum_{i=1}^n (1+\zeta_i) \quad (23)$$

with random variables $\zeta_i \in \{-1,+1\}$. From (23) one obtains the well known exponents $\nu_1 = 1$, $\nu_2 = 0.5$ and the probability distribution for the configurations

$$\text{Prob}\{r_n = k\} = 2^{-n}\binom{n}{k} \quad (24)$$

Both for $d = 2$ and 3 the following typical properties with respect to the parameter η are found for the discharge patterns: high values of η lead to low occupation densities with scarce ramification, with deep penetration of the field into the structure and correspondingly to a broad radial distribution of the charges. The preference for growth in the direction of high field values at the tips is very pronounced. In contrast to this we obtain for low η-values strongly ramified and dense structures with a sharp screening of the exterior field. The corresponding radial densities of occupation ρ_n^{rad} and charge q_n^{rad} defined by

$$q_n^{rad}(r) = \sum_k q_n(k) \delta(|k|-r) \quad (25)$$

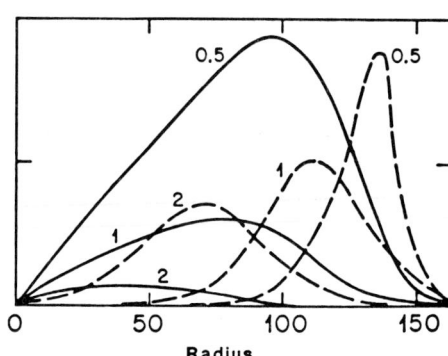

FIGURE 3
Occupation- (ρ_n^{rad}, solid lines) and charge density (q_n^{rad}, dashed lines) in arbitrary units for $\eta = 0.5$, 1 and 2 in $d = 2$.

(and analogous for ρ) are shown in figure 3 for $d = 2$, $\eta = 0.5, 1, 2$, and correspondingly $n \simeq 30000, 10000, 3000$. We can distinguish two regions. The interior, where the structure has achieved its asymptotic limit, is characterized by a typical increase of ρ_n^{rad} with a power $(D-1)$ and the fact that q_n^{rad} is small. The growing or fluctuating zone carries the charges and is accompanied by a decreasing occupation.

The most important result is the nontrivial dependence of the exponents ν_1 and ν_2 from the characteristic parameter η of the probability law. Both ν_1 and ν_2 increase with η. We obtain an inequality $\nu_2 < \nu_1$ with an increasing ratio ν_2/ν_1. Figure 4 illustrates the situation for $d = 3$ (results for $d = 2$ are qualitatively the same). In addition figure 5 shows the double logarithmic plot $\ln<r_n>$ and $\ln<\delta r_n>$ versus $\ln n$ for $d = 2$ and $d = 3$ and $\eta = 1$. The η-dependence of the exponents is clearly continuous with the only possible exception of the value $\eta = 0$. In this case two different definitions of the growth process are possible. Roughly speaking it depends on whether y^η in (10) for $y = 0$ and $\eta = 0$ is defined to be 1 or 0. The definition "1" corresponds to equal weight for each surface bond, independent

of whether the surface bond lies inside a loop of S_n where the field is exactly zero, or not. It can be called "Eden-bond" model. The case "0" corresponds to the limit $\eta = 0$ of our model, but unfortunately it is not treatable with the present numerical methods. However we expect that for high dimensions this difference is irrelevant, whereas in $d = 2$ some care might be needed.

The η-dependence of the growth exponents leads to a new aspect of the controversy in DLA [5], whether the process is governed by one or several length scales. From the mathematical equivalence of DLA with the dielectric breakdown model in the special case $\eta = 1$ [6] one concludes that both have to produce identical results. If ν_2 would tend to ν_1 for large n, we would expect that this happens at an average radius which is smaller for small η (sharp screening) and larger for large η-values (weak screening). We can not observe such a tendency.

Finally table 1 gives an overview over exponents ν, fractal dimension D, the approximate sample sizes n_{max} and the number of samples used to determine average values, for $d = 1,2,3$ and different values of η. We would like to point

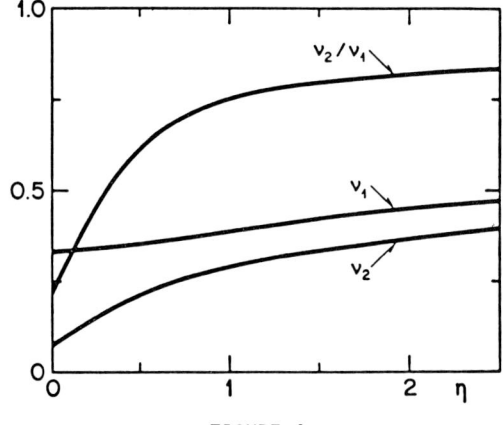

FIGURE 4
Growth exponents ν_1, ν_2 and their ratio as a function of the characteristic parameter η of the probability law.

out that the exponents given in table 1 are determined on the basis of equation (17). Within the limited range we could explore in d = 3 ($r_{max} \sim 25$) an evaluation corresponding[7,8] to DLA leads to slightly different values ($\nu_1 = 0.39$, $\nu_2 = 0.29$, D = 2.55).

From the results presented we can draw the following conclusions:

(i) Exponents and fractal dimension are not universal but depend on the given form of the probability law which reflects certain physical properties of the dielectric breakdown process. Therefore they can be regarded as characteristic numbers of the investigated physical process.

(ii) The inequality $\nu_2 < \nu_1$ indicates that Laplacian fractals tend to build a surface or shape for large n. Structures S_n and $S_{n'}$ with $n \neq n'$ are not similar, only the asymptotic structure S is selfsimilar. A discussion of the growing zone under the aspect of different exponents or length scales is given elsewhere in this volume[9].

TABLE 1
Growth exponents ν_1, ν_2 and fractal dimension D for d = 1,2,3 and several values of η. In addition the approximate sample size n_{max} is indicated and the number of samples evaluated to determine average values.

d	η	ν_1	ν_2	D	n_{max}	nb.
1	all	1	0.5	1		exact
2	0	0.497	0.145	2	20000	20
	0.5	0.52	0.41	1.92	30000	20
	1	0.59	0.50	1.70	10000	20
	2	0.70	0.64	1.43	3000	20
3	0	0.325	0.073	3	20000	20
	0.5	0.36	0.23	2.78	10000	40
	1	0.38	0.28	2.65	4000	40
	2	0.44	0.35	2.26	1500	40

REFERENCES

1. E.E. Kunhardt and L.H. Luessen, Electrical Breakdown in Gases (Plenum Press, New York and London, 1983).

2. T. Hibma, P. Pfluger and H.R. Zeller, Electronic Processes in Polymeric Dielectrics Under High Electrical Fields, in: Electronic Properties of Polymers and Related Compounds, eds. H. Kozmany, M. Mehring and S. Roth (Springer, Berlin, Heidelberg, New York, 1985) p. 317-326.

3. L. Niemeyer, L. Pietronero and H.J. Wiesmann, Phys. Rev. Lett. 52 (1984) 1033.

4. M. Plischke and Z. Racz, Phys. Rev. Lett. 53 (1984) 415.

5. P. Meakin, L.M. Sander and M. Plischke, Z. Racz, Phys. Rev. Lett. 54 (1985) 2053.

6. L. Pietronero and H.J. Wiesmann, J. Stat. Phys. 36 (1984) 909.

7. P. Meakin, Phys. Rev. A27 (1983) 1495.

8. M. Plischke and Z. Racz, Phys. Rev. A31 (1985) 985.

9. L. Pietronero, C. Evertsz, H.J. Wiesmann, Scaling Properties and Growing Zone of Laplacian Fractals, this volume.

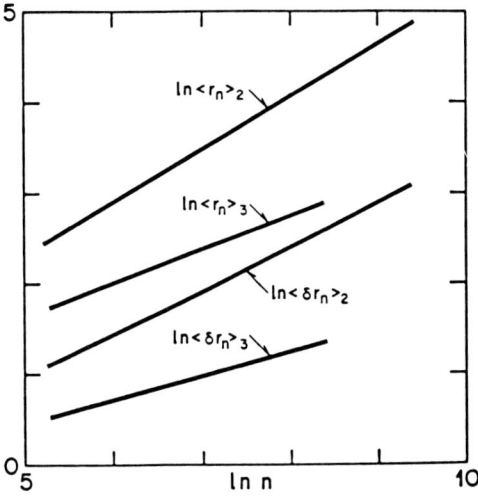

FIGURE 5
Average size $<r_n>_d$ and fluctuations $<\delta r_n>_d$ as a function of n in double logarithmic representation for $\eta = 1$ in d = 2 and 3.

FRACTALS IN PHYSICS
L. Pietronero, E. Tosatti (editors)
© Elsevier Science Publishers B.V., 1986

SCALING PROPERTIES OF GROWING ZONE AND CAPACITY OF LAPLACIAN FRACTALS

L. PIETRONERO[+], C. EVERTSZ[+] and H.J. WIESMANN[++]

[+] University of Groningen, Melkweg 1, 9718 EP Groningen, The Netherlands
[++] Brown Boveri Research Center, CH-5405 Baden, Switzerland

Following the numerical results discussed in the previous paper we discuss why the existence of different length scales in Laplacian Fractals, not only does not lead to any inconsistency, but it actually follows as a natural consequence for this type of problems. We consider also the problem of the total charge (or capacity) of these systems as a function of their size and show that the leading power law only depends on the dimension of the embedding euclidean space. In two dimension an explicit expression for the logarithmic correction is derived and it results in good agreement with numerical results.

1. INTRODUCTION

Laplacian Fractals (LF) have been introduced in the preceding paper[1] as fractal structures whose growth probability is governed by a Laplace equation. Several papers at this Symposium report studies of systems belonging to this class and most of them attempt to elucidate the nature of the growth mechanism[2-6]. A crucial point for further progress appears to be linked to a better understanding of the properties of the growing zone. This is a very controversial topic, the main question being whether the characteristic length governing the thickness of the growing zone scales or not like the size of the system[7,8]. In the preceding paper[1] we have reported numerical results in two and three dimensions for the growing zone exponent ν_2 as a function of the exponent η that relates the local field to the growth probability. These results strongly indicate that the growing zone scales with a different exponent than the radius. However, questions have been raised about the eventual inconsistency of having more than one length scale with respect to the requirement of selfsimilarity[8]. Here we provide evidence that the occurence of two length scales not only does not lead to any inconsistency but it actually follows in a natural way for these type of problems. Such a conclusion is in agreement with that of Plischke and Rácz[2,7]. We also discuss the implications of the fact that both $\nu_1 = 1/D$ and ν_2 explicitly depend on η. Finally we consider another property of the growing zone that is of particular relevance for the dielectric breakdown problem[9,10], the electrical capacity. The question is how does the total charge on the fractal scale with N or R. We find that the leading exponent is only due to the dimensionality d of the embedding euclidean space. The fractal dimension D only affects the prefactor.

2. PROPERTIES OF THE GROWING ZONE

Consider a Laplacian Fractal with its center at the origin. Its fractal dimension D is measured by looking at the number of elements (points or bounds) N(R) contained in a hypersphere of radius R for various of R:

$$N(R) = \int_0^R \rho(r) \, r^{(d-1)} dr \simeq R^D \quad . \qquad (1)$$

This implies that the density $\rho(r)$ scales like

$$\rho(r) \sim r^{-(d-D)} \quad . \qquad (2)$$

Consider now the hypersphere of radius R. This is shown in Fig. 1 for d=2.

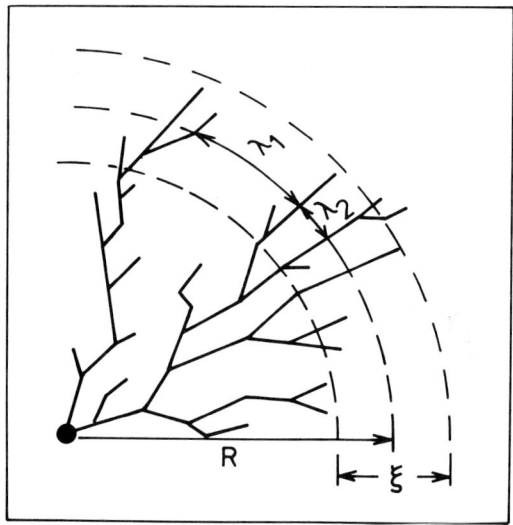

FIGURE 1
Schematic picure of some branches of a Laplacian Fractal. λ_1 and λ_2 indicate the distance between two pairs of branches along the circle of radius R. The growing zone size is characterized by ξ.

The total number of branches intersecting this hypersphere is

$$N(R) = \frac{dN(R)}{dR} \simeq R^{(D-1)} \quad . \qquad (3)$$

We consider here the "bond" version of the problem as in the original dielectric breakdown model[9,10]. All the branches are therefore equally thick and there is no need to introduce an effective branch thickness as for the "site" problem[2,7]. The average area of the hypersurface associated to each branch is

$$<\sigma(R)> = \frac{R^{(d-1)}}{N(R)} \simeq R^{(d-D)} \qquad (4)$$

where the average $<\sigma(R)>$ is taken with respect to all the branches.

This concept is similar to the Voronoi polyedra for disordered systems. The average distance between nearest branches is then

$$<\lambda(R)> \simeq <\sigma(R)>^{1/(d-1)} \simeq R^{(\frac{d-D}{d-1})} \quad . \qquad (5)$$

The meaning of λ is exemplified in Fig. 1 where two examples $\lambda_1(R)$ and $\lambda_2(R)$ are shown. In two dimension $\lambda_i(R)$ is the distance between the i-th branch and the next one along the circle of radius R. From the point of view of the DLA process it is clear that these lengths will be of great importance for the growth process. It is easy to see that the characteristic length $<\lambda(R)>$ in general does not scale like R. In fact the requirement $<\lambda(R)> \sim R$ implies D=1. The simple inspection of the pattern allows therefore to define a length scale that does not scale like R for non trivial cases. This simple fact makes us highly sceptical about the conjecture of a single length scale governing the growth process.

We attempt now to relate the length scale defined by Eq.(5) to the growing properties of the system. The following discussion will be restricted to the d=2 case as illustrated in Fig. 1. Consider a pair of parallel branches separated by the distance λ. It can be shown that the probability distribution for a random walker (we use here the DLA point of view) to penetrate a distance l can be written approximately as[12]

$$P_\lambda(l) \simeq \frac{1}{\lambda} [\ln(\lambda^2/4 + l^2) - \ln(l^2)] \quad . \qquad (6)$$

The characteristic penetration depth is therefore $\zeta \simeq \lambda$. The simplest generalization of this relation in order to obtain a scaling relation is to assume also that

$$<\zeta(R)> \simeq <\lambda(R)> \quad . \qquad (7)$$

Such a relation gives rise however to a number of problems. If we consider in fact the probability distribution given by Eq.(6) and we generalize it to the case of $\eta \neq 1$ we have $[P_\lambda(1)]^\eta$ that still has a characteristic penetration depth of order of λ[12]. This implies that a theory based on Eq.(7) would not be able to take into account for the η dependence of D. On the other hand there is little doubt that for a particular pair of nearest branches $\zeta_i \approx \lambda_i$. This puzzling situation can be resolved by considering a distribution $W[\lambda(R)]$ for the different values of $\lambda(R)$ ranging from zero to $\lambda_{max}(R)$. In this case we have

$$<\lambda(R)> = \int_0^{\lambda_{max}(R)} W[\lambda(R)]\lambda(R)d[\lambda(R)]. \quad (8)$$

but the calculation of the average of $\zeta(R)$ is actually more subtle. In fact a distribution for the values of $\lambda(R)$ implies an extra weighting factor of order of $\lambda(R)$ for the average of $\zeta(R)$ because it is more probable for a random walker to end up in the gulf corresponding to the larger $\lambda_1(R)$ in Fig. 1 rather then in the one defined by $\lambda_2(R)$. We have then

$$<\zeta(R)> \int_0^{\lambda_{max}(R)} W[\lambda(R)] \lambda^2(R) d[\lambda(R)]. (9)$$

It is clear therefore that the introduction of a distribution for $\lambda(R)$ implies in general $<\zeta(R)> \neq <\lambda(R)>$. This may in principle resolve the inconsistencies described above. The problem can be therefore reduced to the determination of self-stabilizing asymptotic distribution $W[\lambda(R)]$. This distribution governs through <u>different relations</u> the fractal dimension D and the interface thickness $\zeta(R)$. It is therefore natural to expect <u>different scaling behaviors</u> for R and $\zeta(R)$. This implies a sort of <u>restricted self-similarity</u> in the sense that, in view of the fact that $\zeta(R)/R$ is not a fixed ratio, the system is selfsimilar only asymptotically for radial transformation with <u>fixed center</u>.

It does not have therefore the properties of homogeneity of the percolation clusters but still there are well defined scaling properties. These conclusions fully agree with those of Plischke and Racz[2,7] and are supported by the studies of the anisotropy of the correlation function[11].

3. ELECTRICAL CAPACITY OF LAPLACIAN FRACTALS

An interesting property to consider in these systems is the electrical capacity as a function of N or R. This question is of course particularly relevant for the dielectric breakdown model where it can have an applied interest. We have seen in the preceding paper how the charge destribution, that can be computed from the local field via a discretized version of Gauss theorem, modulates the growth probability. The question we consider here is what is the total charge on such fractals as a function of their size. We start discussing the two dimensional case for which we illustrate two limiting examples in Fig. 2 where an analytical solution can be obtained.

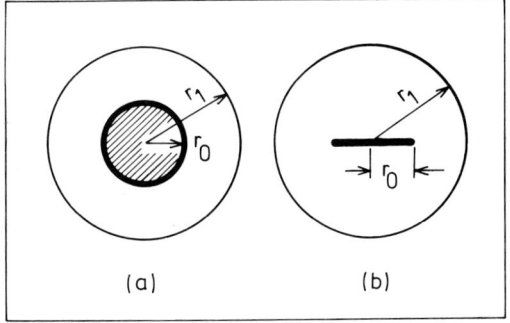

FIGURE 2
Two limiting examples of systems for which the capacity can be computed exactly.

Two dimensions is delicate because due to the logarithmic nature of the Green's function particular care has to be taken for the boundary conditions. In Fig. 2a we have a circle of radius r_0 with the outside boundary at distance r_1, while in Fig. 2b we consider a line of length $2r_0$ with the same boundary conditions. If we consider the circle to include also all its internal points as part of the structure, then these two examples provide limiting cases for all the systems that can be embedded within a circle of radius r_0. The filled circle itself is the most compact of these structures (D=2) while the line is the most tenuous (D=1). These electrostatic problems can be solved by the method of conformal mapping[12]. The charge per unit of voltage results

$$Q_c = 2\pi/\ln(r_1/r_0) \qquad (10)$$

for the circle and

$$Q_s = 2\pi/\ln(2r_1/r_0) \qquad (11)$$

for the line. The analytical behavior is the same for the two cases. Considering that these are limiting cases we can conclude that any fractal structure inscribed in the same circle will be bound above and below by Eqs. (10) and (11). This is more clear if we consider that, given a structure made of N elements with a charge Q(N), if we add an extra element to this structure the total charge (capacity) can only increase

$$\frac{dQ(N)}{dN} \geq 0 \quad . \qquad (12)$$

This result is intuitively clear and can also be derived rigorously[12]. We can generalize this analytical behavior to obtain the N dependence of the total charge on a fractal (in 2-d)

$$Q(N) = 2\pi/\ln[\alpha(D)r_1/<r_0>_N] \qquad (13)$$

where

$$<r_0>_N = A \ N^{1/D} \qquad (14)$$

and $\alpha(D)$ is a coefficient only dependent on D. We have checked a few examples of computer generated fractals and their total charge results in nice agreement with Eq.(13).

We can now turn to the case of euclidean dimensions larger than two. The relation between charge and size can be shown to be of the type[12]

$$Q \simeq R^{d-2} \quad , \qquad (15)$$

and using the aobve arguments of upper and lower bounds for fractal structures we can write in general for a fractal

$$Q(N) = A(D)R^{d-2} = A'(D) \ N^{(d-2)/D} \quad . \qquad (16)$$

In conclusion we have shown that even for a fractal the leading power law of the total charge (capacity) as a function of size is determined only by the dimension of the embedding euclidean space. The fractal dimension only modifies the prefactor for d>2 or the logarithmic term for d=2.

REFERENCES

1. H.J. Wiesmann and L. Pietronero, this volume.

2. M. Plischke and Z. Rácz, this volume.

3. P. Meakin, this volume.

4. L.A. Kurkevic and H. Scher, this volume.

5. G. Rossi, B.R. Thompson, R.C. Ball and R.M. Brady, this volume.

6. R. Jullien and R. Botet, this volume.

7. M. Plischke and Z. Rácz, Phys. Rev. Lett. 53 (1984) 415.
   ibid. C 54 (1985) 2054.

8. P. Meakin and L.M. Sander, Phys. Rev. Lett. C 54 (1985) 2053.

9. L. Niemeyer, L. Pietronero and H.J. Wiesmann Phys. Rev. Lett. 52 (1984) 1033.

10. L. Pietronero and H.J. Wiesmann, J. Stat. Phys. 36 (1984) 909.

11. P. Meakin and T. Vicsek, this volume.

12. L. Pietronero, C. Evertsz and H.J. Wiesmann, unpublished.

AN INFINITE HIERARCHY OF EXPONENTS TO DESCRIBE GROWTH PHENOMENA

Antonio CONIGLIO

Istituto di Fisica Teorica, Mostra D'Oltremare, Pad. 19, 80125 Napoli, Italy
and
Center for Polymer Studies and Department of Physics, Boston University, Boston, Massachusetts 02215, USA

A growth model can be characterized by the set of probabilities $\{p_i\}_{i\in\Gamma}$ that each site at a given time on the external perimeter Γ becomes part of the aggregate. Equations for the set of p_i are given for DLA and other growth models using the electrostatic analogy of the dielectric breakdown model. A scaling approach is developed for the probability distribution and is compared with the voltage distribution in a random resistor and random superconducting network at the percolation threshold. An infinite set of exponents is necessary to fully characterize the moments of the distribution which are related to the surface structure of the aggregate.

1. INTRODUCTION

This talk is based on work done in collaboration with C. Amitrano, L. de Arcangelis, F. di Liberto, P. Meakin, S. Redner, H. E. Stanley and T. Witten. It is complementary to a lecture[1] given at the Cargese School on Growth and Form a week before this Conference. The main ideas introduced in that lecture are reformulated here and developed on the basis of a more general approach to fractal measures presented by Kadanoff[2] at the same school.

What are the relevant parameters to fully describe the essential properties of a growth process? The answer to this question is extremely important in order to be able to understand the complexity and the richness of this exciting field of growth phenomena. It is clear for example that an aggregate cannot be fully characterized by its fractal dimensionality. DLA[3] and percolating clusters in three dimensions have roughly the same fractal dimensionality, yet they have a completely different structure.

A possible way to fully characterize a growth model is by assigning at each time step the growth sites probability distribution (GSPD) $\{p_i\}_{i\in\Gamma}$, where p_i is the probability that site i becomes part of the aggregate. In most cases the growth occurs on the perimeter Γ of the aggregate.

2. SCALING PROPERTIES OF THE GROWTH SITES DISTRIBUTION PROBABILITY

From the GSPD one can obtain not only the static and dynamic quantities of interest but also much more detailed information on the capability of each perimeter site to grow and therefore a better knowledge of the surface structure. In DLA for example the "hottest" sites which are more likely to grow are those at the tips of the cluster, for which the growth probability assumes the highest value, while the very "cold" sites deep inside the fjords are characterized by a very small value of the growth probability.

To characterize the fractal structure of the aggregate we consider the following moments and relative exponents[4-7]

$$Z(q) \equiv \sum_{i\in\Gamma} p_i^q \sim L^{-(q-1)D(q)}, \qquad (1)$$

where L is the size of the aggregate which for example can coincide with the radius. It is convenient to write $Z(q)$ in the following way

$$Z(q) = \sum_p n(p)^q \equiv \sum_p e^{F(p,q)}, \qquad (2)$$

where $n(p)$ is the number of sites with growth probability equal to p and

$$F(p,q) = \ln n(p) + q \ln q. \tag{3}$$

Since for large L $F(p,q)$ is expected to diverge, the sum in (2) can be evaluated by the steepest descent method. If p^* is the value for which $F(p,q)$ has a maximum we have

$$\frac{\partial \ln n(p)}{\partial \ln p}\Big|_{p=p^*} = -q. \tag{4}$$

For each value of q there is a corresponding value $p^* = p^*(q)$. We can make the following scaling *Ansatz*[2]

$$p^* \sim L^{-\alpha(q)} \tag{5}$$

$$n(p^*) \sim L^{f(q)}. \tag{6}$$

Therefore

$$Z(q) \simeq e^{F(p^*,q)} \sim L^{-[q\alpha(q)-f(q)]}, \tag{7}$$

and from (1),

$$(q-1)D(q) = q\alpha(q) - f(q). \tag{8}$$

Note that since p^* is a function of q from (4) we also have $\frac{\partial f}{\partial q} = q\frac{\partial \alpha}{\partial q}$ and therefore

$$\frac{\partial}{\partial q}(q-1)D(q) = \alpha(q), \tag{9}$$

in agreement with the result of Ref. [2]. Therefore given $D(q)$ one can calculate from (8) and (9) $\alpha(q)$ and $f(q)$ and vice-versa. $f(q)$ represents the fractal dimensionality of the set of sites characterized by the growth probability $p = p^*$ and $\alpha(q)$ represents the way p^* scales with L. The concept of density of singularities f and of singularity strength α was introduced in Ref. [6] where however for DLA $\alpha(q)$ and $f(q)$ were taken independently on q.

From the knowledge of $\alpha(q)$ and $f(q)$ we can also predict the scaling behavior of $n(p)$. From (5) we have for large L

$$x \equiv \frac{\ln p^*}{\ln p_{max}} = \frac{\alpha(q)}{\alpha(\infty)}. \tag{10}$$

If $\alpha(q)$ is a monotonic function of q we can invert (10) and obtain $q = q(x)$. Therefore from (6)

$$n(p) \sim L^{\varphi(x)} \qquad x \le 1, \tag{11}$$

where $x = \frac{\ln p}{\ln p_{max}}$ and $\varphi(x) = f(q(x))$ while for $x \gg 1$ $n(p)$ goes to zero very fast. Note that $\varphi(x) = $ const if $f(q) = $ const. Equation (11) expresses the fact that the sites in the aggregate can be divided in different sets, each one characterized by a value $x = \frac{\ln p}{\ln p_{max}}$. Each set with its fractal dimensionality $\varphi(x)$. The "hottest sites" which have more probability of growing correspond to $x = 1$ while the coldest sites are characterized by the miniumum value $x_{min} = \frac{\alpha(-\infty)}{\alpha(+\infty)}$.

In the Eden model the growth probabilities are different from zero on the surface and are all identical $p_i = p^*$. Therefore $n(p^*) \sim L^{d-1}$, since $\Sigma p_i = 1$ $p^* \sim L^{1-d}$. In the Eden model then $D(q) = \alpha(q) = f(q) = d-1$ independent on q. In conclusion and infinite set of exponents is required to describe the fractal structure of DLA. The more intricate the structure, the larger the dispersion in the exponents. As $d \to \infty$, DLA approaches the Eden model where only one exponent is necessary to describe the structure of the aggregate. The presence of an infinite set of exponents is rather different from ordinary critical phenomena as in an Ising model. In this case fractal dimensionality of the critical droplets is the same, independent of their size. It is this fractal dimensionality that dominates, giving rise to one single "gap" exponent that describes all the moments of the order parameters.

3. EQUATIONS FOR GSPD

The exponents (8) have been measured by computer simulations[7] for DLA in $d = 2$ for $k = 2\cdots 8$. This was done by sending N_T random walkers on the DLA aggregate and counting the number of times N_i a given perimeter site i was hit by the walkers. For large values of N_T we have $p_i \simeq \frac{N_i}{N_T}$. From the p_i the set of exponents was determined and indication of a set of independent exponents tending towards $D(\infty) = d_f - 1$ was found.

A different approach was used in Ref. 8 to actually calculate GSDP. Using the electrostatic analogy in the continuum version the density probability $p(x)$ at site x on the surface is given by

$$p(x) = -\vec{n}(x) \cdot \nabla \phi(x) = |E(x)|, \tag{12}$$

where $\vec{n}(x)$ is the normal at the site x on the surface of the aggregate and $\phi(x)$ is the electrostatic potential satisfying the Laplacian equation with the condition that

ϕ const on the conductor and zero at infinity, the constant is chosen in such a way that the total charge on the conductor is equal to unity. The solution for $\phi(x)$ is given by

$$\phi(x) = \int_s \frac{\sigma(x')}{|x-x'|^{d-2}} ds', \qquad (13)$$

for $d > 2$ while for $d = 2$, $|x-x'|^{2-d}$ in the integrand is substituted with $\ln|x-x'|^{-1}$; $\sigma(x') = -\vec{n}(x') \cdot \nabla\phi(x')$ is the charge density. From (12)

$$p(x) = -\vec{n}(x) \cdot \nabla_x \int_s \frac{p(x')}{|x-x'|^{d-2}} ds'. \qquad (14)$$

This equation can be discretized and solved numerically for the $p(x)$, from which the moments have been calculated giving results consistent with the computer simulation results of Ref. 7. Equations for GSPD on lattice have also been given in Ref. 9.

4. VOLTAGE DISTRIBUTION IN RANDOM RESISTOR NETWORK AT THE PERCOLATION THRESHOLD

The theory described in Sec. II holds also for the voltage distribution in a random resistor network at the percolation threshold p_c^5. In fact in view of the analogy between DLA and the dielectric breakdown the properties of the voltage distribution are in fact expected to be very similar to the GSPD in DLA. This is due to the connection of the DLA to the dielectric breakdown model,[10] in which the growth probabilities are directly related to the voltage distribution on the surface of the aggregate.

Consider a d-dimensional hypercubic lattice of size L. Suppose that each bond has a probability p of being active and $1-p$ of being non-active. From percolation theory we know that in the limit of infinite system there exists a percolation threshold p_c, above which an infinite cluster of active bonds is present. Right at p_c the bonds in the spanning configurations, as noted by Stanley[11], may be partitioned in dangling bonds that do not contribute to the electrical resistance and the remaining backbone bonds. The backbone bonds may also be divided into links ("red" bonds), which are singly connected, and the remaining multiply-connected bonds ("blobs"). Whether the blobs could be neglected, as in the Skal-Shklovskii-de Gennes model,[12] or the links, as in the Sierpinski gasket model,[13] is a non-trivial question. It is now well established on the basis of exact results that both links and blobs are critical. The number of links L_1 diverges as $L_1 \sim L^{1/\nu}$ in any dimension when ν is the connectedness length exponent.[14] As d approaches 6, the blobs become less and less important, until they become irrelevant for $d \geq 6$. Strong numerical evidence confirms the validity of this picture.[15]

In order to characterize further the structure of the backbone bonds we associate a unit electrical resistance to each bond and apply a unit voltage at the opposite boundary of the cell connected by the percolating cluster. Each bond is characterized by a voltage drop V equal to the current I flowing through it. The maximum voltage drop $V_{max} = I_{max}$ occurs in the links.

One can further characterize the fractal structure by looking at the moments of the voltage distribution[5]

$$Z(q) = \sum n(v) V^q \sim L^{-(q-1)D(q)}. \qquad (15)$$

These moments have been calculated analytically on a hierarachical model which describes extremely well the properties of the percolating cluster. The expression for $\nu(q-1)D(q)$ (which was called $p(q)$ in Ref. 5) is given by

$$\nu(q-1)D(q) = (q-1) + \frac{q\ln\frac{5}{4} - \ln(1+2^{-q})}{\ln 2}, \qquad (16)$$

which reproduces extremely well the data from computer simulation for 2 dimensional random percolation. From (13) if follows that there is in fact an infinite set of independent exponents describing the moments of the voltage distribution. Physically this means that there is no characteristic value of the voltage which dominates. For each value of q in fact there is a value $V^*(q)$ which maximizes $F(V,q) = n(V)V^q$. In agreement with (5) and (6)

$$V^*(q) = L^{-\alpha(q)}, \qquad n(V^*) \sim L^{f(q)}, \qquad (17)$$

with

$$\nu\alpha(q) = \frac{\ln 5}{\ln 2} - \frac{1}{1+2^{-q}} \qquad (18)$$

$$\nu f(q) = 1 + \frac{q2^{-q}}{1+2^{-q}} + \frac{\ln(1+2^{-q})}{\ln 2}, \qquad (19)$$

which satisfy (8) and (9). It is also found that

$$n(V) \sim L^{\varphi(x)}, \qquad (20)$$

with

$$x = \frac{\ln V}{\ln V_{max}} = \frac{\alpha(q)}{\alpha(\infty)}, \quad (21)$$

and $\varphi(x) = f(q(x))$ where $q(x)$ is given by inverting (21) using (18).

For d above the upper critical dimensionality $d_c = 6$ all exponents coincide with the exponents relative to the links $\nu D(q) = 1$. This result has the physical explanation that above $d = 6$ the blobs are irrelevant and only the links are important. Consequently the moments of the voltage distribution are dominated by only one voltage drop $V = V_{max}$ across the links. On the other hand the maximum dispersion in $D(q)$ occurs at $d = 2$ where the blobs are relatively most important. Note the strong similarity with DLA.

In conclusion, an infinite set of independent exponents is necessary to describe the moments of the voltage distribution. These are related to the structure of the percolating backbone. The more complex the structure the larger the dispersion in the exponents.

The emphasis of this talk has been on the distribution probability of growth sites. This distribution probability has scaling properties similar to the voltage distribution in random resistor networks for which an infinite set of exponents is necessary to fully describe the richness of the structure of the geometrical properties of the aggregate.

I would like to thank my collaborators in this field, namely C. Amitrano, L. deArcangelis, F. diLiberto, P. Meakin, S. Redner, H. E. Stanley and T. Witten.

[1] A. Coniglio, PROCEEDINGS OF CARGESE SUMMER SCHOOL ON GROWTH AND FORMS 1985, eds H. E. Stanley and N. Ostrowsky.

[2] L. P. Kadanoff, PROCEEDINGS OF CARGESE SUMMER SCHOOL ON GROWTH AND FORMS 1985, eds H. E. Stanley and N. Ostrowsky. See also T. C. Halsey, M. H. Jensen, L. P. Kadanoff, I. Procaccia and B. I. Shraiman. Preprint.

[3] T. A. Witten and L. M. Sander, Phys. Rev. Lett. **47**, 1400 (1981).

[4] H. G. E. Hentshel and I. Procaccia, Physica D **8**, 835 (1983).

[5] L. deArcangelis, S. Redner and A. Coniglio, Phys. Rev. B **31**, 4725 (1985).

[6] T. C. Halsey, P. Meakin and I. Procaccia, to be published.

[7] P. Meakin, H. E. Stanley, A. Coniglio and T. Witten, Phys. Rev. A **32**, 2364 (1985), and to be published.

[8] C. Amitrano, A. Coniglio and F. diLiberto, to be published.

[9] L.A. Turkevich and H. Scher, Phys. Rev. Lett. **55**, 1026 (1985).

[10] L. Niemeyer, L. Pietronero and H. J. Wiesman, Phys. Rev. Lett. **52**, 1033 (1984).

[11] H. E. Stanley, J. Phys. A **10**, L211 (1977).

[12] A. S. Skal and B. I. Shklovskii, Sov. Phys. Semicond. **8**, 1029 (1975); P. G. deGennes, La Recherche **7**, 919 (1976).

[13] Y. Gefen, A. Aharony, B. B. Mandelbrot and S. Kirkpatrick, Phys. Rev. Lett. **47**, 1771 (1981).

[14] A. Coniglio, Phys Rev. Lett. **46**, 250 (1981); A. Coniglio, J. Phys. A **15**, 3824 (1982).

[15] R. Pike and H. E. Stanley, J. Phys. A **14**, L169 (1981); H.J. Herrmann and H. E. Stanley, Phys. Rev. Lett. **53**, 1121 (1984).

2-D DIELECTRIC BREAKDOWN BETWEEN PARALLEL LINES

Michael MURAT

School of Physics and Astronomy, Raymond and Beverly Sackler Faculty of Exact Sciences, Tel Aviv University, Tel Aviv 69978, Israel.

The dielectric breakdown model is applied to the simulation of the breakdown of a 2-D dielectric material placed between two lines held at a high potential difference. The resulting discharge patterns closely resemble diffusion limited depositions on surfaces. The resemblance is further confirmed by comparing the geometrical properties of the two structures.

The formation of electric discharge patterns in a material held at a high potential and grounded through a point electrode was recently modelled by Niemeyer, Pietronero and Wiesmann.[1] This dielectric breakdown model leads to the formation of open, tree-like structures that closely resemble diffusion limited aggregates (DLA).[2] The fractal dimensions of the structures generated by both models in 2-dimensions is found to be approximately 1.7. The similarity is justified by comparing the master equation for the DLA problem with the Laplace equation that governs the dielectric breakdown model.[3]

Both models were introduced in a circular geometry in which the seed (in the DLA model) or the grounded electrode (in the dielectric breakdown model) where the growth of the structures initiates, is a single point. DLA model was later applied to the case in which the seed is a plane (or a line in 2-D).[4] This is appropriate for simulating the formation of surface depositions by this model. The fractal dimension was found to be identical to that found in the original geometry. Other geometrical properties of the deposits, like the average height and root-mean-square (rms) thickness were also measured and found to have a scaling behavior with the number of particles in the deposits. The exponents were also reported.

In this paper, we apply the dielectric breakdown model to the case in which a 2-D dielectric material is placed between two lines held at two different potentials. A high potential difference leads to an electric discharge, which we simulate using the dielectric breakdown model. We compare the geometrical properties of the resulting patterns with those of diffusion limited depositions studies by Meakin.[4]

We review briefly the dielectric breakdown model adapted to this geometry. A square lattice represents the dielectric material. The sites at the top (bottom) row of the lattice are assigned a potential V = 1(0). The discharge propagates from the bottom to the top row in a stepwise manner. The discharged sites are assigned a zero potential. At each step the bond to be added to the discharge pattern is determined as follows: Laplace equation

$$\nabla^2 V = 0 \qquad (1)$$

is solved for each site on the lattice using the boundary conditions that V=1 for the top row and V=0 for the bottom row and all the discharged sites. The probability of choosing a given bond connecting a discharged site to one of its neighbors is proportional to the potential difference between the two. A bond is then chosen randomly from among all possible bonds using the above probability distribution. This process is repeated until the discharge pattern reaches the top row.

Such a discharge pattern is shown in Fig. 1. We calculated the fractal dimension of the patterns by first calculating $\rho(r)$, given by the average density of discharged sites at the rth row from the bottom one. This quantity is found to scale as

$$\rho(r) \sim r^{-\alpha} \qquad (2)$$

for rows not too close to the top row. An average of 10 simulations on lattices of size 100 x 100 gives $\alpha = 0.35 \pm 0.05$. The fractal dimension, d_f is then given by

$$d_f = 2 - \alpha \qquad (3)$$

from which we conclude that $d_f = 1.65 \pm 0.05$, in agreement with the DLA value.

We also calculated the behavior of the average and rms height and the rms thickness of the discharge patterns as a function of time steps, t (which is equal to the number of sites belonging to the discharge pattern). These were found to have a power law behavior for long enough times. From this behavior we calculated the exponents ε_i defined by

$$X_i \sim t^{\varepsilon_i} \qquad (4)$$

where X_i is any one of the above characteristics. The exponents were found to depend upon the size, L, of the lattice. Values calculated by extrapolation to infinite size (using 1/L as the extrapolation parameter) are listed in Table I, along with the DLA values.[4] The results are consistent with the suggestion

FIGURE 1
Discharge pattern on a 200 x 200 lattice. The apparently unconnected sections on the two sides of the lattice are actually connected through the use of periodic boundary conditions.

that both models lead to statistically identical structures.

TABLE I: Exponents ε_i for the DLA and the dielectric breakdown model.

Model \ Property	Rms-thickness	Average height	Rms-height
DLA[4]	1.36±0.05	1.45±0.05	-
Dielectric breakdown	1.3 ±0.1	1.45±0.1	1.4±0.1

In conclusion, we have performed simulations of the formation of discharge patterns in a 2-D material placed between two parallel equipotential lines held at a high potential difference. We have characterized the resulting discharge patterns through their geometrical properties. The results indicate that the patterns are statistically identical to diffusion limited depositions on surfaces.

ACKNOWLEDGEMENTS

The author wishes to thank Prof. Amnon Aharony for suggesting this study, and Prof. R. Englman and D. Stauffer for their illuminating remarks. This work was supported in part by grants from the Israel Academy of Sciences and Humanities, by the U.S.-Israel Binational Science Foundation (BSF) and from the Israel AEC Soreq Research Center.

REFERENCES

1. L. Niemeyer, L. Pietronero and H. J. Wiesmann, Phys. Rev. Lett., 52, (1984) 1033. We refer only to the $\eta=1$ case, where η is defined in the above reference.

2. T. A. Witten and L. M. Sander, Phys. Rev. Lett., 47, (1981) 1400.

3. L. Pietronero and H. J. Wiesmann, J. Stat. Phys., 36, (1984) 909.

4. P. Meakin, Phys. Rev., A27, (1983) 2616; P. Meakin, Phys. Rev., B30, (1984) 4207.

FRACTALS IN PHYSICS
L. Pietronero, E. Tosatti (editors)
© Elsevier Science Publishers B.V., 1986

DIELECTRIC BREAKDOWN IN THREE DIMENSIONS

S. SATPATHY

Max-Planck-Institut für Festkörperforschung, Heisenbergstrasse 1,
7000 Stuttgart 80, Federal Republic of Germany

Fractal characters of the dielectric breakdown patterns in three dimensions are studied within the Niemeyer-Pietronero-Wiesmann model by means of direct numerical simulation. Two cases are analyzed: (1) two-dimensional patterns in a three-dimensional Laplace field and (2) three-dimensional patterns in a three-dimensional Laplace field.

1. INTRODUCTION

The formation of a ramified pattern by the pathways of charge-flow during the process of breakdown of a dielectric is well-known through the familiar phenomenon of lightening during a thunderstorm.[1] However, that the pattern may be described as an object of non-integral dimensionality - a "fractal" - has been recognized only recently.[2-6]

Sawada et.al.[3] were perhaps the first to introduce a stochastic model to describe ramifications of dielectric discharge patterns by assigning arbitrarily a priority factor to the growth of tips already present in the growing pattern. This model however produces patterns with the fractal dimensionality[2,7] equal to the Euclidean one. This is not surprising since tips both in the interior of the pattern as well as on its periphery have within this model the same growth rate. Later Niemeyer, Pietronero and Wiesmann (NPW) introduced[4,5] a plausible, stochastic model in which the growth of the pattern depends explicitly on the local electric fields. The resulting patterns from numerical simulations turn out to be fractals and resemble experimental discharge patterns. The simulations performed by NPW were carried out for dielectric breakdown in two dimensions under a two-dimensional Laplace field. Here we present results of simulation for breakdown under a three-dimensional Laplace field. We consider both cases where the breakdown pattern is allowed to grow in three dimensions or where it is restricted to grow on a plane. Examination of the latter case was motivated by the fact that in many laboratory experiments the dielctric discharge pattern is restricted to grow in two dimensions while the field equations are three-dimensional.

2. THE NPW MODEL AND METHOD OF SIMULATION

We simulated dielectric breakdown on a cubic lattice within a sphere of diameter of about 120 lattice units. The surface of the sphere forms an equipotential electrode. The discharge pattern starts from the center and propagates by breakdown of a nearest-neighbour bond, one at a time, adjacent to the already discharged pattern. The probability of breakdown of a bond is proportional to the potential difference across the bond raised to a power η. Our simulations were done for integer values of η between 1 and 4. At each stage of the propagation of the discharge pattern the potentials at the lattice sites V_{ijk} were obtained by iteratively solving the Laplace equation, $\nabla^2 V = 0$, in its discretized form on the cubic lattice, viz.,

$$V_{ijk} = \frac{1}{6}[V_{i+1,j,k} + V_{i-1,j,k} + V_{i,j+1,k} + V_{i,j-1,k} + V_{i,j,k+1} + V_{i,j,k-1}].$$

At each step of the growth of the pattern the potential changes everywhere. However we computed the new potential by keeping the last potential fixed everywhere except in the immediate neighbourhood of the last discharged bond within a volume of 20x20x120 lattice units. Only every fifth step the potential was calculated everywhere inside the sphere. Such a procedure is computationally fast but has no noticeable effect on the fractal dimensionality of the breakdown patterns.

The fractal dimensionality D was obtained from the "mass-length" relation which has the power-law form,
$$N(r) \sim r^D,$$
where $N(r)$ is the total number of discharged bonds enclosed by a sphere of radius r. The linear portion of the ln N ~ ln r plot was least-square-fitted with a straight line the slope of which directly provides an estimate of D.

3. RESULTS

Figs. 1 and 2 respectively show typical computer-generated patterns for two-dimensional and three-dimensional dielectric breakdown. In both cases the Laplace field is three-dimensional. In our simulations we terminated the patterns generally after a size ~5000 or once the pattern touched the sphere boundary. The pattern size becomes smaller with larger values of η and the estimated value of D becomes less accurate. We have however included all results as these are the only available ones at present.

In Fig. 3 we have shown N(r) as a function of r at three different stages of growth of the pattern shown in Fig. 1(a). The fact that additional growth occurs in the exterior region of the pattern is indicated from this figure. A least-square fit provides a fractal dimensionality D = 1.93 for this pattern. A similar analysis for the pattern of Fig. 4(d) pertaining to a three-di-

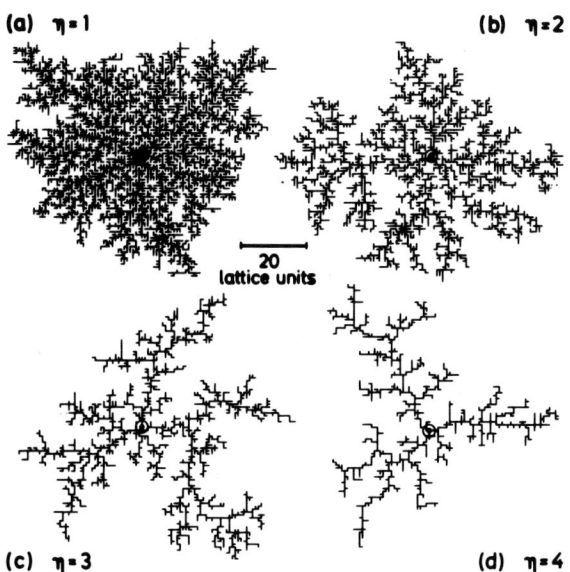

Fig. 1: Typical discharge patterns for two-dimensional dielectric breakdown under three-dimensional Laplace field for η = 1, 2, 3 and 4.

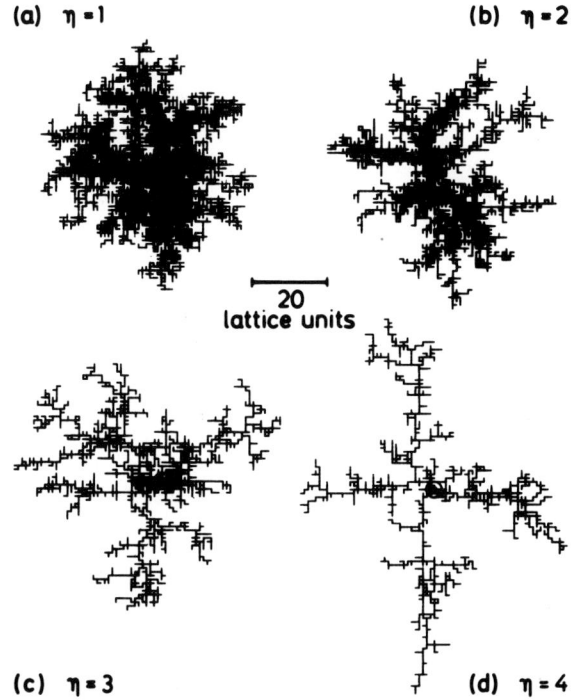

Fig. 2: Projection onto a plane of typical discharge patterns for three-dimensional dielectric breakdown under three-dimensional Laplace field for η = 1, 2, 3, and 4.

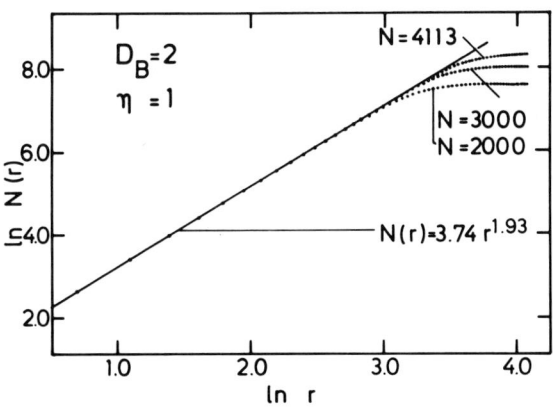

Fig. 3: Total number of discharged bonds N(r) within a sphere of radius r centered at the origin at various stages of growth for the pattern shown in Fig. 1(a). N denotes the total size of the pattern at successive stages of growth. D_B denotes the spatial dimension in which the discharge pattern is allowed to grow.

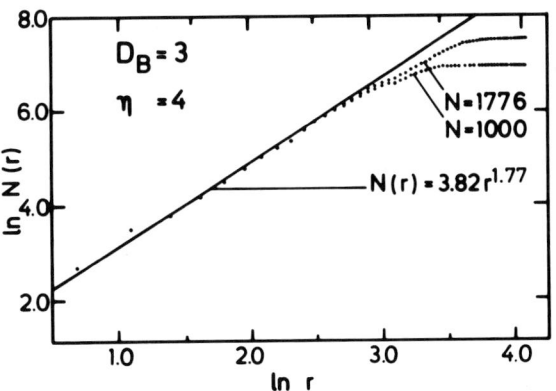

Fig. 4: Same as Fig. 3 for the discharge pattern shown in Fig. 2(d).

mensional breakdown (D_B = 3) under a three-dimensional Laplace field and with η=4 is shown in Fig. 4.

The estimated fractal dimensionali-

Table I: Summary of fractal dimensionality of dielectric breakdown patterns in the NPW model.

D_V/D_B^+	η	$\bar{N}$ %	D
3/3	0	-	3
	1	5000	2.48±0.06
	2	5000	2.11±0.06
	3	2600	1.96±0.08
	4	1750	1.75±0.06
3/2	0	-	2
	1	4500	1.96±0.03
	2	2150	1.82±0.05
	3	1300	1.78±0.06
	4	850	1.69±0.08
2/2*	0		2
	0.5		1.89±0.01
	1	5000	1.75±0.02
	2		~1.6

+ D_V = Dimensionality of Laplace field equation
D_B = Spatial dimension in which the breakdown pattern grows
% $\bar{N}$ = Average size of the computer-generated patterns
* Results for this case are quoted from Ref. 4.

ties and the average pattern sizes are summarized in Table I. These results were determined by averaging over 5-10 different patterns. The estimated errors are the root-mean-square deviations from the mean and do not incorporate possible errors due to finite size of the clusters.

For the value $\eta = 0$ there is no effect of electric field on the growth of the discharge pattern and the model is equivalent to the model of Sawada et. al.[3] with tip priority factor equal to one, i.e., with no priority given to the growth of tips. The resulting pattern has the Euclidean dimensionality. As might be intuitively expected the fractal dimensionality decreases uniformly with increasing value of η as is seen from Table. I. For the value $\eta=1$, i.e., when probability of discharge of a bond is directly proportional to the local electric field, the NPW model is similar to the diffusion limited aggregation[8] (DLA) for which extensive numerical simulations have been performed in the recent past.[8-10] DLA simulations in three dimensions by Richter et. al.[10] provide a value of $D = 2.39 \pm 0.2$. Meakin[9] obtained a more accurate value of $D = 2.51 \pm 0.06$ for the same case. This is to be compared with our value of 2.48 ± 0.06 as seen from Table I.

ACKNOWLEDGEMENT

I thank P.J. Kelly for stimulating discussions.

REFERENCES

1. R. P. Feynman, R. B. Leighton, and M. Sands, <u>Lectures on Physics</u>, Vol. II (Addison-Wesley, London, 1964).

2. B. Mandelbrot, <u>The Fractal Geometry of Nature</u> (Freeman, New York, 1982).

3. Y. Sawada, S. Ohta, M. Yamazaki, and H. Honjo, Phys. Rev. A<u>26</u> (1982) 3557.

4. L. Niemeyer, L. Pietronero, and H. J. Wiesmann, Phys. Rev. Lett. <u>52</u> (1984) 1033.

5. L. Pietronero and H. J. Wiesmann, J. Stat. Phys. <u>36</u> (1984) 909.

6. H. J. Wiesmann, C. Evertsz, and L. Pietronero, this volume; M. Murat, this volume.

7. F. Hausdorff, Math. Ann. <u>79</u> (1919) 157; A. S. Besicovitch, Math. Ann. <u>110</u> (1935) 321.

8. T. A. Witten and L. M. Sander, Phys. Rev. Lett. <u>47</u> (1981) 1400; Phys. Rev. B<u>27</u> (1983) 5686.

9. P. Meakin, Phys. Rev. A<u>27</u> (1983) 1495; Phys. Rev. A<u>27</u> (1983) 604.

10. R. Richter, Z. M. Cheng, L. M. Sander, and T. A. Witten, Bull. Am. Phys. Soc. <u>28</u> (1983) 261.

THE FRACTAL NATURE OF FRACTURE

E. LOUIS*, F. GUINEA+ and F. FLORES+

* Departamento de Física, Universidad de Alicante, Apdo.99, 03080 Alicante, Spain
+ Departamento de Física del Estado Sólido, Universidad Autónoma, Cantoblanco, 28049 Madrid, Spain

Fracture patterns in materials are studied by means of a model which incorporates the equations of elasticity and simple rules of fracture propagation. Different boundary conditions are considered. Self-similar patterns with fractal dimensions around 1.6 are obtained.

1. INTRODUCTION

A wide variety of patterns derived from different growth processes has been shown to have a fractal structure[1-3]. Recently, semiempirical arguments and several measurements have suggested that fracture surfaces in metals may also have a fractal character[4]. Although several theoretical attempts to explore whether fracture patterns have a fractal geometry are being made[5,6], there is a lack of a basic understanding of the problem at least at the level recently achieved for dielectric breakdown[2] (DB) and diffusion limited aggregation[3] (DLA).

In this paper a model which simulates the propagation of fracture in an otherwise perfect mono- or polycrystal (see below) is presented. Fracture phenomena are one of the most intringuing processes in materials science[7], in which many factors, such as material nature, impurities, defects, internal and phase boundaries, temperature, etc., play a crucial role, leading to a large family of fracture mechanisms[7] ranging from ductile fracture to cleavage. Despite this complex phenomenology, it is worthwhile to start with a simple model in which a study of its eventual fractal nature might be feasible.

The results of the present analysis indicate that fracture patterns generated in our simple model have in fact a fractal character, encouraging further work on more realistic systems.

2. MODEL AND METHOD OF CALCULATION

We start from the continuum elastic equations[8],

$$(\lambda+\mu)\partial_i(\sum_j \partial_j u_j) + \mu(\sum_j \partial_j^2)u_i = 0, \qquad (1)$$

where λ and μ are the Lamé coefficients, ∂_i the partial derivative with respect to component i of the r vector and u_i are the components of the strain field. These equations derived within the harmonic approximation are otherwise completely general and allow to study different isotropic materials by varying the ratio between the Lamé coefficients. Different boundary conditions can also be imposed: shear deformation, uniaxial compression, uniform compression, ... The next step is to assume that the probability for a given piece of the sample to break is proportional to a power, η, of the stress acting on it. In this work $\eta=1$ will be used.

We have discretized eqs.(1) by using a triangular lattice with springs joining the nearest neighbour lattice sites. The model can either simulate a monocrystal, in which case springs account for interatomic forces, or a polycrystal whereby the springs describe the grain boundary strength. The continuum limit of this system is characterized by $\lambda=\mu$. Although in this paper we only present results for this simple case, more general cases can also be studied using a more complex unit cell. For instance, a $\sqrt{3}\times\sqrt{3}$ reconstruction can describe an isotropic system with a range of values for the ratio λ/μ, whereas anisotropy can be studied by considering a 2×1 reconstruction.

The discretization of eqs.(1) in the triangular lattice can be easily carried out by using as basis vectors the two lattice vectors $\underline{a}_1=a(1,0)$ and $\underline{a}_2=a(-1/2, \sqrt{3}/2)$. With this basis the equations of motion can be easily written as

$$4u_i(l,m) - u_i(l+1,m) - u_i(l-1,m) -$$
$$u_i(l+1,m+1) - u_i(l-1,m-1) + 2u_j(l,m) -$$
$$u_j(l+1,m+1) - u_j(l-1,m-1) = 0 , \quad (2)$$
$$(i,j = 1,2)$$

where $u_i(l,m)$ is the component i (along the basis vector $\underline{a}_i$) of the displacement vector at the lattice point (l,m). This set of equations are solved iteratively through the whole lattice. Lattices with sizes up to 100×100 were studied. The chosen boundary conditions were uniform expansion and shear deformation. A practical way to impose the boundary conditions through the fracture process is to keep fixed the lattice points at the outermost shell. We start with a lattice whose bounds are stretched by 10% over their equilibrium length. Then a bond is broken and the lattice is allowed to relax to the new equilibrium state using eqs. (1). A bond neighbouring the broken bonds is subsequently broken at random and proportionally to the stress accumulated on it. These steps are repeated until a reasonable number of bonds is broken. We have built up fracture patterns with up to 1200 broken bonds. This number is lower than those usually achieved in DB[2] and DLA[3]. In this concern it should be noticed that

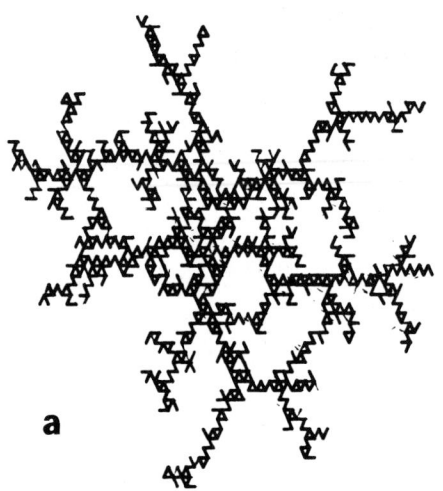

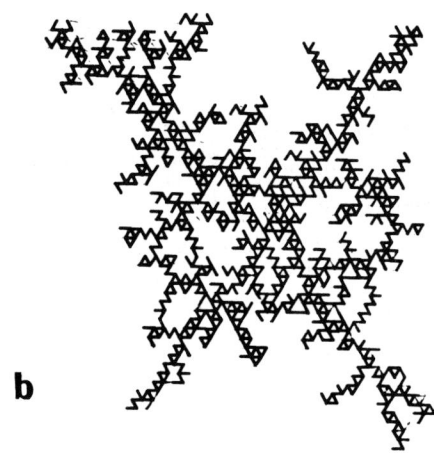

FIGURE 1
Broken bonds: a) expansion, b) shear.

although some similarities between DB and the present work can be found, two main differences make the latter much more time consuming: i) eqs.(2) involve a vectorial field instead of the scalar electrostatic potential[2], and ii) in the present case the whole network, even within the broken zone, relaxes, whereas in DB or DLA

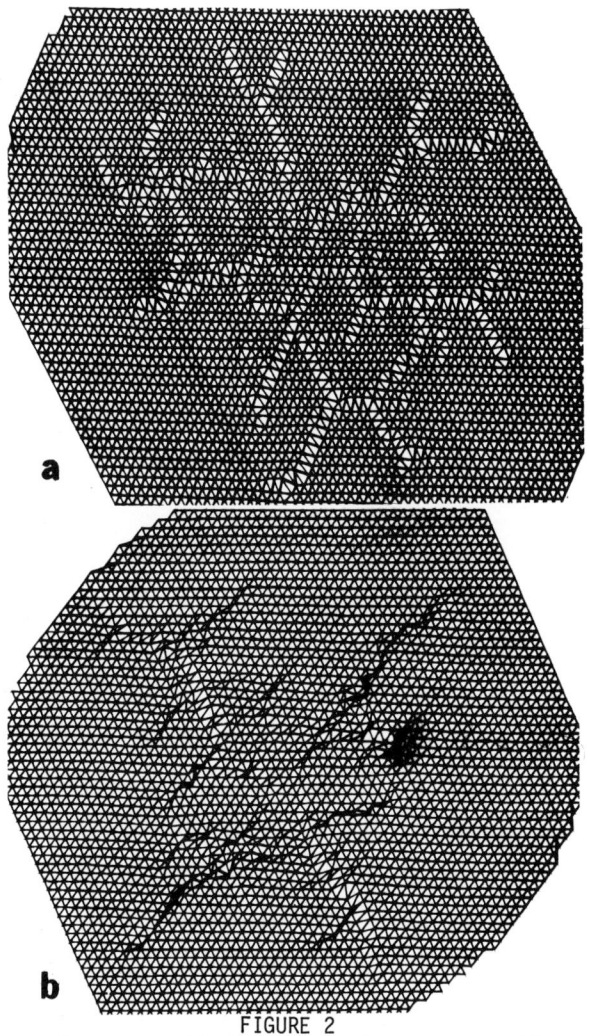

FIGURE 2
Displacements of the atoms: a) expansion, b) shear

the dynamics of the region inside the pattern is frozen.

3. RESULTS AND DISCUSSION

Figs.1a-b illustrate the type of fracture patterns obtained with uniform expansion and shear, respectively. The patterns turn out to be self-similar with a fractal dimension of 1.62±0.05 for expansion, and 1.64±0.05 for shear. The positions of the lattice points associated with the fracture patterns of Figure 1 are shown in Figure 2. The following features should be outlined: i) the lattice relaxes within the region of broken bonds, ii) large deformations are noticed at the active ends of the pattern, and iii) the symmetry imposed by the boundary conditions is clearly visible in the figures, in agreement with a general result which indicates that the strain field for an isotropic medium depends on boundary conditions although not on the elastic constants[7,8].

One of the most appealing results described in the previous paragraph is that fractal dimensionality does not strongly depend on boundary conditions despite the fact that the strain field does. When comparing the fractal dimensions above with those obtained in diffusion limited aggregation and dielectric breakdown in two dimensions, we notice that they are all very similar[2,3].

In conclusion, we have studied the propagation of fracture in a perfect triangular lattice of nearest neighbours springs. The fracture patterns so generated turned out to have fractal character. This could be the beginning of a series of increasingly sophisticated studies on fracture patterns which may help, by introducing this new geometry[1], in the understanding of fracture phenomena in materials.

ACKNOWLEDGEMENTS

Part of the calculations presented in this work were carried out at the Laboratorio de Química Cuántica of the Universidad de Alicante; thanks are due to the Head of the group (Prof.F.Moscardó) for making its computer facilities available to us.

REFERENCES

1. B.B. Mandelbrot, The Fractal Geometry of Nature (W.H.Freeman, San Francisco, 1983)

2. L. Niemeyer, L. Pietronero and H.J. Wiesmann, Phys.Rev.Lett. 52 (1984) 1033, and this Conference

3. T.A. Witten and L.M. Sander, Phys.Rev.B 27 (1983) 5686, and this Conference

4. B.B. Mandelbrot, D.E. Passoja and A.J. Paullay, Nature 308 (1984) 721

5. Y. Termonia and P. Meakin, The Formation of Fractal Cracks in a Kinetic Fracture Model (Preprint)

6. C.W. Lung, these Proceedings

7. R.A. Smith, Ed., Fracture Mechanics Current Status, Future Prospects (Pergamon Press, Oxford, 1979)

8. L.D. Landau and E.M. Lifshitz, Theory of Elasticity (Pergamon Press, Oxford, 1975)

PATTERN FORMATION OF DENDRITIC FRACTALS IN FRACTURE AND ELECTRIC BREAKDOWN

Hideki TAKAYASU

Department of Physics, Faculty of Science, Kyoto University, Kyoto, 606, Japan

By introducing a simple network model, growth process of cracks and electric breakdown patterns is analysed numerically. In both cases, random dendritic fractals are created by a deterministic evolution rule with randomness added only initially. Some resemblance and difference are clarified between these two phenomena.

1. INTRODUCTION

Pattern formation of random fractals may be one of the most interesting problems in non-equilibrium statistical physics now. D.L.A.[1] (diffusion limited aggregation) and electric breakdown patterns[2] have been succesfully simulated by simple models. In these models, random patterns grow stochastically with the probability that is closely related to the solution of Laplace equation. It is found that the random growth with screening effect makes the pattern fractal[3].

Contrary to these stochastic models, I have recently proposed a deterministic model of electric breakdown[4]. In my model, initial small fluctuation of a field variable is enhanced deterministically and random fractal patterns are formed in result. This may indicate that the random growth process is not essential for fractal pattern formation.

In the following section, my electric breakdown model is reviewed briefly. We discuss about brittle fracture of a thin plate in the third section. It is shown that my model of electric breakdown becomes a model of fracture by changing only one parameter.

2. ELECTRIC BREAKDOWN

Let us consider the elementary process of electric breakdown, first. Here, we take the standpoint that the elementary process of electric breakdown is modeled by the following nonlinear and irreversible characteristics of conductivity. Assume that a constant voltage V is applied to a resistor. when V is smaller than a certain critical voltage V_c, the conductivity of the resistor is G. Once V exceeds V_c, then the resistor is considered to be broken and the conductivity is enhanced to λG ($\lambda \gg 1$). This parameter λ denotes the ratio of increase of the conductivity and in an ideal case $\lambda = \infty$. After the breakdown, the resistor keeps the enhanced conductivity regardless of the magnitude of V (see Fig. 1).

Next, we consider a square network of such resistors (see Fig.1 in ref.4). In this network, the electric potential u satisfies the following equation for any lattice point (i,j) :

$$\sum_{k=1}^{4} G_k(i,j)\{u_k(i,j)-u(i,j)\} = 0 \quad (1)$$

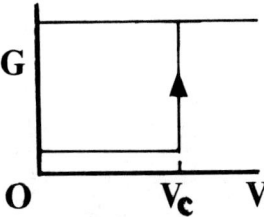

FIGURE 1
The response of the conductivity G with respect to the voltage V.

where the suffix k indicates the direction (see Fig.2) and $G_k(i,j)$ denotes the conductivity of the corresponding resistor. This equation is nothing but a combination of Ohm's law and Kirchhoff's point loop rules. In the case that a lattice point is located on a boundary, we have to put $G_k = 0$ for a missing resistor. For arbitrarily given $\{G\}$ and appropriately given boundary condition of $\{u\}$, eq.(1) for all combination of (i,j) makes a set of linear equation for $\{u\}$, and it can be solved numerically.

Time evolution of this system is defined by the following procedure:

1. Give $\{G\}$ and a boundary condition of $\{u\}$.
2. Solve $\{u\}$ by eq.(1).
3. Check every resistor (except already broken ones). If the broken condition, $|u_k(i,j) - u(i,j)| > V_c$, is satisfied, then let $G_k(i,j) = \lambda G_k(i,j)$.
4. Stop, if no resistor has newly broken in the preceeding procedure. Otherwise, go back to procedure 2 and continue the routine.

It may be obvious from this procedure that we evolve the system by solving eq.(1) and by checking the breakdown condition, repeatedly.

An example of evolution on a 10×10 network is shown in Fig.3. Here, the boundary condition is

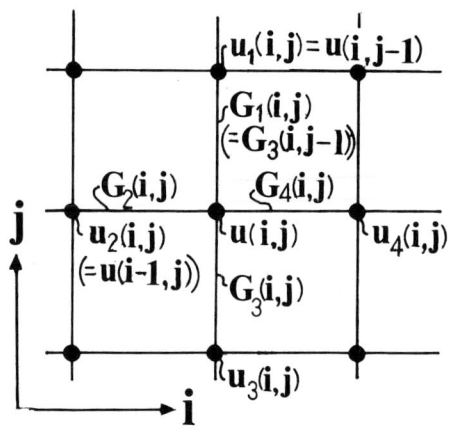

FIGURE 2
The meaning of $u_k(i,j)$ and $G_k(i,j)$. The suffix k denotes up, left, down and right for $k = 1,2,3,4$.

given by

$u(i,0) = 0$ and $u(i,10) = U$ $(i = 1,2,\cdots,10)$ (2)

where U is chosen to be sufficiently large so that at least one resistor breaks, and initial value of $\{G\}$ are given randomly as

$$G_k(i,j) = \bar{G} + G^* Z \qquad (3)$$

where $\bar{G}$ and G^* are constants and Z is a random number distributed uniformly on $[0,1]$. This evolution may be regarded as a kind of dynamical phase transition, that is, a phase transition from an insulated state to a conductive state where the time step acts as a control parameter.

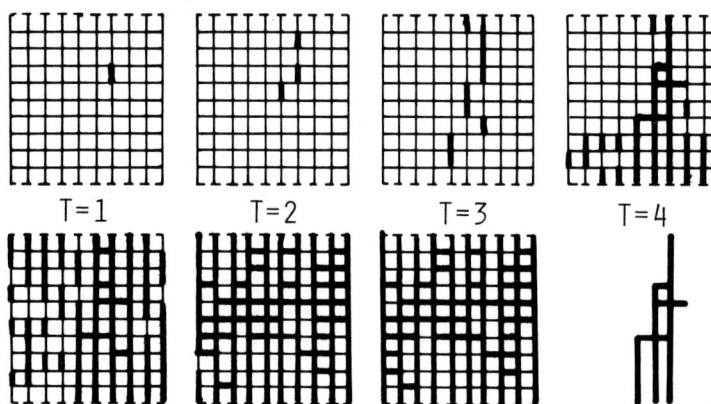

FIGURE 3
An evolution of electric breakdown pattern on a 10 × 10 network. The last figure shows the pattern of the percolation cluster at T = 4. The parameters are $\bar{G} = 2.0$, $G^* = 2.0$, $V_c = 1.0$ and $\lambda = 100$.

The critical point of this phase transition should be the time at which a percolation cluster of broken resistors first appears ($T=4$ in Fig.3). It has been confirmed that critical percolation clusters of this system are fractals with their dimension about 1.6^4.

Although the time step in this model is not a real time, it may be a quite interesting problem to estimate quantitatively the growth rate of the breakdown. By examining 32×32 networks, the following exponential growth was found:

$$N(T) \propto \exp(bT) \quad , \quad b = 0.85 \pm 0.03 \qquad (4)$$

where $N(T)$ denotes the averaged number of broken resistors at the time step T. This relation is vallid for the time step less than the percolation critical time.

The irreversible response of conductivity plays an essential role in this fractal pattern formation. This can be recognized by considering the case that the conductivity has reversible nonlinear character, that is, a step-like response to the voltage. In such cases, percolation phase transitions do not take place and no fractal structure can be found, instead, oscillations generally remain. This is a noteworthy feature of this electric system.

3. FRACTURE

In this section, the story develops quite parallel with that in the preceeding section.

First, let us investigate the elementary process of fracture. Assume the situation that one end of a thin brittle stick is fixed while the other end is free. If the displacement of the free end, d, is less than a critical value, d_c, then its modulus of rigidity is a constant, G. Once d exceeds d_c, the stick may be broken and G may suddenly be reduced to a very small value, εG ($0 \leq \varepsilon \ll 1$), as shown in Fig.4.

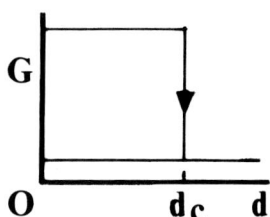

FIGURE 4
The response of the modulus of rigidity G with respect to the displacement d.

Next, we consider a plane square net consisting of such brittle sticks. If we assume the case that displacements at its lattice points are all perpendicular to the plane (anti-plane shear problem), then the equation for the balance of forces conforms to eq.(1), in which $\{u\}$ and $\{G\}$ denotes the displacements and the moduli of rigidity, respectively. This coincidence is not an accident. In a continuum limit, eq.(1) becomes

$$\vec{\nabla} \cdot (G \vec{\nabla} u) = 0 \qquad (5)$$

if G and u are sufficiently smooth. This is a fundamental conservation law and is one of the most popular equations in physics. (Laplace equation is obtained for the special case $G = $ const.)

Time evolution of this brittle net can be carried out by the procedure in the preceeding section if we replace V_c and λ by d_c and ε, respectively. The boundary condition (2) and the initial condition (3) are also available in this case. This boundary condition corresponds to the physical situation that top edge of the square brittle plate is pulled up while the bottom edge is fixed. Owing to these similarities between electric breakdown and brittle fracture, practically, the simulation of electric breakdown turns out to be a simulation of brittle fracture by only changing the parameter λ from the large number to a very small one.

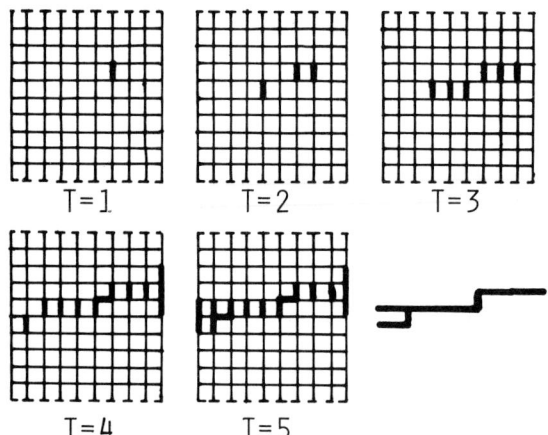

FIGURE 5
An example of evolution of brittle fracture. All conditions except λ are identical to those in Fig.3. Here $\lambda = 0.01 (= \varepsilon)$. The last figure shows the percolation crack at $T = 5$.

In Fig.5, the evolution of fracture under the same condition with the case of Fig.3 is shown. Although the breakbown pattern in Fig.3 grow vertically, in this case it grows horizontally. And the growth stops when the fracture pattern forms a percolation cluster. It may be mathematically quite interesting that these differences originate only in the parameter λ.

Fractal properties of these percolation clusters have been examined and the fractal dimension is estimated as

$$D = 1.65 \pm 0.05 \qquad (6)$$

The averaged number of broken sticks, $N(T)$, follows a power law in this case:

$$N(T) \propto T^a , \quad a = 2.4 \pm 0.2 . \qquad (7)$$

This power law may indicate that the crack growth process has fractal properties not only in its spatial patterns but also in the temporal behaviours.

Contrary to the electric breakdown, the irreversible response is not essential in the fracture. Cracks grow almost identically even if the response of rigidity is replaced by reversible one. This difference can be understood partially if we pay attention to the total energy of the systems. We can find that the total enrgy decreases with time in the fracture while it increases in the electric breakdown. Namely, we may say that the more cracks grow, the more the system is stabilized. Details about the fracture will be published elsewhere[5].

ACKNOWLEDGEMENT

This work was partially supported by the Grant-in-Aid for Research from the Ministry of Education and by the Committee on the Educational Project for Japanese Mathematical Scientists.

REFERENCES

1. T.A. Witten, Jr., and L.M. Sander, Phys. Rev. Lett. 47 (1981), 1400.

2. L. Niemeyer, L. Pietronero and H.J. Wiesmann, Phys. Rev. Lett. 52 (1984), 1033.

3. P. Meakin, Phys. Rev. B28 (1983), 6718; L. Pietronero and H.J. Wiesmann, J. Stat. Phys. 36 (1984), 909.

4. H. Takayasu, Phys. Rev. Lett. 54 (1985), 1099.

5. H. Takayasu, submitted to Prog. Theor. Phys.

COLLAPSE OF LOADED FRACTAL TREES

Sara A. SOLLA

IBM Thomas J. Watson Research Center, Yorktown Heights, New York 10598, USA.

A fractal tree with a statistical distribution of leg strengths provides a hierarchical model for the failure of a structure under an externally applied load. Transfer of load from failed legs to adjacent unbroken regions induces further failure and results in catastrophic collapse of the structure at a critical value of the applied load.

The failure of materials and structures under applied load is a subject of both practical importance and conceptual difficulty. Usual treatments based on continuum elasticity theory do not provide simple tools for discussing the essential nonlinearities of the problem. The purpose of this paper is to propose an alternative approach: a statistical description of the propagation of a region of failure[1]. A loaded fractal tree is used to introduce a scale invariant mechanism for transfer of load which results in catastrophic failure when a critical threshold of the applied load is reached[2].

Consider a simple fractal tree with two legs at an angular separation θ coming down from each node. A vertical load V is applied to the $n = 0$ node at the top. There are 2^n legs at the nth level, connecting each node of order $n - 1$ to two nodes of order n. The height of the nth level is $h_n = h_1 / 2^{n-1}$, a factor of two smaller than that of level $(n - 1)$. The total height of the tree is $H = \sum_{n=1}^{\infty} h_n = 2 h_1$. In the absence of failure, the load on each leg at the nth level is $V_n = V \cos(\theta/2) / 2^n$.

Fluctuations in sizes, shapes, and other internal parameters in the legs of a real structure result in a statistical distribution of values for the failure strength of the nth level legs, V_{nf}. Such variety is modelled here through a Weibull distribution[3]. The failure probability p_n that the load V_n does exceed the failure strength V_{nf} is given by

$$p_n = \text{Prob} (V_{nf} \leq V_n) = 1 - \exp [- (V_n/V_{on})^m], \quad (1)$$

where V_{on} is a reference strength for the nth level legs and the integer m is the order of the distribution. Weibull distributions are often used to represent statistical fluctuations of failure strengths[4].

An essential feature of the model is a scale invariant mechanism for transfer of load: if one nth level leg fails, the load on that leg is transferred to the adjacent nth level leg coming down from the same node of order $n - 1$. The second leg in the pair may suffer an induced failure described by the conditional probability $P_n^{2,1}$ that $V_{nf} \leq 2V_n$ given that $V_{nf} > V_n$. Such conditional probability is easily calculated for a Weibull distribution of the form (1),

$$P_n^{1,2} = 1 - (1 - p_n)^{2^m - 1}. \quad (2)$$

The failure of a pair of nth level legs removes all support from the $n - 1$ order node where they originate, and implies the failure of the $(n - 1)$th level leg going into that node. The probability of failure at the $(n - 1)$ level is then given by

$$p_{n-1} = p_n^2 + 2 p_n (1 - p_n) P_n^{1,2}. \quad (3)$$

The first term describes a direct process due to the failure of both nth level legs in the pair, and the second term describes induced failure and arises when only one of the two nth level legs in a pair fails directly. Substituting (2) into (3) leads to the recursion relation

$$p_{n-1} = 2 p_n [1 - (1 - p_n)^{2^m}] - p_n^2. \quad (4)$$

which describes the propagation of failure up the tree, and is applicable at all levels of the structure.

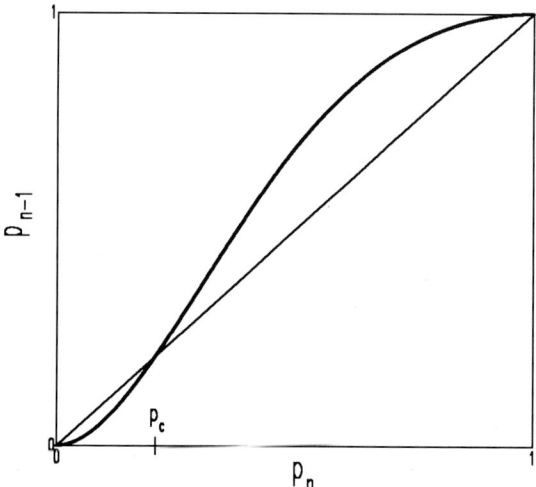

FIGURE 1
Recursion relation for propagation of failure up the tree from equation (4) for the failure probability of equation (1). The curve shown is for $m = 2$.

The functional dependence of p_{n-1} on p_n is shown for $m = 2$ in Figure 1. The fixed point equation $p_{n-1} = p_n$ is satisfied by $p_n = 0$, 1, and p_c. The critical point at $p_c = 0.2063$ describes a transition from a regime in which failed regions remain bound ($p_n < p_c$), into a regime in which the failed regions grow indefinitely and the structure cannot support the external load ($p_n > p_c$). As a specific example, the iteration of the recursion relation (4) with $m = 2$ takes $p_n = 0.1$ into $p_{n-1} = 0.0588$, $p_{n-2} = 0.0218$, $p_{n-3} = 0.0032$. The probability of failure approaches zero as n decreases, and the structure is stable. But for $p_n = 0.6$, the sequence is $p_{n-1} = 0.8093$, $p_{n-2} = 0.9615$, $p_{n-3} = 0.9985$. The probability of failure approaches one as n decreases towards the top of the tree, and the structure collapses.

The catastrophic failure of the structure at p_c is due to a sudden growth of the failed regions as the applied load is increased. This behavior can be characterized by a propagation length L which measures the maximum number of levels that a given fracture can propagate up the tree. The propagation length is finite below p_c, and diverges with a critical exponent ν as p_c is approached from below. For the $m = 2$ case of Figure 1, the collapse of the structure occurs at $p_c = 0.2063$, with $\nu = 1.439$. The corresponding critical load follows from (1), $V_c = 0.4807 V_{on}$, and is considerably less than the mean strength of an nth level leg, $\bar{V} = 0.8862 V_{on}$.

The features of the solution for $m = 2$ discussed above are quite general. As m increases, the distribution (1) of failure strengths becomes narrower, and all legs fail at $V_n = V_{on}$ in the limit $m \to \infty$. For any finite m, however large, the recursion relation (4) yields an S-shaped curve qualitatively similar to that of Figure 1. The critical point at

$$p_c = 1 - (1/2)^{1/(2^m - 1)} \quad (5)$$

corresponds to a critical load

$$V_c = [\ln 2 / (2^m - 1)]^{1/m} V_{on}. \quad (6)$$

Although the critical probability for failure rapidly decreases with increasing m, the failure load is almost independent of m and remains a factor of two smaller than the mean strength

$$\bar{V} = (1/m)! \, V_{on}. \quad (7)$$

The critical exponent ν is almost independent of m for $m > 5$. Results for different values of m are summarized in Table 1.

TABLE 1: Critical parameters

m	p_c	V_c/V_{on}	$\bar{V}/V_{on}$	ν
2	0.2063	0.4807	0.8862	1.439
3	0.0943	0.4626	0.8930	1.368
5	0.0221	0.4676	0.9182	1.328
10	.7 x 10^{-3}	0.4821	0.9513	1.317
20	.7 x 10^{-6}	0.4909	0.9735	1.316
∞	0	0.5	1	1.316

The hierarchical model leading to the recursion relation (4) for the propagation of failure predicts the collapse of the structure through the divergence of a characteristic propagation length. The basic mechanism for this catastrophic behavior is induced failure due to transfer of load. The failure load for the system is almost independent of the spread of failure strengths in the subsystems and lower than their characteristic strength. These results may be applicable to a variety of problems in which the propagation of failure occurs in a hierarchical fashion, due to a correlation between the size of the failed region and the size of the adjacent unbroken region affected by load redistribution.

An interesting possibility is that of preventing the collapse of the structure by incorporating a fraction of very strong legs. Consider a distribution of failure strengths with two characteristic reference strengths, V_{on} and γV_{on}, with $\gamma > 1$. The failure probability for the nth level legs is

$$p_n = \text{Prob}(V_{nf} \leq V_n) = (1 - \alpha)\{1 - \exp[-(V_n/V_{on})^m]\}$$
$$+ \alpha\{1 - \exp[-(V_n/\gamma V_{on})^m]\}. \quad (8)$$

Two additional parameters have been introduced: the fraction α of strong legs and the ratio γ of the characteristic strength of the strong legs to that of the weak legs. The previously considered distribution (1) is recovered for $\alpha = 0$. For $\alpha \neq 0$ the conditional probability $P_n^{1,2}$ can no longer be expressed in terms of p_n in a simple, analytic form. The numerical investigation of the resulting recursion relation as a function of α, γ, and m reveals a regime characterized by two critical points, as illustrated in Figure 2 for $\alpha = 0.15$, $\gamma = 10$, and m = 2.

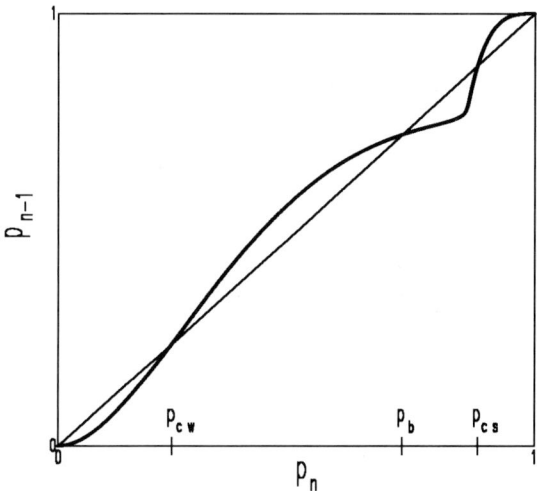

FIGURE 2
Recursion relation for propagation of failure up the tree for the bimodal failure probability of equation (8). The curve shown is for $\alpha = 0.15$, $\gamma = 10$, and m = 2.

The lower critical point at $p_{cw} = 0.2343$ ($V_{cw} = 0.5671 V_{on}$) corresponds to the collapse of the weak legs. A higher threshold at $p_{cs} = 0.8809$ ($V_{cs} = 4.8067 V_{on}$) has to be reached for the collapse of the structure to occur with certainty. The intermediate range $p_{cw} < p_n < p_{cs}$ is controlled by the fixed point at $p_b = 0.7205$. Values of p_n in this range generate a sequence of failure probabilities that approaches p_b as n decreases towards the top of the tree. The value of p_b thus determines the efficiency of the containment mechanism provided by the strong legs. At a fixed value of γ, p_b decreases from one as the fraction α of strong legs increases from zero. The fixed points at p_b and p_{cw}

move towards each other with increasing α, until they merge at $\alpha = \alpha_c$ and disappear. For $\alpha > \alpha_c$ there is no critical point associated with the weak legs and the collapse of the structure is controlled solely by the strong legs. The range $0 < \alpha < \alpha_c$ is characterized by the presence of two critical points at p_{cw} and p_{cs}. For the m = 2, γ = 10 case of Figure 2, $\alpha_c \simeq 0.29$.

To conclude, a loaded fractal tree provides a simple model to investigate the failure of a structure under an externally applied load. The bimodal failure probability considered in equation (8) illustrates one of the possible extensions of the model to incorporate features relevant to the physical systems to which these ideas apply.

ACKNOWLEDGMENTS

This work was done in collaboration with R.F. Smalley, Jr. and D.L. Turcotte, of the Department of Geological Sciences at Cornell University. I thank them both for a very enjoyable interaction.

REFERENCES

1. R.F. Smalley Jr., D.L. Turcotte, and S.A. Solla, J. Geophys. Res. 90 (1985) 1894.

2. D.L. Turcotte, R.F. Smalley Jr., and S.A. Solla, Nature 313 (1985) 671.

3. W. Weibull, J. Appl. Mech. 18 (1951) 293.

4. D.G. Harlow and S.L. Phoenix, Adv. Appl. Probab. 14 (1982) 68.

FRACTALS AND THE FRACTURE OF CRACKED METALS

C.W. LUNG

International Centre for Theoretical Physics, Trieste, Italy and Institute of Metal Research, Academia Sinica, Shenyang, People's Republic of China *

A fractal model for intergranular brittle and ductile fracture surfaces of cracked metals is suggested. It is shown that for small grain size metals, the critical crack extension-force would rise rapidly and faster than the Hall-Petch's relationship due to the increase of the true areas of the irregular fracture surfaces.

1. INTRODUCTION

The fracture surface formed after breaking off is rough and irregular. Mandelbrot et al[1,2] showed that the structure of fracture surfaces of metals was modelled very well by a fractal surface, though metal fractures are only extremely crinkly (down to the limits of their microstructural size range), while fractals are infinitely crinkly. Their experiments in metal fracture showed that the fractal dimension D was very well defined for different specimens of the same metal having similar thermomechanical treatments.

We think that the sizes and orientations of grains in many polycrystalline metals are "irregular" (the distribution of impurities, defects and other internal stress sources are also "irregular"). These may be the physical foundation for may metal fracture surfaces being successfully modelled by fractals.

In this paper, we analyze the grain size effects on the fracture of metals with fractal models.

2. THE CRITICAL CRACK EXTENSION FORCE

On the Griffith theory for perfectly elastic fractures[3], it would have to exceed the work needed to separate the two surfaces, $2\gamma_s$. In Irwin's approach in fracture mechanics, the critical strain energy release rate, i.e. the critical crack extension force, $G_{crit.}$[3] may be written as

$$G_{crit.} = 2\gamma_s \quad \text{(for brittle fracture)} \quad (1)$$

and

$$G_{crit.} = 2\gamma_s + \gamma_p \quad \text{(for quasi-brittle fracture)} \quad (2)$$

where γ_p represents the energy expended in the plastic work necessary to produce unstable crack propagation at the crack tip.

Unlike in glass, the fracture surfaces in metals are rough and irregular. The true areas of the fracture surfaces in metals are actually larger than the data got by macroscopic measurements. The area of the fracture surface per unit thickness of specimen would be $[L(\epsilon)/L_0(\epsilon)] \cdot 1$ (in fracture mechanics, we always simplify the crack as a line in a two-dimensional system). Then, instead of Eq.(1) and Eq.(2), we have (Fig.1a,b).

$$G_{crit.} = 2(L(\epsilon)/L_0(\epsilon))\gamma_s \quad (3)$$

and

$$G_{crit.} = 2(w-a)^{-1}[L_1(\epsilon_1)+L_2(\epsilon_2)]\gamma_s + \gamma_p \quad (4)$$

where ϵ_i's are the yardstick lengths. Other parameters have been shown in Figs.1.

---

* Permanent address.

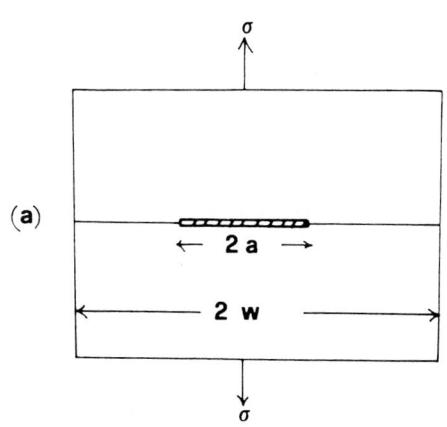

FIGURE 1a
Ideal brittle fracture in glass.

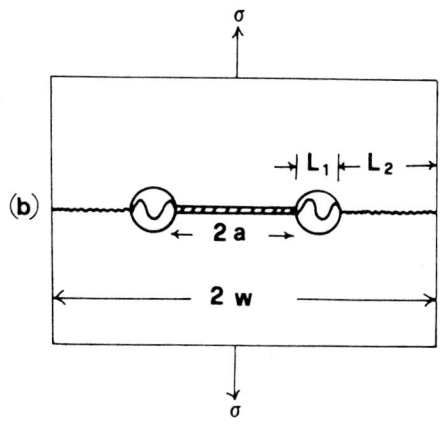

FIGURE 1b
Elastic plastic fracture in metal.

3. A FRACTAL MODEL FOR INTERGRANULAR BRITTLE FRACTURE SURFACES OF METALS

In the intergranular fracture case, the crack would propagate along zigzag grain boundaries. In a smaller scale, the crack would propagate along smaller zigzag subgrain boundaries. In a larger scale, the crack would also propagate along a larger weak passage near by the general direction of crack propagation and which would be formed by irregular distributions of vacancy clusters, micro-voids, inclusions and micro-cracks etc. They are all irregular and can be considered as self-similar and then can be modelled by fractals (Fig.2). In addition, I believe that the fractal dimension D might be well defined for different specimens of the same metal not only having similar thermomechanical treatments [1] but also under the same temperature condition and loading rate of the tensile test.

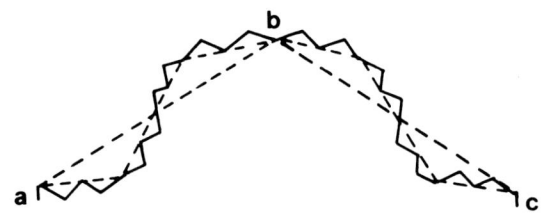

FIGURE 2
New refined zigzag cracks formed in fractal modelled metals.

There are two forms of intergranular brittle fracture (Figs.3a and b). Their fractal dimensions can be estimated by the formal definition

$$D = \log N/\log \left(\frac{1}{r}\right), \qquad (5)$$

where $N = L_i/\varepsilon_{0i}$, $r = \varepsilon_{0i}/L_{0i}$ (Figs.3).

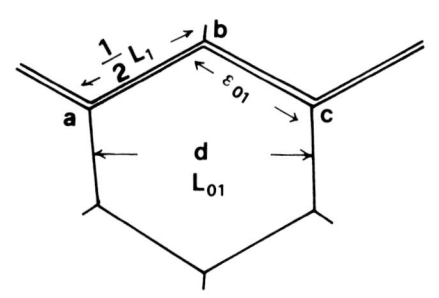

FIGURE 3a

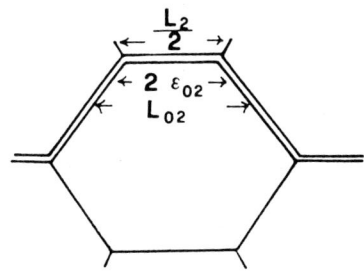

FIGURE 3b
Intergranular brittle fracture.

(a) $N = 2$, $r = \frac{1}{1.732}$, $D = 1.26$,
(b) $N = 4$, $r = \frac{1}{3}$, $D = 1.26$.

Both fractal dimensions of the two forms are 1.26, but the grain sizes are different. The grain size, d,

$$d = L_{01} = 1.73\ \varepsilon_{01} = 3.46\ \varepsilon_{02}, \quad (6)$$

where $L_1 = 2\ \varepsilon_{01}$, $L_{01} = 1.73\ \varepsilon_{01}$, $L_2 = 4\ \varepsilon_{02}$, $L_{02} = 3\ \varepsilon_{02}$. From Eq.(3),

$$G_{crit} = 2\gamma_s [L_i/L_{0i}] = 2\gamma_s (L_{0i}/\varepsilon_{0i})^{D-1} \quad (7)$$

(a) $G_{crit} = 1.73^{0.26} \times 2\gamma_s$
(b) $G_{crit} = 3^{0.26} \times 2\gamma_s$

We may see that case (a) consumes less energy than case (b), then it is preferable. The fractal model in this paper is then based on case (a).

As the grain size is smaller and smaller, the true area of fracture surface becomes larger and larger. According to Mandelbrot[2],

$$L_i(\varepsilon_i) \sim F\ \varepsilon_i^{1-D} \quad (F = L_0^D) \quad (8)$$

$G_{crit} = 2\gamma_s (L_i/L_0) \approx 2\gamma_s\ FL_0^{-1} \varepsilon_i^{1-D} \approx 2\gamma_s d^{-0.26}$

$(1.73^{0.26} \approx 1.1)$

$(FL_0^{-1} = L_0^{D-1} = 1$, for choosing L_0 as a unit length, say 1 cm). Then,

$$G_{crit} = 2\gamma_s \times 10.96 \quad (\text{for } d = 10^{-4}\ cm)$$

$$G_{crit} = 2\gamma_s \times 20 \quad (\text{for } d = 10^{-5}\ cm)$$

$$G_{crit} = 2\gamma_s \times 36.3 \quad (\text{for } d = 10^{-6}\ cm)$$

The term related to γ_s in Eq.(2) is now comparable to or a little smaller than the term related to γ_p (usually $\gamma_p \gtrsim 10\gamma_s$) in brittle fracture; but, it is still not large enough to improve the fracture toughness of materials. However, it might be one of the reasons why the surface energies of metals estimated by low temperature brittle fracture measurements are always higher than by other methods.

4. A FRACTAL MODEL FOR THE INTERGRANULAR DUCTILE FRACTURE SURFACES OF METALS

The fractographic observations on intergranular fracture indicate that the ductile fracture surface is composed of microdimples which are the result of holes forming ahead of the main crack. These holes are thought to initiate in practical alloy steels primarily at the site of precipitated particles in the matrix. Often the large voids in the medium are connected by bands of intense shear, which are formed by dislocation motions. As to our simplified fractal model, plastic deformations in the grains would make the grain boundaries ab and bc to be curves, ab' and b'c. Moreover, ab' and b'c are steeper than ab and bc. An additional angle θ would appear (Fig.4) after loading

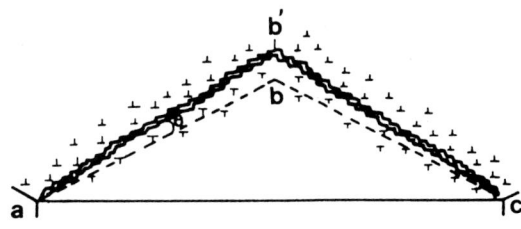

FIGURE 4
The additional angle formed by plastic deformation in the grain.

Now, $L^p = 2\varepsilon$

$$L_0 = 2\varepsilon \cos(30° + \theta) .$$

In this case, $N = 2$, $r = [2\cos(30° + \theta)]^{-1}$

$$D = \log 2/\log[2\cos(30° + \theta)] \quad (9)$$

The value of θ can be estimated as follows

$$\theta = (\rho bL)/L = \rho b , \quad (10)$$

where ρ is the linear density of mobile dislocations. Typical values of total linear density of dislocations range from 10^6-10^7/cm for cold worked crystals to 10^3/cm for annealed crystals. With $b \approx 3 \times 10^{-8}$ cm, the range of θ in Eq.(10) is from 3×10^{-5} (rad.) to $0.03 - 0.3$ (rad.) ($1.7° - 17°$). Then, the fractal dimensions range from 1.26 to 2.23.

Taking $D = 2.23$; then,

$$G_{crit} \approx 2\gamma_s d^{-1.23}$$

$$G_{crit} = 2\gamma_s \times 8.3 \times 10^4 \quad (\text{for } d = 10^{-4} \text{ cm})$$

$$G_{crit} = 2\gamma_s \times 1.4 \times 10^6 \quad (\text{for } d = 10^{-5} \text{ cm})$$

The critical crack extension force estimated by this fractal model would rise rapidly with the decrease of grain size. It rises faster than the Hall-Petch's $d^{-1/2}$ law, if the density of mobile dislocation is high enough. We noted that the grain sizes of almost of the superplastic alloys are very small ($\lesssim 10^{-4}$ cm). This phenomenon probably could be explained by this fractal model.

ACKNOWLEDGMENTS

The author would like to thank Professor Abdus Salam, the International Atomic Energy Agency and UNESCO for hospitality at the International Centre for Theoretical Physics, Trieste, where this work was finished during his stay in the summer of 1985. He would also like to thank Professors S. Lundqvist and B.B. Mandelbrot for their helpful discussions. This work is supported by the Science Fund of the Chinese Academy of Science.

REFERENCES

1. B.B. Mandelbrot, D.E. Passoja and A.J. Paullay, Nature, Vol.308, 19 (1984) 721.

2. B.B. Mandelbrot, The Fractal Geometry of Nature (1983) pp.25, 29, 459.

3. J.F. Knott, Fundamentals of Fracture Mechanics (Butterworths, 1976) pp.109-111.

4. C.W. Lung and H. Gao, Phys. Stat.Sol.(a), 87 (1985) 565.

WHEN ARE VISCOUS FINGERS FRACTAL?

Johann NITTMANN[†], Gérard DACCORD[†] and H. Eugene STANLEY[‡]

[†]Etudes et Fabrication Dowell Schlumberger, B. P. 90, 42003 Saint Etienne, France.
[‡]Center for Polymer Studies and Department of Physics, Boston University, Boston, MA 02215 USA.

We show that the displacement of a highly viscous non-Newtonian fluid by water in linear and radial Hele Shaw cells gives rise to fractal growth. Our experiments with the linear cell indicate that the fractal dimension d_f of the instability structures is a function of the cell width. With increasing width, d_f appears to approach the DLA value of 1.7. This value has been verified in a second independent experiment with a radial cell where boundary effects do not occur. For the Hausdorff dimension of radial viscous fingers we found the value $d_f = 1.70 \pm 0.05$. We also made the first experimental measurement of growth sites, and compare them with DLA simulations. We find that the finger thickness is a linear function of the plate separation. In addition we discuss the application of diffusion limited aggregation and gradient governed growth models to understand the fractal growth of viscous fingers.

1. INTRODUCTION

Recently much interest has focused on both experimental[1-7] and theoretical[8-14] studies on the wide range of physical phenomena which occur when a low viscosity fluid displaces a high viscosity fluid in a Hele Shaw cell. In general the initially straight interface in a linear cell breaks up into a structure which resembles the fingers of a glove — hence the name viscous fingering. Moreover, we have recently discovered physical conditions for which viscous fingers show a very ramified structure. Owing to successive random splitting of the leading finger tips, the viscous fingers continue to grow into branched clusters whose density decreases with time. We find that these structures are fractals and that fractal fingers can be quantitatively characterized by their Hausdorff dimension. The objective of this paper is to define the physical conditions for which fractal viscous fingers develop and to discuss the application of fractal growth model to simulate the grow.

Viscous fingering was first observed in enhanced oil recovery. In order to increase oil production, water is used to displace oil trapped in a porous medium. However, the displacement of a high viscosity fluid by a low viscosity fluid in a medium which provides resistance to flow is unstable and finger-like flow structures develop. Saffman and Taylor[15] have studied this phenomenon in

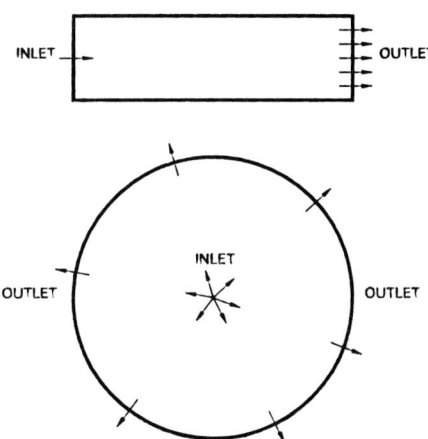

Figure 1: Schematic representation of linear and radial Hele Shaw cells.

a Hele Shaw cell. This apparatus consists of two rectangular glass plates which are seperated by a distance of typically 0.5 mm. The glass plates are sealed off along the longer sides and the fluids can enter through a small opening centered at one of the short ends. The fluids can leave at the opposite end (Fig.1).

The Hele Shaw cell is frequently used to simulate two–fluid flow in a porous medium. For both systems an empirical law exists which relates the flow velocity $\underline{v}$ of a single fluid to the pressure gradient ∇P in the fluid. This relation is called Darcy's law

$$\underline{v} = -\frac{k}{\mu}\nabla P . \qquad (1)$$

Here k and μ are the permeability of the medium and fluid viscosity, respectively. Darcy's law simply states that the flow velocity is proportional to the applied pressure gradient. In a porous medium k will be a function of space, whereas in the Hele Shaw cell k is a constant (It can be shown that $k = b^2/12$, with b being the plate separation).

The early experiments focussed on the displacements of immiscible fluids, for which capillary forces prevent the growth of very small fingers. The ratio of viscous forces to capillary forces can be expressed in form of a non–dimensional number — the capillary number N_{Ca}.

$$N_{Ca} = \frac{\mu v}{\sigma} , \qquad (2)$$

σ is the interfacial tension. Chuoke et al.[16] have shown that the interfacial tension introduces a minimum length scale into the problem. Only pertubations with a wave length λ greater than the critical wave length λ_c can grow, where

$$\lambda_c = 2\pi b(12N_{Ca})^{-1/2} . \qquad (3)$$

Eq.(3) indicates that λ_c is a linear function of the plate separation and decreases for a particular pair of fluids with increasing flow rate.

By repeating the classical Saffman–Taylor experiment with a water–oil system, we noticed that by decreasing λ_c (e.g. by increasing the flow rate) the initialy stable water fingers tip split (Fig.2). It seemed that a further increase of the viscous forces would lead to further ramification. However a further increase of the flow rate was impossible as the experimental apparatus sets limits to the applied viscous forces. We therefore completely redesigned the experiment.

2. EXPERIMENTS

We looked for a two–fluid system with the following properties:
i) essentially zero interfacial tension

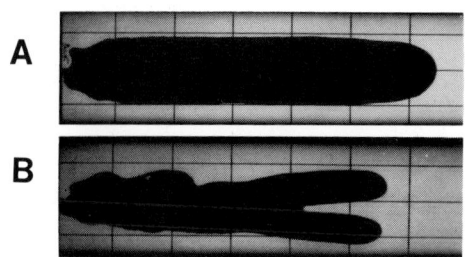

Figure 2: A viscous finger created by water advancing into a Hele Shaw cell filled with a Newtonian fluid, oil. Two different capillary numbers are used: a)$N_{Ca} = 0.08$, b)$N_{Ca} = 0.16$.

ii) a high viscosity ratio
iii) piston–like ("plug–flow") displacement profiles normal to the glass plates.

In order to minimize interfacial tension, miscible fluids were used. As miscible fluids mix, it is important that the time duration of the experiment is short compared to the mixing time. Essentially zero interfacial tension will enable any small fluctuation of the interface to grow. The dominance of randomizing effects over stabilizing effects is extremely important for fractal growth.

The viscosity ratio (viscosity of pushing fluid over viscosity of displaced fluid) is the driving force of the instability, therefore a high viscosity ratio is desirable. The third property prevents "three–dimensional fingering" to occur. Partial displacements are observed in miscible Newtonian fluids which show less ramified instability structures[8].

Our original two–fluid system used[1] was water displacing an aqueous polymer solution of high molecular mass polysaccharide for which the mixing time with water is very long compared with the time of the experiment (diffusion constant $D = 10^{-5}$cm^2/sec). The polymer solution was first placed into the cell and water was then forced through a small opening, either centered at one of the ends (for the linear experiment, Fig.1a) or right at the centre (for the radial experiment, Fig.1b).

The growing structures were filmed with a television camera[6]. The signals were then transferred to a

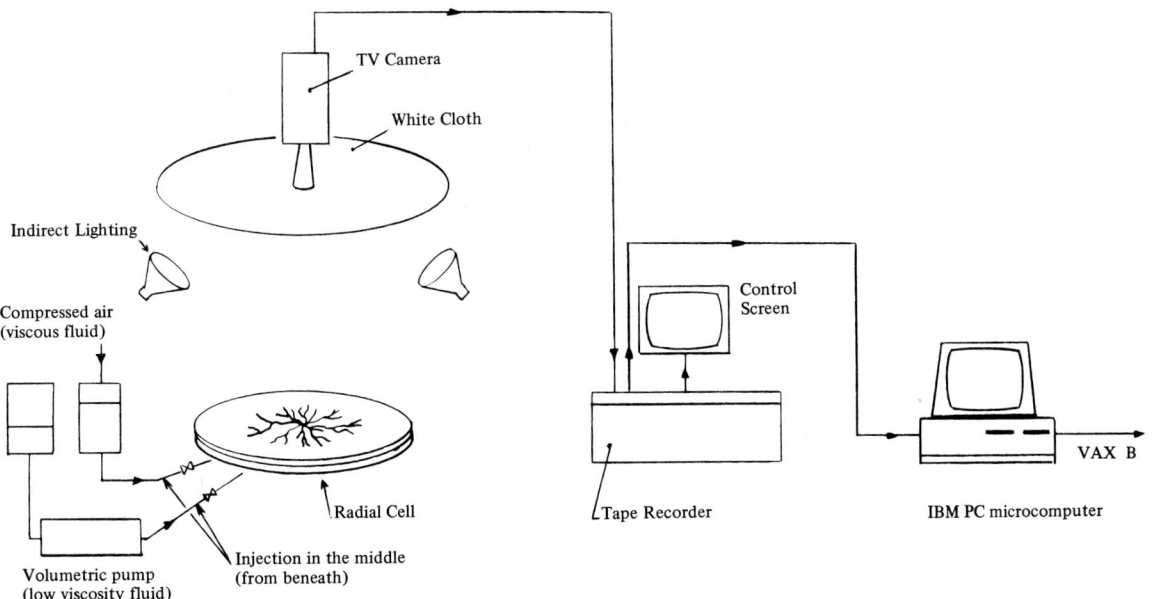

Figure 3: Schematic representation of the experimental set up.

digitizing system which recorded the filmed structures on to a 256 × 256 grid. The digitized structures were then transferred onto a VAX 11/785 for further analysis. Fig.3 gives a schematic representation of the experimental setup.

3. RESULTS OF EXPERIMENTS AND ANALYSIS

3.1. Determination of fractal Dimension

In Figs.4 and 5 we show two typical results obtained when water displaces a high molecular mass polymer solution. Fig.4 is taken from a linear cell of 1 m length and a height of 10 cm. Fig.5 shows the same result obtained in a radial cell. The plate separation is in both cases 0.5 mm. The quantitative analysis of the digitized viscous fingers has been carried out with several methods:

i) Sandbox method[17]. About every lattice point in the fingering structure, one forms an imaginary $L \times L$ square box. The number $N(L)$ of finger points inside the box is counted and averaged over all possible lattice points as center. A double logarithmic plot of $N(L)$ versus L has slope d_f.

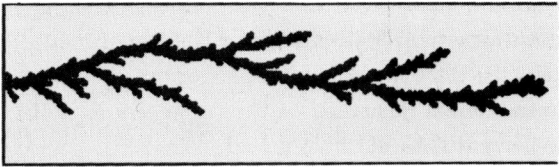

Figure 4: Typical viscous fingers created by water advancing into a linear Hele Shaw cell filled with a polymer solution (scleroglucan).

Figure 5: Typical viscous fingers created by water advancing into a radial Hele Shaw cell filled with a polymer solution (scleroglucan).

ii) Density–density Correlation function[18]. Every lattice point on the structure is the center of a circle with radius R. Then the number of finger points which lay on the circle perimeter are counted. This number divided by the total number of fluid points in the cluster scales with R^{d_f-d}. d is the euclidian dimension.

iii) Radius of Gyration[19]. Here several photographs of the finger are taken at successive stages of growth. For each photograph, R_g is measured and plotted against finger mass. The slope is $1/d_f$.

For the fingering structure in Fig.4 we obtained $d_f = 1.4 \pm .04$. However this value depends very strongly on the height of the model[1]. With increasing model height the fractal dimension increased. It is evident that lower and upper boundary of the cell constrained the growth of the instability. Simultaneous numerical studies indicated that with increasing cell height d_f seems to approach the diffusion limited aggregation[18-20] (DLA) value of $d_f = 1.7$. In order to obtain the fractal dimension without growth limiting boundaries we designed a second cell of radial geometry. Fig.5 is a typical structure obtained with a cell of 1 m diameter[3]. Using the three methods mentioned above we obtain[3] for the fractal dimension of viscous fingers in radial geometry (Fig.6a–c)

$$d_f = 1.70 \pm 0.05$$

Previous[1] and recent new experiments[2,3] indicate that this value of d_f does not depend on the flow rate (range studied: 0.4–40 ml/min) nor on the polymer concentration (range tested: 2.4–12 g/l). This value of d_f is further reproducible for different types of polymer (tested: scleroglucan, a guar gum derivative and a hydroxyethyl cellulose). In addition fractal structures with similar d_f have been also obtained with a latex fluid which is a suspension of polymer spheres (size = 0.1 micron) in water.

3.2. How do viscous fingers grow ?

To investigate where new growth is added during a finite interval of time, we take two successive photographs of the growing viscous fingers (the time interval is 1 sec). By substracting the earlier photograph

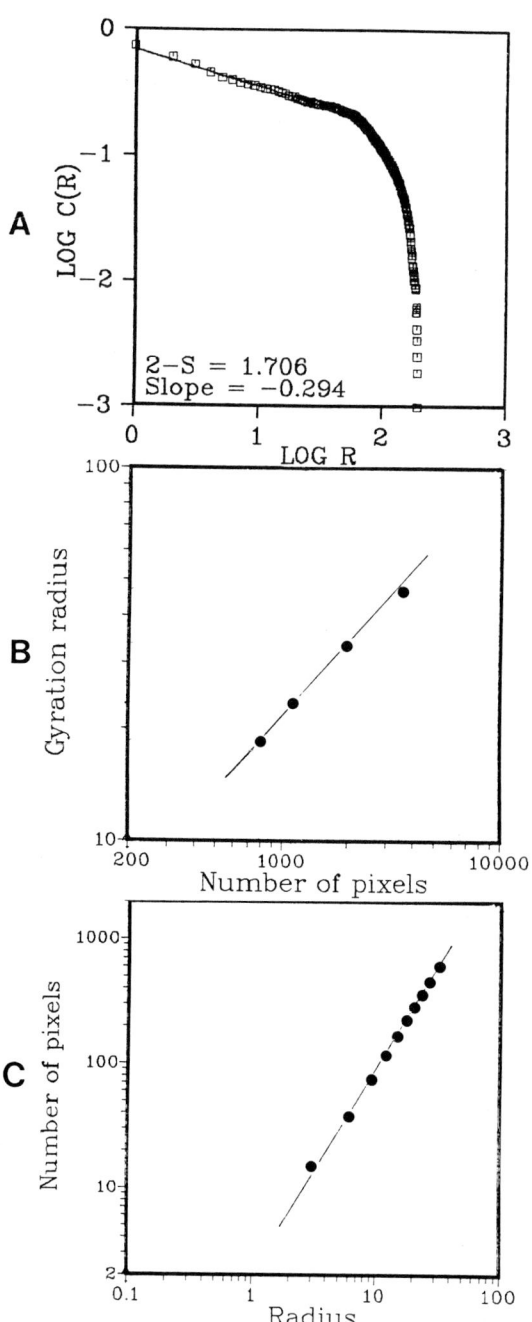

Figure 6: Determination of the fractal dimension: a) density–density correlation function method (slope = 1.70), b) radius of gyration method (initial slope 1.6), c) sandbox method (slope = 1.70).

from the later one (Fig.7a) we obtain the "growing sites" of the cluster. Fig.7b is quantitatively similar to equivalent studies of large DLA–clusters[19]. Both structures show that growth occurs only on the tip and almost never in the fjords. Please notice that an almost indistinguishable growth process has occurred for such different physical mechanisms as—aggregation and viscous fingering. Quantitatively, Meakin found a slope of -1.1 for his density correlation analysis. We found the value of -1.3 ± 0.1 for our relative small viscous finger clusters[3]. Viscous fingers grow by successive irregular splitting of the leading finger tips. This is different from crystal growth where dendrites show frequently stable tips with regular side branches[21].

3.3. Fingerwidth versus plate separation

As we pointed out, we observe a well defined (i.e. constant) finger thickness during the growth process. As a first step to understand the reason for this phenomenon we have studied how the finger thickness changes with the plate seperation distance b. In the Chuoke et al. theory of the critical finger width for immiscible fluids, the finger thickness depends linearly on the plate separation (Eq. 3). Paterson has recently proposed a theory for miscible fluids. He makes viscous dissipation mechanism responsible for enabling only wavelengths above a certain threshold value to grow. He finds also a linear dependence of the finger thickness on the plate separation. However he predicts that the finger thickness does not depend on the flow rate.

Fig.8 shows our results for several plate separations. We find within experimental accuracy a linear dependence. We measured for the slope the value 4.6 — a value close to that proposed by Paterson for Newtonian fluids. This close proximity and the fact that our finger thickness does not depend on the flow rate support Paterson's theory. However, dissipation mechanism in polymer solutions will be different from that in Newtonian fluids.

Figure 7: The digitized "growth region" of a typical radial viscous finger structure (a), obtained by substracting the images of the same finger photographed at slightly different times. The fact that the growth regions (b) are only at the tips is a striking indication of the high degree of screening present.

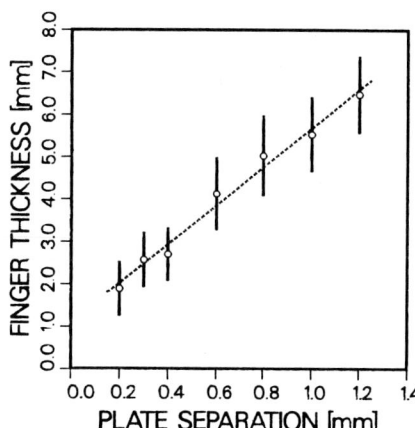

Figure 8: Dependence of the finger width ("critical wavelength") upon the separation b of the Hele Shaw plates for the non–Newtonian fluid scleroglucan (shear–thinning exponent $n' = 0.15$).

4. STOCHASTIC MODELLING

4.1. Diffusion Limited Aggregation Model of RVF

As mentioned above radial viscous fingers (RVF) and the DLA show quantitative agreement in the fractal dimension and qualitative agreement in the growing site analysis. Moreover, visual comparison of the Fig.5 and a DLA structure (Fig.9) shows further similarities in the growth:

i) Spoke–like main branches grow from the centre and form the skeleton of the structure.
ii) As the structure grows some of these main branches tip bifurcate into two main branches.
iii) On all main branches side branches grow, some of them tip split again.
iv) The angle of side branching is considerably less than 90°.

Besides these striking similarities we also see certain differences distinguishing RVF from DLA:
i) fractal fingers have straighter main branches.
ii) the tip split angle is smaller than in DLA and has a narrower distribution.
iii) there is more "foliage" growing on the DLA spokes compared to the relative barren impression of the RVF spokes

Figure 9: A typical DLA cluster made of 10 000 points.

Why are RVF structures so similar to DLA clusters? Can we explain the differences?

For DLA it can be shown[18,20] that the probability $\phi(r,t)$ of a random walker to be at the spatial position r at time t follows a Laplace equation,

$$\nabla^2 \phi = 0 . \quad (4)$$

The growth rate $\underline{u}$ of a cluster is proportional to the local probability gradient between a cluster site and a neighbour site,

$$\underline{u} \propto \underline{\nabla}\phi . \quad (5)$$

The flow of two Newtonian fluids in a Hele Shaw cell follows similar equations. If we assume that one fluid is considerably more viscous than the other, the fluid potential or fluid pressure can be considered to be constant in the less viscous fluid[9]. The normal velocity of the interface will be equal to the local pressure gradient,

$$u_n = -\underline{n} \cdot \underline{\nabla} P . \quad (6)$$

With the incompressible condition $\nabla u = 0$, this leads to a Laplace equation for the pressure field,

$$\nabla^2 P = 0 . \quad (7)$$

Eq. (6) differs from (5) as it represents the Mean Field Limit (MFL) of Eq. (5). Tang[11] has shown that indeed the MFL of time reversed DLA produces stable Saffman–Taylor fingers, which are not fractal. The

MFL introduces a smoothing of the growth profile similar to surface tension. However surface tension might not necessarily prevent fractal growth. It introduces, however, a length scale below which the growth is certainly not fractal. To see fractal growth in the presence of interfacial tension the experimental apparatus will have to be of a size much larger than the critical wavelength of growth. For many two–fluid systems this is rather difficult to achieve practically. Our two–fluid system has a priori no interfacial tension because the two fluids mix. The fact that a well defined finger thickness is present, might be caused by some residual interfacial tension or by other so far not understood physical phenomena. The similarities between DLA and RVF indicate that in the limit $l_{finger} \ll l_{apparat}$ viscous finger growth follows Eq. (5).

Our considerations have so far focused only on Newtonian fluids. The next question we will ask is how does the presence of a non–Newtonian fluid change Eq. (6) and (7)? The non–Newtonian fluids used in our experiments are shear thinning fluids, where the viscosity μ decreases with increasing shear rate $\dot{\gamma}$ ($\dot{\gamma} = -dv/dz$, with z being the coordinate axis normal to the glass plates). For the range of flow rates applied, the logarithmic relation between shear rate and viscosity is linear. Frequently, these fluids are referred to as powerlaw fluid.

$$\mu = c\dot{\gamma}^{n'-1} \qquad (8)$$

where c and n' are characteristic constants of each polymer solution[22]. n' is the shear thinning index which was for the fluids we used between 0.4 and 0.1. Eq. (8) modifies Eqs. (6) and (7) to

$$u_n = -\underline{n} \cdot (|\nabla P|^{m-1} \underline{\nabla} P) \qquad (9)$$

$$\underline{\nabla}(|\nabla P|^{m-1} \underline{\nabla} P) = 0 \qquad (10)$$

where $m = 1/n'$. It is not obvious how the system of Eqs. (9) and (10) can be represented by a random walk model. Moreover, if fluids with a finite viscosity ratio are assumed, then the pressure gradient in the less viscous fluid cannot any longer be neglected. Again the DLA model would need some major modifications.

DLA has also the purely phenomenological drawback that cluster growth occurs by aggregation, whereas in RVF growth originates from the centre of the structure. A similar fractal growth morphology has been found recently for the dielectric breakdown phenomenon by Niemeyer et al.[23]. They have proposed a model which can simulate fractal growth without using random walkers.

4.2. Gradient Governed Growth Model (GGG)

The dielectric breakdown model (DBM) represents an alternative way of solving Eqs. (4) and (5). As in DLA it is assumed that the less viscous fluid has zero pressure gradient. The first step consists in finding a numerical solution for the Laplace equation in the high viscosity fluid. This is done by a finite difference discretisation of Eq. (4) with Dirichlet boundary conditions. Starting off with an analytic solution at time zero, after each step of advance Eq. (4) is solved by an iterative procedure.

The second step concerns the application of Eq. (5). After the distribution of the pressure field between interface and boundary is known, the local pressure gradients between any potential growth site and neighbour site is calculated. Proportional to the local gradient one new site is chosen randomly. Then the finite difference representation of Eq. (4) is solved again and the cycle is closed.

The structures obtained have a fractal dimension slightly larger than DLA. A direct comparison of computer time shows that DLA grows clusters considerable faster than DBM. However DBM has the big advantage of being easily extendable to more complex systems[23]. One interesting question to ask is how does a finite viscosity ratio influence the growth. It is clear that the interface profile of a high viscosity fluid displacing a low viscosity fluid is stable. Therefore a crossover from fractal to non-fractal growth must occur as a function of the viscosity ratio. For this purpose Sherwood and Nittmann[24] have extended the DBM model to a general gradient governed growth (GGG) model. In Figs.10a–d we show radial simulations of unstable structures for viscosity ratio $\eta = 0.0001, 0.01, 0.1, 10$. ($\eta$ is the vis-

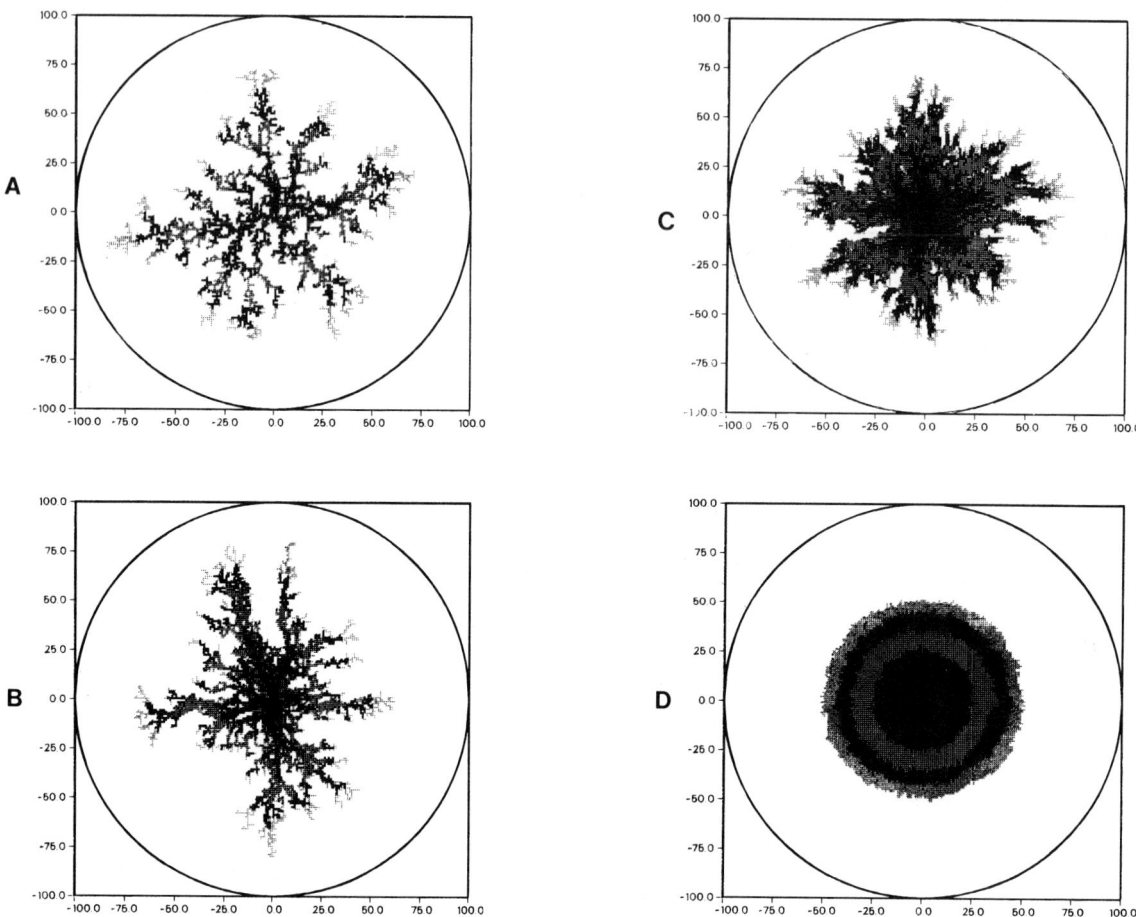

Figure 10: Typical structures obtained with the gradient governed growth model. The compactness of the structure is a function of the viscosity ratio η: a) $\eta = 0.0001$, $d_f = 1.90 \pm 0.05$, b) $\eta = 0.01$, the structure is non-fractal, c) $\eta = 0.1$, d) stable case, the highly viscous fluid is displacing the low viscosity one, $\eta = 10$.

cosity ratio of displacing fluid over displaced fluid). As a function of the viscosity ratio we observe a cross over of the fractal dimension. For $\eta = 0.0001$ we find[24] $d_f = 1.90 \pm 0.05$. For $\eta = 0.01$ the structure is non-fractal. The crossover has occurred owing to the fact that the pressure in the low viscosity fluid is not constant anymore but has a slope with the same sign as in the highly viscous fluid. This gives the sites in the fjords of the structure a better chance to grow.

We would like to return now to the problem of representing the non-Newtonian aspect of our displaced fluid accurately. Eqs. (4) and (5) have the considerable advantage over (9) and (10) that they are linear equations. The non-Newtonian character has made Eq. (9) highly non-linear and a solution procedure is not yet available. However from studying the analytic solution of (9) and (10) for a system at time zero, we are able to come to certain conclusion about the growth process. The solution of Eqs. (9) and (10) is

$$P(r) = \frac{R^{(m-1)/m} - r^{(m-1)/m}}{R^{(m-1)/m} - 1} \quad \text{for} \quad m \neq 1 \quad (11)$$

$$P(r) = 1 - \frac{\log r}{\log R} \quad \text{for} \quad m = 1, \quad (12)$$

with the boundary conditions $P = 1$ at $r = 1$ and $P = 0$ at $r = R$. The case $m = 1$ corresponds to Newtonian fluids, DLA and DBM. In that case the growth

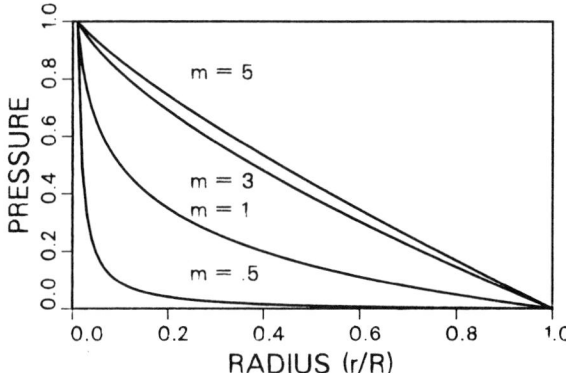

Figure 11: Dependence of the pressure field $P(r)$ of the distance r from the centre of a radial Hele Shaw cell. Shown are several curves for different values of the parameter $m = 1/n'$. The Newtonian fluids analyzed previously concern the case $m = 1$. Experimentally, we used high-viscosity fluids with n' in the range 0.1-0.4 so $m > 1$.

rate is initially proportional to $1/r$. It is interesting to notice that also a non-Newtonian fluid follows the same growth law $1/r$. Let us consider the case $m > 1$ (a similar argument holds for the case $m < 1$): although u, owing to Eq. (9), now is a function of the pressure gradient to the power m, which would indicate a more violent growth at the tips where the gradients are generally high, the pressure distribution is flatter owing to Eq. (10). To demonstrate that, we have plotted several pressure fields for $m = 0.5, 1, 3$ and 5 in Fig.11. Hence we believe that instability structures obtained with non-Newtonian fluids should not differ very much from those obtained with Newtonian fluids. This, however, can only be proved by a numerical solution of Eqs. (9) and (10).

5. Summary

We have shown that fascinating fractal structures can develop in laminar fluid flow. We have evidence to believe that these structures are caused by randomizing processes winning over stabilizing forces. Fluctuation at the interface of the two miscible fluids can grow chaotically without the confining influence of surface smoothing forces. The growth is governed by the global

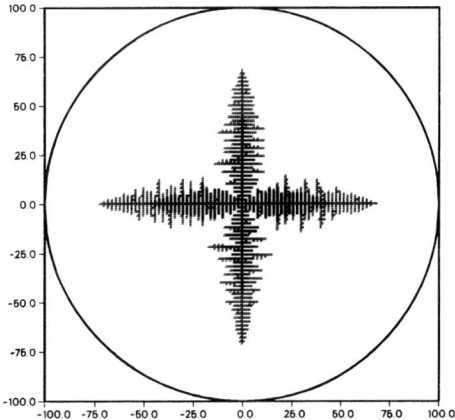

Figure 12: A simulation of viscous fingering in an etched Hele Shaw cell. The introduction of spatial anisotropy and surface smoothing has transformed the initialy disordered DLA-type structure into a highly ordered four-fold structure.

pressure field, and the competition between neighbouring interface elements to grow proportionally to the local pressure gradient.

Fractal fingers have also been found in unstable two-fluid displacements in porous media. Lenormand and Zarcone[25] and recently Chen and Wilkinson[26] have found DLA-type fingering in two-dimensional porous medium models in the case that viscous forces dominate over capillary forces. Chen and Wilkinson suggest the randomness in the pore size distribution as the origin of the fractal behaviour. To test this they performed studies on models with narrower pore size distributions and find in the limit of only one pore size a highly ordered structure. In particular their Figs.1a and 2a resemble our Fig.12 which has been obtained by introducing anisotropy and surface smoothing into the gradient governed growth model. The objective was to model the growth of viscous fingers during the presence of spatial anisotropy. Lenormand[6] and Ben-Jacob et al.[7] have recently obtained snow crystal-type structures in a Hele Shaw cell by etching channels into one of the two plates. Fig.12, although highly ordered at first glance, has a fractal dimension of $d_f \simeq 1.64$

and work is in progress to understand these interesting findings.

All these very recent developments make us hope that our discovery of fractal viscous fingers is just the tip of an iceberg beneath which lies new understanding of the growth and form of random and ordered structures.

Acknowledgement

We would like to thank our colleagues from Dowell-Schlumberger Research, R. Lenormand, F. Rondelez, J. D. Sherwood, E. Touboul, and Dr. P. H. Gaskell (Leeds University) for helpful discussions and M. Bourlion for technical assistance.

References

1. J. Nittmann, G. Daccord and H. E. Stanley, Nature 314 (1985) 141.
2. G. Daccord, J. Nittmann and H. E. Stanley, Fractal viscous fingers: Experimental results, in: On Growth and Form, eds. H. E. Stanley and N. Ostrowsky (Martinus Nijhoff Pub., Dordrecht 1985) pp 175–182.
3. G. Daccord, J. Nittmann and H. E. Stanley, submitted to Phys. Rev. Lett.
4. J. Maher, Phys. Rev. Lett. 54 (1985) 1498. T. Maxworthy, preprint.
5. C. W. Parks and G. M. Homsy, Phys. Fluids 28 (1985) 1583.
6. R. Lenormand (private comunication).
7. E. Ben Jacob, R. Godbey, N. D. Goldenfeld, J. Koplik, H. Levine, T. Mueller and L. M. Sander, Phys. Rev. Lett. 55 (1985) 1315.
8. L. Paterson, Phys. Rev. Lett. 52 (1984) 1621.
9. L. Paterson, Phys. Fluids 28 (1985) 26.
10. L. P. Kadanoff, J. Stat. Phys. 39 (1985) 267.
11. C. Tang, Phys. Rev. A 31 (1985) 1977. S. Liang, preprint.
12. D. Bensimon, preprint.
13. G. Tygravson and H. Aref, J. Fluid. Mech. 154 (1985) 287. A. J. DeGregoria and L. W. Schwartz, preprint.
14. D. A. Kessler and H. Levine, preprint.
15. P. G. Saffmann and G. I. Taylor, Proc. Roy. Soc. A 245 (1958) 312.
16. R. L. Chuoke, P. Van Meurs and C. J. Van der Poel, J. Petrol. Tech. 11 (1959) 64.
17. L. Forrest and T. A. Witten, J. Phys. A 12 (1979) L109.
18. T. A. Witten and L. M. Sander, Phys. Rev. Lett. 47 (1981) 1499.
19. P. Meakin, H. E. Stanley, A. Coniglio and T. Witten, Phys. Rev. A32 (1985) 2364. See also the recent review: P. Meakin, Computer simulations of growth and aggregation processes, in: On Growth and Form, eds. H. E. Stanley and N. Ostrowsky (Martinus Nijhoff Pub., Dordrecht 1985) pp 81–110.
20. T. A. Witten and L. M. Sander, Phys. Rev. B27 (1983) 5685.
21. H. Honjo, S. Ohta and Y. Sawada, Phys. Rev. Lett. 55 (1985) 841.
22. R. B. Bird, R. C. Armstrong and D. Haffager, in: Dynamics of Polymeric Liquids, Vol.1, (Wiley 1977).
23. L. Niemeyer, L. Pietronero and H. Wiesmann, Phys. Rev. Lett. 52 (1984) 1033.
24. J. D. Sherwood and J. Nittmann, J. Physique 47 (1986) in print.
25. R. Lenormand and C. Zarcone, Phys. Chem. Hydro. 6 (1985) 497.
26. J.-D. Chen and D. Wilkinson, preprint.

Part V (b)
IRREVERSIBLE GROWTH MODELS: DIFFUSION-LIMITED AGGREGATION, DENDRITIC GROWTH, EDEN MODEL AND CLUSTER-CLUSTER AGGREGATION

FRACTALS IN PHYSICS
L. Pietronero, E. Tosatti (editors)
© Elsevier Science Publishers B.V., 1986

SOME RECENT ADVANCES IN THE SIMULATION OF DIFFUSION LIMITED AGGREGATION AND RELATED PROCESSES

Paul MEAKIN

Central Research and Development Department, E. I. du Pont de Nemours and Company, Experimental Station, Wilmington, Delaware 19898

Large numbers (hundreds) of large ($\sim 10^5$ sites or particles) two dimensional DLA clusters have been generated on a variety of lattices using improved algorithms. For the case of the square lattice our results are consistent with the theoretical results of Turkevich and Sher (D=5/3). For the case of the hexagonal and triangular lattices the simulations do not agree with this theory but the discrepancy can be understood in terms of the concepts contained in this theory. Our results suggest that the fractal dimensionality of DLA models with high symmetry belongs to the same universality class as off-lattice DLA but that models of low symmetry are non-universal. The effects of sticking anisotropy and diffusional anisotropy as well as lattice anisotropy are explored.

1. INTRODUCTION

In recent years a very rapid growth of interest in non-equilibrium growth and aggregation processes has taken place. In particular, models which lead to the formation of fractal[1] structures have been extensively studied. Much of this work was stimulated by the Witten-Sander[2] model for diffusion limited aggregation (DLA) and this remains the most important model for the exploration of non-equilibrium growth and aggregation processes. Most of the initial enthusiasm for the Witten-Sander model was generated by the interesting scaling and universality properties[3,4] associated with the structures generated by this model. More recently interest has also been sustained by the application of DLA and related models to electro-deposition[5,6], the morphology of thin films[7], fluid-fluid displacement processes[8-10], dielectric breakdown[11], dendritic growth[12-14] and other phenomena. A variety of theoretical approaches has been explored.[15-17] While all of these theories give results which are in good agreement with computer simulation for the dependence of the fractal dimensionality (D) on the Euclidean dimensionality (d) of the lattice or embedding spaces they are difficult to evaluate and cannot easily be extended to other related models.

More recently a new theoretical approach has been developed by Turkevich and Sher[19,20] based on the ideas that the fractal dimensionality is determined by the strength of the singularities in the site occupation probabilities for the DLA growth process and that the strength of these singularities is determined by the structure of the lattice used in the simulation. Related ideas have been developed by Ball et. al.[21] and by Halsey et. al.[22]. Unlike previous theories the methods employed by Turkevich and Sher can easily be extended to other related models and lead to specific predictions which can be tested via computer simulation. Perhaps the most surprising prediction of the Turkevich-Sher model is that the fractal dimensionality D is non-universal in the sense that it depends not only on the dimensionality of the lattice used in the simulation but also on the structure of the lattice. In particular, Turkevich and Sher predict that the fractal dimensionality should be 5/3 for a two dimensional square lattice and 7/4 for two dimensional hexagonal and triangular (honeycomb) lattices. One of the main objectives of the work described here is to test these predictions.

In order to distinguish between the predicted fractal dimensionalities of 1.667 and 1.750 and that found from large scale off-lattice simulations (1.715±0.002)[23] large numbers (hundreds) of large clusters (on the order of 10^5 particles or sites) must be generated to reduced statistical uncertainties and approach the asymptotic (large cluster mass) limit.

2. MODELS

In the Witten-Sander DLA model, particles (frequently represented by occupied lattice sites) are added one at a time to a growing cluster or aggregate of particles via random walk (Brownian) trajectories. Particle trajectories are started outside of the region occupied by the cluster and are terminated when the particle contacts the cluster and growth occurs.

Efficient algorithms for simulating the formation of two dimensional off-lattice and semi-lattice DLA aggregates have been described recently.[24] In these algorithms the mobile particle undergoes an off-lattice random walk and is allowed to take large steps when it is a large distance from any occupied site or particle in the cluster. In the off-lattice model, aggregation occurs when the particle trajectory first causes it to "contact" a particle already incorporated into the growing aggregate. In the semi-lattice model, the growing cluster consists of occupied lattice sites and growth occurs whenever a particle (of zero diameter) enters an unoccupied lattice site which has an occupied nearest neighbor. At no stage in the simulation is a "particle" allowed to enter an occupied lattice site. Two new algorithms have been developed in connection with the work described in this paper. In one of these models the mobile particle is transferred to the nearest lattice site and undergoes a random walk on the lattice whenever it is within a prescribed distance from any occupied lattice site (typically 6-7 lattice units). Growth occurs whenever the particle enters an unoccupied site with an occupied nearest neighbor. Thus in the vicinity of the cluster this model works in the same way as an ordinary lattice model for DLA.[2,3] At greater distances from the cluster the particle undergoes an off-lattice random walk. The step size is increased with increasing distance from the cluster but the step length is not large enough to allow the particle to penetrate more than one lattice unit into the region in which the on lattice walk occurs. Since this model should give results which are undistinguishable from those obtained from a model employing completely on lattice random walks it is called a "lattice model" for DLA. This model is illustrated in Figure 1.

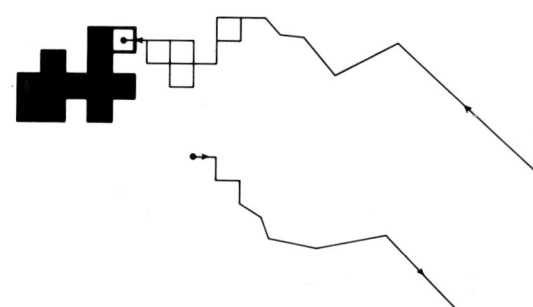

FIGURE 1
This figure illustrates the lattice model for diffusion limited aggregation. A possible trajectory at an early stage in a simulation is shown. Near to the cluster the mobile particle moves by means of an on-lattice random walk.

In the second model the simulation proceeds in the same way as an off-lattice simulation.[24]

However, after the particle has been added to the cluster it is rotated about the center of mass of the contacted particle on the cluster until it occupies the nearest unoccupied lattice site. This model is illustrated for the case of a square lattice in Figure 2. This model is called a semi-lattice-2 model to distinguish it from the earlier[24] or semi-lattice-1 model. Both of the models described above have been implemented for the square lattice, the three coordinate triangular or honeycomb lattice and the six coordinate or hexagonal lattice.

SEMI-LATTICE MODEL VERSION II

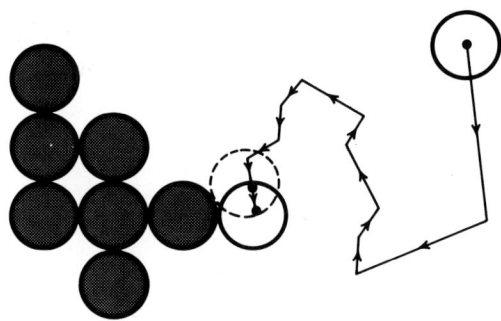

FIGURE 2
Semi-lattice model 2 on a square lattice. The dotted circle shows the position at which the particle first sticks to the cluster. This particle then rotates about the center of the contacting particle in the cluster until it occupies the nearest unoccupied position on a lattice (solid circle).

3. RESULTS

Figure 3 shows some of the results obtained from over 300 clusters generated using the two dimensional lattice model (see above) and a similar number of clusters obtained from semi-lattice model 1[24] for the case of a square lattice. Each of the clusters contains 80,000 occupied lattice sites. The fractal dimension-

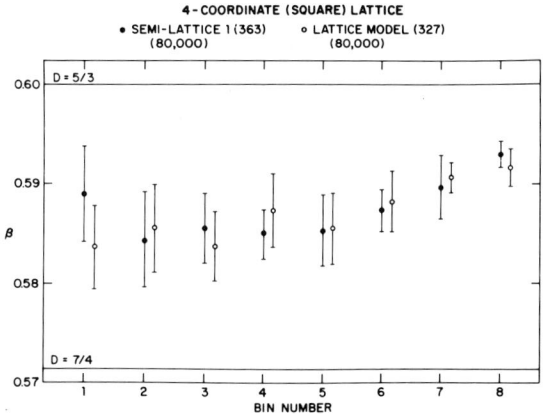

FIGURE 3
The dependence of the radius of gyration exponent (β) on cluster mass obtained from lattice model and semi-lattice model 1 simulations on a square lattice. In this figure (and in similar figures) the number of clusters (363 for the semi-lattice model and 327 for the lattice model) and their sizes (80,000 occupied lattice sites) are shown. The horizontal lines indicate the value of the radius of gyration exponents predicted by Turkevich and Sher[16] for the square lattice (D_β = 5/3, β = 0.6) and for the hexagonal lattice (D_β = 7/4, β = 0.5714). In this and other figures the error bars are 95% confidence limits.

ality D_β is estimated from the dependence of the radius of gyration on the number of occupied lattice sites (N) or mass of the clusters. The results are divided into "bins" corresponding to clusters containing 40,000 - 80,000 sites (N(max)/2 - N(max)) for bin 8, 20,000 - 40,000 sites (N(max)/4 - N(max)/2) for bin 7, 10,000 - 20,000 sites for bin 6 etc., where N(max) is the number of occupied lattice sites for the maximum cluster size. The effective fractal dimensionality D_β[25] is determined for the clusters in each bin by least squares fitting straight lines to the dependence of $\ln(R_g)$ on $\ln(N)$.

Figure 3 shows the values obtained for the exponent β obtained assuming that

$$R_g \sim N^\beta \qquad (1)$$

The effective fractal dimensionality D_β is equal to $1/\beta$. For small cluster masses β has a value of 0.585 (very similar to that found for clusters of all sizes from the off-lattice model).[23,26] However, for larger cluster masses the effective value for β increases and our results are consistent with a limiting value of 0.6 corresponding to one of the predictions of Turkevich and Sher ($D_\beta = 1/\beta = 5/3$).

Figure 4 shows similar results obtained from simulations of diffusion limited aggregation on a triangular lattice (or on the <u>bonds</u> of a honeycomb lattice). In this case the exponent β is almost independent of the range of cluster sizes used to estimate its value. There is no indication that the effective fractal dimensionality ($D_\beta = 1/\beta$) is approaching a value of 7/4 as predicted by Turkevich and Sher. In fact if there is any variation with cluster mass at all D_β seems to be decreasing with increasing cluster mass.

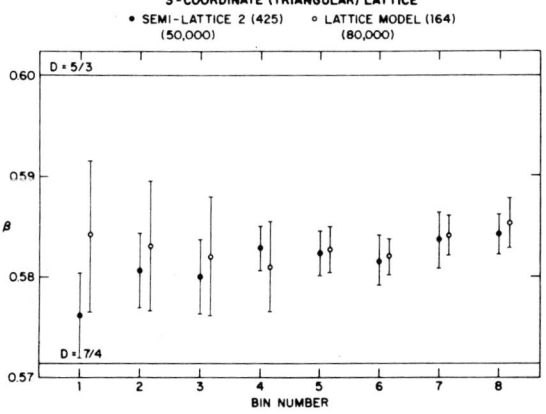

FIGURE 4
Dependence of the radius of gyration exponent β, with increasing cluster mass for DLA simulations carried out on a triangular lattice. In this case the cluster sizes generated using the semi-lattice and lattice models are not equal.

Figure 5 shows a similar set of results obtained using a six coordinate (hexagonal lattice). In this case there is no discernable dependence of β or D_β on the cluster mass and the numerical values of these quantities are very similar to those obtained from off-lattice simulations. These results are not in good agreement with the predictions of Turkevich and Sher for this lattice ($D_\beta = 7/4$).

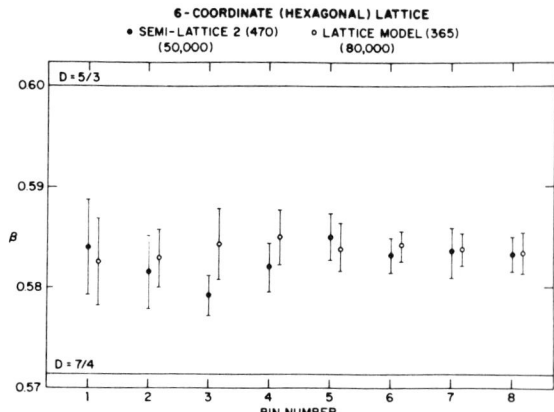

FIGURE 5
This figure shows that the exponent β and the corresponding effective dimensionality D_β are insensitive to cluster size for simulations carried out on a hexagonal lattice using both lattice and semi-lattice models for diffusion limited aggregation.

4. RESULTS FROM OTHER MODELS

Several other models have been developed in order to investigate the effects of anisotropy on DLA. Figure 6 shows a modification of the hexagonal "lattice" model in which the steps in the on lattice random walk near to the cluster are only permitted in 3 of the 6 possible directions. This model leads to clusters which have a "triangular" shape. Figure 7 shows how the radius of gyration exponent β depends on cluster mass for this model. In this case the increase in β with increasing cluster mass is quite pronounced and it seems that the limiting

Simulation of diffusion limited aggregation

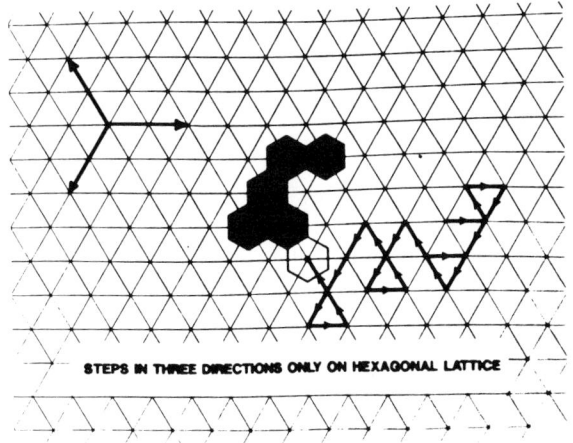

FIGURE 6
The hexagonal lattice model for diffusion limited aggregation with random walk steps in 3 directions only. A typical trajectory resulting in aggregation is shown. The shaded hexagons belong to the growing cluster and the open hexagon indicates the new occupied site added as a result of the trajectory shown. Further away from the cluster, off-lattice random walks are used and long steps are permitted.

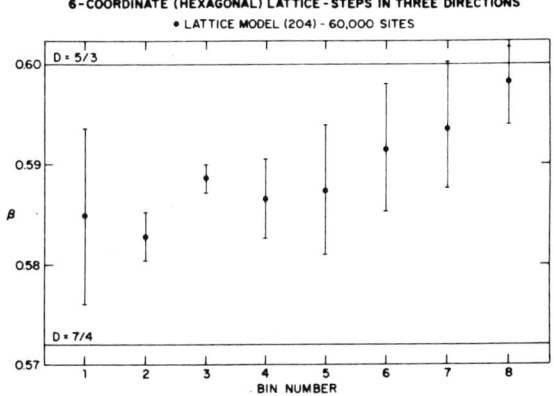

FIGURE 7
Results obtained for the radius of gyration exponent (β) from the two dimensional hexagonal lattice DLA model with random walk steps in three directions only.

($N \rightarrow \infty$) value will exceed 0.6 (i.e. $D_\beta < 5/3$).
Simulations have also been carried out on hexagonal lattices in which sticking of the mobile particle is only allowed in 3 of the six possible directions. This model is illustrated in Figure 8. Figure 9 shows a randomly selected cluster which has been generated using this model. Similar simulations have been carried out using a semi-lattice model in which the mobile particles are rotated so as to form bonds with the contacting particles on the cluster in three directions only. After contact the particle is rotated about the center of the contacting particle in the cluster until the new bond is oriented parallel to the closest of 3 possible directions which are oriented at angles of 120° with respect to each other. The clusters obtained from this model have an appearance very similar to those obtained from the "lattice" model (Figure 9).

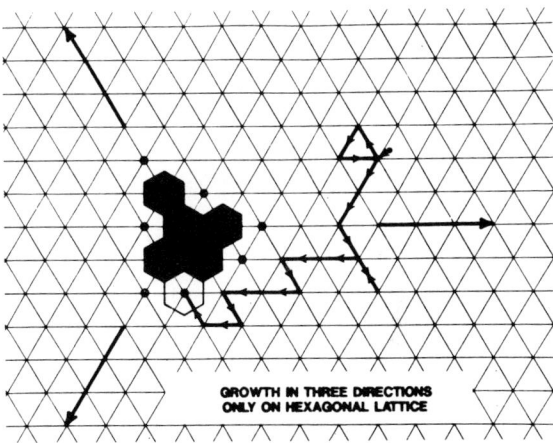

FIGURE 8
This figure shows an early stage in a DLA simulation using a hexagonal lattice with sticking in 3 directions only. The unoccupied surface sites at which growth can occur are indicated by small open circles. The mobile particle can occupy the other surface sites but does not stick. At no stage is the mobile particle allowed to enter occupied (shaded) sites on the cluster.

Figure 10 shows how the exponent β depends on the range of cluster masses used to estimate its value. This exponent increases strongly with increasing cluster size and may approach a limiting value of 2/3. In this case, the internal structure of the clusters does not seem to be fractal-like and it is not appropriate to interpret the radius of gyration exponent, β, in terms of a fractal dimensionality (D_β).

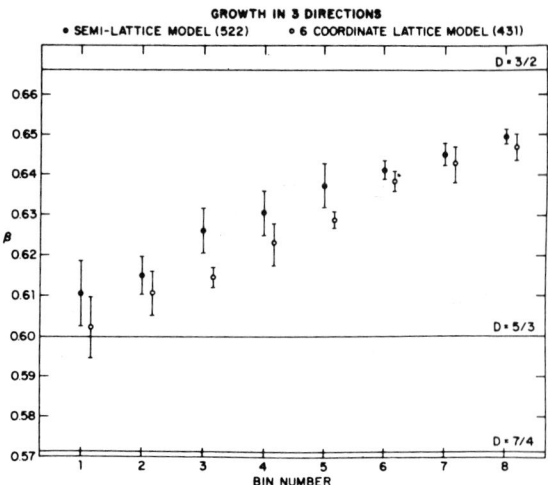

FIGURE 10
Results obtained for the radius of gyration exponent β for lattice model and semi-lattice model simulations of DLA with sticking in three directions only.

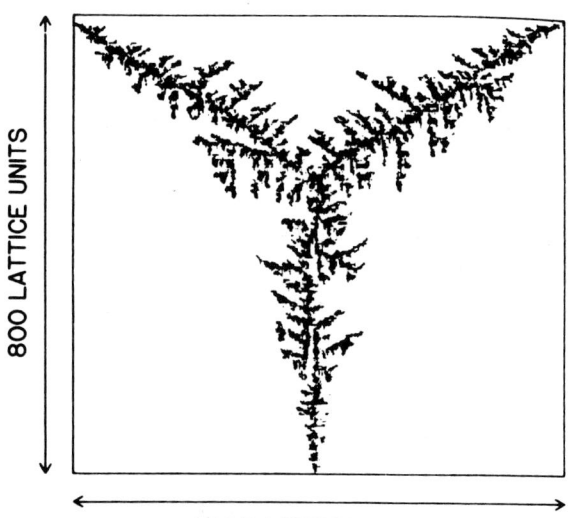

FIGURE 9
A typical 40,000 site cluster obtained using the model depicted in Figure 8. (DLA on a hexagonal lattice with sticking in 3 directions only).

5. DISCUSSION

One of the main objectives of this work is to check the theoretical predictions of Turkevich and Sher for DLA on two dimensional square, triangular and hexagonal lattices. For the two dimensional square lattice our results are consistent with the predictions of Turkevich and Sher (D=5/3) but for the other lattices we find D≈1.71 instead of 1.75. In these cases the statistical uncertainties are much smaller than the difference between the theoretical and simulation results and there is no indication of significant finite size effects. Despite the disagreement between theory and simulation our results can be rationalized using the ideas contained in the theory of Turkevich and Sher. According to this theory the growth of a DLA cluster is controlled by the most rapidly growing regions (the tips of the cluster). The "rate of growth" of the DLA clusters (and their fractal dimensionality) is then controlled by the strength of the singularities in the distribution of growth probabilities which are associated with these tips. It is assumed that the strength of the singularities is the same as that for a regular object having the shape of the lattice. Thus the strength of the singularities is determined by the angle (θ) associated with the lattice structure (θ=90° for the square lattice and 120° for the hexagonal and triangular[27] lattices). According to this theory the fractal dimensionality is given by

$D = (3\pi - \theta)/(2\pi - \theta)$

The stronger the leading singularity in the distribution probability (the smaller the angle θ) the smaller will be the fractal dimensionality.

For the case of off-lattice simulation, the fractal dimensionality of about 1.71 can be associated with an effective angle of about $108°$[16,19] (the angle associated with a pentagon). The singularity associated with the distribution of growth probabilities in off-lattice DLA is stronger than that computed for DLA on hexagonal and triangular lattices ($\theta=120°$). Consequently, it is not unreasonable to expect that DLA on these lattices will grow with a fractal dimensionality of 1.71 instead of 1.75 since stronger singularities are associated with completely random off-lattice DLA than are associated with the lattice. This leads to a picture in which DLA on lattices of high symmetry ($\theta \geq 108°$) lead to structures in the same universality class (same fractal dimensionality) as off-lattice DLA. For lattices of lower symmetry the fractal dimensionality depends on the lattice structure and these models belong to a different universality class than off-lattice DLA.

The results obtained by introducing other forms of anisotropy into the DLA model (see above) indicate that these conclusions can probably be extended to these cases as well. For the case of anisotropic sticking probabilities on a square lattice, Ball et. al.[21] have shown that compact needle-like structures are formed in which the growth in the "easy" direction can be described by the scaling relation $X \sim N^{2/3}$ and the growth in the "hard" direction by $Y \sim N^{1/3}$.

The theory of Turkevich and Sher is supported by the observation that large DLA clusters grown on a square lattice take on a diamond-like shape characterized by an angle of $90°$.[24,28] However, we do not know if this shape is stable or evolves into a "cross" shaped structure with a possibly non-fractal internal structure. The results shown above for anisotropic sticking probabilities and those obtained by Ball et. al. suggest that this could happen. It is also possible that the decrease in the effective value for D_β with increasing cluster size is associated with the "distortion" to a diamond shape and that D_β will return to a value of about 1.71 when this process is complete. Much larger scale simulations and a much more careful analysis of the simulation results will be needed to resolve these questions.

It should also be noted that the effective value for D_β is essentially equal for all of the models studied in this paper for small cluster sizes. This explains why no indications of non universality in the fractal dimensionality of DLA were found in earlier work with small clusters.[2,4]

REFERENCES

1. B. B. Mandelbrot, The Fractal Geometry of Nature, W. H. Freeman and Company (San Francisco 1982).

2. T. A. Witten and L. M. Sander, Phys. Rev. Lett. 47 (1981) 1400.

3. P. Meakin, Phys. Rev. A27 (1983) 604; A27 (1983) 1495.

4. T. A. Witten and L. M. Sander, Phys. Rev. B27 (1983) 5686.

5. R. M. Brady and R. C. Ball, Nature 309 (1984) 225.

6. M. Matsushita, M. Sano, Y. Hayakawa, H. Honjo, and Y. Sawada, Phys. Rev. Lett. 53 (1984) 286.

7. W. T. Elam, S. Wolf, S. A. Sprague, D. V. Gubser, D. VanVechten, G. G. Barz, Jr. and P. Meakin, Phys. Rev. Lett. 54 (1985) 701.

8. L. P. Kadanoff, preprint

9. J. Nittmann, G. Daccord and H. E. Stanley, Nature 314 (1985) 141.

10. C. Tang, Phys. Rev. A31 (1985) 1977.

11. L. Niemeyer, L. Pietronero and A. T. Wiesmann, Phys. Rev. Lett. 54 (1985) 1346.

12. T. Vicsek, Phys. Rev. Lett. 53 (1984) 2281.

13. T. Szep, T. Cserti and T. Kertesz, preprint.

14. E. Bern-Jacob, R. B. Godbey, J. E. Mueller and L. M. Sander, Bull. Am. Phys. Soc. 30 (1985) 269.

15. H. Gould, F. Family and H. E. Stanley, Phys. Rev. Lett. 50 (1983) 686.

16. M. Muthukumar, Phys. Rev. Lett. 50 (1983) 839.

17. M. Tokuyama and K. Kawasaki, Phys. Lett. 100A (1984) 337.

18. H. G. E. Hentschel, Phys. Rev. Lett. 52 (1984) 212.

19. L. Turkevich and H. Sher, preprint

20. L. Turkevich and H. Sher (these proceedings)

21. R. C. Ball, R. M. Brady, G. Rossi and B. R. Thompson, preprint

22. T. C. Halsey, P. Meakin and I. Procaccia, preprint

23. P. Meakin in "On Growth and Form. Proc.", 1985 Cargese NATO ASI, H. E. Stanley and N. Ostrowsky, Editors, Martinus Nijhoft, Amsterdam.

24. P. Meakin, J. Phys. A18 (1985) L661.

25. H. E. Stanley, J. Phys. A10 (1977) L211.

26. P. Meakin and L. M. Sander, Phys. Rev. Lett. 54 (1985) 2053.

27. Although the triangular lattice has three fold symmetry the environments of adjacent lattice sites are not equivalent but are related by a 180° rotation. Consequently, growth processes on a triangular lattice behave as if the lattice had six fold rather than three fold symmetry.

28. R. M. Brady and R. C. Ball, CECAM Workshop, Orsay 1984 (unpublished).

INTERNAL ANISOTROPY OF DIFFUSION-LIMITED AGGREGATES

Paul MEAKIN

Central Research and Development Department, Experimental Station, E.I. du Pont de Nemours and Company, Wilmington, DE 19898, USA

Tamás VICSEK

Department of Physics, Emory University, Atlanta, GA 30322, USA[*]

The correlations within a layer at a distance R from the origin of large diffusion limited aggregates were investigated by determining the tangential correlation function $c_R(\theta)$ as a function of the angle θ. We find that for $\theta \ll 1$ $c_R(\theta)$ decays algebraically with an exponent $\alpha_\perp = 0.41 \pm 0.04$ which is different from the corresponding value obtained for the radial correlations, indicating that the aggregates are internally anisotropic. This fact has an effect on the results obtained for the fractal dimension in the form of the appearance of a correction-to-scaling term.

1. INTRODUCTION

Diffusion-limited aggregation (DLA) has been introduced by Witten and -Sander[1] in order to provide a simple model for a broad class of growth processes in which diffusion is the rate limiting step and the rearrangement of the material within the clusters is not allowed. The DLA clusters are large, randomly ramified objects having a fractal dimension[2], D, less than the Euclidian dimension of the space in which the aggregation process takes place.

The fractal dimension is a global property of a cluster and it does not provide an insight into the structural details of the aggregates. On the other hand, we expect that the diffusion-limited mechanism by which the DLA clusters are grown has a very specific impact on the correlations and interrelation of branches inside a large aggregate.

In this paper we investigate the internal structure of diffusion-limited aggregates by determining the density-density correlations in a layer at a distance R from the origin of the cluster, $c_R(\theta)$, as a function of the angle θ measured from the origin of the clusters, with θR being the distance separating to sites in the layer.

2. RESULTS FOR $c_R(\theta)$

In our simulations large two-dimensional DLA clusters (consisting of 50,000-100,000 particles) were generated using improved algorithms[3]. Given the coordinates of the particles within an N particle aggregate we calculate the tangential density-density correlation function $c_R(\theta)$ in a layer of width ΔR being at a distance R from the origin

---

[*]Permanent address: Research Institute for Technical Physics, Budapest, Ujpest 1, Pf.76, H-1325, HUNGARY

(Figure 1.) using the expression

$$c_R(\theta) = N^{-1} \sum_{\theta'} p_R(\theta+\theta') \, p_R(\theta') \qquad (1)$$

where $p_R(\theta)=k$ if there are k particles in a box of size $\Delta R \Delta \theta$ at the point (R,θ) and $p_R(\theta)=0$ otherwise. The summation in expression (1) is taken over θ' values increased by a fixed small $\Delta \theta'$ from $\theta'=0$ to $\theta'=\pi$.

Expression (1) was first used to evaluate data obtained for 168 clusters grown on a square lattice with off lattice random walk particle trajectories[3] and containing of 80,000-100,000 particles. Figure 2 shows the results of these large scale simulations. (The curves are not smooth because of the statistical errors.) Unlike the radial correlation function the tangential correlation functions displayed in Figure 2 have a pronounced minimum and a second maximum which become sharper as the distance of the layer from the origin R (in which $c_R(\theta)$ is determined) is increased. The position of the second maximum is approximately at $\ln(\theta_{min}) \simeq 0.45 \simeq \ln(\pi/2)$ corresponding to an anisotropic structure with four large branches and an overall shape having the symmetry of a square.

As expected, the angular correlation function for not too large values of θ decays as a power law with some exponent $\alpha_\perp$. We determined this exponent for several values of R in order to take into account finite-size effects. From Figure 4 one can see that $\alpha_\perp$ for R<200 decreases with increasing R. On the other hand for 200<R<400 the $\alpha_\perp$ values seem to saturate about 0.39. Assuming that the latter tendency is valid for much larger radii as well, our extrapolated estimate

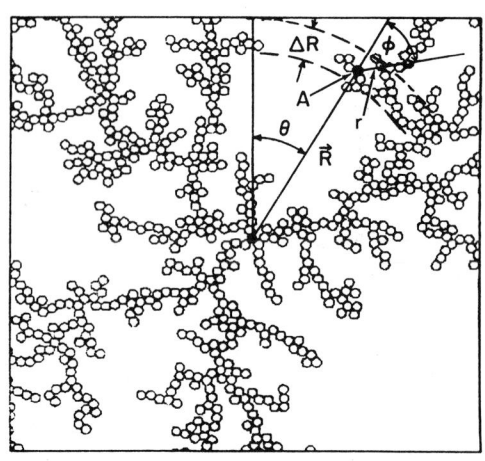

FIGURE 1
Central part of an off lattice diffusion-limited aggregate. The tangential correlations were determined in a layer of width R at a distance R from the origin as a function of the angle . The arguments of the orientation dependent correlation function $c(r,\phi)$ at the point A are the distance from this point r and the local angle ϕ.

for DLA clusters grown on a square lattice is $\alpha_\perp = 0.39 \pm 0.04$. This value is definitely different from $\alpha = 0.29 \pm 0.02$ which describes the algebraic decay of the ordinary radial correlation function[1,3] $c(r) = N^{-1} \sum \rho(r'+r)\rho(r')$.

It is natural to assume that we obtained diamond shaped clusters because the aggregation process was biased by the underlying square lattice. Therefore, we carried out additional simulations using a completely off lattice model. In Figure 3 the angular correlations in several layers of large nonlattice aggregates are shown. These data were obtained from 44 clusters consisting of 50,000 particles. As before the angular correlation function has a minimum, but no well defined maximum can be observed on these curves. The linear regions of the plots describe a

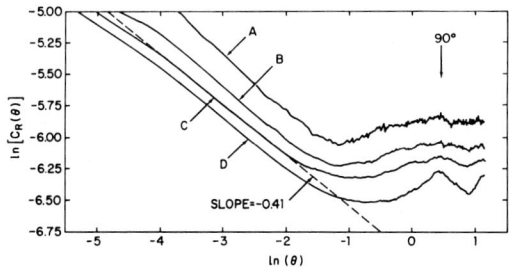

FIGURE 2
The tangential correlation function $c_R(\theta)$ as a function of the angle θ for different values from the origin R. These results were obtained for 168 diffusion-limited aggregates grown on a square lattice and containing of 80,000-100,000 particles. In order to get curves a-d the data were averaged over the intervals $\delta R = R \pm 0.05R$ with a) R=100; b) R=200; c) R=300 and d) R=400. The second maximum at $\theta \simeq \pi/2$ indicates that the overall shape of the clusters is biased by the underlying square lattice

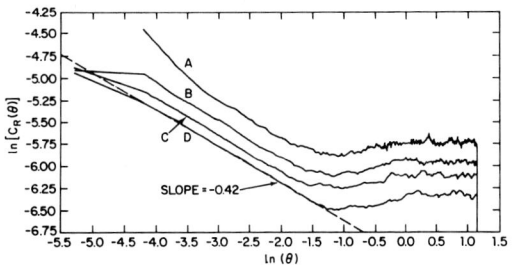

FIGURE 3
Tangential correlations in off lattice diffusion-limited aggregates. Courves a-d were obtained for 44 clusters of 50,000 particles by making an average over the intervals $\delta R = R \pm 0.05R$ with a) R=75; b) R=150; c) R=225 and d) R=300. The position of the minimum of these correlation functions does not seem to change with the increasing distance R from the origin, showing that the aggregates are built up from a few large tree-like branches. The slope of the straight line drawn through the courve d is $\alpha_\perp = 0.42$

power law decay of the angular correlation function with some exponent $\alpha_\perp$. In Fig. 4 we give the α values obtained for various R. The dependence of α on R in the off lattice case is weaker and its value is fluctuating about 0.41 for R >60 suggesting that in the R→∞ limit $\alpha_\perp = 0.41 \pm 0.04$.

3. DISCUSSION

According to our numerical results, the correlations around a given particle belonging to a large aggregate and being at point $\vec{R}$ decay as a function of the particular direction in which the density-density correlations are measured. In order to demonstrate the effects of this fact on the fractal dimension D and on the radial correlation function $c(r)$ determined in the usual way, we make an assumption for the correlation function $c(r,\phi)$ depending both on r (denoting the distance from the given particle) and the local angle ϕ (with $\phi=0$ along the radial direction $\vec{R}$, as it is shown on Figure 1). This assumption should represent our data suggesting that for not too large values of r the correlations in the direction parallel to $\vec{R}$ decay as $r^{-\alpha_\parallel}$, while in the direction perpendicular to R decay as $r^{-\alpha_\perp}$. For this purpose we use the expression

$$c(r,\phi) = f(r,\phi) \, (r^{-\alpha_\parallel}\cos^2\phi + \rho_1 r^{-\alpha_\perp}\sin^2\phi) \quad (2)$$

where ρ_1 is a constant. Equation (2) becomes an approximation if we assume that the supposedly slowly varying function $f(r,\phi)$ is equal to a constant ρ_0.

As a consequence of (3) the fractal dimension is given by $D = \lim_{a\to\infty}\{\ln(N(a))/\ln(a)\} = \lim_{a\to\infty} 2 - \alpha_\parallel + D_1(a) = 2 - \alpha_\parallel$ with a correction-to-scaling term $D_1(a) = \ln\{\rho_0 \pi(1+\rho_1 a^{\alpha_\parallel - \alpha_\perp})\}/\ln(a)$. Therefore,

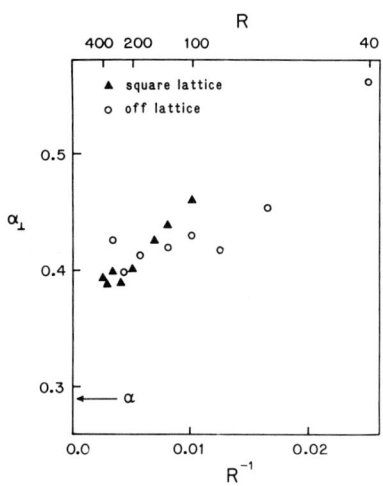

FIGURE 4
Finite size scaling plot of the estimates for $\alpha_\perp$ versus R^{-1} for the square lattice and the off lattice cases. The data represent the slopes of the straight lines giving the best least-square fit to our results for the tangential correlations on the interval $0.035<\theta<0.135$. A recent estimate (Ref. 3.) for the radial correlation function exponent, α, is also shown for comparison

the radial distribution function (which is an average of $c(r,\phi)$ over the angles ϕ) and the value of the fractal dimension D are determined by the correlations along the radial directions in the aggregate and the fact that $\alpha_\parallel < \alpha_\perp$ results only in the appearance of a lower order term diminishing as $a \longrightarrow \infty$. This correction-to-scaling effect is probably one of the reasons for the overestimates of the exponent obtained from the early simulations of DLA clusters consisting a considerably smaller number of particles than in the present study.

ACKNOWLEDGEMENTS

One of the authors (T. V.) thanks Fereydoon Family for useful discussions. The research conducted at Emory University was supported by grants from the Emory University Research Fund, and by the National Science Foundation grant no. DMR-82-08051.

REFERENCES

1. T.A. Witten, Jr. and L.M. Sander, Phys. Rev. Lett. 47, 1400 (1981); T.A. Witten, Jr. and L.M. Sander, Phys. Rev. B27, 5686 (1983).

2. B.B. Mandelbrot, The Fractal Geometry of Nature (Freeman, San Francisco, 1982).

3. P.Meakin, unpublished

GROWING INTERFACE IN DIFFUSION-LIMITED AGGREGATION AND IN THE EDEN PROCESS

M. PLISCHKE and Z. RÁCZ*

Physics Department, Simon Fraser University, Burnaby, B.C., Canada, V5A 1S6

Growth is described in terms of the motion of an active zone defined as that region of the cluster which collects the newly arriving particles. It is shown that the active zone can be specified by two parameters, the average deposition radius of the N-th particle ($\bar{r}_N \sim N^\nu \sim N^{1/D}$ where D is the fractal dimension of the infinite cluster) and the width of the active zone which scales independently ($\xi_N \sim N^{\bar{\nu}}$ with $\bar{\nu} < \nu$). It is argued that in case of diffusion-limited aggregation (DLA) a small difference in $\nu - \bar{\nu}$ might arise because of the intrinsic anisotropy of DLA clusters. More transparent is the Eden model where the appearance of the second length-scale is traced back to the non-equilibrium dynamics of the process.

1. INTRODUCTION

It must be quite clear now from the preceding lectures[1] that there are several growth processes such as diffusion limited aggregation[2] (DLA) which produce scale invariant structures and that a basic theoretical question is how to calculate the quantities characterizing this scale invariance, how to calculate e.g. the fractal dimension of the resulting structures.

You might have also noticed in the previous talks[3,4] that a common feature of these growth processes is that growth occurs in the outer regions of the clusters. The reason for this might be screening as in case of DLA or exclusion as in case of the Eden model[5]. The main point, however, is that there always exists an outer "active" zone[6] which collects all the newly arriving particles. This active zone moves outward and leaves behind a frozen-in structure the properties of which we would like to calculate. Clearly, since the frozen-in part is built in the active zone, to understand its properties we must understand the properties of the active zone. This provides the basic motivation for studying the surface structure of growing clusters. (Some other incentives more directly related to experiments are listed in Ref. 3 and 4).

The first step in understanding something is always the description of the object of interest. So the first problem is how to characterize the active zone. As a simple approach[7] one might try to investigate only the radial (spherically averaged) properties of the cluster and ask the following question. What is the probability P(r,N)dr that the N-th particle is attached at a distance r from the centre of mass of the existing N-1 particle cluster. If the "active-zone" picture is valid then the probability density P(r,N) is expected to be zero both in the frozen-in zone ($r \ll R_N$) and far-away from the cluster ($r \gg R_N$ where R_N is the characteristic radius of the N-particle cluster). As can be seen from Fig. 1 this expectation is verified for two-dimensional (d=2) DLA. Similar pictures could be shown for other dimensions and other models[6,8] as well.

A remarkable feature of Fig. 1 is that the

---

*On leave from the Institute for Theoretical Physics, Eötvös University, H-1088 Budapest, Hungary.

solid lines are Gaussians and they fit the Monte Carlo data with great accuracy. This means that, as far as the spherically averaged properties are concerned, the growth process is completely characterized by two parameters, namely by the *average deposition radius* $\bar{r}_N$ (the center of the Gaussian) and by the *width of the active zone* ξ_N (the width of the Gaussian). This result seems to be valid quite generally. It holds for DLA and the Eden model in dimensions d=2 and 3 and for various generalizations[8] of the Eden model in d=2.

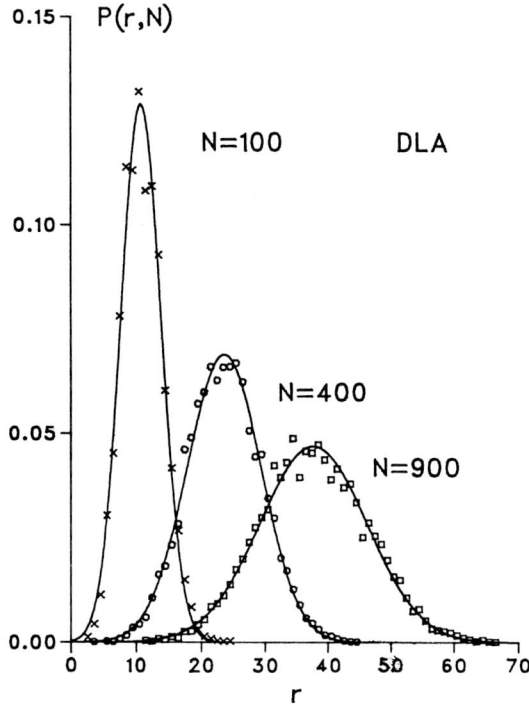

FIGURE 1
Probability density $P(r,N)$ as obtained from Monte Carlo simulations of DLA in d=2. The solid lines are Gaussian fits. For details see Ref. 6.

The description of the growth process is reduced now to understanding the functions $\bar{r}_N$ and ξ_N. Since $\bar{r}_N$ is practically the radius of the N-particle cluster, for large N we must have

$$\bar{r}_N \sim N^\nu \sim N^{1/D} \qquad (1)$$

where D is the fractal dimension of the frozen-in structure. In general, one expects some kind of scaling for ξ_N as well

$$\xi_N \sim N^{\bar{\nu}} \qquad (2)$$

and then the study of the large N behavior of the cluster is simplified to the study of the two exponents ν and $\bar{\nu}$.

The theory of equilibrium critical phenomena of isotropic systems would suggest that actually we have only one parameter since there is only one characteristic length in those systems and thus $\bar{\nu} = \nu$. The point I would like to make in this lecture is that growing clusters, in general, are neither isotropic nor are they in equilibrium thus the appearance of an extra characteristic length should not be surprising; $\bar{\nu} \neq \nu$ is the rule rather than an exception.

2. DIFFUSION-LIMITED AGGREGATION IN d=2

The question of whether $\bar{\nu} = \nu$ or $\bar{\nu} \neq \nu$ has not been settled in this case. Monte Carlo (MC) simulations on relatively small clusters[6] indicate $\bar{\nu} < \nu$ but the effective value of $\bar{\nu}$ seems to be increasing with cluster size[9] and for the largest clusters (50000 particles) $\nu - \bar{\nu}$ is of the order of 0.05. I would like to argue, however, that the increasing effective value of $\bar{\nu}$ does not mean that $\bar{\nu} \to \nu$ in the $N \to \infty$ limit since a difference of $\nu - \bar{\nu} \approx 0.05$ is expected[10] to arise from the small anisotropy observed in DLA clusters.[11]

The argument is based on the fact that a diffusing field penetrates an absorbing structure a distance ℓ if the structure contains gaps of size ℓ. For DLA this means that the penetration depth of the incoming particles and, consequently, the width of the active zone ξ can be calculated as the average distance between neighboring branches at the deposition radius

(Fig. 2). The average distance between the branches can be found in the following way.

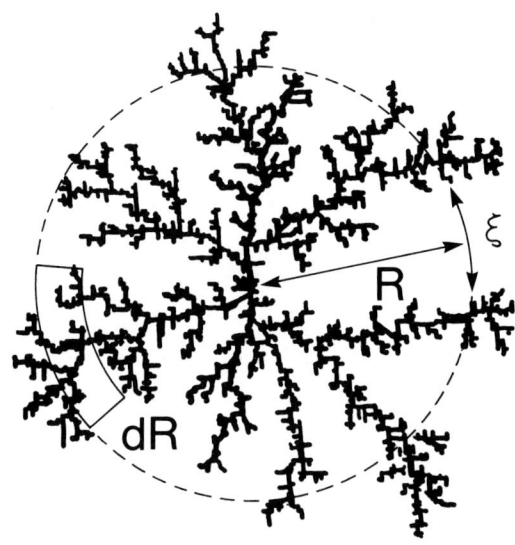

FIGURE 2
DLA cluster with the average radius of deposition drawn as a dotted circle. The width of the active zone ξ is calculated as the average distance between neighboring branches.

First we count the number of branches at the deposition radius $\bar{r}_N$. Since the density of particles $\rho(\bar{r}_N)$ scales as $\rho \sim \bar{r}_N^{-d+D}$ the number of particles in a shell of width dr is $dN \sim \bar{r}_N^{D-1} dr$. The number of particles dN_1 belonging to a given branch in the shell is also given by $dN_1 \sim \bar{r}_N^{\alpha} dr$ but α is not known. If the cluster is porcupine like and consists of one dimensional branches (extreme anisotropy) then $\alpha=0$ whereas in case of isotropic branches $\alpha=D-1$. In general, one expects $0<\alpha<D-1$. Next, we find the number of branches n as $n \sim dN/dN_1 \sim \bar{r}_N^{D-1-\alpha}$ and distributing this number of points evenly on a d-dimensional sphere of radius $\bar{r}_N$ one obtains the average distance between the branches

$$\xi \sim \bar{r}_N^{(d-D+\alpha)/(d-1)} \qquad (3)$$

One can see that $\bar{\nu} = \nu$ holds only in the isotropic ($\alpha=D-1$) case and small anisotropy might produce a small difference between ν and $\bar{\nu}$. Anisotropy of DLA clusters has actually been measured[11] and from those results one might infer[10] $\alpha \approx 0.59$ leading to the estimate $\bar{\nu} \approx 0.51$. This result shows that if anisotropy is not a finite-size effect then $\nu - \bar{\nu} \approx 0.07$.

The phenomenon of two distinctly diverging length-scales produced by anisotropy is not new. It can be observed in equilibrium critical phenomena as well (for a simple example see directed percolation[12]). The next example will demonstrate, however, that there is a more general and more distinct cause for the appearance of a second length-scale in growth processes, namely the nonequilibrium nature of the process.

3. EDEN MODEL IN d=2

There are several slightly different prescriptions to grow Eden clusters. Here we shall consider version C introduced by Jullien and Botet.[13] In this version an occupied surface site of the cluster is chosen randomly and a new particle is added equiprobably to one of the adjacent empty sites. The results of our MC simulations of this process are shown on Fig. 3. The data for the radius of gyration ($R_g \sim N^{\nu} \sim N^{1/2}$) demonstrates convincingly that the clusters are compact[14] and it is also quite clear that the other length-scale, the width of the surface zone, $\xi \sim N^{\bar{\nu}}$ diverges with an exponent $\bar{\nu}$ different from ν.

In order to understand the "roughening" of the surface of Eden clusters it is convenient to study the growth process in a strip geometry[13] i.e. to restrict the growth to a strip of width L. For initial conditions a substrate may be used in which all the sites up to height $h_0=0$ are occupied (Fig. 4). The strip geometry provides a convenient separation of the control parameters. The width of the strip, L, and the average height of the deposit $\bar{h}$, or, in appropriate units, the time of the growth $t \sim \bar{h}$ can be varied independently. This is quite different from the "circular" geometry usually considered[6,15,16] where the cluster is grown from a seed particle and a single parameter N controls both the "height" which might be considered to be the mean radius $R \sim N^{1/2}$ and the "strip-width" which is equivalent to the circumference at the mean radius $2\pi R \sim N^{1/2}$. The curvature effects are, of course, expected to disappear for $N \to \infty$. Thus the scaling properties of Eden clusters in "circular" geometry can be obtained from those in strip geometry provided $t \sim \bar{h} \sim L \sim N^{1/2}$ is chosen.

The separation of control parameters is useful because it makes us free to consider limits which are simple but inaccessible in "circular" geometry. Especially important is the $t \to \infty$ limit since the growth process becomes stationary for large t and the surface properties of the moving front become time independent.[13] The hope is that this stationary state can be more readily treated by analytical methods and then the $t \sim \bar{h} \sim L \sim N^{1/2}$ limit may be obtained by studying relaxation towards the stationary state.

At present, however, we can report only on MC simulations. Jullien and Botet[13] found that the width of the surface zone obeys scaling in the following form

$$\xi(L,t) \sim L^{\alpha} G(t/L^z) \qquad (4)$$

where $\alpha = 0.50 \pm 0.03$ and $z = 1.7 \pm 0.3$. The

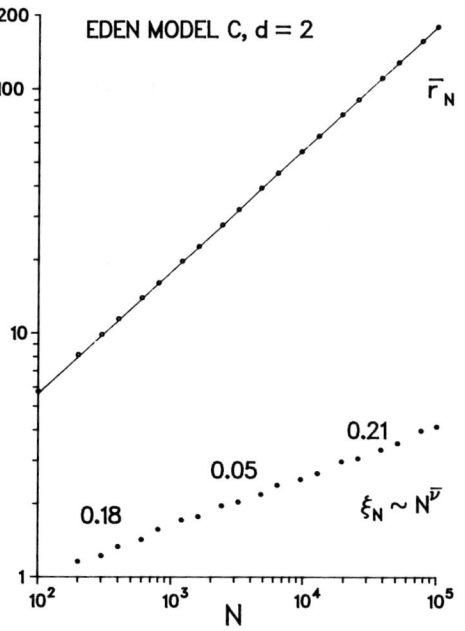

FIGURE 3
Average radius of deposition $\bar{r}_N$ and width of the active zone ξ_N of Eden clusters obtained from MC data on 1000 clusters in version C of the growth process. The solid line denotes the compact-cluster result $\bar{r}_N \sim (N/\pi)^{1/2}$. The numbers above the lower curve are the effective-exponent estimates for ξ_N in the corresponding decade of N.

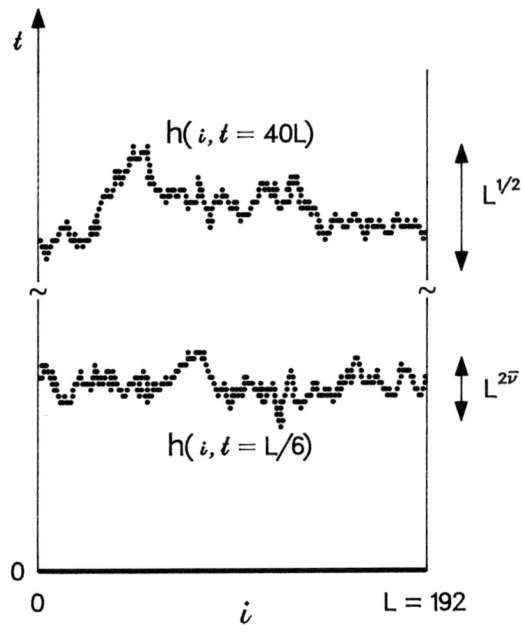

FIGURE 4
Eden growth in strip geometry. The black dots denote the surface sites at various stages of development of the cluster.

scaling function $G(x)$ is obviously finite, $G(x) \to G(\infty) \neq 0$, for $x \to \infty$ while in the less obvious limit $x \to 0$ it has been conjectured[13] to behave as $G(x) \sim x^{\alpha/z}$ (then the L-dependence in ξ cancels for small t). The small x behavior of $G(x)$ is important in finding the $t \sim L \sim N^{1/2} \to \infty$ limit and thus obtaining the exponent $\bar{\nu}$:

$$\xi(L,L) \sim L^\alpha G(L^{1-z}) \sim L^{\alpha/z} \sim N^{\alpha/(2z)} \sim N^{\bar{\nu}} \quad (5)$$

and consequently

$$\bar{\nu} = \alpha/(2z) \quad . \quad (6)$$

Since the scaling form (4) suggests interesting analogies with dynamical critical phenomena and since the unchecked $x \to 0$ limit is important in obtaining eq. (6) we undertook a more detailed study of the surface by expanding it into Fourier modes and investigating the static and dynamic properties of these modes.

To use Fourier analysis, a single valued function for the surface height is needed. Although the number of overhangs and holes are very limited in version C of the Eden model, the surface occasionally consists of more than one occupied site above a particular point of the substrate. A way to overcome this problem is to define a locally averaged height $h(i,t)$ for substrate point i at time t:

$$h(i,t) = \frac{1}{n_s(i,t)} \sum_{j=1}^{n_s(i,t)} h_j(i,t) \quad (7)$$

where $h_j(i,t)$ is the height of the j-th surface site in column i, $n_s(i,t)$ is the number of surface site in that column, and t is measured in number of particles deposited per substrate site. Now the Fourier analysis of this "smoothed out" surface is carried out by defining

$$\hat{h}(q,t) = \frac{1}{\sqrt{L}} \sum_{j=1}^{L} [h(j,t) - \bar{h}(t)] e^{iqt} \quad (8)$$

where $q = \pm 2k\pi/L$, $k = 1,2,\ldots,L-1$ and $\bar{h}(t)$ is the average height $\bar{h} = \sum_j h(j,t)/L$. In terms of $\hat{h}(q,t)$ one can write

$$\xi^2(L,t) = \frac{1}{L}\sum_{j=1}^{L} [h(j,t) - \bar{h}(t)]^2 = \frac{1}{L}\sum_q |\hat{h}(q,t)|^2 \quad (9)$$

Thus we see that the square of the amplitude of the Fourier modes determines the width of the surface.

Our MC results[16] for $|\hat{h}(q,t)|^2$ show that in the steady state ($t \to \infty$) this function is practically independent of L and $|\hat{h}(q,\infty)|^2 \sim q^{-2}$. This immediately establishes that

$$\xi^2(L,\infty) \sim \int_{2\pi/L}^{\pi} dq/q^2 \sim L \quad (10)$$

a result which is in agreement with $\alpha = 0.50 \pm 0.03$ obtained in Ref. 13. The finding $\xi \sim \sqrt{L}$ reminds one of the properties of interfaces in roughening and Ising models.[17,18] The similarity, however stops at the static aspects of the phenomena. Our MC simulations reveal[16] that $|\hat{h}(q,t)|^2$ obeys dynamic scaling[19]

$$|\hat{h}(q,t)|^2 \sim \frac{1}{q^2}[1-g(q^z t)] \quad (11)$$

but the dynamic critical exponent $z = 1.55 \pm 0.15$ is different from those ($z \geq 2$) found for the dynamical generalizations of the roughening[17] and Ising[20] models. The scaling function $g(x)$ is exponentially small for $x \to \infty$ while $1-g(x) \sim x$ for $x \to 0$. This implies that substituting (11) into (9) one obtains the scaling form (4) for $\xi(L,t)$ with $\alpha = 0.5$ and $z = 1.55 \pm 0.15$ and the behavior $G(x) \sim x^{\alpha/z}$ of the scaling function for $x \to 0$ is not a conjecture any more but a consequence of (11).

The scaling properties of the surface modes of the Eden model closely resemble the properties of modes in a thermodynamic system at a critical point. Further studies of these modes will hopefully help in constructing a phenomenological field theory which will yield the exponents α and z through an appropriate renormalization group treatment.

Clearly, the identification of z with the dynamical critical exponent (11) and the connection between the surface exponent $\bar{\nu}$ and the dynamic exponent z proves our point that a second length-scale arises naturally as a consequence of the nonequilibrium nature of the process. We believe, the connection between $\bar{\nu}$ and z is quite general and this connection will help in clarifying the means by which the classification of growth processes might be achieved.

REFERENCES

1. See preceding lectures in this volume.

2. T.A. Witten and L.M. Sander, Phys. Rev. Lett. 47, 1400 (1981).

3. See lecture by H.J. Wiesmann, C. Evertsz and L. Pietronero in this volume.

4. See lecture by P. Meakin in this volume.

5. M. Eden, Proceedings of the Fourth Berkeley Symposium on Mathematical Statistics and Probability, edited by F. Neyman (University of California, Berkeley, 1961) Vol. IV, p. 223.

6. M. Plischke and Z. Rácz, Phys. Rev. Lett. 53, 415 (1984), Z. Rácz and M. Plischke, Phys. Rev. A31, 985 (1985).

7. For more complicated definitions of the surface zone see P. Meakin and T.A. Witten, Phys. Rev. A28, 2985 (1983), P. Meakin, Phys. Rev. B30, 4207 (1984), Phys. Rev. A32, 453 (1985) and also Ref. 4.

8. A. Bunde, H.J. Herrmann, A. Margolina and H.E. Stanely, preprint 1984.

9. P. Meakin and L.M. Sander, Phys. Rev. Lett. 54, 2053 (1985).

10. M. Plischke and Z. Rácz, Phys. Rev. Lett. 54, 2054 (1985).

11. P. Meakin and T. Vicsek, Phys. Rev. A, in print, M. Kolb, in this volume.

12. W. Kinzel and J.M. Yeomans, J. Phys. A14, L163 (1981).

13. R. Jullien and R. Botet, Phys. Rev. Lett. 54, 2055 (1985), J. Phys. A in print.

14. D. Dhar, Phys. Rev. Lett. 54, 2058 (1985).

15. H.P. Peters et. al. Z. Phys. B34, 399 (1979).

16. M. Plischke and Z. Rácz, preprint 1985.

17. J.D. Weeks and G.H. Gilmer, Adv. Chem. Phys. 40, 157 (1979).

18. D. Jasnow, Rep. Prog. Phys. 47, 1509 (1984).

19. P.C. Hohenberg and B.I. Halperin, Rev. Mod. Phys. 49, 435 (1977).

20. R. Bausch, V. Dohm, H.K. Janssen and R.K.P. Zia, Phys. Rev. Lett. 47, 1837 (1981).

FRACTALS IN PHYSICS
L. Pietronero, E. Tosatti (editors)
© Elsevier Science Publishers B.V., 1986

STICKING PROBABILITY SCALING IN DIFFUSION-LIMITED AGGREGATION

Leonid A. TURKEVICH and Harvey SCHER

The Standard Oil Company, Corporate Research Center, 4440 Warrensville Center Road, Cleveland, Ohio 44128

Scaling relates the DLA Hausdorff dimension D to the perimeter occupancy probabilities of the cluster tips. On a 2d square lattice, we find D=5/3 for DLA, D=2 for the Eden model, and D=2-η/3 for the Brown-Boveri dielectric breakdown model of degree η. With uniaxial anisotropy D=3/2 for DLA. The results are extended to d≥2. We find no upper critical dimension for DLA, although D → d-1 for large d. Our values for D are fully consistent with Meakin's Cartesian lattice simulations for 2≤d≤6. With uniaxial anisotropy, the upper critical dimension is d=3, above which D=d-1.

1. INTRODUCTION

This paper concerns the importance of occupation probabilities in diffusion-limited aggregation (DLA)[1,2]. These are the probabilities for a random-walker to end up on the growing perimeter sites of a DLA cluster. In Section 2, the occupancy probabilities for growth sites are related[3] to the fractal dimension D of the cluster. A simple scaling relation[3] implements the physics that all the growth takes place at, and is driven by, the tips. This reduces DLA to classical electrostatics, with other models as innoccuous variants. In order to calculate D, we need only the scaling, with cluster mass N, of the occupancy probability, $P_{max}(N)$, of the maximally extending tips. In Section 3, we utilize an ansatz[3] that the large-scale structure of the cluster determines the strength of the tip cusps (wedges). This permits a simple electrostatic calculation of the scaling of $P_{max}(N)$, in d=2 via conformal transformations. The theory predicts a mild nonuniversality (lattice dependence) of D. In Section 4, we extend[4] these results to higher dimensions. Consistent with the earliest speculations[2], our theory yields no upper critical dimension for DLA on Cartesian lattices. The results are in good agreement with Meakin's large Cartesian lattice simulations[5,8]. With uniaxial anisotropy, the upper critical dimension is reduced to d=3.

2. RELATION OF THE OCCUPANCY PROBABILITIES TO D

Consider a DLA cluster of N particles. The N+1st random-walker is launched from infinity and may land on any one of the perimeter sites of the cluster (indexed by i) with probability p_i. These occupancy probabilities are not uniform, as they would be for the Eden model[6], but rather depend intimately on the diffusive (fractal) character of the random-walk. In particular, there is a far greater probability for the random-walker to adhere to a tip perimeter site than to a perimeter site along a fjord deep within the cluster. It is this screening that is embodied in our scaling treatment.

If the DLA cluster were fully developed to its extremeties, it would have a nominal radius $r_o = N^{1/D} a$, where a is the lattice constant. However, the interior of the cluster is screened (with screening length ξ): additional random-walkers are only captured by perimeter sites in the "active zone". Thus the DLA cluster is fully developed (i.e. with the number of particles within radius r scaling as $N(r) \sim r^D$) only up to radius $r_- = r_o - \xi$. In order to still accommodate all the N particles of the cluster, the cluster must extend further, to a radius $r_+ > r_o$ (Figure 1, top).

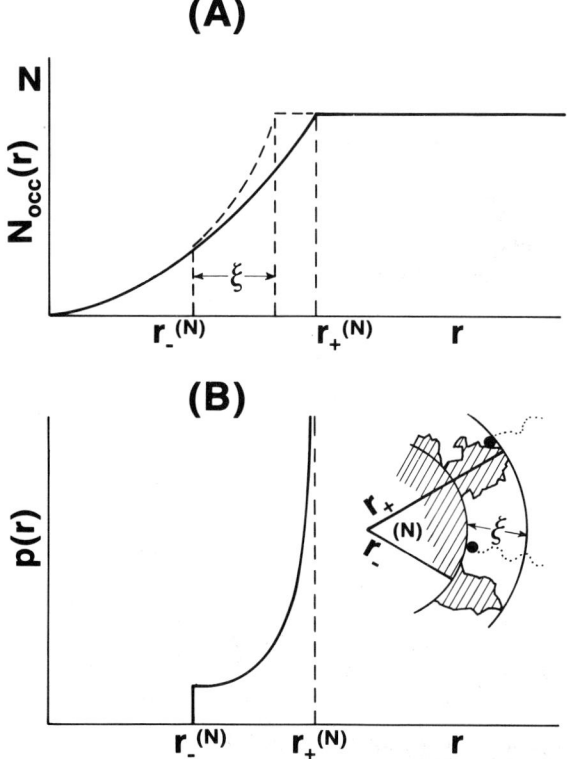

FIGURE 1
A) Schematic of the number of occupied sites N(r) within distance r from the centre of mass. The cluster is fully developed for $r < r_-$ with a lower density edge region for $r_- < r < r_+$. A cluster with no edge region would extend to a nominal radius r_o. B) Schematic of the occupancy probability density $p_N(r)$. No growth occurs for $r < r_-$. $p_N(r)$ is singular at r_+, reflecting maximal growth probability at the tips. Inset: two possible random-walkers, one adhering near a cluster tip $r \sim r_+$, the other adhering deep within the edge region.

Self-similarity ensures that all the lengths, r_o, r_+, ξ, scale similarly with N.

We now consider DLA clusters specifically in d=2. The occupancy probabilities may be expressed as a probability density, namely, $p_N(r)dr$ is the probability that the N+1st random-walker lands in the annulus of width dr at radius r. Clearly $p_N(r)$ is maximal at the tips; in fact, $p_N(r \to r_+)$ diverges (Figure 1, bottom).

This divergence in the occupancy probability density determines D. Given an N-cluster, the probability that the N+1st random-walker will land at the outermost portion of the cluster (the annulus at $r_+^{(N)}$) is

$$P_{max} = \int_{r_+^{(N)} - a}^{r_+^{(N)}} p_N(r)dr , \quad (1)$$

in which case the cluster grows: $r_+^{(N+1)} = r_+^{(N)} + a$. There is zero probability for this walker to land within the fully developed region ($r < r_-$)--this is just the effect of screening. The probability that the N+1st random-walker lands within the active zone $r_-^{(N)} < r < r_+^{(N)}$ (but <u>not</u> at the tips) is $1 - P_{max}$, but in this case the cluster does not grow: $r_+^{(N+1)} = r_+^{(N)}$. Combining these alternatives,

$$r_+^{(N+1)} = P_{max}(r_+^{(N)} + a) + (1 - P_{max}) r_+^{(N)}, \quad (2)$$

which, for large N, becomes

$$\frac{dr_+^{(N)}}{dN} = P_{max}(N) \, a . \quad (3)$$

Equation (3) quantifies the physics that the growth of a DLA cluster is driven by the occupancy probability at the tips. Concomitantly, Eq.(3) also expresses the fact that, with growth, $p_{N+1}(r)$ (the occupancy probability density for the N+2nd random-walker) is altered from $p_N(r)$, while without growth, $p_{N+1}(r) \simeq p_N(r)$. Finally, since $r_+ \sim N^{1/D} a$, the singularity in the occupancy probability density $p_N(r \to r_+)$, as the tip is approached, completely determines the fractal dimension D.

As we require $p_N(r)$ only at the tips, the intricacies of the random, ramified cluster interior are irrelevant--<u>all the growth takes place at the tips</u>. We thus may utilize <u>any</u> object possessing the same cusp structure as a DLA cluster. This is the fundamental feature of our approach to the probability scaling: D does not depend on the cluster's irregular structure but rather on its more regular growth. We thus can utilize an object with the symmetry of the lattice--for the square lattice, a right-angled diamond--in other words, we consider a random-walker in the presence of a square array of absorbing traps. We compute the occupancy probabilities as a function of distance δs away from the corners of the square.

In the infinite time limit, the random-walk diffusion problem reduces[7] to $\nabla^2 u=0$, where u is the concentration field of the random-walker. As the perimeter sites are perfect traps, u=0 along the cluster perimeter. Since random-walkers are isotropically launched from infinity, there is the additional boundary condition that u=1 for $r=R_\infty$. The occupancy probability density $p(\vec{r})$ is just proportional to the flux $\vec{\nabla}u$ of random walkers at perimeter site $\vec{r}$. Thus to obtain the probability density $p(\vec{r})$, we in principle solve the electrostatics problem $\nabla^2 \Phi=0$ for a conducting cluster ($\Phi=0$) inside an infinite radius conducting circle ($\Phi=1$)

$$p(\vec{r}) = |\vec{E}(\vec{r})| / \int_\Pi |\vec{E}(\vec{r})| \, ds \qquad (4)$$

where the electric field $\vec{E}=-\vec{\nabla}\Phi$, and where the probability density is normalized over the cluster perimeter Π. In addition, one may consider[7] a whole class of models where (4) is generalized to

$$p(\vec{r}) = |\vec{E}(\vec{r})|^\eta / \int_\Pi |\vec{E}(\vec{r})|^\eta ds \qquad (5)$$

For $\eta=0$ we recover the Eden model[6], $\eta=1$ DLA and $\eta \neq 1$ a class of dielectric breakdown models considered by the Brown-Boveri group[7].

3. THE LARGE-SCALE STRUCTURE ANSATZ

The electrostatics problem of a randomly-branched conducting cluster within a conducting cylinder is too difficult to solve exactly. Our scaling equation of growth (3), however, only requires the scaling behavior of the probability density <u>at the tips</u> of the cluster. The interior of the cluster is screened, and thus the scaling of P_{max} should be insensitive to the details of the ramified, random interior. In order to obtain the scaling of P_{max}, it suffices to consider <u>any object with the same cusp structure</u> as the DLA cluster.

For a cluster grown on a 2d square lattice, the easy growth directions are precisely oriented along the lattice axes. The cusp structure of the cluster inevitably reflects the symmetry of the lattice. In fact, for the large square-lattice simulations of Meakin[8] and of the Cambridge group[9], the clusters acquire a definite diamond shape (a square rotated $\pi/4$ from the Cartesian axes). We thus propose, as an <u>ansatz</u>, that the cusp structure of a DLA cluster grown on a square lattice is identical to the cusp structure of a diamond outline of sites on the lattice.

The <u>ansatz</u> enables us to calculate, via conformal transformation, the electrostatic potential near the tip of a DLA cluster. Near the tip of a square of length L,

$$E \sim L^{-2/3} \delta s^{-1/3} \qquad (6)$$

where δs measures the distance to a tip. Normalizing with (4) over a screening length $\xi^{(N)}$, we obtain the probability density

$$p_N(r) \sim \frac{2}{3} \xi^{(N)\,-2/3} (r_+^{(N)} - r)^{-1/3}. \quad (7)$$

Using (1), we obtain the occupancy probability at a tip, $P_{max} \sim (a/\xi)^{2/3}$. The scaling equation of growth (3) yields $N \sim (r/a)^{5/3}$, i.e. $D=5/3$ for 2d DLA on a square lattice.

We have verified[3] the consistency of the diamond cusp structure <u>ansatz</u> by examining the numerical scaling of $P_{max}(N) \sim \xi^{-2/3} \sim N^{-2/5}$, using small realizations of DLA cluster grown from a CTRW recipe. Similar scaling ($P_{max} \sim N^{-.39}$) is observed by Meakin[10] with large ($N \sim 5 \times 10^4$) simulations.

The extension of these 2d square lattice results to the more general class of η-models[7] is straightforward. Normalizing (6) according to (5),

$$p_N(r) \sim \xi^{\eta/3-1} (r_+ - r)^{-\eta/3}, \quad (8)$$

where (1) yields $P_{max} \sim (a/\xi)^{1-\eta/3}$. Integrating the scaling equation of growth (3), $N \sim (r/a)^{2-\eta/3}$, i.e. $D_\eta = 2-\eta/3$. This trivially yields $D_0=2$ for the Eden cancer model[6], $D_{1/2}=11/6$ and $D_2=4/3$ for the respective the Brown-Boveri dielectric breakdown models[7] on a 2d square lattice. These last results are in excellent agreement with the simulations of Meakin[8] ($D_\alpha=1.86\pm0.02$, $D_\beta=1.92\pm0.05$, for $\eta=1/2$, and $D_\alpha=1.44\pm0.02$, $D_\beta=1.39\pm0.10$ for $\eta=2$). For $\eta > 3$, the probability distribution (8) ceases to be normalizable, indicating the impossibility of sustaining a fractal structure under such an extreme growth rule[11].

Our <u>ansatz</u> of the large scale structure of the cluster determining the cusp structure may be applied to DLA grown on other lattices. For the 2d triangular lattice we would expect a DLA cluster "in the large" to resemble a hexagon, with cusp angle $2\pi/3$. An argument identical to that used for the square lattice yields $D=7/4$. Meakin's latest numerical results[8] for clusters grown on a hexagonal lattice indicate $D_\beta=1.71$, consistent with that for off-lattice DLA, with stringent enough error bars to statistically (but not systematically) rule out $D=1.75$. However, no discernable hexagonal structure has emerged in these clusters. Either the clusters ($N \sim 10^3$) are not yet large enough to properly identify D, or a hexagon, in fact, never emerges, indicating that the stronger singularity present for off-lattice DLA dominates the singularity generated by the triangular lattice (whereas the still stronger square-lattice singularity drives the 2d square-lattice DLA away from the off-lattice fixed-point to the diamond shape).

We remark that the off-lattice results[8] $D=1.71$ may be qualitatively understood as an averaging over coordination of the accreted random-walkers. The two possible 2d isotropic lattices nicely bracket the off-lattice result: $D=5/3$ (4-fold coordination) and $D=7/4$ (6-fold coordination). It is amusing to note that a hypothetical 5-fold coordinated structure yields $D = 12/7 = 1.7143...$ Expressed in this language, an adequate theory for off-lattice DLA will provide information on the local coordination of the particles in the cluster.

Finally the above results may be extended to anisotropic (but still 4-fold coordinated) lattices, with oblique angle β. A straightforward exercise yields $D=(3\pi-\beta)/(2\pi-\beta)$, which is stringently bounded: $1.50 < D < 1.67$. The (weak) nonuniversality of this result may be understood by "unskewing" the oblique lattice. Naively, we should recover isotropic Cartesian lattice DLA, but the "unskewing" tampers with the distribution of sources at infinity, uniaxially concentrating the source of random-walkers. This prediction is sufficiently dramatic as to warrant testing by simulation. For the case of extreme uniaxial anisotropy, the general class of η-models[7] should have $D_\eta = 2-\eta/2$, with unsustainable fractal structures for $\eta > 2$.

IV. HIGHER DIMENSIONS

We now apply these geometric ideas to higher dimnensions[4]. For simlicity we consider only Cartesian lattices. In d=3 we must integrate the probability distribution away from the tips to obtain

$$P_{max} = \int_{r_+ -a}^{r_+} p_N(r) r dr \quad . \quad (9)$$

The scaling equation of growth (3) is unchanged The large-scale strucutre ansatz motivates the solution of the electrostatics of an octahedron (dual to the simple cubic lattice) held at $\Phi=0$ inside a conducting sphere (at R_∞) held at $\Phi=1$. As the divergence of the electric field near the edges is weaker than at the corners, we solve for the dominant singular behavior near the tip of a rectilinear cone.

With azimuthal symmetry, $\Phi=R(r)\theta(\theta)$; the radial part $R\sim r^\nu$, i.e. $E\sim r^{\nu-1}$, whence (9) yields $D=2+\nu$. The separation constant ν appears in the polar (Legendre) equation

$$\frac{1}{\sin\theta}\frac{d}{d\theta}\left(\sin\theta\frac{d\theta}{d\theta}\right) + \nu(\nu+1)\theta = 0 \quad , \quad (10)$$

and is determined by the boundary condition that the cone be held at constant potential[12]

$$P_\nu(\cos\beta) = F(-\nu,\nu+1;1;z) = 0 \quad , \quad (11)$$

where $P_\nu(\cos\theta)$ is the Legendre function of order ν, $F(\alpha,\beta;\gamma;z)$ is the hypergeometric function, and where $z=(1-\cos\beta)/2$. Note that the cone angle β is measured exterior to the cone from its axis (i.e. $\beta=3\pi/4$ for a rectilinear cone).

The extension to higher dimensions is straightforward[4]. To obtain P_{max}, the probability distribution must be integrated away from the tips of a hypercone

$$P_{max} \sim \int_{r_+ -a}^{r_+} p_N(r) \, d^{d-1}r \quad . \quad (12)$$

Using hyperspherical coordinates and assuming hyperazimuthal symmetry, $\Phi=R(r)\theta(\theta)$; the radial part $R\sim r^{\nu-d+3}$, whence (12) yields $D=2+\nu$. The separation constant ν appears in the polar equation

$$\frac{1}{\sin^{d-2}\theta}\frac{d}{d\theta}\left(\sin^{d-2}\theta\frac{d\theta}{d\theta}\right)+(\nu-d+3)(\nu+1)\theta=0 \quad (13)$$

and is determined by the boundary condition that the hypercone be at constant potential

$$F(d-3-\nu,1+\nu;\frac{d-1}{2};z) = 0 \quad . \quad (14)$$

The radius of gyration exponent $\beta=1/D$ is displayed in Figure 2, as a function of Euclidean dimension d. Also plotted are the

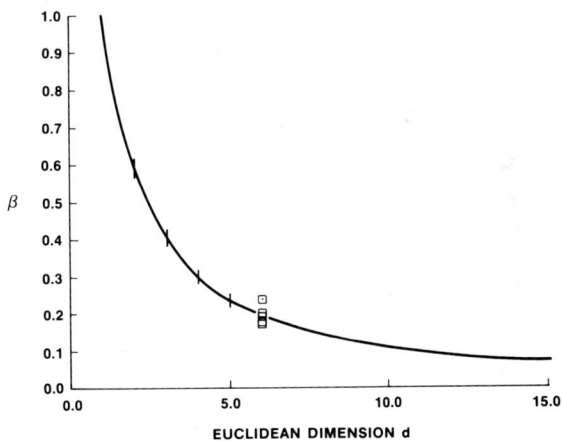

FIGURE 2
Radius of gyration exponent β for DLA clusters grown on d-dimensional Cartesian lattices. Data is from Meakin [5] (vertical lines indicate statistical confidence limits and squares are individual clusters).

simulation results of Meakin[5] for Cartesian lattices of $2 \leq d \leq 6$. The agreement of the simulation results with the theory is remarkable.

Figure 3 displays the Hausdorff dimension itself, plotted as the deviation of D from its mean-field value (the causality lower bound[13]) $d-1$. We find that $D \to d-1$ for large d but that

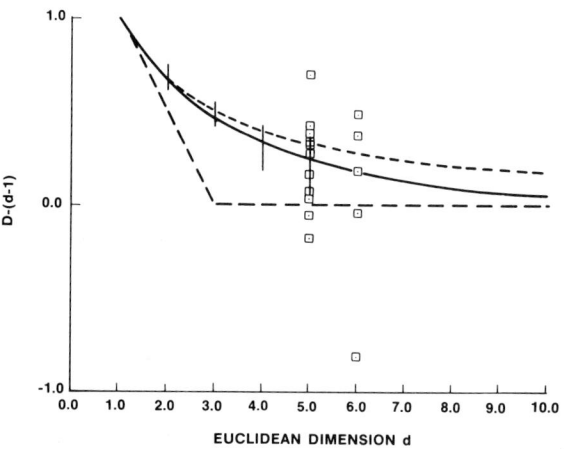

FIGURE 3
Deviation of the Hausdorff dimension D from its mean-field value (causality lower bound [13]) $d-1$. The upper dashed line is the mean-field conjecture [15]. Data (squares and vertical lines) is from Meakin [5]. The lower dashed line is the case of uniaxial anisotropy.

$D > d-1$; thus there is no upper critical dimension of DLA on <u>isotropic</u> Cartesian lattices. The lack of upper critical dimension had been conjectured from the outset[2] and is consistent with the observed[14] lack of dimensional dependence for the mass-scaling of the minimum path length along DLA clusters. Also shown (dotted line on Figure 3) is the conjecture[15], based on mean-field arguments, that $D=(d^2+1)/(d+1)$. We remark that our results (14) for D approach the causality lower bound $d-1$ faster than the "mean-field" formula. We have also plotted on Figure 3 the results of Meakin's Cartesian lattice simulations[5]. The vertical solid lines indicate the statistical confidence limits Meakin has placed on the Hausdorff dimension D for the appropriate Euclidean dimension; where these statistical confidence limits are unavailable, the data squares[5] represent Hausdorff dimensions for individual clusters. The plotted simulation values have been obtained by inverting Meakin's radius of gyration exponent β. For the general class of η-models[7], $D_\eta - d = \eta(D_1 - d)$, with unsustainable fractal growth for $\eta > 1/(d-D_1)$. In particular, the Brown-Boveri $\eta=2$ dielectric breakdown model should not possess fractal solutions for $d \geq 3$.

We finally examine the effect of uniaxial anisotropy on these results, namely their lattice dependence. This is easily effected by changing the exterior angle β of the hypercone. In the extreme uniaxial limit ($\beta \to 0$), $D=(d+1)/2$ for $d \leq 3$ and $D=d-1$ for $d \geq 3$. Thus for such anisotropic lattices, the Hausdorff dimension is reduced to the mean-field value for $d \geq 3$, even though there is no upper critical dimension for the isotropic lattices.

V. SUMMARY

In summary, we have presented a scaling formulation of DLA. We focus, at each stage of the growth, on the perimeter occupancy probabilities for capturing an incident random-walker. Growth occurs predominantly at the cluster tips, and is controlled by the occupancy probability P_{max} of these maximally extending tips. The scaling of P_{max} determines the Hausdorff dimension D of the cluster, and may be obtained using the singular part of the probability density for a regular object with the same cusps as exhibited by the DLA cluster tips. We find $D=5/3$ for DLA on a 2d square lattice and $D=7/4$ on a 2d triangular lattice, thus evincing mild nonuniversality. We have extended our solution to higher dimensions and find no upper critical dimension for DLA on Cartesian lattices: $D \to d-1$. However with uniaxial anisotropy, the upper critical dimension is reduced to $d=3$.

REFERENCES

1. T.A. Witten, Jr. and L.M. Sander, Phys. Rev. Lett. 47, 1400 (1981); P. Meakin, Phys. Rev. A 27, 604, (1983.)

2. T.A. Witten, Jr. and L.M. Sander, Phys. Rev. B 27, 5686 (1983).

3. L.A. Turkevich and H. Scher, Phys. Rev. Lett. 55, 1026 (1985).

4. L.A. Turkevich and H. Scher, Phys. Rev. A, to be published.

5. P. Meakin, Phys. Rev. A 27, 1495 (1983).

6. M. Eden, in Proc. Fourth Berkeley Symp. on Mathematical Statistics and Probability, ed by J. Neyman (U. of California Press, Berkeley, 1961), Vol. 4, p. 223; H.P. Peters, D. Stauffer, H.P. Hölters, K. Loewenich, Z. Phys. B 34, 399 (1979).

7. L. Niemeyer, L. Pietronero, H.J. Wiesmann, Phys. Rev. Lett. 52, 1033 (1984).

8. P. Meakin, in On Growth and Form, A Modern View, ed. by H.E. Stanley and N. Ostrowski (M. Nijhoff, Dordrecht, to be published, 1985).

9. R.C. Ball and R.M. Brady, to be published.

10. P. Meakin, priate communication.

11. We thank L.M. Sander for a question at this conference, which prompts this observation.

12. J.D. Jackson, Classical Electrodynamics, 2nd edition (Wiley, New York, 1975), pages 94-8; R.N. Hall, J. Appl. Phys. 20, 925 (1949).

13. R.C. Ball and T.A. Witten, Jr., Phys. Rev. A 29, 2966 (1984).

14. P. Meakin, I. Majid, S. Havlin and H.E. Stanley, J. Phys. A 17, L975 (1984).

15. M. Muthukumar, Phys. Rev. Lett. 50, 839 (1983); M. Tokuyama, K. Kawasaki, Phys. Lett. 100A, 337 (1984).

CONE ANGLE PICTURE AND ANISOTROPY IN DLA CLUSTER GROWTH

Giuseppe ROSSI, Bernard R. THOMPSON, Robin C. BALL and Robert M. BRADY

TCM Group, Cavendish Laboratory, University of Cambridge, Madingley Road, Cambridge CB3 OHE, United Kingdom.

Computer simulation results for DLA clusters grown with anisotropic sticking rules provide evidence in support of a new model of DLA cluster growth. The model relates the exponents controlling the growth of a DLA cluster to its overall shape. It predicts universal exponents in the case of uniaxial anisotropic sticking. It also gives good estimates for the fractal dimension of ordinary DLA.

1. INTRODUCTION

We present here a new model, hereafter referred to as the *cone angle* picture, to describe the growth of DLA clusters[1,2]. It consists of a set of scaling arguments which provide a link between the large scale features of the cluster (especially the shape of the tips) and the exponents which characterize its growth. The picture is based on a deterministic growth model but it attempts to account in an averaged (effective) way for the stochastic nature of the process.

We have tested the ideas on which our model is based by studying the properties of DLA clusters grown with anisotropic sticking rules[3] on a two dimensional square lattice. The rules we used in our computer simulation to grow these clusters are as follows: a particle is launched from a site chosen at random on a circle surrounding the cluster and performs a random walk[4] until it reaches a site adjacent to the cluster; the walker sticks (and a new particle is launched) if the left or right nearest neighbor of this site are occupied; otherwise, it only has a probability p of sticking. If the particle does not stick it keeps walking, but it cannot move onto occupied sites. The walker is killed (and a new particle is launched) if it ever wanders more than 200 cluster radii away from the center of the cluster.

The results of our simulation provide strong confirmation for the cone angle picture. Our results suggest that for *any* value of p smaller than 1 the characteristic lengths of the cluster in the easy (x) and hard (y) direction of growth increase respectively as $N^{2/3}$ and $N^{1/3}$ in the limit $N \to \infty$ (N is the number of particles in the cluster). This means that for any applied uniaxial anisotropy the cluster will eventually grow into a compact object (in the sense that the area covered by the cluster grows linearly with N) with a rodlike shape. Arguments relying upon the cone angle picture predict this result.

Furthermore if the considerations employed in the case of anisotropic sticking are continued to ordinary DLA (p = 1) one obtains approximate estimates for the fractal dimension D which turn out to be close to the values obtained by computer simulation: in particular in two dimensions the cone angle picture predicts D = 5/3.

2. CONE ANGLE PICTURE

Start by considering an ordinary DLA cluster made up of N particles and having extremal radius R. The probability $\phi(\vec{r})$ of finding a diffusing particle in a certain site $\vec{r}$ outside the cluster obeys the Laplace equation $\nabla^2 \phi(\vec{r})=0$

subject to the absorbing boundary condition $\phi=0$ on sites adjacent to the cluster. The flux of diffusing particles onto a certain absorber is proportional to the normal derivative (with respect to the surface of the cluster) of ϕ at the point where the absorber is located.

The electrostatic analog of this situation is the problem of a charged conductor having the shape of the cluster. In particular the electric field (or the surface charge density) at a point on the conductor surface is the analog of the flux of particles onto that point. One can use the analogy to formulate in a familiar language problems such as finding for example (dN/dR) as a function of R: in fact (dN/dR) = (dN/dt)/(dR/dt), where the total flux of particles onto the cluster (dN/dt) corresponds to the total charge, and the rate of growth of the extremal tip (dR/dt) is proportional to the flux of walkers onto the tip, i.e. to the charge density at the tip.

Since DLA clusters are stochastic fractals one expects $N \sim R^D$, i.e. (dN/dR) $\sim R^{D-1}$; by comparing this expression with the result for (dN/dR) obtained from the electrostatics one should be able to get an estimate as to the value of D. For example, in two dimensions since D is not an integer, one expects the solution of the electrostatic problem to yield a non integer power behavior for (dN/dR) as a function of R.

The cone angle picture consists of substituting the electrostatic problem associated with the complicated fractal geometry of a DLA cluster with a simpler problem associated with a solvable geometry, which however, yields the appropriate power behavior in R for (dN/dR). It turns out, at least in the case of clusters grown with an anisotropic sticking rule, that the simple solvable geometry introduced in this way is directly related to the large scale shape of the cluster.

In two dimensions the simplest geometry that gives (dN/dR) as a non integer power of R is that of a cone. In fact consider the solution of the Laplace equation for a conducting infinite cone of exterior half angle β (see fig. 2): it is given by

$$\Phi(r,\theta) = C \cdot r^{\pi/(2\beta)} \cdot \cos(\pi\theta/2\beta) \qquad (1)$$

where C is a normalization factor. Thus the steady state flux of random walkers onto the cone edge at a distance ρ from its tip is

$$u(\rho) = (\pi/(2\beta)) \cdot C \cdot \rho^{\pi/(2\beta)-1} \qquad (2)$$

We obtain (dN/dt) by integrating $u(\rho)$ from $\rho=0$ up to some large cutoff at $\rho \sim R$ and (dR/dt) by

FIGURE 1
A cluster with $66 \cdot 10^3$ particles grown with p = 1/3. The segment on the bottom right is 500 lattice units in length.

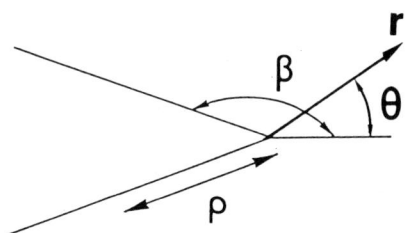

FIGURE 2
Cone angle geometry.

integrating up to a small cutoff at $\rho \sim a$ (a is the size of the diffusing particle or lattice spacing and hereafter is taken to be unity). Thus from (2) we find $dN/dt \cong C \cdot R^{\pi/(2\beta)}$ and $dR/dt \cong C$ so that

$$dR/dN \cong R^{-\pi/(2\beta)}. \qquad (3)$$

It appears that to obtain a non integer power behavior for (dN/dR) geometries involving sharp angles (such as cones and polygons) are needed: smooth geometries (such as circles) will only give integer powers. In other words one needs the singular structure associated with a sharp angle to reproduce the non analytic features of the stochastic fractal.

Up to this point no attempt has been made to relate the angle β to the geometrical features of the cluster. In fact eq. (3) can be regarded as just the definition of an effective angle β_{eff} related to D by

$$D = 1 + \pi/(2\beta_{eff}) \qquad (4)$$

and as it stands our picture has no predictive power as to the value of D.

In the case of DLA clusters grown with an anisotropic sticking rule there appears to be a natural interpretation for the cone angle introduced above. In fact it is easy to identify in these clusters an approximate overall diamond shape and it is tempting to look at this diamond geometry as the one relevant to the determination of the exponents. In what follows we shall show that it is sensible to do so; i.e. that on this basis predictions for the exponents can be made which agree well with our computer simulation results.

3. PREDICTIONS OF THE CONE ANGLE PICTURE

One can use a Schwarz-Christoffel mapping to treat the electrostatic problem associated with a diamond geometry[3,5]. However, the same result can be obtained by applying naively the type of argument which led to eq.(3) to the diamond tips in the x and y directions (see fig. 3). One obtains the following equations for the rate of growth of the tip to tip distances X and Y of a diamond shaped cluster:

$$dX/dN = A \cdot R^{-\pi/(2\beta_x)} \qquad (5)$$

$$dY/dN = B \cdot R^{-\pi/(2\beta_y)} \qquad (6)$$

Here $R = (1/2)(X^2+Y^2)^{1/2}$ is the length of each side of the diamond, A and B are scaling functions bounded of order 1 (they may exhibit slowly varying dependence on X, Y and p); since $\beta_x = \pi - \arctan(Y/X)$ and $\beta_y = \pi/2 + \arctan(Y/X)$

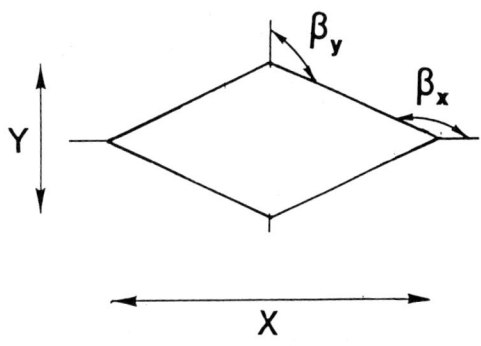

FIGURE 3
Diamond geometry

the exponents in the r.h.s. of eqs. (5) and (6) depend *only on the aspect ratio* Y/X.

It should be stressed that eqs. (5) and (6) are continuum equations: they refer to the mean growth rates of the length and the width of the cluster[5].

For a DLA cluster grown with p < 1 one expects $\beta_x > \beta_y$ so that (dX/dN) > (dY/dN), in other words eqs. (5) and (6) indicate that for DLA clusters grown with an anisotropic sticking rule X *grows faster than* Y. Therefore as N increases the cluster will grow ever more elongated: eventually it will reach a rodlike shape characterized by $\beta_x = \pi$ and $\beta_y = \pi/2$. Then eqs. (5) and (6) become $dX/dN \sim X^{-1/2}$ and $dY/dN \sim X^{-1}$, so that in the limit $N \to \infty$ one has

$$X \sim N^{2/3}$$
$$Y \sim N^{1/3} \qquad (7)$$

and the area covered by the cluster grows linearly with N.

In the case of ordinary DLA (p=1) there is no distinction between the X and Y direction of growth: if one extrapolates to this case the description used for the anisotropic case one has A = B and $\beta_x = \beta_y = 3\pi/4$ throughout the growth, which in turn implies $R \sim N^{3/5}$, namely D = 5/3. (Note however, that single ordinary DLA clusters do not exhibit any obvious diamond shape). This result compares satisfactorily with the most recent simulation results[2,6] D = 1.71 ± .01.

This sort of argument can be generalized to dimension higher than two (see ref. 3). For example, for d = 3 one gets D ≅ 2.46 and for d = 4 one has D = 10/3, again in good agreement with available computer simulation results[7].

4. SIMULATION RESULTS FOR DLA CLUSTERS GROWN WITH AN ANISOTROPIC STICKING RULE

Using the prescription described in the introduction, we grew by computer simulation at least seven clusters of up to 10^5 particles for each of seven different values of p, ranging from p = 1/50 to p = 2/3. (Further results of this simulation are discussed in ref.5).

The exponents which control the growth of these clusters can depend on the number of particles N and on the probability p of sticking in the y direction. Consider for example the root mean square length X_g and width Y_g of the clusters (these quantities are the anisotropic analogs of the radius of gyration). One can measure the exponents D_x and D_y by assuming

$$X_g \sim N^{1/D_x}$$
$$Y_g \sim N^{1/D_y} \qquad (8)$$

If the cone angle picture described above is correct one expects (i) that for N sufficiently large, D_x and D_y will approach 3/2 and 3 respectively, independent of the value of p, and (ii) that D_x and D_y will be two universal functions of the aspect ratio Y/X. In order to test this possibility we measured D_x, D_y and Y/X as functions of N for each cluster. For each set of clusters relative to a certain value of p we obtained the averages $\langle D_x \rangle$ and $\langle D_y \rangle$ at given values of N and plotted it against the average aspect ratio $\langle Y/X \rangle$ relative to that value of N. The result of this procedure is shown in figure 4. Data from clusters grown with different values of p turn out to lie to a good approximation on the same curve suggesting that (ii) is indeed fulfilled; also it appears that D_x and D_y reach their asymptotic values as the clusters grow more and more elongated.

If one uses eq. (5) and (6) with the further hypothesis that the prefactors A and B are p independent, one can obtain theoretical predictions for the form of D_x and D_y as functions of (Y/X). In fact consider for

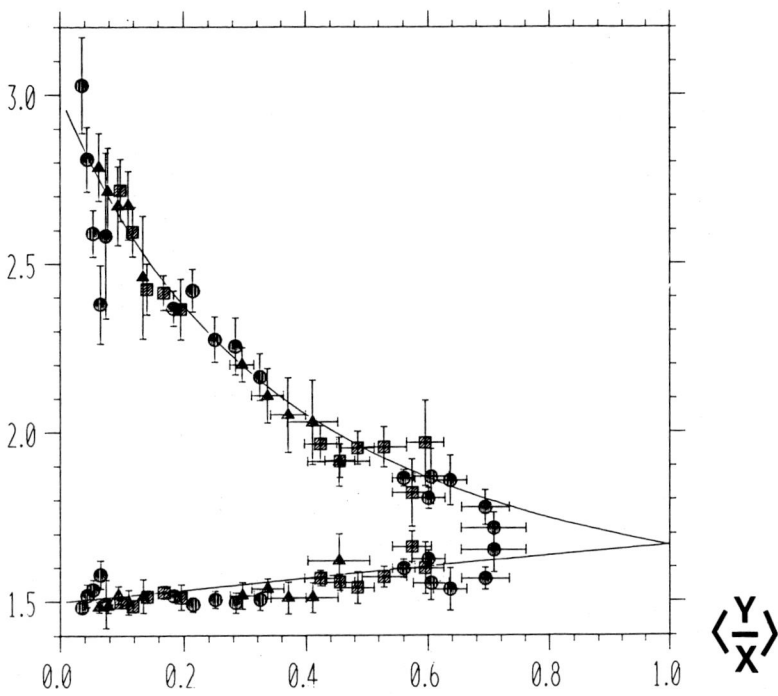

FIGURE 4
Measured values of D_x (lower data) and D_y (upper data) vs. $\langle Y/X \rangle$. Different groups of symbols correspond to different p: from left to right p = 1/50 (circles), p = 1/20 (triangles), p=1/10 (squares), p = 1/5 (circles), p = 1/3 (triangles), p = 1/2 (squares), p = 2/3 (circles). Data of different p represented by the same symbols do not overlap. The continuous curves are theoretical predictions.

example eq.(5): for (Y/X) close to some value (Y_0/X_0) one can expand the exponent

$$\nu_x(Y/X) = \pi/(2 \cdot \beta_x(Y/X)) \quad (9)$$

in a Taylor series around Y_0/X_0; neglecting terms of order $((Y/X) - (Y_0/X_0))$ one gets

$$dX/dN \cong A \cdot X^{-\nu_{xo}} \quad (10)$$

where ν_{xo} denotes the value of the exponent ν_x at $(Y/X) = (Y_0/X_0)$: therefore

$$D_x(X_0/X_0) = \nu_{xo} + 1 \quad (11)$$

and a similar treatment of eq. (6) gives

$$D_y(Y_0/X_0) = (\nu_{xo} + 1)/(\nu_{xo} - \nu_{yo} + 1) \quad (12)$$

These approximate values of D_x and D_y give the continuous curves plotted in fig. 4 (note that they do not depend on the values of A and B). Again the agreement with the simulation results is remarkable.

It should be stressed that to obtain our results we have made appeal to the applied uniaxial anisotropy but *not* to the type of lattice on which the clusters are grown. This is true both for the exponents of anisotropic DLA in the $N \to \infty$ limit (which we believe to be exact) and for our estimates of the fractal dimension of ordinary DLA (which can only be regarded as approximate). In fact our anisotropic clusters were grown on a square lattice: however, we believe that their rough overall diamond shape (which we used in our arguments) is a consequence of the applied uniaxial anisotropy[8] and would be realized to the same extent for clusters grown off lattice or on lattices other than square.

5. CONCLUSION

In summary our cone angle picture relates DLA exponents to the large scale shape of the cluster. Anisotropic sticking rules allow one to give a natural definition of what is meant by large scale shape so that quantitative predictions can be made. In particular, the *universal* behavior $X \sim N^{2/3}$, $Y \sim N^{1/3}$ (as $N \to \infty$) predicted on the basis of our picture agrees with *all* the available evidence from computer simulation.

ACKNOWLEDGEMENTS

G. Rossi would like to thank the SERC (UK) and B.R. Thompson the NSERC (Canada) for financial support. B.R. Thompson also acknowledges support from King's College Cambridge and the Cambridge Commonwealth Trust.

REFERENCES

1. T.A. Witten and L.M. Sander, Phys. Rev. Lett. 47 (1981) 1400. T.A. Witten and L.M. Sander, Phys. Rev. B27 (1983) 5686.

2. P. Meakin and L. M. Sander, Phys. Rev. Lett. 54 (1985) 2053.

3. R.C. Ball, R.M. Brady, G. Rossi and B.R. Thompson, Phys. Rev. Lett. 55, in print.

4. The algorithm we used is described in R.M. Brady and R.C. Ball, J. Phys. A, in print.

5. B.R. Thompson, G. Rossi, R.C. Ball and R.M. Brady, Growth of anisotropic DLA clusters, this volume.

6. P. Meakin, Bull. Amer. Phys. Soc. 30 (1985) 222.

7. P. Meakin, Phys. Rev. A27 (1983) 1495

8. P. Meakin presented at this meeting results for cluster grown with different anisotropy symmetries.

GROWTH OF ANISOTROPIC DLA CLUSTERS

Bernard R. THOMPSON, Giuseppe ROSSI, Robin C. BALL and Robert M. BRADY

TCM Group, Cavendish Laboratory, Madingley Road, Cambridge CB3 0HE, United Kingdom.

Equations describing the evolution of DLA clusters grown with anisotropic sticking rules are derived. Their predictions are compared with simulation data and good agreement is found.

This paper first describes results of computer simulations of DLA[1] cluster growth in two dimensions with anisotropic sticking rules[2,3]. Then equations describing the growth of these clusters according to the cone angle picture[3] (assuming an overall diamond shape) are derived. Finally we compare the solutions of these equations with results from the simulations; there is remarkably good agreement. This paper should be read after reference 2 in this volume. Reference 2 describes the cone angle picture of DLA cluster growth and the algorithm used for growing DLA clusters with anisotropic sticking rules.

In figure 1 a cluster grown with p = 1/5 is shown at various stages of its growth. (Here p is the sticking probability in the y direction; the sticking probability in the x direction is unity.) One sees that the cluster becomes more elongated (its aspect ratio Y/X decreases) and more compact as the number of particles N increases. Also the large scale boundary of the cluster becomes more clearly defined as it becomes larger.

We grew at least 7 two dimensional clusters of 50000 or more particles for each of the following values of p: 1/50, 1/20, 1/10, 1/5, 1/3, 1/2 and 2/3. We measured the tip to tip distances X and Y for each cluster (i.e., the difference in lattice units between the maximum and minimum abscissae and ordinates) as a function of the number of particles N in the cluster. In figure 2 the average aspect ratio $\langle Y/X \rangle$ is plotted as a function of N for sets of clusters grown with different values of p. We see that for each value of p this quantity decreases as N increases. The cone angle picture predicts that for all values of p less than unity $\langle X \rangle \sim N^{2/3}$ and $\langle Y \rangle \sim N^{1/3}$ and so $\langle Y/X \rangle \sim N^{-1/3}$ and $\langle X \rangle \cdot \langle Y \rangle \sim N$ in the limit $N \to \infty$. Thus the cluster becomes a compact rodlike object as $N \to \infty$. In figure 2 we see that the asymptotic slope is close to the theoretical value -1/3 for small values of p already at $N \sim 10^4$. In figure 3 the aspect ratio $\langle Y/X \rangle$ is plotted as a function of $\langle X \rangle$. The cone angle picture predicts that $\langle Y/X \rangle \sim \langle X \rangle^{-1/2}$ in the limit $\langle X \rangle \to \infty$. Again there is good agreement with theory for small values of p.

We now use cone angle arguments to derive the evolution equations for a DLA cluster grown with anisotropic sticking rules. This allows one to compare the theory with the simulation results away from the asymptotic regime.

DLA clusters grown with anisotropic sticking rules have a striking overall diamond shape (this is easily seen in figure 1) and it is natural to model them as perfectly absorbing diamonds. Problems with this model are discussed below. At this level the growth of a

FIGURE 1
A cluster grown with anisotropic sticking rules (p = 1/5) is shown at various stages of growth. From top to bottom, N = 5000, 10000, 20000, 30000, 40000 and 50000 particles. The line segment at the top is 500 lattice units long.

cluster is completely described by dX/dN and dY/dN, the rates at which the length and width of the cluster grow as particles are added to it. To find the rate of growth of X and Y we first consider the steady state flux of particles, u_x and u_y, onto the x and y tips of the cluster. We will later identify u_x and u_y with the time rates of growth of the length and width of the cluster (diamond), dX/dt and dY/dt. To find u_x and u_y we use a Schwarz-Christoffel

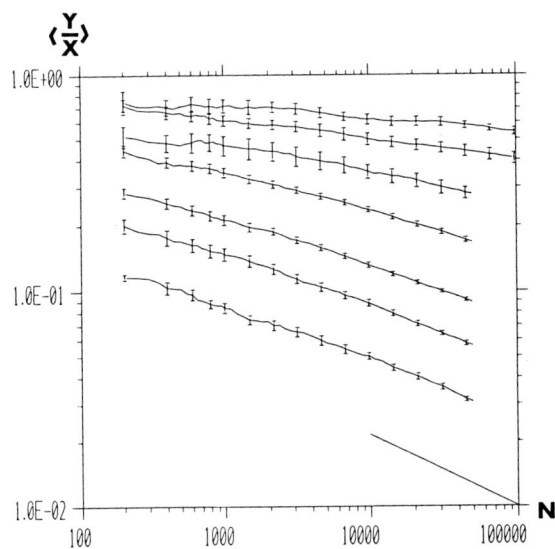

FIGURE 2
Average measured aspect ratio $\langle Y/X \rangle$ as a function of N. Error bars are shown at selected data points. The curves correspond to p = 2/3, 1/2, 1/3, 1/5, 1/10, 1/20 and 1/50 going from top to bottom. The line segment has the predicted asymptotic slope -1/3.

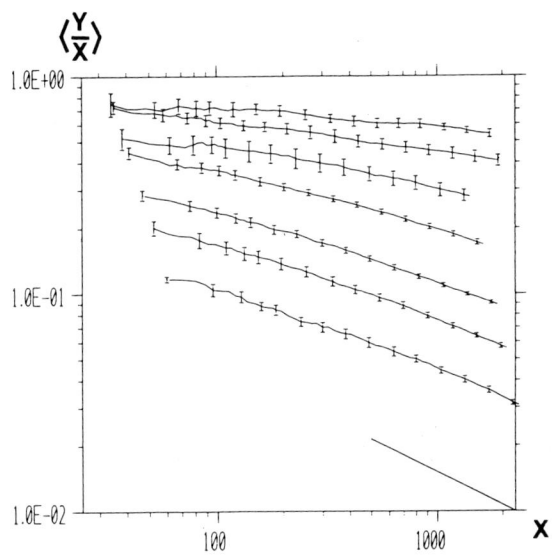

FIGURE 3
Average measured aspect ratio $\langle Y/X \rangle$ as a function of $\langle X \rangle$. Error bars are shown at selected data points. The curves correspond to p = 2/3, 1/2, 1/3, 1/5, 1/10, 1/20 and 1/50 going from top to bottom. The line segment has the predicted asymptotic slope -1/2.

mapping to solve $\nabla^2 \phi = 0$ with the boundary conditions $\phi = 0$ on the diamond boundary and $\phi(\underline{r}) \to \ln r$ as $r \to \infty$. We find that

$$u_x \sim C \cdot (1/R) \cdot (\rho_x/R)^{(\pi/2\beta_x)-1} \quad (1)$$

$$u_y \sim C \cdot (1/R) \cdot (\rho_y/R)^{(\pi/2\beta_y)-1} \quad (2)$$

Here (see figure 3 in reference 2) $R = (1/2) \cdot (X^2+Y^2)^{1/2}$ is the length of each side of the diamond, X and Y are its tip to tip dimensions, $\beta_x = \pi - \arctan(Y/X)$, $\beta_y = (\pi/2) + \arctan(Y/X)$ and ρ_x and ρ_y are the distances along the edge of the diamond from its x and y tips. We also find that the total particle flux onto the diamond, $dN/dt \sim C$, is independent of R by Gauss's theorem.

We impose a short distance cutoff at $\rho \sim a$ where a is the tip radius or lattice spacing in order to eliminate the divergence in the flux onto the tips of the diamond. This is because the diffusing particles have finite size: thus a cluster tip must have a finite curvature rather than the infinite curvature of the sharp points of a diamond which gives rise to the divergences in the fluxes u_x and u_y at the tips of the diamond. We now note that if we multiply the steady state fluxes onto the tips of the diamond by the particle diameter the result is the time rates at which the tips advance, $(1/2) \cdot dX/dt$ and $(1/2) \cdot dY/dt$. Thus we have

$$dX/dN = (dX/dt) \cdot (dt/dN) = A \cdot R^{-\pi/(2\beta_x)} \quad (3)$$

$$dY/dN = (dY/dt) \cdot (dt/dN) = B \cdot R^{-\pi/(2\beta_y)} \quad (4)$$

The coefficients A and B are bounded of order 1 but may be slowly varying functions of X and Y and of p.

A number of observations should be made about these equations. (i) Modelling a DLA cluster

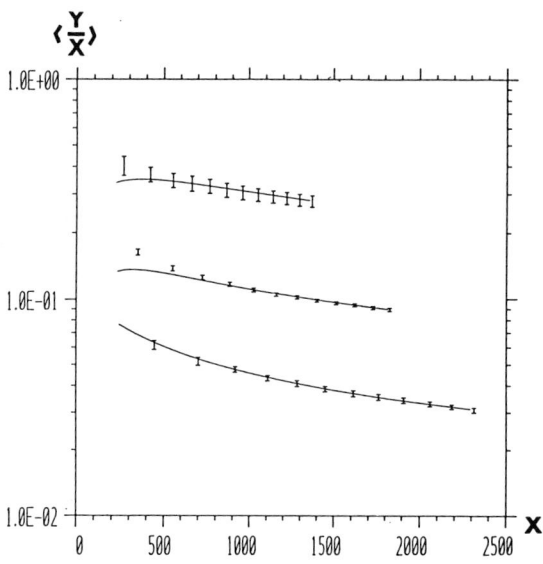

FIGURE 4
Plot of two parameter fit of equation (5) to simulation data for ⟨Y/X⟩ as a function of ⟨X⟩. The continuous curve is from integrating equation (5) and the error bars are from the simulation data. The curves correspond to $p = 1/3$, $1/10$ and $1/50$ going from top to bottom.

grown with anisotropic sticking rules as a perfectly absorbing diamond means that we have replaced the problem of describing the stochastic growth of an irregularly shaped object with a simple deterministic growth problem. We are not looking at the growth of the entire diamond but at the way its tips advance. Note that the diamond shape is not in fact preserved under our assumptions. (ii) We have assumed that DLA clusters grown with anisotropic sticking probabilities can be modelled as perfect absorbers although incoming particles do not necessarily stick on first contact when $p < 1$. This is because any particle will stick near its first contact point with the cluster. The average distance between the first and final contact points is independent of the size of the cluster and thus sticking is local on the scale where our equations are valid. (iii) It should finally be noted that these are continuum equations; they apply to the mean growth rates of the length and width of a cluster.

To test equations (3) and (4) by comparing their predictions with simulation results we noted that N can be eliminated from equations (3) and (4) to give

$$d(Y/X)/dX = (B/A) \cdot (1/X) \cdot R^{-\pi/(2\beta_y) + \pi/(2\beta_x)} - (Y/X^2) \quad (5)$$

We integrated this equation numerically. We assumed B/A to be constant with its value determined by requiring that the curve Y/X versus X pass through the simulation data points at the X values corresponding to N = 10000 and 50000 particles. This was done for $p = 1/3$, $1/10$ and $1/50$. The results of this fitting procedure are shown in figure 4. The continuous curves in figure 4 are the results of integrating equation (5) and the error bars are from the simulation data of figure 3. The fit is remarkably good in view of the approximations and assumptions used in deriving equations (3) and (4). The fit between equation (5) and the simulation data appears to worsen for higher values of p. We expect any fitting procedure to work better for smaller values of p since the asymptotic regime is reached more rapidly as p decreases.

REFERENCES
1. T.A. Witten and L.M. Sander, Phys. Rev. Lett. 47 (1981) 1400. T.A. Witten and L.M. Sander, Phys. Rev. B27 (1983) 5686.

2. G. Rossi, B.R. Thompson, R.C. Ball and R.M. Brady, Cone angle picture and anisotropy in DLA cluster growth, this volume.

3. R.C. Ball, R.M. Brady, G. Rossi and B.R. Thompson, Phys. Rev. Lett. 55, in print.

CONTINUUM DLA: RANDOM FRACTAL GROWTH GENERATED BY A DETERMINISTIC MODEL

Leonard M. SANDER

University of Michigan, Department of Physics, Ann Arbor, Michigan, 48109-1120

We consider from a theoretical and experimental point of view the role of noise in diffusion-limited aggregation. We present a deterministic model which shows fractal growth from random initial has many of the features of DLA.

1. INTRODUCTION

Irreversible aggregation of particles into clusters has recently attracted a good deal of attention. It is now seen as one of the ways in which fractals can arise from irreversible processes. Witten and Sander[1,2] introduced a model of this sort, the diffusion-limited aggregation (DLA) model which has been intensively studies. It is very simple: random walking particles of finite size accrete to form an aggregate by wandering in from far away, one at a time, and sticking to a point-like nucleation center or to the particles that have already accreted. Extensive computer investigations have shown that complex, branched fractals are produced.

In order to put the DLA model in context it is appropriate to consider a general phenomenology which has begun to emerge in recent years for other kinetic processes which exhibit a transition from equilibrium to non-equilibrium behavior. Typically, and Rayleigh-Bernard convection is a well-known example, we can isolate three regimes of dynamics: 1) near-equilibrium, ii) pattern-forming, and iii) disorderly. For the Rayleigh-Bernard problem, these correspond to the regimes of heat-conduction, formation of convection rolls, and tubulence.

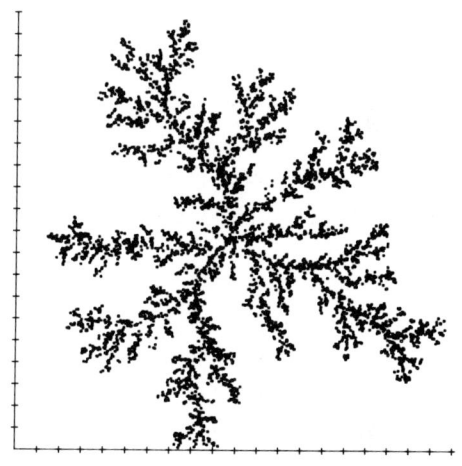

FIG. 1
A DLA cluster

We will argue here that we can identify these regimes for diffusion-limited processes, as well. This will give insight into the essential features of the DLA model which has, so far, substantially resisted analytical explanation despite its extreme simplicity.

The relationship between DLA and dendritic solidification has been known for some time.[1] We will examine this in more detail, below.

For the moment, it suffices to say that we can identify the diffusion of latent heat away from a growing crystal with the motion of random walkers toward the aggregate. Clearly, the near-equilibrium crystal whose shape is given by the Wulff construction corresponds to regime i) above. The beautiful patterns of snowflakes are in regime ii). Disorderly, chaotic crystal growth, which we will identify with DLA in a certain limit is, we claim, an example of iii), and gives rise to fractal patterns.

In the next section we will discuss some specific realizations of DLA-like processes, all of which can be described in a common framework. Then we will discuss the role of noise and anisotropy in the process in light of a recent experiment of Ben-Jacob, et al.[3] In the last section we will present a numerical treatment of an interface-dynamics model for diffusion-limited processes which achieves random fractal growth without external noise.

2. REALIZATIONS OF THE DIFFUSION-LIMITED PROCESS

The most literal experimental example of DLA that we know of is the electrolytic deposition of metals on a small electrode. The first experiment of this type to be analyzed in terms of fractal growth was that of Brady and Ball,[4] whose system was shown to be very like three-dimensional DLA. Subsequent work by Matsushita, et al.[5] gave an example of two-dimensional growth. The analogy with the process described above is immediate: the ions of copper or zinc diffuse by random motions in the electrolyte under the proper conditions until they stick to the deposit. However, since the deposit turns out to be polycrystalline with a crystallite size which is macroscopically small but contains many atoms, it is probably appropriate to identify the crystallite size with the particle size in the original model.[4]

For future reference we can write down a set of equations which describe the average growth under these conditions. For the density of ions, $u(r,t)$, we have a steady-state diffusion equation with absorbing boundary conditions:

$$\nabla^2 u = 0 \qquad (1)$$
$$u|_S = 0 + \text{cutoff}, \qquad (2)$$

supplemented by a growth condition on the boundary:

$$v_n \sim n \cdot \nabla u|_S , \qquad (3)$$

where v_n is the normal velocity of growth of a point on the boundary.

Another realization of the process is dielectric breakdown.[6] In this case we can identify u, above, with the electrostatic potential in the material. Eq. (2) corresponds to the fact that the breakdown channel is an approximate equipotential, with the cutoff given by its characteristic size. Finally, Eq. (3) represents the physical assumption that the rate of further breakdown is proportional to the electric field at the channel.

By using Eq. (3) as a probability in a simulation, it was shown[6] that the process is, in fact, equivalent to DLA. In this case, as above, we can identify a macroscopic cutoff and explicitly recognize the role of discreteness and noise (corresponding to the shot noise of the arriving particles) in the process. However, in this, and the cases to be described below, we will have to explain how the shot noise manages to produce the large structures of a fractal pattern without averaging out eventually.

For the solidification problem we can identify u with $T_m - T$, where T is the undercooled melt and T_m is the melting temperature. In the limit of very slow growth the latent heat (whose generation at the surface is represented by Eq. (3)) diffuses away according to Eq. (1). (In the general case, there is a time-dependent term on the left side of Eq. (1)). However, the cutoff in the system is not explicitly related to crystallite size. Ordinary surface tension provides a cutoff even in the case of the growth of a single crystal. It is tempting to conclude that because, in this case, there is no evident form of shot noise or discreteness, we should never get into regime iii) without some large external source of shot noise, but will always find patterned growth and never fractals. This seems, at first glance to correspond to the common occurrence of dendrites such as snowflakes. We will argue, below, that the conclusion is not correct, and that solidification physics is capable of producing fractals in the right conditions even without external noise.

The most explicit experimental demonstration of this fact is not in a solidification experiment, but in a certain kind of fluid flow. Paterson[7] pointed out that the set of equations written above were exactly those of the viscous fingering system of Saffman and Taylor.[8] In this case, we consider the displacement of a viscous fluid flowing either in a porous medium or between closely spaced, parallel plates (a Hele-Shaw cell). In these systems the flow velocity is proportional to the gradient of the pressure in the fluid:

$$v \sim \nabla P . \qquad (4)$$

If the viscous fluid is displaced by another, whose viscosity may be neglected (and thus flows at more or less constant pressure) a complex interface pattern results. These are the viscous fingers.

We can take $u = P - P_0$, where P_0 is the constant pressure of the intruding fluid. Then for incompressible flow, Eq. (1) follows from Eq. (4). Eq. (2) holds for the pressure jump at the fluid boundary with the cutoff supplied by the usual surface-tension boundary condition:

$$u|_S = -d \kappa \qquad (5)$$

where d is the capillary length and κ the curvature. Eq. (2) is evidently the proper boundary condition for Eq. (4).

Thus, this fluid flow system has a special significance for the understanding of DLA because there is no discreteness in the system above the scale of individual molecules. If complex, branching patterns can be produced here which resemble DLA aggregates, and are fractal, then we will have gained a valuable insight into the growth process.

In fact, most investigations of the phenomenon do not produce branched structures because they are done in the usual Hele-Shaw geometry of a long channel with closed sides. The walls of the channel serve to direct the flow and give rise to a single finger. However, Paterson[9] showed experimentally that a radial geometry (with the less viscous fluid injected in the center of plates with open edges) does not have this feature, and that branching patterns can be made. Ben-Jacob, et al.[3] took this subject up again, and demonstrated that the fractal dimension of the patterns was approximately that of DLA. Thus particle aggregation without particles seems possible.

3. INSTABILITIES, NOISE, AND ANISOTROPY

We are led, on the basis of the above, to conjecture that discreteness and noise are not essential to produce random fractal patterns of the DLA type. To be sure, some randomness must be present in the system to give rise to the ensemble of final aggregates. However, we suggest that the process of growth is essentially deterministic, and that the initial stages of growth, when the shot noise is substantial on the scale of the aggregate, serve as a kind of initial condition for the subsequent stages. The random initial conditions serve to nucleate instabilities which grow and interact to produce the fractal pattern. In fact, this is the point of view taken in Ref. 3, where, on the basis of experimental observations, it is suggested that the instability which is relevant is <u>tip-splitting</u> of the viscous fingers in the absence of a directing mechanism such as the side-walls of a Hele-Shaw cell.

In the dendritic growth problem, patterns of the snowflake type arise because of another directing mechanism, namely crystalline anisotropy.[10,11] Most previous investigations of this fact have been in the opposite limit to the one relevant to DLA where the right side of Eq. (1), which we have set to zero, denominates the physics, and introduces another length into the problem, the diffusion length. Because of the existence of this new scale, fractal scaling is not possible for this sort of solidification, but tip-splitting is, and apparently occurs for small anisotropy.

An experimental demonstration that even in the limit of infinite diffusion length (i.e., when Eq. (1) holds) this sort of transition occurs has also been given in Ref. (3). This is done by inscribing a lattice on one of the plates of the radial Hele-Shaw cell. A clear transition between tip-splitting, for small flow rates, and directed snowflake-like patterns, for rapid flow, has been seen. Thus a puzzle of patterned crystal growth seeming to predominate has, at least, a tentative resolution. Crystalline anisotropy is too large, in most cases, to allow fractal patterns to be seen.

4. CONTINUUM DLA

We were led, on the basis of the foregoing, to reinvestigate the theoretical basis of fractal aggregation in a way that allows a clear separation of the roles of noise in creating initial conditions and directing the growth. To this end we have introduced a formulation in terms of the motion of an interface fed by the random walkers. It is convenient to rewrite the equations introduced above in the following form:

$$\nabla^2 u = 0 \tag{6a}$$
$$v_n = -n \cdot \nabla u|_s / 4\pi \tag{6b}$$
$$u(R_0) = 0 \tag{6c}$$
$$u(x_s) = 1 - \kappa(x_s) \tag{6d}$$
$$u(x_{int}) = 1 \tag{6e}$$

The field u is held constant at zero at some large distance R_0 from the interface and at 1 in the interior of the region enclosed by the interface, $\kappa(x_s)$ represents the curvature of the interface at a point x_s, and v_n its normal velocity.

The form of the cutoff here is appropriate for the viscous fingering problem, but it is not clearly applicable to DLA. In fact, the boundary condition for DLA simulations involves only a short-distance cutoff corresponding to the particle size. We will return to this point. However, we do need some cutoff--otherwise the interface develops unphysical cusps.[12] A solution to Eq. (6) by direct means (say, by relaxation methods) would be difficult. However, it has recently been shown[13] that a more efficient solution can be given by converting to the integro-differential equation:

$$1 + (1/4\pi) \int dx_s' \kappa(x_s') \partial G((x_s, x_s'))/\partial n'$$
$$= \int dx_s' G(x_s, x_s') v_n(x_s') \tag{7}$$

Here $G(x,y)$ is the Green's function for the 2-d Laplace equation. The integral on the left hand side of Eq. (7) is the potential due to a dipole layer of strength $-\kappa/4\pi$, which ensures a discontinuous jump in the u field from 1 in interior to $1-\kappa(x_s)$ on the interface. We invert Eq. (7) numerically to obtain v_n by discretizing and converting to a matrix equation. Then we step the interface forward in time.

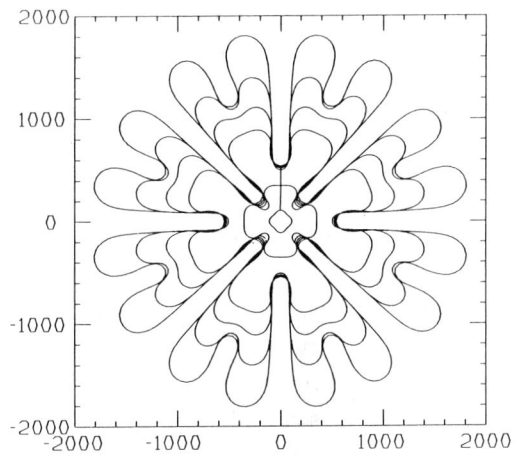

FIG. 2
Numerical solution of Eq. (6).

The result of a simulation is shown in Fig. 2 where a branched pattern has formed from asymmetric initial conditions. Unfortunately we cannot, for numerical reasons, carry the solution far enough to see whether it resembles the branched, wispy structure of DLA. However, suppose we arbitrarily modify Eq. (7):

$$1+(1/4\pi)\int dx \kappa^N \partial G/\partial n$$
$$= \int dx G v_n \qquad (8)$$

The motivation for this replacement is clear for DLA since, in that case, large curvatures, $\kappa > 1/a$, are forbidden, but smaller ones have no effect. For large N we approach this limit, and the rate of tip-splitting is greatly increased. For fluid flow, however, Eq. (6d) is physically correct.

We can now address the question of the kinds of structures that Eq. (6) generates for various boundary conditions given by Eq. (8). Values of N up to 5 are sufficiently large to elucidate the problem. Beginning with a four-fold symmetric initial condition with R=20 and δ_m=0.05 we solve Eq. (6), using Eq. (8), for N=5. We obtain the results displayed in Fig. 3. The existence of a ramified structure is clear with successive tip splittings as is the similarity between the numerical results and the patterns observed in a Hele-Shaw cell in experiments of Ben-Jacob et al.[3] and Paterson.[9] We think that if we could grow the structure further, we would obtain the wispy structure of DLA. Note, for example, that the thickness of the branches in Fig. 3 does not seem to increase

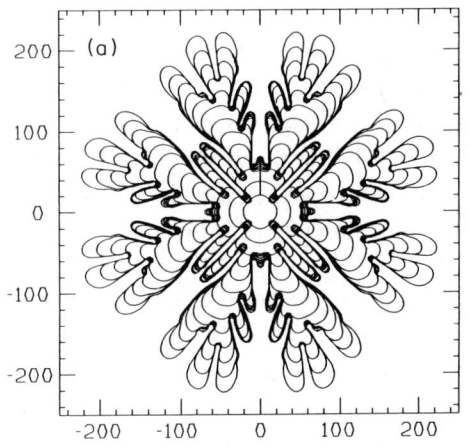

FIG. 3
Solution of Eq. (6) for N=5.

as fast as the size of the structure. We believe, though we cannot prove, that eventually even for N=1 we will approach the fractal limit.

We can calculate the fractal dimension of the objects in Fig. 4 by measuring the scaling of the area with the radius of gyration, R_g:

$$A \sim R_g^D \ . \qquad (9)$$

We find D=1.72 for the patterns in Fig. 3. For N=3 the situation is similar.

Finally, we can reexamine the role of imposed anisotropy. If we model the anisotropy[10,11] by modifying the boundary condition:

$$u(x_s) = 1 - \kappa - v_n f(\theta) \quad (10)$$
$$f(\theta) = \beta(1-\cos m\theta).$$

This modifies Eq. (7)

$$1+(1/4\pi)\int dx \kappa \partial G/\partial n = \int dx[G-(f/4\pi)\partial G/\partial n]v_n \quad (11)$$

Starting with an initial condition of R=200, δ_m=0.05 and m=4 with an anisotropic strength β=0.004 we obtain a structure with a parabolic tip and the side branching typical of crystalline dendrites in this case.[10,11] The tip velocity is observed to oscillate with periodicity defined by the onset of the side branches, and the tip-splitting instability is surpassed.

SUMMARY

In summary, we have tried to show that a model closely related to DLA is sufficiently unstable to give fractal growth, and that the explicit of the aggregation process does not seem to play an essential role. The numerical results for the fractal dimension and the general appearance of the structures obtained are reminiscent of computer simulations of DLA and experiments of the viscous fingering process. We have tried to shed light on the problem with the continuum approximation of DLA. Note that even if one declines to accept our risky assumption that the N=1 case is similar to the others, we have demonstrated that for some short range cutoffs we can make DLA-like objects.

DLA is produced by the proliferation of instabilities. Once an unstable mode starts growing, its growth is deterministic though, presumably, sensitive to initial conditions. In short, the model produces its own noise.

ACKNOWLEDGEMENTS

The calculations in this paper were done in collaboration with E. Ben-Jacob and R. Ramanlal.[14] We would like to thank R. Ball and D. Kessler for helpful conversations. Supported by DoE Grant DE-FG02-85ER45189.

REFERENCES

1. T.A. Witten and L.M Sander, Phys. Rev. Lett. 47, 1400 (1981).
2. T.A. Witten and L.M. Sander, Phys. Rev. B 27, 5686 (1983).
3. E. Ben-Jacob, R. Godbey, N. Goldenfeld, J. Koplik, H. Levine, T. Mueller, and L. Sander, Phys. Rev. Lett. 55, 1315 (1985).
4. R. Brady and R.C. Ball, Nature 309, 225 (1984).
5. M. Matsushita, M. Sano, Y. Hayakawa, H. Honjo and Y. Sawada, Phys. Rev. Lett. 53, 286 (1984).
6. L. Neimeyer, L. Pietronero, and H. Weisman, Phys. Rev. Lett. 52, 1033 (1984).
7. L. Paterson, Phys. Rev. Lett. 52, 1621 (1984).
8. P. Saffman and G. Taylor, Proc. Roy. Soc. A 245, 312 (1958).
9. L. Paterson, J. Fluid Mech. 113, 513 (1981).
10. R.C. Brower, D.A. Kessler, J. Koplik and H. Levine, Phys. Rev. A 29, 1335 (1984).
11. E. Ben-Jacob, Nigel Goldenfeld, J.S. Langer and Gerd Schon, Phys. Rev. A 29, 330 (1984).
12. B. Shraiman and D. Bensimon, Phys. Rev. A 30, 2840 (1984).
13. D.A. Kessler, J. Koplik and H. Levine, Phys. Rev. A 30, 2820 (1984).
14. L. Sander, P. Ramanlal and E. Ben-Jacob, Phys. Rev. A (in press).

FRACTALS IN PHYSICS
L. Pietronero, E. Tosatti (editors)
© *Elsevier Science Publishers B.V., 1986*

FORMATION OF SOLIDIFICATION PATTERNS IN AGGREGATION MODELS

Tamás VICSEK

Department of Physics, Emory University, Atlanta, GA 30322, USA

Generalizations of the diffusion-limited aggregation model are considered in order to simulate pattern formation during solidification. The two-dimensional clusters grown on a seed particle are initially circular but at later stages the process crosses over into dendritic growth. The effects of an anisotropic surface tension are studied by assuming that the sticking probability of the particles depends on the local orientation of the interface. Directional solidification is simulated by the deposition of particles undergoing biased random walks. Linearly stable patterns are generated if the basic features of the directional solidification experiments are taken into account. The resulting patterns are very similar to those observed experimentally.

1. INTRODUCTION

The formation of patterns by growing interfaces is one of the main processes in a wide range of phenomena in science and technology. Such behavior is exhibited during solidification, when the crystalline phase is growing in supersaturated vapor or undercooled melt.[1] Examples for formation of solidification patterns include the evolution of a snowflake in the atmosphere or directional solidification in a number of metallurgically important situations.[2]

The process of solidification is described by nonlinear partial differential equations and both the analytical and the numerical treatment of these equations are extremely difficult. As a result, many of the questions concerning patterns formation have not so far been satisfactorily answered. One possible way to examine the above questions is the study of model systems which produce patterns.[3-7]

To simulate the behavior of the solidification front in the presence of non-local driving forces one needs new efficient numerical methods and models. The diffusion-limited aggregation (DLA) model of Witten and Sander[2] seems to be particularly appropriate for treating the effects of the non-local diffusion field. In a recent paper a simple generalization of the DLA was introduced[3], where the stabilizing force of the surface tension was accounted for through a local curvature dependent sticking probability. In this paper the above generalized DLA model will be used to generate solidification patterns.

2. MODEL

The following rules[3] will be used to simulate the process of solidification:
i) Random walks by the particles (as in DLA),
ii) Sticking to the surface of the

Permanent and present address: Research Institute for Technical Physics, Budapest, Ujpest 1, Pf. 76, H-1325, HUNGARY growing cluster with a probability depending on the local interface curvature,

iii) Relaxation to a position with the highest number of occupied nearest neighbors.

The first rule simulates the effects of a non-local diffusion field as it was discussed by Witten and Sander[8] and by Kadanoff.[9] This destabilizing force is compensated by the surface tension which is taken into account by rule ii. (The growth is slowed down at places with large curvature.) The third rule is needed in order to get compact clusters with a low density of local defects (holes). A detailed description of the model can be found in Ref.3.

Application of the above model to the solidification problem has a number of advantages. The numerical method is simple and effective. Relatively complex geometries can be generated easily. Second, the fluctuations which are always present in a thermodynamical system (and play an important role during the growth process) are included in a natural way through the random walks of the particles. Finally, the model can easily be modified in order to take into account various experimental conditions. For example, the effects of an anisotropic surface tension or a temperature gradient imposed upon the system can be directly simulated.

3. DENDRITIC GROWTH

In order to simulate dendritic growth the model described in Sec. 2 was used with a single seed particle on a square lattice. The process starts with a growing, nearly circular cluster since at this stage the surface area minimizing effect of the curvature dependent sticking probability dominates the growth.

As the process goes on the initially structureless behavior crosses over into dendritic growth as it is demonstrated in Figure 1, where we show the number of surface sites, N_s, versus the number of sites in the cluster, N, in a log-log plot. For relatively small sizes the slope of the straight line connecting the data is approximately equal to 1/2 in accordance with the growth of a circular cluster. At later stages, however, the number of surface sites becomes linearly proportional to N (resulting in a slope nearly equal to 1) which corresponds to the development of dendrites.

The effects of an anisotropic surface tension can be investigated in the present model by introducing a sticking probability, p_{an}, depending on the local slope of the surface.

In Figure 2 a cluster of 25,000 particles is shown which was generated using an anisotropic surface tension enhancing the growth along the main

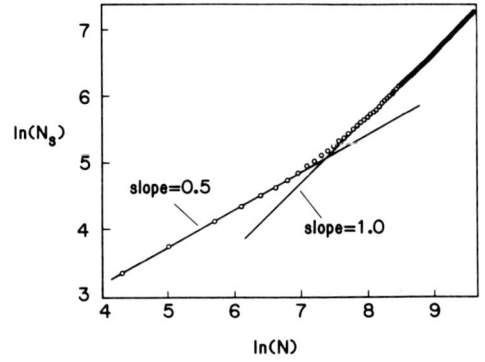

FIGURE 1
Dependence of the number of surface sites, N_S, on the number of particles in the cluster, N. The change in the slope of the curve indicates the crossover from compact to dendritic growth

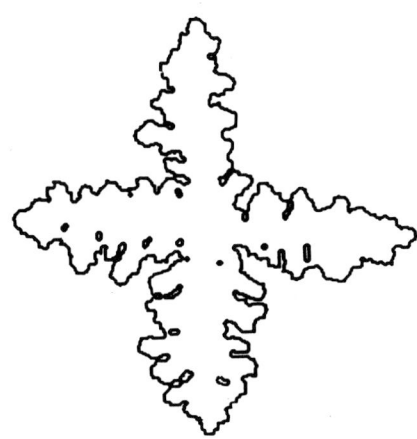

FIGURE 2
The effect of a sticking probability depending on the local orientation of the surface. This figure shows a cluster of 25,000 particles generated using a condition for the sticking probability which enhances growth along the axes of the square lattice. Only the surface sites (those which have less than four occupied nearest neighbors) are plotted

axes of the square lattice. This pattern exhibits a number of properties typical of dendritic solidification and it appears to be stable in the sense of the absence of any tip splitting. However, this cluster is still not as regular as most of the solidification patterns observed in the experiments. This is due to the fluctuations which are relatively large for small surface tension. For larger values of A more regular patterns are expected to grow in this model. On the other hand, we had to keep A relatively low in order to have larger curvatures and more complex patterns in our medium scale simulations.

4. DIRECTIONAL SOLIDIFICATION

In this section we consider a version of the model in which the particles

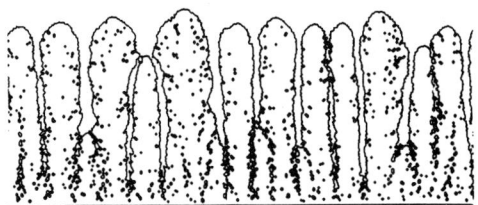

FIGURE 3
This pattern was generated using biased random walks with a ratio of down- to upward jumps of R=1.1. The insert shows the experimental results of Heslot and Libchaber[11] on the directional solidification of succinonitrole

are deposited onto a line instead of a single particle in order to simulate the conditions of directional solidification experiments.[10,11] During these experiments the working material (usually a long rod or a thin strip) is drawn with a given velocity through a fixed temperature gradient.

To account for the fact that the working material is moving we introduce a biased random walk by increasing the probability of jumping into the direction of the interface or "downward", p_{down}, with regard to the probability of jumping "upward", p_{up}. The simulations for several values of the ratio $R=p_{down}/p_{up}$ result in patterns which are more regular for R>1 than in the unbiased case. In fact, these patterns look very similar to those observed in the experiments of Heslot and Libchaber[11] on directional solidification of thin samples of succinonitrole. This is demonstrated in Figure 3, where both the simulation and the experimental results are shown.

5. CONCLUSIONS

Several generalizations of the diffusion-limited aggregation model have been applied during solidification. It has been shown that the aggregation model proposed in Ref. 3. produces patterns which have properties consistent with the experimental observations. These patterns emerge from an antirely random process which leads to either fractal objects or to nearly regular solidification patterns depending on the value of a parameter corresponding to the surface tension.

ACKNOWLEDGMENTS

The author thanks F. Family, L. Kadanoff and A. Libchaber for useful discussions. This research was supported by grants from the Emory University Research Fund and the National Science Foundation (Grant No. DMR-82-08051).

REFERENCES

1. J.S. Langer, Rev. Mod. Phys. $\underline{52}$, 1 (1980).
2. See, e.g. Crystal Growth, edited by B.R. Pamplin (Pergamon, New York, 1975).
3. T. Vicsek, Phys. Rev. Lett. $\underline{53}$, 2281 (1984).
4. E. Ben-Jacob, N. Goldenfield, J.S. Langer and G. Schon Phys. Rev. Lett. $\underline{51}$, 1930 (1983), and Phys.Rev. A 29, $\underline{330}$ (1984).
5. R.C. Brower, D.A. Kessler, J. Koplik and H. Levine, Phys. Rev. Lett. $\underline{51}$, 1111 (1983), and Phys. Rev. A $\underline{29}$, 1335 (1984).
6. D.A. Kessler, J. Koplik and H. Levine, Phys. Rev. A $\underline{30}$, 2820 (1984).
7. J. Szep, J. Cserti and J. Kertesz, J. Phys. A18 (1985) L413.
8. T.A. Witten and L.M. Sander, Phys. Rev. Lett. $\underline{47}$, 1400 (1981), and Phys. Rev. B $\underline{27}$, 5686 (1983).
9. L.P. Kadanoff, to be published
10. K.A. Jackson, in Solidification (American Society for Metals, Metals Park, Ohio, 1971).
11. F. Heslot and A. Libchaber, Phys. Scr. T9, 126 (1985).

SCALING PROPERTIES OF THE SURFACE OF THE EDEN MODEL

R. JULLIEN and R. BOTET

Laboratoire de Physique des Solides, Bât. 510, Université Paris-Sud, Centre d'Orsay, 91405 Orsay, France

The surface of the Eden model is investigated numerically by finite size scaling, using a strip-geometry. Three different versions are studied and it is shown that the one mostly used previously exhibits strong finite-size corrections. One finds that the surface thickness σ takes the scaling form $\sigma(\ell, h) \sim \ell^\alpha \, f(\frac{h}{\ell^\gamma})$ where ℓ is the width of the strip and where $h = N/\ell$, N being the number of particles in the cluster. $f(x) \to$ cste for $x \to \infty$ and $f(x) \sim x^{\alpha/\gamma}$ for $x \to 0$. In two dimensions one finds $\gamma = 0.50 \pm 0.03$ and $\gamma = 1.7 \pm 0.3$.

Many irreversible growth models[1] (Eden model, DLA process, ballistic model,...) have been introduced in a 'spherical' geometry with a spherical seed which plays the role of growth germ at the center of coordinates of an infinite lattice. Another way to study these models is to use the strip-geometry and finite size scaling. This method has been applied to study DLA process[2], ballistic model[3], and recently Eden models[4].

In d dimensions of space, it consists in an infinite strip with cylindrical periodic boundary conditions, and whose section is a (d-1)-dimensional hypercube of size ℓ. At the beginning of the process, all the sites of the strip are occupied up to height $z = 0$. Then the growing process starts from this hyperplane.

Three versions of the Eden model are introduced here:

In version A, we consider all unoccupied sites adjacent to the surface, with the same probability, and we choose randomly one of these sites. Then a new particle is added on this site. This version ('Eden model for physicists') has been widely studied in the past[5].

In version B, we consider all open bonds (joining an occupied site to an unoccupied one) with the same probability, and we choose randomly one of these bonds. Then a new particle is added on the empty edge site. This version corresponds to the original Eden model[6].

In version C, we consider all occupied sites of the surface with the same probability and we choose randomly one of these sites. Then we consider all open bonds starting from this site, with the same probability and we choose randomly one of these bonds.

All the three versions are different in the sense that, while the sets of allowed sites are the same for a given configuration, the statistical weight (probability to choose) one of these sites is in general different in the three cases.

We present here some results in d = 2. All the three models lead to compact clusters (fractal dimension equal to the dimension of space) of density 1.

We see on figure 1 the last top rows of typical clusters with $\ell = 96$. Below these rows, in the three cases, there is a full strip of height 40 ℓ.

If the fractal dimension is trivial, this is not the case for the thickness of the surface. It is defined by the formula

$$\sigma^2 = \frac{1}{n_s} \sum_i (z_i - \bar{z})^2$$

and

$$\bar{z} = \frac{1}{n_s} \sum_i z_i$$

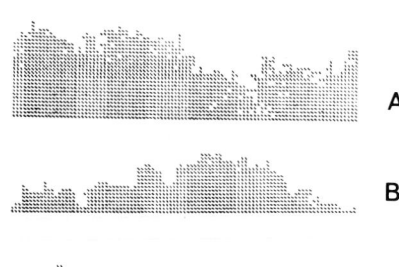

FIGURE 1
Typical two-dimensional examples with $\ell = 96$. The figure only shows the last top rows, containing surface sites.

where the sums cover the n_s surface sites labelled by index i, and z_i denotes the height of site i.

This surface depends on two characteristic lengths : the width ℓ of the strip and the effective height $h = N/\ell$ of the cluster, where N is the total number of particles.

For large ℓ and h, general scaling arguments imply that σ takes the scaling form[7] :

$$\sigma(\ell, h) \sim \ell^\alpha f(h/\ell^\gamma)$$

with :

$f(x) \to$ constant for $x \to \infty$
$f(x) \sim x^\beta$ for $x \to 0$

and $\gamma = \alpha/\beta$ to insure independence of σ versus ℓ for $\ell \gg h$. This scaling relation implies the existence of a steady state for $h \gg \ell$, in which we must have :

$$\sigma \sim \ell^\alpha$$

A direct estimation of the exponent α gives in case C :

$$\alpha = 0.50 \pm 0.03$$

The other cases (especially case A) show strong finite size corrections[4]. For A, the value of α goes through a minimum $\alpha \simeq 0.37$ between $\ell = 24$ and $\ell = 48$, then increases slowly. A similar poor convergence for model A has already been noted by Meakin and Witten[5].

The same global features appear in the Log-Log plot of $\sigma(\ell, \frac{1}{6}\ell)$ versus ℓ (see ref. (4) for details) and lead to :

$$\beta = 0.30 \pm 0.03$$

which, combined with the preceding result for α, gives :

$$\gamma = \frac{\alpha}{\beta} = 1.7 \pm 0.3$$

These values for α and γ are very close to the values found analytically for a related growth model[8] : $\alpha = 1/2$, $\gamma = 2$. But while the values of α are in good agreement. This is not the case for γ. Even if our value of β seems not compatible with the value $\beta = 1/4$ of some exactly solved growth models, it is really difficult to conclude and much larger simulations are needed (recent simulations[9], up to 2^{24} sites cannot decide more clearly).

REFERENCES

1. For a review see : 'Kinetics of Aggregation and Gelation', eds. F. Family and D. P. Landau (North Holland, 1984)

2. R. Jullien, M. Kolb and R. Botet, J. Physique 45, 395 (1984)
Z. Racz and T. Vicsek, Phys. Rev. Lett. 51 2385 (1983)
L. Turban and J. M. Debierre, J. Phys. A 17, L 289 (1984)

3. P. Gelband and P. N. Strenski, J. Phys. A, 18, 611 (1985)

4. R. Jullien and R. Botet, Phys. Rev. Lett., 54, 2055 (1985)
R. Jullien and R. Botet, J. Phys. A, 18, 2279 (1985)

5. H. P. Peters, D. Stauffer, H. P. Hölters and K. Loewenich, Z. Phys. B 34, 399 (1979)
P. Meakin, J. Colloid and Interface Sci. 96, 415 (1983)
P. Meakin & T.A. Witten, Phys. Rev. A 28, 2985 (1983)
M. Plischke and Z. Racz, Phys. Rev. Lett., 53, 415 (1984)

6. M. Eden, in 'Proceedings of the Fourth Berkeley Symposium on Mathematical Statistics and Probabilities', ed. F. Neyman (Univ. of California Press, Berkeley, 1961), vol. IV p. 233

7. F. Family and T. Vicsek, J. Phys. A **18**, L 75 (1985)

8. D. Richardson, Proc. Camb. Phil. Soc. **74**, 515 (1973)
   D. Dhar, Phys. Rev. Lett. **54**, 2058 (1985)

9. P. Freche, D. Stauffer and H. E. Stanley, submitted to J. Phys. A Letters.

FRACTALS IN PHYSICS
L. Pietronero, E. Tosatti (editors)
© Elsevier Science Publishers B.V., 1986

CLUSTER AGGREGATION

R. BOTET, R. JULLIEN and M. KOLB

Laboratoire de Physique des Solides, Bât. 510, Université Paris-Sud, Centre d'Orsay, 91405 Orsay, France

We introduce the clustering of clusters process as a model to describe aggregation of colloidal or aerosol particles in the low concentration regime. We investigate modifications of the parameters of the model. In particular, various reversible models of aggregation are detailed and we discuss possible relations between them (universality classes).

1. INTRODUCTION

The formation of aggregates by clustering of isolated particles plays an important role in numerous scientific areas[1]. In colloids and aerosols, aggregates show a self-similar geometrical structure, well-defined on many length-scales[2,3]. A beginning for a theoretical explanation has been found using numerical simulations. We do not yet know why these aggregates are fractal but we begin to know how they are fractal and how their fractal dimension is related to physics.

In the following, we introduce the model of irreversible cluster-cluster aggregation (Cl-Cl) which gives a realistic description of colloid and aerosol aggregation.

2. THE MODEL OF CLUSTERING OF CLUSTERS

This model has been proposed independently by Meakin[4] and by us[5]. We describe here only the 3-dimensional version of this model since it is the most useful for direct comparison with experiments[6]. Nevertheless all dimensions less than 6 have been numerically studied[7] as well as dimensions larger than the upper critical dimension[8] : $d_c = \frac{\text{Log} 36}{\text{Log} \frac{3}{2}} = 8.8...$ The model is defined as follows : in a box L x L x L with periodic boundary conditions, which is a portion of an infinite cubic lattice, we put randomly N_0 identical particles. Each of these particles follows a random walk on the lattice. The model simulates an infinite medium, statistically homogeneous, with an initial density N_0/L^3 of monodisperse brownian particles.

When two of these particles collide (i.e. when they are on two neighbouring sites of the lattice) they stick irreversibly and form a small rigid dimer. This cluster also follows a random walk trajectory with a velocity which is a function of its mass. When two clusters collide, they stick and form a larger rigid cluster, and so on. Since sticking is irreversible, the number of clusters decreases with time and the process ends when there remains only one single cluster in the box.

In this version, the underlying lattice forbids continuous rotation of the clusters along their trajectory. In the simulation the clusters have not been allowed to rotate. Off-lattice simulations have been done with a generalization of the random walk, and they show that clusters have the same fractal structure if this rotation is reasonable[9].

Likewise, the relation between velocity and mass is unimportant for the geometrical features, if it is realistic, that is if the velocity is not an increasing function of the mass[10]. Of course, the size distribution of the clusters as a function of time depends on this relation[11].

Figure 1 shows a typical cluster of 1024 particles, grown by this process in a box of size 70 x 70 x 70.

When averaging over a large number of clusters, Log-Log plot of radius of gyration versus mass shows a scaling relation of the form :

$$N \sim R^D$$

for all length scales greater than a few monomer radii. D is the fractal dimension and its value is $D = 1.78 \pm 0.05$. This value is very close to the experimental values of Forrest and Witten[2]

FIGURE 1
Typical 3-dimensional cluster of 1024 particles, grown by Cl-Cl process on a lattice.

$D = 1.8 \pm 0.1$ on quenched Fe vapors, and of Weitz and Oliveria[3] : $D = 1.75 \pm 0.05$ on gold colloids.

3. PHYSICAL PARAMETERS OF THE MODEL

It is interesting to note how few parameters are needed to recover the experimental results.

The starting conditions are : monodisperse particles at low concentration.

The diffusion conditions are : brownian diffusion.

The sticking conditions are : once two clusters collide, they stick irreversibly and form a new rigid cluster.

What happens when we change some of the parameters of the model ?

4. MODIFYING THE STARTING CONDITIONS

If we allow polydispersity of the monomers (balls with a distribution of radii) the same kind of simulations show that the resulting aggregates are still fractal objects, with the same fractal dimension as in the monodisperse case[12]. These numerical results have been successfully compared with recent experiments on polydisperse Fe aggregates[13]. Moreover, as time goes on, the effective density :
$$\sum_{i=\text{all clusters}} \frac{R_i^3}{L^3} \sim \bar{R}^{3-D}$$
becomes of order unity, since the mean cluster radius $\bar{R}$ increases with time. Then the brownian screening is inefficient because of entanglement. We are in the so-called kinetic gelation regime and the fractal dimension typical of this regime is $D = 1.75 \pm 0.07$[14] (in d = 2).

5. MODIFYING THE DIFFUSION CONDITIONS

We can imagine a molecular (instead of brownian) diffusion of the clusters, where the mean free path is only limited by collisions between clusters. Here also the screening is less efficient than in the brownian case and on average the clusters penetrate more deeply into each other. The fractal dimension of the resulting clusters is $D = 1.91 \pm 0.03$[15].

6. MODIFYING THE STICKING CONDITIONS

Here, three parameters can be modified : the probability of sticking when colliding, the rigidity of the clusters and the irreversibility of the sticking process.

If we let the sticking probability tend to 0, we obtain the chemical model. In this case, two clusters must collide many times before sticking. But once stuck, irreversibility implies that clusters can not break. This chemical model has been studied numerically[16] and experimentally[17] (the height of the repulsive barrier between two gold colloids can be varied chemically). The fractal dimension of the resulting clusters is $D = 2.00 \pm 0.06$ numerically and $D = 2.01 \pm 0.10$ experimentally.

Some aspects of restructuring by deformation of the clusters during aggregation process, have been studied by Meakin and Jullien[18] and are not discussed here.

The last point to investigate is irreversibility. If, in the above model, we allow breaking of clusters so that a steady state exists (for example : each bond has a finite life time) the fractal dimension of the clusters[19] is very close to the fractal dimension of lattice animals[20].

The problem which arises is to know if the two models (reversible Cl-Cl and lattice animals) lead to the same universality class. For the moment, only the fractal dimension is available to test this correspondance. What about other kinetic aggregation-fragmentation models at equilibrium ?

Reversible Eden model has been studied by Stauffer[21] in another context (percolation). It is a model of connected aggregates where we remove a particle (if this does not disconnect the cluster) and then put this particle at a point (randomly chosen) of the unoccupied surface of the cluster (Eden-type aggregation), and so on. Stauffer found the same fractal dimension as for lattice animals.

Reversible DLA model has been studied by two of us[22]. In this version of the particle-cluster aggregation[23], we start from any connected cluster without loops, on a lattice. We take randomly a particle in the set of the singly connected ones. We break the corresponding bond, then the particle begins a random walk until it sticks to the cluster again (DLA-type aggregation).

To have a loopless structure, we decide that sticking arises when the diffusive particle, say A, reaches a site occupied by a particle of the cluster, say B. Then particle A backs up to its last brownian step and we decide it is connected only to particle B of the cluster (A can have several neighbours, but only one bond (between

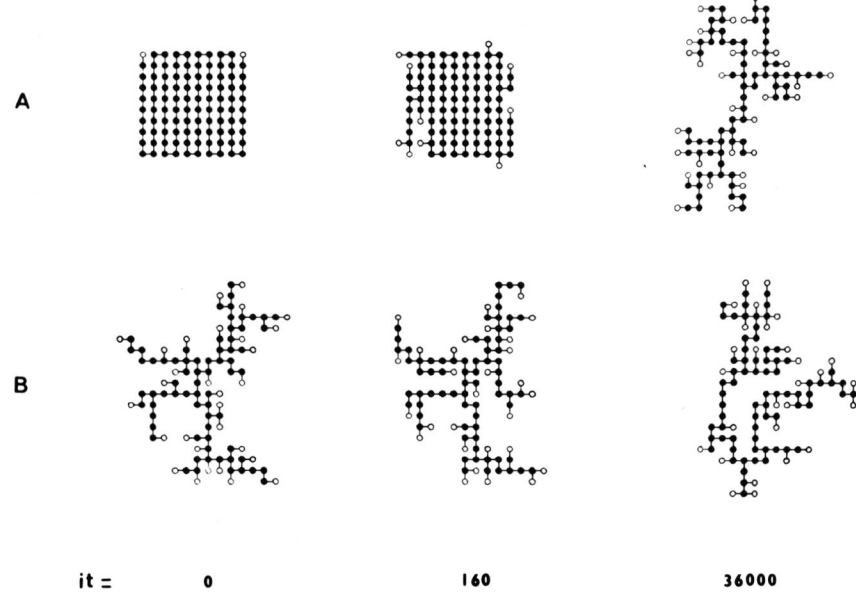

FIGURE 2
Sketch of reversible DLA process acting on a compact loopless cluster (A) or a typical loopless DLA cluster (B) of 100 particles. The number of iterations is indicated below the figures, and grows from left to right.

A and B) is formed). This is a process already introduced by Kadanoff[25] for DLA.

Starting from any initial configuration leads to the same statistical distribution of clusters for very large times. In this steady state, the fractal dimension is found to be very close to that of lattice animals. The four reversible models (lattice animals, reversible Cl-Cl, reversible Eden, reversible DLA) are all different, in the sense that the statistical weight of a given finite cluster is different in each model. This does not mean, however, that the scaling properties for very large cluster are different. The fractal dimensions suggest that all these reversible models belong to the same universality class (lattice animals). But so far no theoretical arguments have supported this result.

We acknowledge the collaboration and discussions with H. Hermann and P. Meakin. This work has been supported by an ATP C. N. R. S. and by the CCVR, Palaiseau.

REFERENCES

1. A review on aggregation process can be found in 'kinetics of Aggregation and Gelation', eds. F. Family and D. P. Landau (North Holland 1984)
2. S. R. Forrest and T. A. Witten Jr., J. Phys. A 12, L 109 (1979)
3. D. A. Weitz and M. Oliveria, Phys. Rev. Lett. 52, 1433 (1984)
   D. A. Weitz, M. Y. Lin and C. J. Sandroff, Surface Sci., 158, 147 (1985)
4. P. Meakin, Phys. Rev. Lett., 51, 1119 (1983)
5. M. Kolb, R. Botet and R. Jullien, Phys. Rev. Lett., 51, 1123 (1983)
6. Recent experiments have been done in a two-dimensional confined space (layer). They are described in : A. Hurd and D. Schaefer, Phys. Rev. Lett. 54, 1043 (1985). A possible explanation of their results has been given in : R. Jullien, to be published
7. R. Jullien, M. Kolb and R. Botet, J. Physique Lett. 45, L 211 (1984)
   P. Meakin, Phys. Lett. A 107A, 269 (1985)
8. R. C. Ball and T. A. Witten, Phys. Rev. A 29, 2966 (1983)
   R. C. Ball, J. Stat. Phys. 36, 873 (1984)
   S. P. Obukhov, 'Kinetically aggregated clusters', preprint (1984)
   R. Botet, J. Phys. A 18, 847 (1985)
9. P. Meakin, J. Chem. Phys. 81, 4637 (1984)
10. R. Botet, R. Jullien and M. Kolb, Phys. Rev. A 30, 2150 (1984)
    P. Meakin, J. Colloid and Interface Sci. 102, 491 (1984)
11. R. Botet and R. Jullien, J. Phys. A 17, 2517 (1984)
    M. Kolb, Phys. Rev. Lett. 53, 1653 (1984)
    P. Meakin, T. Vicsek and F. Family, Phys. Rev. B 31, 564 (1985)
12. J. P. Chevalier, C. Colliex, M. Tencé, R. Jullien and R. Botet, 'Fractal structure of polydisperse iron aggregates : STEM analysis and numerical simulations', in preparation
13. J. P. Chevalier, C. Colliex and M. Tencé, 'Analysis of digitalized STEM micrographs : Application to the calculation of the fractal dimension of iron aggregates', poster presented to : Colloque annuel de la Société Française de microscopie électronique, Stransbourg 28-31 mai 1985
14. M. Kolb and H. J. Herrmann, J. Phys. A 18, L 435 (1985)
15. P. Meakin, J. Colloid and Interface Sci. 102, 505 (1984)
    P. Meakin, Phys. Rev. A 29, 997 (1984)
    R. C. Ball and R. Jullien, J. Physique Lett. 45, L 1031 (1984)
16. R. Jullien and M. Kolb, J. Phys. A 17, L 639 (1984)
    M. Kolb and R. Jullien, J. Physique Lett. 45 L 977 (1984)
17. D. A. Weitz, J. S. Huang, M. Y. Lin and J. Sung, 'The limits of the fractal dimension for irreversible kinetic aggregation of colloids', preprint (1985)
18. P. Meakin and R. Jullien, J. Physique 46, 1543 (1985)
19. M. Kolb, 'Reversible diffusion limited cluster aggregation', preprint (1985)
20. H. P. Peters, D. Stauffer, H. P. Hölters and K. Loewenich, Z. Physik B 34, 399 (1979)
    B. Derrida and L. de Sèze, J. Physique 43, 475 (1982)
    V. Privman, F. Family and A. Margolina, J. Phys. A 17, 2837 (1984)
21. D. Stauffer, Phys. Rev. Lett. 41, 1333 (1978)
22. R. Botet and R. Jullien, 'Diffusion limited aggregation with disaggregation', preprint (1985)
23. For a review, see ref. 1 and proceedings of Geilo ('Scaling phenomena in disordered systems', ed. R. Pynn (1985)), Les Houches ("Finely divided matter', ed. M. Daoud (1985)), and Cargese ('On growth and forms' eds. H. E. Stanley and N. Ostrowsky, Martinus Nighoff publishers (1985)).
24. Note that DLA clusters off-lattice have no loops. So loops are an artefact of the lattice.
25. Leo P. Kadanoff, J. Stat. Phys. 39, 267 (1985)

ANISOTROPY IN CLUSTER AND PARTICLE AGGREGATION

M. KOLB

Laboratoire de Physique des Solides, Bât. 510, Université de Paris-Sud, 91405 Orsay, France
and
Institut für Theorie der Kondensierten Materie, Freie Universität Berlin, Arnimallee 14, 100 Berlin 33, West Germany

Anisotropy both due to the lattice and due to the growth mechanism is analysed for diffusion limited particle resp. cluster aggregation. In cluster aggregation, there is merely anisotropy of the amplitude due to the lattice, whereas particle aggregation has different scaling powers in the direction of growth and perpendicular to it. The correlation-function exponents differ by $\Delta A = 0.16 \pm 0.05$ in two dimensions.

Though many different growth models have been concerned and investigated in recent years and numerous scaling exponents have been determined, the basic understanding of what determines these properties has not progressed very much[1]. After some early efforts to give a qualitative description of the Witten-Sander[2] or diffusion limited particle aggregation model (PA) have not provided a sufficiently clear picture, one resorts to study the model in more detail with numerical methods to eventually be able to draw a complete description of its relevant features.

A question which has been asked since the model was originally conceived is whether the radially inward flux of particles in spherically grown PA leads to anisotropy effects despite the fact that visually small portions of the aggregate appear isotropic. The question has been raised anew by observed lattice anistropies for very large clusters[3]. This would imply that the scaling picture, which usually is applied to such clusters in analogy with static critical phenomena, would have to be refined.

One criteria, which both is simple and decisive is the anisotropy of correlation functions. Hence, the correlation functions of aggregation clusters have been measured as a function of the angle with the center of the cluster. Care has been taken to avoid measuring spurious effects due to finite size, incomplete growth and statistical uncertainty. The same analysis has also been performed on cluster-cluster aggregates[4] (CA) to be sure the measured, small effect is characteristic of particle aggregation.

The result is that PA has a small but clear anisotropy of the correlation function : along the direction of growth, the correlation falls off with a power which is weaker by $\Delta A = 0.16 \pm 0.05$ than perpendicular to the direction of growth[5]. No such effect is found for CA, which is expected if one attributes the anisotropy to the growth mechanism ; CA does not have a fixed center of growth as usually two clusters of comparable size aggregate. In both cases, PA and CA, there is additionally an amplitude anisotropy for the correlations due to the underlying square lattice. All calculations were done in two dimensions.

PA clusters of 5000 particles and CA clusters grown hierarchically (with 4096 particles) were analysed as shown in Fig. 1. R is the radius of the cluster, $\vec{\rho}$ the vector from the origin to a point on the cluster and $\vec{r}$ the direction in which the correlation function is

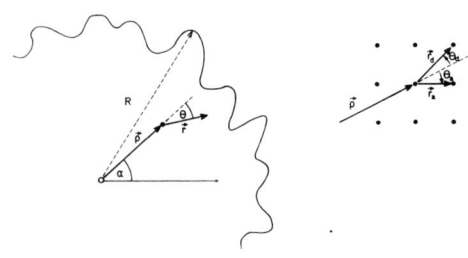

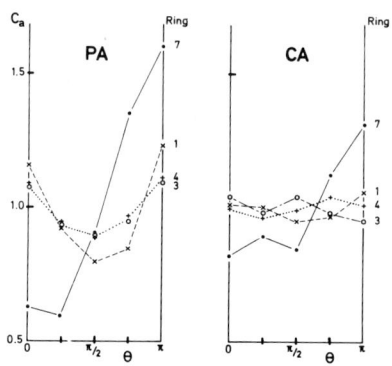

FIGURE 1
Definition of the angular correlations $c(\vec{\rho},\vec{r}) = c(r,\theta)$. From the origin (the seed particle for PA, the particle closest to the centre of mass for CA) $\vec{\rho}$ denotes the coordinates of a particle of the cluster. The correlation functions $c(\vec{\rho},\vec{r}) = \langle n(\vec{\rho}+\vec{r})n(\vec{\rho})\rangle = \langle n(\vec{\rho}+\vec{r})\rangle$ is the average density n at $\rho+r$ and only depends on r and θ (as shown on the left). On the lattice, $c(\rho, r)$ also depends on the angle α between $\vec{\rho}$ and one axis of the lattice, and $\vec{r}$ is restricted to points on the lattice. Axial (θ_a) and diagonal (θ_d) correlations are calculated here (as indicated on the right).

FIGURE 2
Axial correlations $c_a(r = 6, \theta)$ (normalized by the average of c_a over angles). They are calculated for all points on the cluster and then are averaged separately over concentrical rings with the same number of particles around the origin (ring 1-ring 7). Initial and surface effects clearly modify the angular dependence of c_a. PA is shown on the left, CA on the right.

measured. Off lattice, in the scaling region where

(1) $1 \ll r \ll \rho \ll R$

the correlations can only depend on $r = |\vec{r}|$ and θ, the angle of $\vec{r}$ with respect to the growth direction $\vec{\rho}$. As the clusters are grown on a square lattice, both axial and diagonal correlations are calculated separately to eliminate possible effects due to the lattice. In Fig. 2, the results are shown for PA and CA, for axial correlations. The data of all the angles have been grouped into five segments from forward ($\theta = 0$) to backward ($\theta = \pi$). Measurements have been made separately in seven concentrical rings around the seed particle, whereby rings 3 and 4 best represent the scaling region, eq. 1. Rings 1 and 7 show clear deviations.
In Fig. 3, axial and diagonal correlations in the scaling region show that with increasing r, the anisotropy between $\theta = 0, \pi$ and $\theta = \pi/2$ grows for PA. For CA there is no such effect.

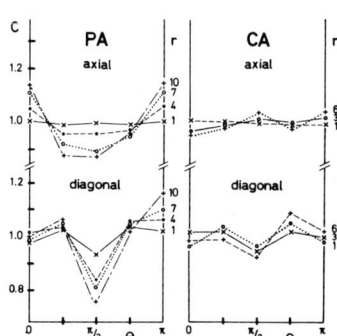

FIGURE 3
Axial and diagonal correlations $c(r, \theta)$ (normalized) for PA and CA as a function of the angle θ with r as a parameter. The results are obtained deep inside the cluster, where equation (1) is valid (the correlations are averaged over the points of the fourth ring alone). The results for PA (left) differ markedly from CA (right). They suggest different scaling behaviour parallel and perpendicular to the growth direction, for PA only. Comparing the axial correlations (top) with the diagonal correlations (bottom) also show, that the lattice changes the amplitudes of the correlations, both for PA and for CA. The statistical error of the points is less than 0.04. For the diagonal correlations r is labelled in units of $\sqrt{2}$ times the lattice spacing.

To get a quantitative estimate of the anisotropy, the ratio $c_\|/c_\perp = r^{-(A_\|-A_\perp)} = r^{\Delta A}$ is shown in a log-log plot in Fig. 4. ($c_\| = c(r, \theta = 0, \pi)$, $c_\perp = c(r, \theta = \pi/2)$). Both axial and diagonal correlations support an anisotropy of PA of $\Delta A = 0.16 \pm 0.05$ but not for CA.

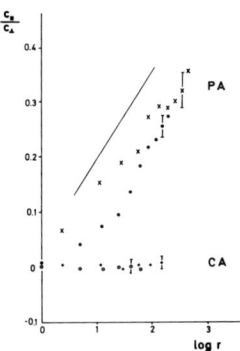

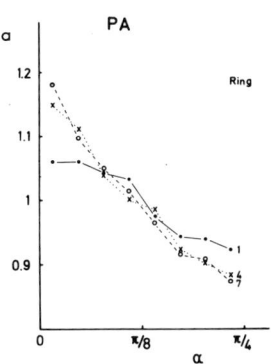

FIGURE 4

Log-Log plot of the ratio $\dfrac{c_\|}{c_\perp} = \dfrac{c(r,\theta=0)+c(r,\theta=\pi)}{2c(r,\theta=\pi/2)}$ versus r for PA to determine the exponent $\Delta A = A_\perp - A_\| = 0.16 \pm 0.05$. The straight line has slope 0.16. The curves from axial and diagonal correlations gradually approach the same slope for large r. The same ratio for CA suggests $\Delta A = 0$. The scaling region breaks down for $r \gtrsim 30$ ($r \gtrsim 20$) for axial (diagonal) correlations, primarily as the edge of the cluster is reached (PA). The symbols stand for axial (•) and diagonal (x) correlations for PA and for axial (O) and diagonal (+) correlations for CA.

Finally, the observed shape anisotropy[3] of the clusters in PA can be quantified by calculating the average Radius $R^2(\alpha, N)$ of the (N+1)th particle aggregating at an angle α with respect to one lattice axis. To have a measure that does not depend on N explicitly, the ratio $a = \dfrac{R^2(\alpha,N)}{R^2(N)}$ is calculated. $R^2(N)$ is the angular average of $R^2(\alpha,N)$. For increasing rings (1 - 7) the anisotropy grows steadily, as shown in Fig. 5. In fact, the data indicates that the anisotropy becomes indefinitely more pronounced with increasing cluster size. The measure used here is quite sensitive, as the clusters visually still appear isotropic for the sizes considered.

FIGURE 5

$R(N,\alpha)$ is the average radius of the Nth particle aggregating at an angle α on a square lattice. It is shown normalized $a(\alpha) = R^2(N,\alpha)/R^2(N)$ where $R^2(N)$ is the angular average of $R^2(N,\alpha)$, and averaged separately for increasing rings, for PA, in the interval $0 \leq \alpha \leq \pi/4$. It clearly shows the anisotropy of the shape of the clusters. The statistical errors are less than 0.03.

REFERENCES

1. For a review see the abstracts of the workshop on "Kinetic models for cluster formation" (september 17-28, 1984, CECAM Orsay); R. Jullien, M. Kolb, H. Herrmann and J. Vannimenus eds., J. Stat. Phys. 39, 241 (1985).

2. T. A. Witten and L. M. Sander, Phys. Rev. Lett. 47 (1981) 1400

3. R. Brady and R. Ball, unpublished. They suggest that large PA clusters on a square lattice visually have the shape of a diamond

4. P. Meakin, Phys. Rev. Lett. 51 (1983) 1119
   M. Kolb, R. Botet and R. Jullien, Phys. Rev. Lett. 51 (1983) 1123

5. M. Kolb, J. de Physique Lett. 46, L 631 (1985). P. Meakin and T. Vicsec and independently R. Voss also found cluster anisotropy for PA, with similar methods.

FRACTALS IN PHYSICS
L. Pietronero, E. Tosatti (editors)
© *Elsevier Science Publishers B.V., 1986*

REVERSIBILITY IN CLUSTER AGGREGATION

M. KOLB

Laboratoire de Physique des Solides, Bât. 510, Université de Paris-Sud, 91405 Orsay, France
and
Institut für Theorie der Kondensierten Materie, Freie Universität Berlin, Arnimallee 14, 1000 Berlin 33, West Germany

Reversibility is introduced systematically into diffusive cluster-cluster aggregation. The scaling analysis of large clusters suggest that reversibility destroys the features of irreversible clustering leading to fractal dimensions D = 1.57 (2.03) in two (three) dimensions independent of the kinetics. These values are consistent with purely static cluster statistics. The cluster size distribution on the other hand is dependent on the kinetics.

What makes theoretical investigations of growth processes difficult is the irreversibility. The usual methods of equilibrium statistical mechanics are not easily generalised to situations far from equilibrium. On the other hand, the belief is that it is this feature that leads to the rich structure of new scaling properties observed in many growth processes[1-3].

This motivates the study of growth where the irreversibility is partially relaxed. Does one retain the properties of irreversible aggregation, do new scaling phenomena appear or are the properties those of the well known static cluster models[4-5] ? There is also experimental interest in partial reversibility, as fragmentation is a commonly observed process[6-9]. The answer suggested by cluster cluster aggregation with bond breaking as proposed here is that the geometrical fractal properties of the clusters become numerically comparable to static models, completely independent of the kinetics. This suggests that any reversibility destroys the scaling features of cluster cluster aggregation completely.

The model considered is that of clustering of clusters[2]. Particles diffuse independently and stick whenever they touch each other. These dimers also diffuse and stick when they touch.

This way larger and larger clusters grow, as the bonds are permanent and irreversible. The new feature that is now added to this model is that each bond has a finite (but large) lifetime . Irreversible clustering corresponds to $\tau = \infty$. When a bond breaks, a cluster falls apart into two smaller clusters (if there are no loops). It is now assumed that these two clusters diffuse independently of each other, as if they had never been bonded. The bonds are supposed to break independently of each other : per unit time every bond is broken with a probability equal to $1/\tau$. Starting initially with single particles and no bonds, the aggregation process dominates at first . As during the growth more and more bonds form, some of them start to break. Eventually there are sufficiently many bonds such that the time to form one new bond in the aggregation process is equal to the time when one of the many existing bond breaks. A dynamic equilibrium is reached. The early stages of the process correspond to irreversible clustering, the long time behaviour to reversible clustering. This latter case will be studied here. The average cluster size in this region increases with increasing τ. In a scaling analysis, the fractal dimension of large clusters is determined by calculating their radius of gyration as a function of their mass (the mass is varied by varying the average bond lifetime τ). The results indicate that the reversibility destroys the fractal aspect of irreversible clustering. The properties now appear to be those of static cluster statistics (lattice animals) irrespective of the kinetics of the diffusive motion of the clusters and sticking probabilities, both in two and three dimensions. Loop formation also is irrelevant in the process.

The model investigated numerically both in two and three dimensions is a lattice version of cluster aggregation. Particles and clusters move randomly on a periodically bounded hypercubic lattice of length L. Initially N_0 particles are placed randomly on the lattice, not occupying the same sites. Then the particle jump randomly to nearest neighbor sites. If two particles sit on nearest neighbor sites, a bond forms between them and they move together from now on as a dimer. Similarly, when clusters touch each other (nearest neighbors), a rigid bond forms. A cluster always moves as a rigid object. While for simulations off lattice the probability to form loops is vanishing, on the lattice loops may form when two clusters touch simultaneously in two points. Two cases have been considered to show that loop formation does not influence the results. In the loopless version, one simply pick one pair of touching particles at random (if several particles of the clusters touch) and places a single bond between them. In the looped version all possible bond are formed. On the other hand, a cluster then does not necessarily fall apart if a bond is cut. Cutting a bond then either breaks up a cluster or just breaks a loop.

Qualitatively, the following analysis can be given for the case of low cluster concentration. D denotes the fractal dimension of the clusters and d the spatial dimension. Furthermore the diffusing velocity of the clusters is mass dependent, $v(m) \sim m^\alpha$. Using the mean field expression[10] for the time to pair up, $\tau_a \sim m^{1 - \alpha - (d-2)/D}$ and the average time for one bond to break $\tau_f \sim \tau/m$ (the subscripts stand for aggregation resp. fragmentation, the equilibrium condition $\tau_a = \tau_f$ determines the typical cluster mass m as a function of τ, $m_{eq} \sim \tau^{1/(2 - \alpha - (d-2)/D)}$. The characteristic time which determines the onset of the equilibrium aggregation is $t_{eq} = \tau/m_{eq}$. For times $t \ll t_{eq}$, the aggregation process is irreversible, for $t \gg t_{eq}$ it is in equilibrium. Here, the latter regime has been simulated for different values of τ and hence m.

Figure 1 shows the crossover between the two

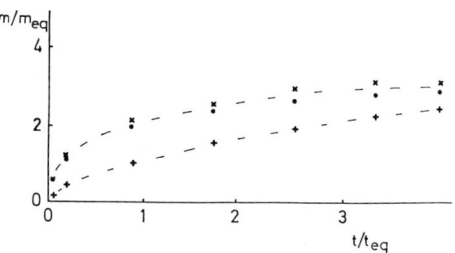

FIGURE 1
Crossover from growth (irreversible clustering) to equilibrium (reversible clustering) aggregation. The average mass m (normalised by m_{eq}) is plotted vs. time t (normalised by t_{eq}). For comparison, $\alpha = -2$, $m_{eq} = 6$ (.), $\alpha = -2$, $m_{eq} = 12$ (x) and $\alpha = 0$, $m_{eq} = 12$ (+) are shown. The data is an average over 100 separate simulations in two dimensions.

two regimes in reduced coordinates, m/m_{eq} versus t/t_{eq}, for different m_{eq} and α. For a given α, the points for different m lie on the same curve, indicating that this is a scaling function. In Fig. 2, the radius of gyration R is shown in a log-log plot versus the mass m. Both aggregation with and without loops and the data taken from an average over all clusters and only from the largest cluster show straight lines with a slope corresponding to D = 1.57 ± 0.06

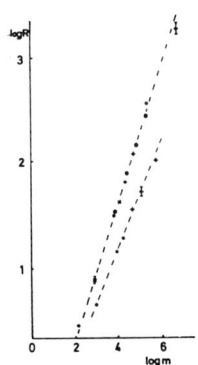

FIGURE 2
Log-log plot of the radius of gyration R vs. the mass m. The upper curve is for d = 2, the lower for d = 3. Data is shown for α = -2 (.) and α = 0 (x) for the average over all the clusters and for α = -2 (+) for the largest cluster, for loopless clusters. Clusters with loops are shown for α = -2 (o) average). The estimated fractal dimension is D = 1.57 ± 0.06 (2.03 ± 0.05) in d = 2 (3). The data is an average over 2000 measurements in the steady state regime.

(2.03 ± 0.05) in two (three dimensions). The kinetics of the diffusion, through the parameter α, is varied and also does not influence the measured value of D within the error bars. Another way to change the kinetics is to introduce a sticking probability : a bond is formed only with probability p < 1 when two particles touch. Simulations with p = 0.05 show that this modification also leaves the fractal dimension at the values quoted above.

The conclusion of these measurements under widely varied conditions is that reversibility has a very drastic effect on the growth. The fractal dimension changes from the value of irreversible clustering, leaving the clusters more compact, but insensitive to the kinetics. Neither the loop structure, nor the diffusivity or the sticking conditions seem to matter. This suggests that as soon as there is reversibility only configurational (static) aspects determine the fractal properties. This is supported additionally by the fact that D is consistent with the corresponding values for static lattice animals.

The kinetics nevertheless leaves its trace when considering the cluster size distribution. The number of clusters of mass m, N(m), when put in scaling form, $N(m) = \bar{m}^{-2} p(m/\bar{m})$, shows a dependence on the kinetics. $\bar{m}$ is the weight averaged mass. The scaling function p(x) is independent of $\bar{m}$ ($\bar{m}$ large) and is shown in Fig. 3. $\bar{m}$ is proportional to m_{eq}.

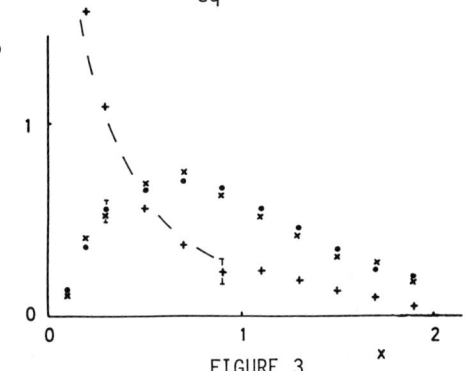

FIGURE 3
Reduced cluster-size distribution p(x) for d = 2 and loopless clusters. The distribution does not depend on m_{eq} but on α. For α = -2, m_{eq} = 6 (.) and m_{eq} = 12 (x) and for α = -1, m_{eq} = 12 are shown.

Its shape in the monodisperse (α negative large) regime shows a much broader maximum than in the corresponding irreversible process[10]. This is due to the random breakup mechanism which favors polydispersity.

REFERENCES

1. T. Witten and L. M. Sander, Phys. Rev. Lett. 47, 1400 (1981)

2. P. Meakin, Phys. Rev. Lett. 51, 1119 (1983)
   M. Kolb, R. Botet and R. Jullien, Phys. Rev. Lett. 51, 1123 (1983)

3. For a general introduction see "Kinetics of aggregation and gelation", F. Family and D. P. Landau, eds. (North Holland 1984)

4. T. Lubensky and J. Isaacson, Phys. Rev. Lett. 41, 829 (1978) ; Phys. Rev. A 20, 2130 (1979)
   G. Parisi and N. Sourlas, Phys. Rev. Lett. 46, 871 (1981) ; U. Glaus, preprint

5. D. Stauffer, Phys. Repts 54, 1, (1979) ;

"Introduction to Percolation", Taylor and Francis, London 1985
6. P. J. Blatz and A. V. Tobolsky, J. Phys. Chem. 49, 77 (1945) ; V. S. Nanda and R. K. Pathria, J. Chem. Phys. 30, 27 (1959) ; E. M. Hendriks, Z. Phys. B 57, 307 (1984)
7. C. Allain and B. Jouhier, J. de Physique Lett. 44, L 421 (1983) ; P. Richetti, J. Prost and P. Barois, J. de Physique Lett. 45, L 1137 (1985) ; C. Camoin and R. Blanc, J. de Physique Lett. 46, L 67 (1985) ; numerically, rotational readjustment was investigated by P. Meakin and R. Jullien, J. de Physique, J. de Physique 46, 1543 (1985)
8. M. Herbst and J. H. Goldstein, p. 53 in ref. 3
9. B. Vincent and S. G. Whittington, Surface and Colloid Science 12, 1 (1982)
10. M. Kolb, Phys. Rev. Lett. 53, 1653 (1984)

FILM ON AGGREGATION PROCESSES

M. KOLB

Laboratoire de Physique des Solides, Bât. 510, Université de Paris-Sud, 91405 Orsay, France
and
Institut für Theorie der Kondensierten Materie, Freie Universität Berlin, Arnimallee 14, 1000 Berlin 33, West Germany

A film has been produced to illustrate several mechanisms of growth processes. The film is based on simulations of theoretical models. It shows how different fractal like structures appear depending on the different growth mechanisms.

There are two reasons for making a movie on aggregation phenomena. First, the processes in nature which lead to disordered fractal structures can be modelled by rather simple stochastic rules. Secondly, much of the theoretical understanding comes from numerical experiments (Monte-Carlo simulations). These simulations actually execute in some simplified way the movements of particles in nature.

The purpose of the film is pedagogical. The slogan that one picture can say much more than a thousand words is particularly valid for growth processes.

In order to achieve some cohesion, four important growth models have been chosen which experimentally cover a very wide range of applications but which theoretically are closely related, i. e. they can be throught of as special cases of one general aggregation model. The four models illustrated are : 1) the Eden model which may be used to describe surface growth like in biological systems. 2) The Witten-Sander or diffusion limited particle aggregation model, which has been invoked to model lightning, electrodeposition and fluid-fluid displacement phenomena, 3) Diffusion limited cluster aggregation or clustering of clusters which describes aggregation in aerosols and colloids and 4) Chemical clustering, also used in colloidal systems.

The reason these models are closely related is that two parameters can be used to characterise all four of them in terms of generalised cluster cluster aggregation. One parameter is the mechanism of aggregation : particle-cluster or cluster-cluster. This can be controlled by a diffusion constant which depends on the cluster size. The second parameter determines whether diffusion or chemical reaction dominates the irreversible bond formation between clusters. A sticking probability is the variable that theoretically distinguishes the two cases.

In the film the Eden model is shown first, with the visible feature that growth from a center leads to concentrical rings like one observes when cutting the trunk of a tree. The structure is compact, the surface roughness decreasing gradually compared with the size of the aggregate. The dynamics of the growing surface is illustrated by showing the active zone.

Secondly, the Witten-Sander model is grown - in the same spherical geometry as the Eden model. Growth on two different length scales shows simultaneously the (approximate) scale invariance and effects of the underlying lattice which influences the allover shape of the object. The active surface in this growth process has an entirely different structure than the Eden model, reflecting the importance of the diffusion and showing that the aggregate looks like a fractal. Using a zoom, the invariance of the object viewed on different length scales is demonstrated. In this model, the geometry in which the aggregate is grown, determines its shape. As an example, this model has been grown on a surface, which looks strikingly similar to pictures of metal oxide enclosed in rock. Finally, when introducing a sink, where particles are absorbed, one can study the properties of a single branch of this treelike structure.

Thirdly, the clustering of clusters model is shown, where an assembly of randomly moving particles form larger and larger objects with very ramified, stringy features. This shows explicitly, why for example smoke particles have a whisky appearance. Important information on the kinetics of the experiments can be obtained from the cluster size distribution. In the film, its qualitative features can be seen because of the color coding of the clusters according to size.

The theoretical connection with the Witten-Sander model can be demonstrated in an amusing way by inverting the law relating cluster mass and diffusion constant : now large clusters diffuse faster than small ones. The bigger they get, the greedier they become !

Two variants of this model describe other physical situations. When the cluster concentration becomes large, gel formation sets in which, in contrast to the usual modelling in terms of percolation, is determined by the kinetics. Also, when the bond formation is not irreversible, the system reaches a dynamic equilibrium, with clusters looking somewhat more compact.

The last of the models illustrated in chemical (reaction limited) clustering. As in actual colloidal systems, the model is shown for the case of diffusing clusters, but where the sticking probability to form a bond (when two clusters touch) is so low, that it is the reaction rate rather than the diffusion that determines the growth. The resulting clusters again have visibly different features from the diffusive situation : they are also more compact (though still fractal).

The film shows exclusively two dimensional simulations of the above aggregation processes (the basic concepts are identical to those in three dimensions and it turned out that projections resp. sections of three dimensional realisations do not illustrate the processes very well). The colors have been used systematically to show dynamic properties : as time goes on, the colors cover the spectrum of the rain bow !

Technically, the movie was generated as follows : on a computer, a series of subsequent configurations are generated by simulating one of the above growth processes. For the Eden model, a new configuration consists of the previous one where a few hundred particles have been added at the surface. For cluster aggregation, the time between two configurations is such that most clusters have diffused by about the size of a particle.

These configurations are then sent to a high resolution color screen, one by one. After each frame is completely drawn on the screen, a camera placed in front of it takes one picture. Then the next configuration is transmitted to the screen and so on. For every second or film 25 frames have to be exposed.

For those interested, the film is ca. 25 minutes long and all the aggregation processes are explained in detail. No expertise in the field is necessary to follow the way the structures grow. Copies are available upon request, either in 16 mm or in video (color).

This film has been made with the help of ZEAM (production), the department of geography, ZEDAT, all at Freie Universität Berlin and ZUSE Zendum Berlin. Fotos were provided by Lemma, Pietronero, Rasz and Weitz. Support was also given by Pathé Marconi (EMI Records), Paris. Special thanks go to U. Hall and K. A. Penson. Financial assistance by the DFG, Bonn is acknowledged.

FIELD THEORY APPROACH TO THE EDEN MODEL AND DIFFUSION LIMITED-AGGREGATION

Luca PELITI

Dipartimento di Fisica, Università "La Sapienza", Piazzale A. Moro 2, I-00185 Roma (Italy) and GNSM - CNR, Unità di Roma

ZHANG Yi-Cheng[*]

Physics Department, Brookhaven National Laboratory, Upton, NY 11973 USA

Field theories of growth models, such as the Eden model and diffusion-limited aggregation, are introduced by means of the Doi-Grassberger-Scheunert Fock space formalism for classical objects.

Kinetic aggregation models are presently the subject of intensive investigations. The most popular models of this kind are the Eden model [1] and the Witten-Sander model of diffusion limited aggregation [2,3]. A large part of this investigation is carried out by computer simulation [4] although different theoretical approaches are not lacking. A few exact results for the Eden and DLA models in the limit of many dimensions have indeed been obtained by Parisi and Zhang [5] and by Vannimenus et al. [6].

One of the main purposes of current research on these models is to identify the mechanism at the origin of the remarkable self-similarity of the aggregates and a way to calculate their fractal dimension. On the basis of the experience of the theory of critical phenomena, one would expect field theoretical techniques to appear as a powerful computational tool in such a project. Indeed, a few field theoretical approaches to birth-death processes on a lattice closely resembling aggregation models have appeared in the literature. Field theoretical approaches to birth-death processes with and without memory have recently been considered. In refs [7-9] a field theoretical treatment of the Schlögl model [10] of chemical reactions (related to directed percolation) is introduced and related to the Reggeon Field Theory (RFT) of high energy physics [11]. Cardy [12] has then introduced immunization effects in Schlögl's model and has treated the corresponding non--Markovian process by a field theory similar to that describing the "true" self-avoiding walk (TSAW;[13]). This analysis has been reexamined by Cardy and Grassberger [14] and by Janssen [15] who came to the conclusion that the static properties of this model are in the same universality class as the statistics of percolation clusters. On the other hand, Parisi and Zhang [16] have introduced, on the basis of heuristic

---

[*] Supported by U.S. Department of Energy under contract No. DE-AC02-76CH00016.

considerations, field theories describing the Eden and DLA models. Their conclusion in that the Eden model is described by a RFT with local interactions, deep in the ordered region, whereas DLA corresponds to a generalized RFT with nonlocal interactions. Shapir and Zhang [17] have considered a systematic Hamiltonian approach to the Eden model, deriving a field theory which is local in space, but nonlocal (non-Markovian) in time. We show that a single well-cut tool is sufficient to derive field theoretical descriptions of both Eden and DLA models in a systematic way. The tool is the Fock space formalism for classical objects, first introduced by Doi [19] and more recently reformulated by Grassberger and Scheunert [20].

The method involves essentially expressing the evolution equations of the model as a master equation of the form

$$\frac{d|\phi(t)>}{dt} = L|\phi(t)> \qquad (1)$$

where $|\phi(t)>$ is the macroscopic state of the system, identified by the probabilities $\phi(\underline{n};t)$ of finding the system in the microscopic state defined by the set of occupation numbers $\underline{n} = \{n_{\vec{r},\alpha}\}$, where $\vec{r}$ identifies sites of a D-dimensional lattice and α identifies the species of particles which may be present of the system. The evolution operator L (the Liouvillian) is expressed in terms of annihilation $a_{\vec{r},\alpha}$ and creation $\pi_{\vec{r},\alpha}$ operators defined by

$$a_{\vec{r},\alpha}|\{n_{\vec{r}',\alpha'}\}> = n_{\vec{r},\alpha}|\{\ldots,n_{\vec{r},\alpha}-1,\ldots\}>, \qquad (2)$$

$$\pi_{\vec{r},\alpha}|\{n_{\vec{r}',\alpha'}\}> = |\{\ldots,n_{\vec{r},\alpha}+1,\ldots\}>, \qquad (3)$$

These operators satisfy the usual commutation relation:

$$[a_{\vec{r},\alpha},\pi_{\vec{r}',\alpha'}] = \delta_{\vec{r}\vec{r}'}\delta_{\alpha\alpha'} \qquad (4)$$

A Hilbert space structure is introduced for the space of macroscopic states such that the Hermitean conjugate $a^+_{\vec{r},\alpha}$ of the annihilation operator $a_{\vec{r},\alpha}$ is given by:

$$a^+_{\vec{r},\alpha} = \pi_{\vec{r},\alpha} - 1. \qquad (5)$$

If L is written as a normal product, i.e. as a linear combination of monomials in the creation and annihilation operators, with the annihilation operators on the right of the creation ones, the corresponding path integral is identified by the Lagrangian

$$\mathcal{L} = \sum_{\vec{r},\alpha} i\hat{\psi}_{\vec{r},\alpha}\dot{\psi}_{\vec{r},\alpha} - L[i\hat{\underline{\psi}},\underline{\psi}] \qquad (6)$$

as a function of the two real, classical fields $\hat{\underline{\psi}} = \{\hat{\psi}_{\vec{r},\alpha}\}$, $\underline{\psi} = \{\psi_{\vec{r},\alpha}\}$. The dot denotes a derivative with respect to time. The function $\mathcal{L}(i\hat{\psi},\psi)$ is obtained from the normal product expression of the Liouvillian L by substituting $i\hat{\psi}_{\vec{r},\alpha}$ for $a^+_{\vec{r},\alpha}$ and $\psi_{\vec{r},\alpha}$ for $a_{\vec{r},\alpha}$ and taking into account eq. (5).

Let us now consider the following formulation of the Eden model [1]. While in the original Eden model exactly one particle was added at each time step, we consider that each member of the aggregate may add a particle at one of its neighboring sites, if it is empty, with a certain rate μ. The "time" of the original Eden model is therefore proportional to the aggregate mass of this model. It is easy to write down a normal product expression for the Liouvillian of such a process:

$$L = \mu \sum_{\vec{r}} \sum_{\vec{e}} \sum_{k=0}^{\infty} \frac{(-1)^k}{k!} (\pi_{\vec{r}} - 1) \pi_{\vec{r}+\vec{e}} \pi_{\vec{r}}^k a_{\vec{r}+\vec{e}} a_{\vec{r}}^k \quad (7)$$

The factor

$$\gamma(\pi_{\vec{r}}, a_{\vec{r}}) = \sum_{k=0}^{\infty} \frac{(-1)^k}{k!} \pi_{\vec{r}}^k a_{\vec{r}}^k \quad (8)$$

ensures that no particle is added to sites which are already occupied. The sum over $\vec{e}$ runs over all nearest neighbor vectors, i.e. over all vectors which lead from one site to one of its nearest neighbors. By applying the above scheme we are led to the Lagrangian

$$\mathcal{L} = \sum_{\vec{r}} \left\{ i\hat{\psi}_{\vec{r}} \dot{\psi}_{\vec{r}} - \mu \sum_{\vec{e}} \sum_{k=0}^{\infty} \frac{(-1)^k}{k!} (i\hat{\psi}_{\vec{r}})(i\hat{\psi}_{\vec{r}+\vec{e}}+1) \right.$$
$$\left. (i\hat{\psi}_{\vec{r}}+1)^k \psi_{\vec{r}+\vec{e}} \psi_{\vec{r}}^k \right\} \quad (9)$$

To make this expression more transparent we isolate the terms which are lowest order in the fields. We obtain:

$$\mathcal{L} = \sum_{\vec{r}} \{ i\hat{\psi}_{\vec{r}} \dot{\psi}_{\vec{r}} - \mu \sum_{\vec{e}} [i\hat{\psi}_{\vec{r}} \psi_{\vec{r}+\vec{e}} + (i\hat{\psi}_{\vec{r}})(i\hat{\psi}_{\vec{r}+\vec{e}}) \psi_{\vec{r}+\vec{e}}$$
$$-i\hat{\psi}_{\vec{r}} \psi_{\vec{r}+\vec{e}} \psi_{\vec{r}}] + \text{terms higher order in the fields} \}. \quad (10)$$

We may rearrange the terms explicitly written down in eq. (10) to obtain:

$$\mathcal{L} = \sum_{\vec{r}} \{ i\hat{\psi}_{\vec{r}} \dot{\psi}_{\vec{r}} - \mu q i\hat{\psi}_{\vec{r}} \psi_{\vec{r}} - \mu i\hat{\psi}_{\vec{r}} \Delta\psi_{\vec{r}} - \mu(i\hat{\psi}_{\vec{r}})^2 \psi_{\vec{r}} +$$
$$+ \mu i\hat{\psi}_{\vec{r}} \psi_{\vec{r}}^2 - \mu i\hat{\psi}_{\vec{r}} \Delta(i\hat{\psi}_{\vec{r}} \psi_{\vec{r}}) - \mu i\hat{\psi}_{\vec{r}} \psi_{\vec{r}} \Delta\psi_{\vec{r}} + \ldots \}, \quad (11)$$

where q is the coordination number of the lattice and Δ is the discrete Laplacian:

$$\Delta\psi_{\vec{r}} = \sum_{\vec{e}} (\psi_{\vec{r}+\vec{e}} - \psi_{\vec{r}}). \quad (12)$$

The first two lines of eq. (11) correspond to a RFT with negative "mass", i.e. deep in the ordered region. Neglecting higher order terms such as those appearing in the third line of this equation one recovers RFT in the continuous limit. Let us remark that our field theory, eq. (9), is local (Markovian) in time and quasilocal in space, only involving derivatives up to second order. Moreover no fields except the density field ψ and its conjugate $i\hat{\psi}$ appear in it.

Let us now dwell on the DLA model, which we formulate as follows. There is a steady (but small) flow of diffus<u>ing</u> (D) particles and an aggregate of A particles. If a D particle finds itself in a site which does not contain A particles, but which is nearest neig<u>h</u>bor to a site containing one of them, it may turn into an A particle and stop, with a certain rate μ. Let us indicate by ϕ, ψ the density fields of D and A particles respectively, and by $i\hat{\phi}$, $i\hat{\psi}$ the corresponding conj<u>u</u>gate ones. One obtains therefore the Lagrangian:

$$\mathcal{L} = \sum_{\vec{r}} \left\{ i\hat{\phi}_{\vec{r}} \dot{\phi}_{\vec{r}} + i\hat{\psi}_{\vec{r}} \dot{\psi}_{\vec{r}} - \omega \sum_{\vec{e}} (i\hat{\phi}_{\vec{r}+\vec{e}} - i\hat{\phi}_{\vec{r}}) \phi_{\vec{r}} \right.$$
$$+ \mu \sum_{\vec{e}} (1+i\hat{\psi}_{\vec{r}+\vec{e}})(i\hat{\psi}_{\vec{r}} - i\hat{\phi}_{\vec{r}}) \psi_{\vec{r}+\vec{e}} \phi_{\vec{r}} \quad (13)$$
$$+ \mu \sum_{\vec{e}} \sum_{k=1}^{\infty} \frac{(-1)^k}{k!} (1+i\hat{\psi}_{\vec{r}+\vec{e}})(i\hat{\psi}_{\vec{r}} - i\hat{\phi}_{\vec{r}})$$
$$\left. (1+i\hat{\psi}_{\vec{r}})^k \psi_{\vec{r}}^k \psi_{\vec{r}+\vec{e}} \phi_{\vec{r}} \right\}.$$

We have not written explicitly the source terms at infinity which ensure the steady incoming flow of D particles. If the interaction terms proportional to $i\hat{\psi}_{\vec{r}+\vec{e}}$ are neglected, one recovers the mean field theory of DLA proposed by Nauenberg [20].

We have shown in conclusion that the techniques of Doi and Grassberger and Scheunert allow for a derivation of field theories for the most popular models of irreversible aggregation.

We thank Y. Shapir for illuminating discussions.

REFERENCES

[1] Eden M., 1961, in Neyman, J. (ed.), *Proceedings of the 4th Berkeley Symposium on Mathematical Statistics and Probability* (Berkeley: University of California Press). Vol. IV, p. 223.

[2] Witten T.A. and Sander L.M., Phys. Rev. Letts. 47 (1981) 1400.

[3] Witten T.A. and Sander L.M., Phys. Rev. B27, (1983) 5686.

[4] Family F. and Landau D.P. (eds.), 1984, *Kinetics of Aggregation and Gelation* (Amsterdam: North-Holland).

[5] Parisi G. and Zhang Y.C., Phys. Rev. Letts. 53 (1984) 1791.

[6] Vannimenus J., Nickel B. and Hakim V., Phys. Rev. B30 (1984) 391.

[7] Grassberger P. and De La Torre A., Ann. Phys. (NY) 122 (1979) 373.

[8] Cardy J.L. and Sugar R.L., J. Phys. A: Math. Gen. 13 (1980) L423.

[9] Janssen H.K., Zeit. Physik B42 (1981) 151.

[10] Schlögl F., Zeit. Physik 253 (1972) 147.

[11] Amati D., Ciafaloni M., Marchesini G. and Parisi G., Nucl. Phys. 448 (1976) 483.

[12] Cardy J.L., J. Phys. A: Math. Gen. 16, (1983) L709.

[13] Obukhov S.P. and Peliti L., J. Phys. A: Math. Gen. 16 (1983) L167.

[14] Cardy J.L. and Grassberger P., J. Phys. A18 (1985) L267.

[15] Janssen H.K., 1985, Zeit. Physik B58 (1985) 311.

[16] Parisi G. and Zhang Y.C., 1985, Brookhaven preprint, to appear in J. Stat. Phys.

[17] Shapir Y. and Zhang Y.C., 1985, Lett. J. de Phys. June Issue.

[18] Doi M., J. Phys. A: Math. Gen. 9 (1976) 1465.

[19] Grassberger P. and Scheunert M., Fortschritte der Physik 28 (1980) 547.

[20] Nauenberg M., Phys. Rev. B78 (1983) 449.

SPREADING OF EPIDEMIC PROCESSES LEADING TO FRACTAL STRUCTURES

Peter GRASSBERGER

Physics Department, University of Wuppertal, Wuppertal, W.-Germany

Population growth models are reviewed that can model epidemics and lead to fractal clusters. All known such models allow also for shrinking of the epidemic in addition to its growth, and are related to either directed or undirected percolation. In the case of short-range infection, Monte Carlo simulations lead to very precise estimates of critical parameters of ordinary percolation. Epsilon-expansions of dynamic critical exponents are discussed, both for short- and long-range infections.

1. INTRODUCTION

In this talk I shall discuss cluster growth models which can serve as models for the spread of epidemics, in populations with low mobility and with not too long ranged infections [1,2,3]. Such models can - with only minor modifications - be applied also to a variety of other phenomena, from forest fires to the formation of stars. For consistency, I shall mainly speak of epidemics.

The simplest model is Malthus' exponential growth. Incorporating saturation effects leads, if spatial distribution is not considered, to the Verhulst model. If spatial clustering is taken into account, it leads to the Eden process [4]. The latter leads to compact clusters, so I will not say much about it (although there might be non-trivial scaling laws associated with the boundary of these clusters [5]).

More realistic models must take into account that the cluster of infected individua does not always grow. For instance, ill individua are less fit to defend themselves, and thus might be replaced by off-spring of healthy ones.

Schematically, we can write this as a "reaction"

$$\text{ill + healthy} \rightarrow \begin{cases} \text{ill + ill} \\ \text{(infection, rate a)} \\ \text{healthy + healthy} \\ \text{(death and replacement, rate b<a)} \end{cases} \quad (1)$$

This has been proposed as a model for growth of tumors [6]. It again does not lead to fractal clusters, but again we should expect that the surface of the cluster shows interesting and non-trivial scaling laws similar to those found in ref. [5].

In order to obtain fractal clusters, it seems that one must take into account recovery or death as single-"particle" reactions,

$$\text{ill} \rightarrow \text{healthy or dead} \quad (2)$$

The main difference between eqs.(1) and (2) is that under eq.(1) a cluster of ills can shrink only along its perimeter, while under eq.(2) it can develop internal holes. There are two qualitatively different models:

I. The "simple epidemic with recovery" [1] characterized by only two types of individua, and by the following reactions taking place between neighbours:

ill + healthy → ill + ill
                     (infection, rate a)
        ill → healthy
                    (recovery, rate b)

When the ratio a/b is above a critical value, we can have an endemic situation, i.e. a stationary state with non-zero density of ills. The critical behaviour at its threshold is, as we shall see, described by directed percolation in $d+1$ dimension, where d is the number of space dimensions.

II. The "general epidemic" or "epidemic with removal" [1], with a third state of "removed" (i.e. immune or dead) individua, and with

ill + susceptible → ill + ill
                       (infection, rate a)
        ill → removed
                    (death or recovery with immunization, rate b)

Notice that the reservoir of susceptibles is not re-filled. Thus, every epidemic in a finite population dies out sooner or later. In an infinite population, however, an epidemic can spread forever as a solitary wave, leaving behind it a cluster of immunes. If these do not move later on, they form a (undirected) percolation cluster.

Thus we see that all known epidemic models leading to fractal structures are related to percolation.

In sec. 2 we shall elaborate on the last two models, and shall present Monte Carlo results. Field theory and epsilon expansions for them will be studied in sec. 3, and generalizations (mainly to long-range infections) will be discussed in sec. 4.

2. RELATION TO PERCOLATION AND MONTE CARLO SIMULATIONS

2.1. To see the connection between model (I) and directed percolation, take a square lattice in 1+1 - dimensional space-time oriented in the following way:

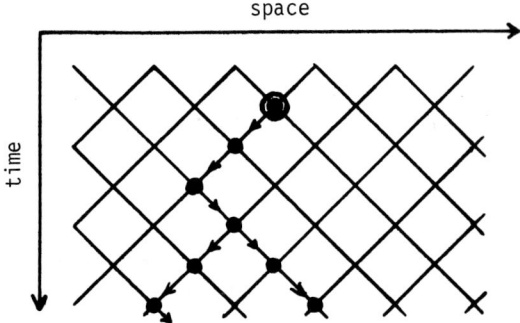

Each individuum occupies one lattice site, and illness lasts exactly one time step. Incubation or latency times are neglected. If we assume that each ill individuum infects both neighbours independently and with probability p, we see that a single ill seed generates a directed bond percolation cluster. If, on the other hand, infection hits always either both neighbours or none, we have site percolation. The following is a typical space-time pattern generated at $p \simeq p_c$ in 1 dimension of space, with all sites originally infected:

Critical exponents of this model have been computed in a variety of ways, e.g. by finite-size scaling [7], series expansions [8], and Monte Carlo [9]. At $p = p_c$, the fractal dimension of the cluster of ills at any fixed time is given by

$$d_F = d - \beta/\nu, \qquad (3)$$

which gives [7,8] $d_F = .749 \pm .001$ for $d=1$ and $d_F = 1.119 \pm .004$ for $d=2$. In the one-dimensional case, this suggests that $d_F = 3/4$ exactly, but nothing is known exactly in spite of the simplicity of the model. In all dimensions ≥ 1, typical clusters are disconnected (as in the above picture for $d=1$).

2.2. For the model with immunization, we employ again the same lattice. We see immediately that the process dies always in 1 dimension:

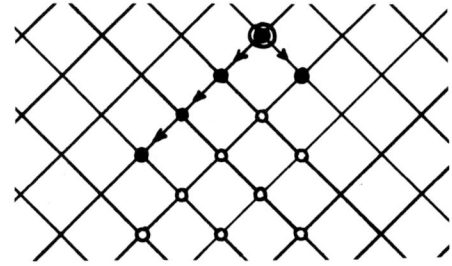

For $d \geq 2$, it can survive forever if $p > p_c$,

and the immune sites at $t=\infty$ form a connected cluster. Clearly, the statistics of these clusters are those of (undirected) bond percolation [10].

Monte Carlo simulations are most efficient if one starts with an entire infected hyperplane and lets the infection spread away from it. A typical pattern generated in this way (in $d=2$) is shown in the following figure. There, removed sites are grey, and ill ones are black, and the infection spreads from bottom to top:

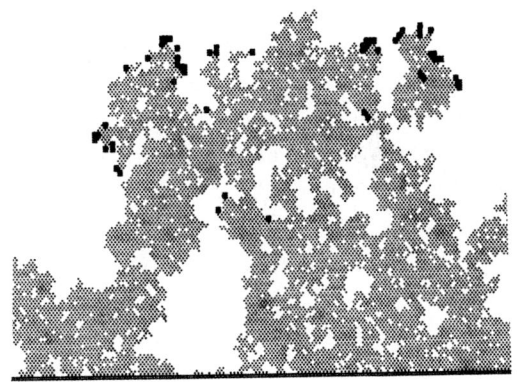

In $d=2$, p_c and the static critical exponents are exactly known, and there is just one independent kinetic exponent, e.g. the exponent ν_t defined by $t_{corr} \sim |p-p_c|^{\nu_t}$. By measuring the average number of ill sites and their average distance from the starting line as functions of t, we found [11] $\nu_t = 1.509 \pm .004$ which might suggest $\nu_t = 3/2$ exactly (other estimates [12,13] have larger errors). In higher dimensions, Monte Carlo simulations [12,14] yield p_c and static critical exponents with roughly the same precision as other methods. For the kinetic exponent, we found [14] $\nu/\nu_t = .725 \pm .006$ ($d=3$) and $.625 \pm .01$ ($d=4$). For a more detailed comparison with other estimates of critical parameters, see ref. [14] (a discrepancy in p_c for d=4 reported in a previous version of [14] was due to a programming error).

3. FIELD THEORY AND -EXPANSIONS

Although field theories of epidemic processes can also be obtained from Martin-Siggia-Rose type arguments [15], a more systematic deviation uses a field theory for particle processes developped by Doi and others [16].

For the epidemic without immunization, one introduces a field operator $\psi(x)$, together with its conjugate $\psi^+(x)$, which play the roles of annihilation and creation operators for infected individua. The density of the latter, when there was one infectious at $(\vec{x},t)=(0,0)$, is given by a Green's function

$$<0|\psi^+(\vec{x})\, e^{LT}\psi(0)|0> \quad (4)$$

Here, the Liouvillean $L = \int d\vec{x}\, L(\vec{x})$ contains the following terms

$$\begin{aligned}L = &-D\vec{\nabla}\psi^+\cdot\vec{\nabla}\psi - b\psi^+\psi \\ &\text{(diffusion + recovery)} \\ &+ a(1+\psi^+)\psi^+\psi \\ &\text{(infection)} \\ &- c(1+\psi^+)\psi^+\psi^2 \\ &\text{(saturation due to finite density of susceptibles.)}\end{aligned} \quad (5)$$

Already some time ago, this theory has been studied in detail, under the name of "reggeon field theory" [17]. Its upper critical dimension is $d_c=4$. Epsilon-expansions ($\varepsilon=d_c-d$) and loop expansions [17] give results in good agreement with other methods.

In the case of immunization, one introduces a second pair of field operators $\phi(x)$ and $\phi^+(x)$. They describe the holes in the distribution of susceptibles, and thus do not propagate in space. The Liouvillean is now

$$\begin{aligned}L = &-D\vec{\nabla}\psi^+\cdot\vec{\nabla}\psi \\ &\text{(diffusion)} \\ &+ a(1+\psi^+)\psi^+\psi \\ &\text{(infection)} \\ &+ c(1+\psi^+)\phi^+\psi \\ &\text{(creation of holes)} \\ &- c'(1+\phi^+)\psi^+\psi\phi \\ &\text{(encounter between infection and holes)}\end{aligned} \quad (6)$$

This time, the upper critical dimension is $d_c=6$. Epsilon-expansions for static exponents agree with those obtained from Φ^3-theory [18]. For the kinetic exponent, we get $\nu_t=1+\varepsilon/28+O(\varepsilon^2)$ [15,19] Numerical agreement with exponents obtained by other methods is reasonable [14].

4. LONG-RANGE INFECTION AND OTHER MODIFICATIONS

4.1. In realistic situations, infection rarly passes between nearest neighbours only. A better approximation, suggested in ref. [2], might be an infection probability which decreases with distance like some inverse power,

$$\text{infection prob.} \sim r^{-\alpha-d}$$

If $\alpha > 2$, this has no effect on the critical exponents. If $\alpha < 2$, one finds that the diffusion terms in eqs(5) and (6) have to be replaced in Fourier space by terms

$$\int dk\, k^\alpha\, \psi^+(k)\, \psi(k) \quad (7)$$

Otherwise said, the interacting Brownian walks have to be replaced by interacting Levy flights [20]. Exponents ν and ν_t have now to be defined via geometric averages,

$$\langle \log. \ell_{corr} \rangle \sim -\nu \langle \log|p-p_c| \rangle ,$$
$$\langle \log. t_{corr} \rangle \sim -\nu_t \langle \log|p-p_c| \rangle \quad (8)$$

Upper critical dimensions are now $d_c=2\alpha$ (without removal) and $d_c=3\alpha$ (with removal). Epsilon-expansions, with $\varepsilon=d_c-d$, can again be obtained straightforwardly. For process (I) (no removal; directed percolation) we find

$$\nu_t = 1+\frac{2\varepsilon}{9\alpha}, \quad \nu = \frac{1}{\alpha}\left(1+\frac{4\varepsilon}{9\alpha}\right),$$
$$\beta = 1-\frac{\varepsilon}{3\alpha}, \quad d_F = \alpha - \frac{2\varepsilon}{9} ; \quad (9)$$

for process (II), we find similarly

$$\nu_t = 1+\frac{5\varepsilon}{32\alpha}, \quad \nu = \frac{1}{\alpha}\left(1+\frac{\varepsilon}{4\alpha}\right),$$
$$\beta = 1-\frac{\varepsilon}{4\alpha}, \quad d_F = 2\alpha - \frac{\varepsilon}{2} . \quad (10)$$

In this latter case, the static exponents had already been given in ref.[18].

A problem is now that these exponents are not analytic at $\alpha=2$, compare ref.[21]. Accordingly the range of applicability ot these expansions is doubtful [22].

For $d>d_c$, there can be more than one infinite cluster. Thus, the probability P(p) that a given site is the source of an inifinite epidemic when $p>p_c$ (i.e., that it belongs to some infinite cluster) is larger than the density $\rho(p)$ of infected resp. removed sites in a given infinite cluster:

$$P(p) \sim (p_c-p)^\beta, \quad \rho(p) \sim (p_c-p)^{\beta'}, \quad (11)$$
with $\beta' > \beta$.

Consider the model with immunization. There, the exponents ν_t, β, and the spreading dimension $\hat{d}$ [11] are "intrinsic", i.e. independent of the lattice and of the exponent α, for $d>d_c$:

$\nu_t = \beta = \hat{d}/2 = 1$. On the other hand, the exponents ν, β', and d_F do depend on the lattice and on α: $\nu = 1/\alpha$, $\beta'=d/\alpha-2$, $d_F = 2\alpha$. These results hold also for the short-range case, with α replaced by 2.

4.2. Real epidemics are of course much more complicated then the simple models discussed above. There are e.g. latency periods, incubating times, carriers, vectors, etc.[1]. All these effects do not change critical exponents provided they do not involve infinite time or length scales. They do, however, influence the range where critical scaling laws are expected to hold. Critical behaviour might, however, be changed if immunity is provided only for finite times, which are distributed according to an inverse power.

Finally, we can consider the case whe where immunes move around diffusively, thus the shape of their cluster. Since diffusion is slow compared to the evolution of the epidemic process, there should exist intermediate times where it is irrelevant, with a cross-over to a structureless cluster at time $t\to\infty$.

As a last example, we can consider epidemics (such as some helminthic infections) where the degree of illness (number of helminths per individuum) can vary over a broad range, and changes only by multiple infection [23]. In this case, it might be that the fractal cluster if infected individua has to be replaced by a fractal measure with different order - α dimensions [24]. Each of the critical exponents would then be replaced by an entire hierarchy, similarly to the hierarchies discussed by Coniglio [25].

REFERENCES

1. N.T.J. Baily, The Mathematical Theory of Infectious Diseases (Griffin, London, 1975)
2. D. Mollison, J. Roy. Stat. Soc. B39 (1977) 283
3. T. M. Liggett, Interacting Particle Systems (Springer, New York, 1985)
4. M. Eden, in Proc. fourth Berkeley Symp. on Mathem. Statistics and Probability vol IV, p. 233, ed. F. Neyman (Univ. of California Press, Berkeley 1961)
5. R. Jullien and R. Botet, J. Phys. A 18 (1985) 2279
6. T. Williams and R. Bjerknes, Nature 236 (1972) 19
7. W. Kinzel and J. M. Yeomans, J. Phys. A14 (1981) L163
8. R. Brower, M. A. Furman, and M. Moshe, Phys. Lett. 76B (1978) 213
9. P. Grassberger and A. de la Torre, Ann. Phys. (N.Y.) 122 (1979) 373
10. P. Grassberger, Math. Biosci. 62 (1983) 157
11. P. Grassberger, J. Phys. A18(1985) L 215
12. Z. Alexandrowicz, Phys. Lett. 80A (1984) 284
13. R. Pike and H.E. Stanley, J. Phys. A10 (1981) L169
    D.C. Hong and H. E. Stanley, J. Phys. A16 (1983) L475, L525
    S. Havlin and R. Nossal, J. Phys. A17 (1984) L427
    R. Rammal, J. C. Angles d'Auriac, and A. Benoit, J. Phys. A17(1984) L 491
14. P. Grassberger, to be published in J. Phys. A
15. K. H. Janssen, Z. Phys. B58(1985)311
16. M. Doi, J. Phys. A9(1976) 1456,1479
    P. Grassberger and M. Scheunert, Fortschr. Phys. 28(1980) 547
    L. Peliti, Univ. Rome preprint
17. M. Moshe, Phys. Rep. C37 (1978) 255
18. R. G. Priest and T.C. Lubensky, Phys. Rev. B13 (1976) 4159
    D. J. Amit, J. Phys. A9 (1976) 1441
19. J. L. Cardy and P. Grassberger, J. Phys. A18 (1985) L267
20. B.B. Mandelbrot, The Fractal Geometry of Nature (Freeman, San Francisco, 1982)
21. M. E. Fisher, S.-K. Ma, and B.G. Nickel, Phys. Rev. Lett. 29(1972) 917
22. W. K. Theumann and M.A. Gusmao, Phys. Rev. B31 (1985) 379
23. K. Dietz, private communication
24. P. Grassberger, Phys. Lett 97A (1983) 222; 107A (1985) 101
25. A. Coniglio, these proceedings

FRACTALS IN PHYSICS
L. Pietronero, E. Tosatti (editors)
© Elsevier Science Publishers B.V., 1986

RANDOM RAIN SIMULATIONS OF DENDRITIC GROWTH

B.CAPRILE, A.C.LEVI and L.LIGGIERI

Università di Genova, Dipartimento di Fisica and Gruppo Nazionale di Struttura della Materia del CNR, Via Dodecaneso 33, 16146 Genova, Italy

Two-dimensional growth simulations are described for a "random rain" model, where the candidates for sticking approach the growing cluster along random straight lines. Both isotropic growth from a central seed and growth on a base line on to which the "atoms" fall obliquely from a parallel line are studied. The resulting clusters appear to be highly ramified, although less so than for DLA, and their Hausdorff-Besicovitch dimension is considered. Integro-differential equations for the local density as a function of position and time are also derived. Further the simulation is modified by including different physical effects, namely: 1) evaporation; 2) surface tension (in the form of differential sticking probabilities); 3) heat diffusion in the solid; 4) surface diffusion along the border. In this way a partially realistic picture of two-dimensional crystal growth is approached.

1. INTRODUCTION

Although dendritic crystal growth has been for a long time a model for the generation of ramified objects, microscopic simulations of this phenomenon are far from abundant and the theory has remained to a large extent macroscopic[1]. Diffusion-limited aggregation (DLA) was shown[2] to bear a mathematical similarity with crystal growth; but dendritic crystals are much more compact and less ramified (except in special cases) than DLA clusters. Besides, DLA simulations are fairly expensive computationally (because of the cost of random walk). Finally, since in our laboratory atom-surface scattering is studied, we were more interested in growth from vapour than in growth from solution, which would be best modeled by DLA.

For all these reasons, we chose to perform simulations where the "atoms", candidates for sticking, "rain" on to the growing cluster along random straight lines rather than along the Brownian paths characteristic of DLA. We call the resulting model the "random rain" (RR) model.

2. THE RANDOM RAIN MODEL

The RR model is not altogether new. It has a long history, starting from Marjorie Vold[3] and involving the work of Sutherland[4]; recently, it has been considered in simulations by Bensimon et al.[5] and two of the authors[6]. In the present work two geometries were considered, both in two dimensions. In the former, first a seed is placed at the centre; then the "atoms" candidates for aggregation are made to start inwards from the circumference of a large circle, starting from random points and moving along random cords. In the latter, the "atoms" are made to fall on to a line, in random directions, from a parallel line and to stick when they meet either the base line or the growing cluster. A representative example of a cluster grown in the former geometry is shown in Figure 1. The RR procedure produces ramified structures (although less so than in DLA), because the probability to stick on a branch is higher than near the centre.

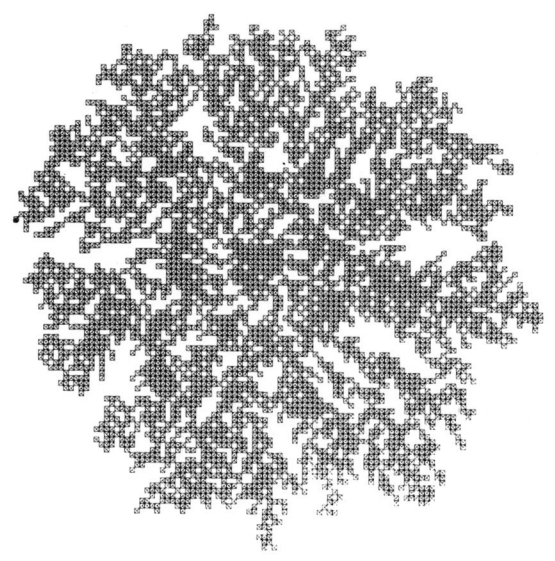

FIGURE 1
Cluster obtained by simple RR attachment on a square lattice.

3. HAUSDORFF-BESICOVITCH DIMENSION

3.1. Growth from a seed

The Hausdorff-Besicovitch dimension D of a cluster was measured by counting "atoms" within circles of increasing radius, as shown in Fig.2. The result is D=1.86±.03. However, both extensive simulations by Meakin[7] and a mathematical argument by Ball and Witten[8] indicate that the H.-B. dimension of RR clusters is trivial, i.e. D=2. Thus the value 1.86 is presumably not final, and should increase slowly to 2 with increasing cluster size.

3.2. Growth on a line

Here D was measured by counting "atoms" within strips of length 1 and increasing width z, according to the formula $N(z) \sim 1z^{D-1}$. The result is again D=1.86±.02, but again we expect D to approach 2 for larger clusters.

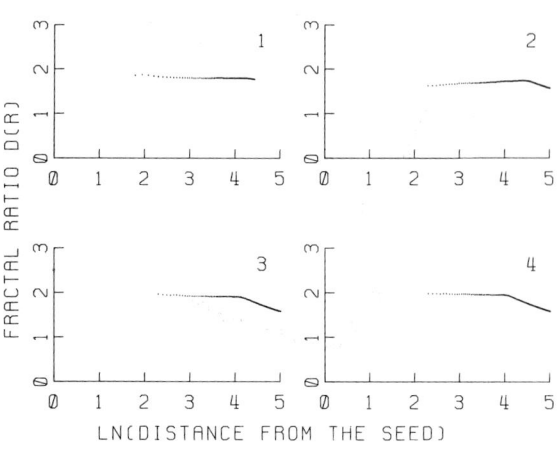

FIGURE 2
Evaluation of the H.-B. dimension according to the algorithm $D(r)=\ln[A_1 N(r)/\pi]/\ln r$ where A_1 is the unit cell area and $N(r)$ the number of occupied sites within a circle of radius r, for the clusters shown in (1) Fig.1; (2) Fig.3; (3) Fig. 4 and (4) Fig. 5.

4. INTEGRO-DIFFERENTIAL EQUATIONS

The mean radial density $\varphi(r,t)$ of a cluster growing from a seed in the RR model increases with time according to the integro-differential equation[6]:

$$\frac{\partial \varphi}{\partial t} = A\, \varphi \int d\psi (R^2 + r^2 - 2Rr \cos \psi)^{-\frac{1}{2}} \times$$

$$\times \exp\left\{\int \ln[1-\varphi(r',t)]\,dl\right\} \quad (1)$$

where (R,ψ) are polar coordinates of the source point of the circumference of the large circle and dl is the differential path length. Eq.(1) holds provided the cluster is very large in comparison to the lattice spacing. The exponential describes the opacity of the cluster.

Eq. (1) belongs to an interesting class of evolution equations whose simplest instance is:

$$\frac{\partial \varphi}{\partial t} = [f(x)+g(x)\varphi(x,t)]\exp\left\{-\int_x^\infty \varphi(x',t)dx'\right\} \quad (2)$$

which can be easily solved analytically in the cases: a) $g=0$; b) $g=1$, $f=0$. When $t\to\infty$, φ tends to a function φ_∞ whose integral diverges at long distances. The opacity of the outer parts of the cluster becomes infinite, so that at each point x the density φ stops increasing after a certain time $\bar{t}(\bar{t}$ depends on x).

5. EVAPORATION

Sticking without evaporation corresponds to infinite chemical potential difference $\Delta\mu$ between fluid and solid. In real life $\Delta\mu$ is finite and atoms evaporate. This situation is simulated by allowing random detachment of atoms from the periphery of the cluster. Subsequently they move along straight lines until they either meet another part of the cluster and stick again or disappear far from the cluster.

6. HEAT PROPAGATION AND DIFFERENTIAL STICKING

The previous simulations lack many physical properties that are of importance in real crystal growth. Among these probably the most important are:
a) surface tension;
b) heat diffusion in the solid;
c) macroscopic crystal symmetry
d) surface diffusion.

Attempts have been made in more recent simulations to account for these effects.

FIGURE 3
Cluster obtained on a hexagonal lattice. The sticking probability of an "atom" is fixed as 0.02, 0.3 or 1 if its attachment gives the fulfilment of a segment, a triangle or a hexagon.

a) The effects of surface tension are partly considered, simply by favouring attachment at points having many neighbours. Taking an underlying hexagonal lattice, a differential sticking probability is assumed by assigning higher probability for sticking when the added "atom" completes a triangle and still higher when it completes a hexagon. By favouring linear borders, this procedure simulates surface tension.
b) Latent heat is produced when an "atom" sticks, and is propagated through the cluster, with heat flux proportional to temperature difference between neighbouring cells. Excess heat is radiated away out of the plane.

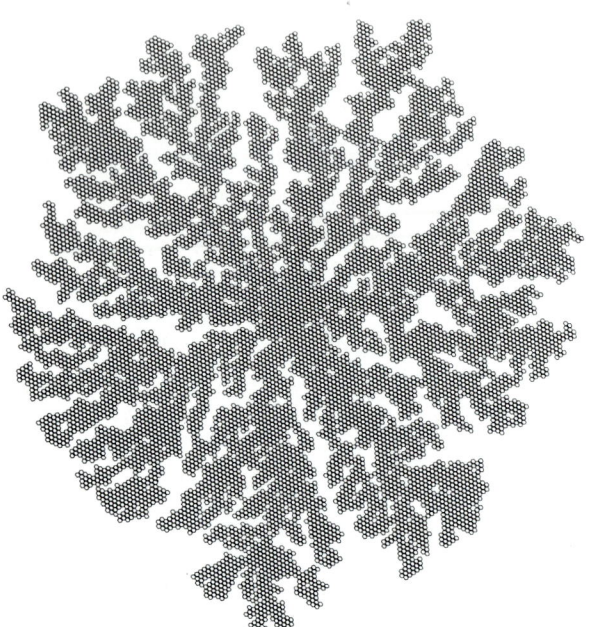

FIGURE 4
Cluster obtained introducing 1) differential sticking probabilities; 2) heat transport inside the cluster; 3) heat radiation out of the cluster; 4) "surface" diffusion of the sticking "atom".

H.-B. dimensions are evaluated in Fig. 2: the apparent D tends to increase as more physical effects are included.

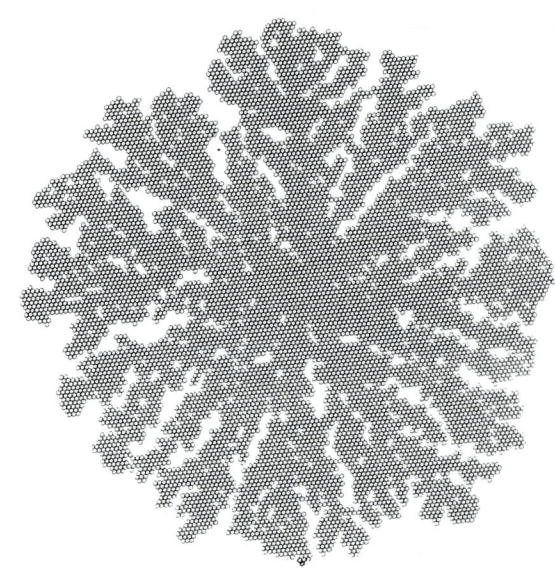

FIGURE 5
Same as Fig. 4, but including evaporation.

c) Macroscopic hexagonal crystal symmetry is not achieved in the present simulations; however the physical information concerned should be again related to differential sticking.

d) Surface diffusion is included in the simulations by allowing the "atom" to wander along the perimeter of the cluster, until it finds a site where sticking is favoured either because of neighbour number (a) or because of ease to dissipate heat (b).

Simulations differ by including one or more of the effects (a) - (d) and/or evaporation. The examples shown are clusters grown including differential sticking only (Figure 3), or adding heat propagation and surface diffusion (Fig. 4) and evaporation (Fig. 5). The corresponding

REFERENCES

1. J.S.Langer, Rev. Mod.Phys. 52 (1980) 1.

2. T.A.Witten and L.M.Sander, Phys.Rev.Letters 47 (1981) 1400; Phys.Rev. B27 (1983) 5686.

3. M.T.Vold, J.Colloid Sci. 18 (1963) 684.

4. D.N.Sutherland, J.Colloid Sci 22 (1966) 300; 25 (1967) 373.

5. D.Bensimon, E.Domany and A.Aharony, Phys.Rev. Letters 51 (1983) 1394.

6. A.C.Levi and L.Liggieri,Surf.Sci.148(1984) 212.

7. P.Meakin, private communication.

8. R.C.Ball and T.A.Witten, Phys.Rev. A 29 (1984) 2966.

EXPERIMENTAL STUDY OF TWO DIMENSIONAL AGGREGATION

Catherine ALLAIN, Michel CLOITRE

Laboratoire d'Hydrodynamique et Mécanique Physique UA CNRS/857, ESPCI, 10 rue Vauquelin, Paris 75005, France

In this paper, we present experimental investigations of the two dimensional aggregation of small spheres floating on an air-water interface. The attractive interactions between particles, here capillary effects, are caracterized quantitatively. We show their importance on the evolution of the system. The structures are found to be self-similar with a fractal dimension D=1.65±0.05.

1. INTRODUCTION

The formation of clusters of particles is a very common process involved for instance in the flocculation of colloids, coagulation of aerosols or in radicalar chemical reactions. In addition to computer simulations leading to different models of irreversible aggregation[1-3], several experimental studies of clustering have been reported recently[4-7]. In this paper, we present recent results obtained on an experimental system exhibiting two dimensional aggregation[8].

2. EXPERIMENT

Our system consists of uniform spherical particles floating on an air-water interface. The particles used are polypropylen or wax balls of 3.1 mm in diameter; the box where the experiment takes place is a 600mm× 600mm square. The random velocity gradients are low: $G < 10^{-2} s^{-1}$; the influence of these velocity gradients on the aggregation phenomenon has been investigated in our laboratory by Blanc et al[9] who studied the clustering process of the same particles in a shear flow.

The equilibrium of a ball at the air-water interface results from a balance between its weight, the Archimede's force and the surface tension forces. Because of the capillary attraction, when two particles come into contact, they stick irreversibly. At the beginning of the experiment the particles are distributed randomly on the surface. The position and motion of the particles are recorded on a photographic film.

3. INTERACTIONS BETWEEN PARTICLES

The bonding between balls is only due to capillary forces. Caracterization of this intersphere interaction is acheived through the measurement of deformations of the air-water interface around a particle. We used a method based on a moire technique. A periodic pattern of lines is projected optically on the surface S. After reflection on S, it is photographed and analyzed by optical filtering; we determine isocurvature and isoclinal lines.

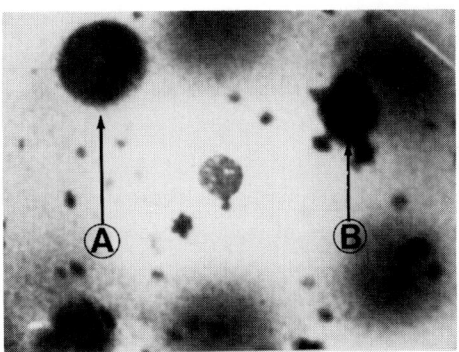

FIGURE 1
Deformation zones around two different particles (x0.3)

In figure 1, the black circles surrounding the different particles A and B represent the regions of the surface where the radius of curvature is lower than 60 m (far from the particles it is infinite). The diameters of these zones, d_A=50 mm and d_B=30 mm, are related to the ranges of the capillary effects due to A and B. In figure 2, clusters (made with A type particles) are surrounded by black fringes; these are the isoclinal lines.

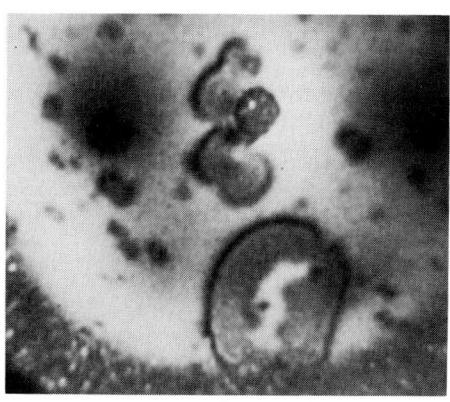

FIGURE 2
Isoclinal lines around clusters (×0.3)

Clearly, the deformation zones around the clusters are not isotropic and, qualitatively, newly arriving particles or clusters almost always stick following lines of greatest slope. This can explain their ramified structure.

4. RESULTS

A first result concerns the existence of two different regimes of aggregation, according to the initial concentration in particles ϕ[8]. In the dilute regime ($\phi \approx$ 1-3%), single cluster is obtained only after a very long time (12 hours). When ϕ becomes greater than a critical concentration ϕ_c, a cluster joining two opposite sides of the box appears after a finite time.

We observe that the value of the critical concentration ϕ_c is increased when the attractive interaction is lowered; for instance, we find for particles A $\phi_c(A) \approx 8\%$ while for particles B $\phi_c(B) \approx 12\%$. Picture 3 has been taken just before gelation, 20 mn after the beginning of the experiment; the initial concentration was $\phi \approx 10\%$.

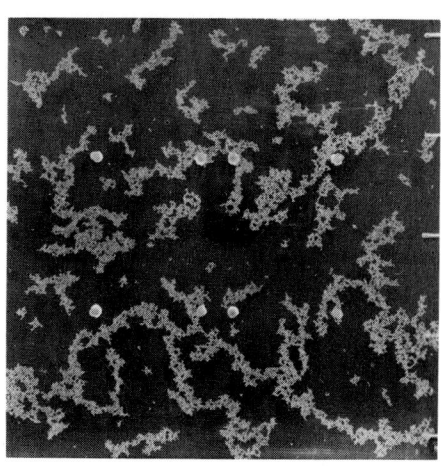

FIGURE 3
Picture taken just before gelation : two links only are missing to form an infinite cluster.

To illustrate the self-similarity of the aggregates and to determine the Hausdorff's dimension, we count the number of particles in a given cluster, N, as a function of its diameter Δ, for each of the aggregates observed in a picture such as figure 3. As a determination of Δ, we take the mean of the longest linear dimension of a cluster and the length on a perpendicular direction. A logarithmic plot of LogN versus LogΔ shows that $N(\Delta)$ conforms to a power law variation, $N(\Delta) \approx \Delta^D$, over nearly three decades[8]. We find $D \approx 1.65 \pm 0.05$ for experiments in the gelation regime. Each cluster also exhibits self-similarity. For instance, in the case of the infinite cluster, we have measured the mass M(L) embedded in boxes of linear size L centered on a given point. We find D=1.7±0.1; this value is consistent with the preceding determination. It is noteworthy that D is greater than the values calculated in computer simulations of cluster-cluster aggregation where D≃1.4 [2,3]. We think that this increase may be attributed to internal rearrangements and to intra-cluster aggregation due to the flexibility of the branches of the clusters. Moreover, we have observed that the clusters become more and more compact as capillary interactions are decreased.

We are now investigating the importance of capillarity on clustering. For instance, another related effect is the existence of a segregation process when the aggregation experiments are performed with two categories of particles whose wetting properties are different. Figure 4 is the final stage of an experiment performed using A and B particles with the initial concentrations : $\phi(A) \simeq 5\%$ and $\phi(B) \simeq 10\%$.

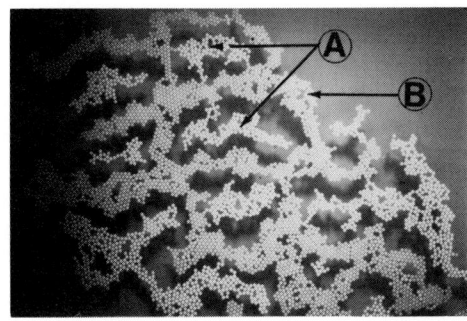

FIGURE 4
Picture showing the final stage of an experiment performed with two types of particles, A and B.

Because of the surface properties of A and B, interactions A-B are repulsive while A-A and B-B are attractive. We observe the formation of a large cluster of particles B interpenetrated by small clusters of particles A. We note that, probably because of the existence of repusive interactions A-B, the cluster composed of particles B is more compact than usually observed in experiments performed with particles B alone.

5. CONCLUSIONS

In this paper, we have presented some recent results concerning an experimental system exhibiting two dimensional aggregation. We have given a quantitative caracterization of the inter-particles interactions which are responsible for clustering. Experimentally, we have shown the existence of a critical concentration above which gelation is observed. The clusters have been found self-similar with a fractal dimension equal to 1.65±0.05 in the gelation regime. Finally, we have indicated the importance of the capillary interactions on the final structure of the clusters; for instance, a segregation process is observed when particles of different wetting properties are mixed.

REFERENCES

1. T.A. Witten and L.M. Sander, Phys.Rev.Lett. 47(1981)1400.

2. M. Kolb, R. Botet and R. Jullien, Phys. Rev. Lett. 51(1983)1123.

3. P. Meakin, Phys. Rev. Lett. 51(1983)1119

4. D.W. Shaefer and K.D. Keefer, Phys.Rev.Lett. 53(1984)1383.

5. D.W. Shaefer, J.E. Martin, D. Cannel and P. Wiltzius, Phys. Rev. Lett. 52(1984)2371.

6. P. Richetti, J. Prost and P. Barois, J.Phys. Lett.45(1984)L1137.

7. D.A. Weitz and M. Oliveira, Phys.Rev.Lett.52 (1984)1433.

8. C. Allain and B. Jouhier, J.Phys.Lett. 44 (1983)L421.

9. R. Blanc and C. Camoin, J.Phys.Lett. 46(1985) L67.

Part VI
KINETICS OF CLUSTERING

FRACTALS IN PHYSICS
L. Pietronero, E. Tosatti (editors)
© Elsevier Science Publishers B.V., 1986

KINETICS OF CLUSTERING IN IRREVERSIBLE AGGREGATION

M.H. ERNST

Institute for Theoretical Physics, University of Utrecht, P.O. Box 80.006, 3508 TA Utrecht, The Netherlands

The analytic results are reviewed that have been obtained for the cluster size distribution on the basis of Smoluchowski's coagulation equation. The behavior of the solution depends strongly on the coagulation rate constants K_{ij} for the model considered. The following subjects are discussed: exact results on completely and partially solved models and on existence and uniqueness; the possible occurrence of a gelation transition; asymptotic solutions (i) in the scaling limit $k, t \to \infty$ with $k/s(t)$ kept fixed where $s(t)$ is the mean cluster size; (ii) large-k-behavior at a fixed time t and (iii) large-t-behavior at a fixed k-value.

1. INTRODUCTION

Smoluchowski's coagulation equation has been widely used to describe aggregation phenomena in many fields of science and technology since its introduction in 1914, and comprehensive reviews of the older literature have been given by Drake [1] and Friedlander [2], and of more recent literature by Ziff [3] and the author [4]. However, in the last few years many new analytic results have been obtained, that give a fairly complete picture of the size distribution $c_k(t)$ in the whole (k,t)-plane and justify a new review.

Consider a suspension of aggregating particles suspended in a host fluid or gas, which are continuously growing as a result of pairs of particles coming into contact and adhering or bonding to form clusters or polymers.

For a statistical description of its macroscopic properties we need the distribution of clusters over different cluster sizes, i.e. the concentration $c_k(t)$ with $k = 1,2,...$ as a function of time. The zeroth moment $M_0(t)$ of this distribution determines the osmotic pressure (moments are defined as $M_n(t) = \Sigma_k k^n c_k(t)$); the first moment $M_1(t) = M$ determines the total mass or total concentration of units; the average cluster size, $s(t) = M_2(t)/M_1(t)$, determines the viscosity and light scattering intensity of the suspension.

Thus, we are looking for a description of the coupled system of chemical reactions; $A_1 + A_1 \to A_2$, $A_1 + A_2 \to A_3$,... $A_2 + A_2 \to A_4$,... $A_i + A_j \to A_{i+j}$, where $A_1, A_2...A_k$ represent respectively monomers, dimers and k-mers. The reactions are assumed to be irreversible, because dissociation or fragmentation is assumed to be absent.

In such reactions one can distinguish at least two characteristic times: the diffusion time t_D and the reaction time t_R. The first time is needed by two particles to come into contact via diffusion. It takes in general a time t_R before a chemical bond is formed, which depends on the reaction or sticking probability per unit time P_0, where $t_R \sim 1/P_0$ as $P_0 \to 0$.

Simplifications in the description occur if both time scales are very different. Here we distinguish two limiting cases: (i) *diffusion-limited aggregation* (DLA) with $t_D \gg t_R$, where at each collision a bond is formed instantaneously ($P_0 = 1$); (ii) *reaction-limited aggregation* (RLA) with $t_R \gg t_D$ corresponding to a low sticking probability P_0.

In these limiting cases we can give a description at the time scale of the slow process in the form of a set of coupled reaction equations for $c_k(t)$, containing a set of coagulation rate constants K_{ij}. The rate of a specific (ij)-reaction is $K_{ij} c_i c_j$, where it is assumed that the probability to find an i-cluster and a j-cluster are statistically uncorrelated (stosszahlansatz). We thus obtain Smoluchowski's coagulation equation, describing the time evolution of the concentrations $c_k(t)$:

$$\dot{c}_k = \tfrac{1}{2} \sum_{i+j=k} K_{ij} c_i c_j - c_k \sum_{j=1}^{\infty} K_{kj} c_j \qquad (1.1)$$

This nonlinear set of first order differential equations (k=1,2,...) has to be solved for a general initial condition $c_k(0)$, where the monodisperse initial condition $c_k(0) \sim \delta_{k1}$ is physically the most relevant one.

What are the further limitations on the validity of the coagulation equation? The equation has the typical structure of a mean field approximation, where the particles react irrespectively of their mutual distance. The spatial fluctuations and correlations between the positions of the reacting particles have been neglected.

In many applications of (1.1), such as in aerosol coagulation [1,2] it is customary to consider the size distribution c(x,t)dx as a function of a continuous size-variable x, and a continuous version of the coagulation equation is being used. In this review I restrict myself to a discrete size variable. Most asymptotic results discussed here apply to the continuous version of the coagulation equation as well.

Returning to the coagulation equation I note that the mean cluster size $s(t)$ keeps increasing and aggregation proceeds until all monomers form a single macroscopic aggregate. The cluster size distribution $c_k(t)$, at fixed k, approaches the trivial stationary solution $c_k(\infty) = 0$ for $k = 1,2,\ldots$. During this aggregation process the mass is conserved, as expressed by:

$$M_1(t) = \sum_{k=1}^{\infty} kc_k(t) = M = 1 \quad (1.2)$$

where $kc_k(t)/M$ is the probability that a monomer is contained in a k-mer. The unit volume is chosen such that the total density M equals unity. The total number of clusters $M_0(t)$ is, of course, not conserved.

The conservation law can be simply derived by multiplying Smoluchowski's equation with k^α, summing over all k and freely interchanging the order of summation. As long as this interchange is allowed one finds the general moment equations:

$$\dot{M}_\alpha = \tfrac{1}{2} \sum_{i,j} c_i c_j K_{ij} [(i+j)^\alpha - i^\alpha - j^\alpha] \quad (1.3)$$

with α arbitrary. For $\alpha = 1$ one finds mass conservation. One of the most interesting properties of this equation is the violation of mass conservation after a finite time, as may occur in certain classes of coagulation models, and corresponds to a gelation transition. Mathematical peculiarities obtained from Smoluchowski's equation for certain coagulation kernels K_{ij}, where $M_0(t)$ would go negative or $M_2(t)$ would diverge after a finite time, were considered before as limitations on the time-interval, on which the Smoluchowski equation was physically meaningful. The proper interpretation of these peculiarities as being signs of a gelation transition was not so long ago given by Lushnikov [6] and Ziff [7].
The plan of the paper is as follows. In section 2 the different coagulation kernels K_{ij} are divided into three classes with different asymptotic properties. Section 3 deals with the few exactly solved models for K_{ij}, with partially solved models and rigorous results on existence and uniqueness. In section 4 the criteria for gelation are discussed. In the remaining sections I study the asymptotic properties of the solution $c_k(t)$: in section 5 the scaling limit, where $k,t \to \infty$ with $k/s(t)$ kept fixed ($s(t)$ is the mean cluster size); in section 6 the large-k-behavior at fixed k, and in section 7 the large-t-behavior at fixed k. The different asymptotic solutions have common regions of validity. I close with a brief discussion.

2. CLASSIFICATION OF COAGULATION PROCESSES

Before formally classifying different types of coagulation kernels I discuss some representative examples. A typical example of a DLA-process is Brownian coagulation. If the clusters are compact objects, the collision frequency K_{ij} can be calculated by solving the diffusion equation in the presence of an absorbing sphere [2]. This yields $K_{ij} = 4\pi D_{ij} R_{ij}$ with $D_{ij} = D_i + D_j$ and $R_{ij} = R_i + R_j$, valid for dimensionality $d = 3$. More generally one has $K_{ij} \sim D_{ij}(R_{ij})^{d-2}$ for $d > 2$. For $d = 2$ the above diffusion problem does not have a stationary solution. In these expressions R_k is the radius of gyration of a cluster, with $R_k \sim k^\nu$ ($k \to \infty$). For a compact cluster $\nu = 1/d$. If Einstein's formula for the coefficient D_k applies for these clusters, one has $D_k \sim 1/R_k$.

Examples of RLA-processes are reactions occurring in a continuously stirred tank reactor, in which the slow diffusion process is eliminated by stirring. Other examples are polymerization reactions. For such reactions Flory estimates the sticking probability P_0 as 10^{-9} for monomer-cluster-reactions and 10^{-13} for cluster-cluster-reactions [8].
In RLA-reactions the coagulation rate constant K_{ij} will be the larger the more reactive groups – denoted here by A's – are accessible on a cluster. To illustrate this I consider Flory's RA_f-model ("repeated A_f-units"). Here each monomer has f A's, where f is the functionality or coordination number; polymers are formed through AA-bonds; a k-mer has no cycles and therefore contains $\sigma_k = (f-2)k+2$ unreacted A's, and two polymers are assumed to be completely interpentrable. Then the coagulation constant is proportional to the number of possible AA-pairs on both clusters, i.e. $K_{ij} \sim \sigma_i \sigma_j$ which reduces to $K_{ij} \sim f^2 ij$ for large clusters.

In case the clusters are compact – as opposed to fractal – they have "holes" or "fjords" of a characteristic length and two clusters penetrate one another over a finite distance. Now only the A's in a surface layer are accessible, their number being $\sigma_k \sim k^\omega$ with $\omega = 1-1/d$. This yields for large clusters the estimate $K_{ij} \sim (ij)^\omega$. In this surface interaction model the dimension d of the space, in which the cluster is embedded, appears only through the surface exponent ω.

How these arguments should be modified to describe RLA of fractal clusters is not clear. In table 1 I have collected a representative set of coagulation kernels used in the literature.

In order to discuss the (asymptotic) properties of the different coagulation models it is convenient to classify them according to two properties: (i) how does a *large* cluster react with another *large* cluster, specified by an

K_{ij}	process
$\left(i^{1/3}+j^{1/3}\right)\left(i^{-1/3}+j^{-1/3}\right)$	Brownian coagulation in the continuum regime [1] ($\mu = -1/3$)
$\left(i^{1/3}+j^{1/3}\right)^2\left(i^{-1}+j^{-1}\right)^{\frac{1}{2}}$	Brownian coagulation in free molecular flow [1] ($\mu = -1/2$)
$\left|i^{1/3}-j^{1/3}\right|\left(i^{1/3}+j^{1/3}\right)^2$	coagulation by gravitational settling [2] ($\mu = 1/3$)
$\left(i^{1/3}+j^{1/3}\right)^3$	coagulation in shear flow [2] ($\mu = 0$)
ij	branched polymer model with AA-bonds [7] ($\mu = 1$)
$(ij)^\omega$ ($\omega = 1-1/d$)	same with surface interactions [9]
$i+j$	branched polymer model with A-B-bonds [7] ($\mu = 0$)
$i^\omega + j^\omega$	same with surface interactions [9]
1	linear polymer model [7] ($\mu = 0$)
$(i+j)^\omega$	mathematical model [10] ($\mu = 0$)
$i^\mu j^\nu + j^\mu i^\nu$	same with $\nu > \mu$ [11]
$(ij)^\mu (i+j)^{\nu-\mu}$	same with general ν, μ
$j^{2\omega}\delta_{ij}$	hierarchical model [12] ($\mu = +\infty$)
$j^\omega \delta_{i1} + i^\omega \delta_{j1}$	addition model [13] ($\mu = -\infty$)

Table 1. *Homogeneous coagulation kernels*

exponent λ, and (ii) how does a *large* cluster react with a *small* one, specified by an exponent μ. To that purpose, I define

$$K_{j,aj} \sim j^\lambda \quad (j \gg 1, a \text{ fixed})$$
$$K_{ij} \sim i^\mu j^\nu \quad (j \gg i; \lambda = \mu + \nu) \quad (2.1)$$

Since the average number of reactive sites on a cluster cannot increase faster than its size, I impose the physical restriction $\lambda \leq 2$ and $\nu \leq 1$, but no restrictions are imposed on μ. The upper-limits can only be reached if the clusters are fully penetrable. I do not know of any further physical or mathematical restrictions on μ and λ that limit the region of acceptable models in fig. 1.
As illustrated in table 1, most coagulation kernels, used in the literature, are homogeneous

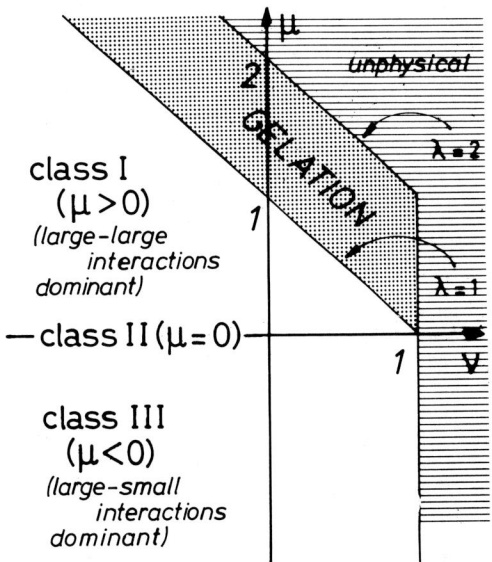

Fig. 1. Homogeneous coagulation kernels $K_{ij} \sim i^\mu j^\nu$ ($j \gg i; \lambda = \mu + \nu$) are divided into classes with different asymptotic behavior. Class I contains gelling and non-gelling models.

functions of i and j - at least for large cluster sizes. Since I shall be focussing on large size properties of $c_k(t)$, I consider kernels that are asymptotically homogeneous, i.e.

$$\lim_{a \to \infty} a^{-\lambda} K_{ai,aj} = K(i,j) \quad (2.2)$$

In the actual presentation I restrict myself to kernels that are strictly homogeneous

$$K_{ij} = a^{-\lambda} K_{ai,aj} = K(i,j) \quad (2.3)$$

The degree of homogeneity λ is the most important parameter, but some properties of $c_k(t)$ are also determined by the parameter μ. I therefore distinguish class I with $\mu > 0$, class II with $\mu = 0$ and class III ($\mu < 0$), as indicated in fig. 1. Typical for class I are RLA-reactions, such as polymerization; typical for class III are the DLA-reactions occurring in aerosol coagulation, e.g. Brownian coagulation.

3. EXACT RESULTS

3a. *Exactly solved cases*

In separate subsections I briefly review (i) exactly solved models; (ii) partially solved models for which either (a) the size distribution can be calculated sequentially, but not the moments, or (b) only a single moment can be calculated and (iii) rigorous results on existence, uniqueness and positivity.

The solution for arbitrary initial conditions is only known for K_{ij} = constant [1], i+j [14] and ij [15-17]. For the last model the global solution for general initial conditions was only found a few years ago [16].
For monodisperse initial conditions $c_k(o) = \delta_{k1}$ the global solution is known for the bilinear kernel $K_{ij} = Aij+B(i+j)+C$ [1,18-20]. For instance, for the case $C \neq o, A = B = o$ one has $c_k(t) = t^{k-1}(1+t)^{-k-1}$, where C = 2 by choosing proper units of time. However, for the case $A \neq o$ the known solution [15] was only valid in a finite time interval. Lushnikov [6] and Ziff [7] seem to have been the first to realize that a phase transition occurs in this model. This can be understood most easily by studying the behavior of the mean cluster size, $s(t) = M_2(t)/M_1(t)$. For the simplest of such models, $K_{ij} = ij$, the moment equations (1.3) yield

$$\dot{M}_o = -\tfrac{1}{2} \quad ; \quad \dot{M}_1 = o \quad ; \quad \dot{M}_2 = M_2^2 \qquad (3.1)$$

implying that the mean cluster size becomes infinite after a finite time $t_c = 1/M_2(o)$. In this manner one obtains the behavior of the moments for $t < t_c$ as illustrated in fig. 2. The physical interpretation of this singularity is the occurrence of a phase transition at t_c, called gelation. What happens as $t \uparrow t_c$ can be understood by calculating with the help of (1.1) the mass flux $J(L,t)$ from clusters with size $k < L$ to clusters with $k > L$, yielding [7,16]:

$$J(L,t) = \sum_{k=1}^{L} k\dot{c}_k = -\sum_{k=1}^{L} \sum_{j=L-k+1}^{\infty} k^2 j\, c_k c_j \qquad (3.2)$$

As long as c_k is exponentially bounded for large k, the limiting value $J(\infty,t)$ vanishes and the total mass is conserved, $M_1(t) = 1$. This happens for $t < t_c$. However, if c_k behaves algebraically, $c_k \sim k^{-\tau}$ ($k \to \infty$), the large-L-limit of the mass flux can be easily estimated by power counting [17,21] and yields $J(L,t) \sim L^{5-2\tau}$. As we shall see below, the exact solution has $\tau = 5/2$ for $t \geq t_c$. Thus, $J(\infty,t) \neq o$ for $t \geq t_c$. This implies that there is a finite *non-vanishing mass flux from finite size clusters (sol particles) to the infinite cluster (gel)*, and the total (sol)mass contained in finite clusters, $M_1(t) = \sum_{k=1}^{\infty} kc_k(t)$ is no longer constant, but starts to decrease. The gel has a vanishing concentration, but a non-vanishing mass, the so-called gel fraction $G(t) = 1 - M_1(t)$, which is the order parameter of this phase transition.

As an illustration I simply quote the size distribution for the monodisperse initial condition $c_k(o) = \delta_{k1}$ [16,18]

$$c_k(t) = \begin{cases} (k^{k-2}/k!)t^{k-1} e^{-kt} & (t \leq 1) \\ (k^{k-2}/k!)e^{-k}/t & (t \geq 1) \end{cases} \qquad (3.3)$$

where $t_c = 1$ is the gel point. The pre-gelation solution was first given by McLeod [15]. It is more instructive to look at the asymptotic behavior of this solution for large k, i.e.

$$c_k(t) \simeq \begin{cases} (2\pi)^{-1/2} k^{-5/2} (te^{1-t})^k & (t \leq 1) \\ (2\pi)^{-1/2} k^{-5/2} t^{-1} & (t \geq 1) \end{cases} \qquad (3.4)$$

These results show that the size distribution is exponentially bounded, $c_k < \exp(-\alpha k)$ as $k \to \infty$, in the pre-gelation stage (sol phase), and has an algebraic tail $c_k \sim k^{-5/2}$ ($k \to \infty$) in the post-gelation stage (gel phase). Consequently, all moments are bounded in the sol phase, whereas in the gel phase $M_\alpha(t)$ with $\alpha \geq 3/2$ is divergent and mass conservation is violated, i.e. $M_1(t) = 1/t$.

The behavior (3.4) of the size distribution at large k implies that the generating function $f(x,t) = \sum kc_k(t) \exp(kx)$ is regular in $x = o$ for $t < t_c$ and has a square root branch point in $x = o$ for $t > t_c$ on account of the algebraic tail, $c_k \sim k^{-5/2}$.

For more general kernels $K_{ij} = Aij+B(i+j)+C$ and monodisperse initial conditions also pre- and post-gelation solutions have been found [19, 20,22]. One can add monomer sources to the co-

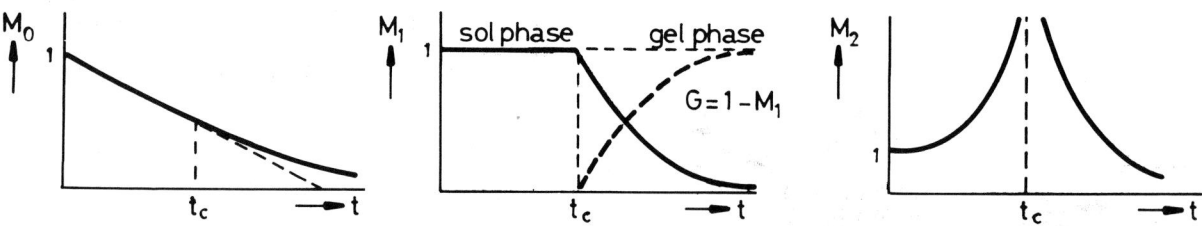

Fig. 2. *Behavior of the first three moments in the model* $K_{ij} = ij$.

agulation equation for this model [23-25], included fragmentation [26-28], extend the bilinear kernels to copolymerization models with different types of monomers [29], include in the bilinear rate constants K_{ij} not only size-dependence, but also shape-dependence, such as the number of branch points in a cluster [30].

There also exists an exact mapping [31] of the solution of Smoluchowski's equation for model I with $K_{ij} = (ai+b)(aj+b)$ on model II with $K_{ij} = a(i+j)+b$ and vice versa. If $c_k(t)$ is a solution of model I and $n_k(t)$ of model II then:

$$n_k(t) = a(ak+b)t_c \, e^{-at} \, c_k\!\left(t_c(1-e^{-at})\right) \quad (3.5)$$

Here t_c is given in terms of initial distributions $n_k(o)$ or $c_k(o)$. This transformation maps the pre-gelation solution for $o < t < t_c$ of the gelling model I on the global solution for $o < t < \infty$ of the non-gelling model II.

3b. *Partially solved cases*

For special coagulation-kernels K_{ij} the size distribution can be calculated sequentially, but the moments cannot be determined explicitly. The first model has been discussed by Lushnikov and Piskunov [32]. Their coagulation kernel has the general form $K_{ij} = i\sigma_i + j\sigma_j$, where σ_k is an arbitrary positive function with $\sigma_k \leq Ck$ $(k \to \infty)$. The substitution $c_k(t) = \nu_k(t) \exp(kM_0(t))$ together with mass conservation $M_1(t) = 1$ (for gelling systems this is only applicable for $t < t_c$) transforms the coagulation equation into

$$\dot{\nu}_k = \sum_{i+j=k} i\sigma_j \nu_i \nu_j - \sigma_k \nu_k$$

This set of equation can be solved sequentially for general initial conditions, starting with $k = 1, 2, \ldots$. Thus $\nu_1(t) = \nu_1(o) \exp(-\sigma_1 t)$, etc. If $\gamma(x;t) = \Sigma \nu_k(t) \exp(kx)$, then $M_0(t)$ can be solved from $\dot{M}_0(t) = \gamma(M_0(t);t)$.
The second example was given by Ernst et al. [33] and concerns the kernel $K_{ij} = \sigma_i + \sigma_j$ with the same constraints on σ_k as above. Here the substitution

$$\nu_k(\tau) = c_k(t)/M_0(t)$$

$$d\tau = M_0(t)dt$$

transforms the coagulation equation (3.1) into

$$\dot{\nu}_k = \sum_{i+j=k} \sigma_j \nu_i \nu_j - \sigma_k \nu_k$$

with the constraint $\Sigma \nu_k = 1$. This set can also be solved sequentially for general initial conditions, and one recovers $M_0(t)$ and the original time variable from mass conservation.

In the second part of this subsection I consider some models constructed by Ziff [7], where one moment can be calculated exactly, which may lead to an exact value of the gel point for a gelling model.

In model (a), $K_{ij} = Ai^\beta j^\beta/[(i+j)^\beta - i^\beta - j^\beta]$, with degree of homogeneity $\lambda = \beta < 2$ the moment $M_\beta(t)$ can be calculated exactly from (1.3) with the result $M_\beta(t) = M_\beta(o)/[1-\tfrac{1}{2}AM_\beta(o)t]$. If $\beta > 1$ and $A > o$, this is a class-I-model ($\mu=\beta-1; \nu=1$) and the β-th moment diverges within a finite time $t_c = 2/(AM_\beta(o))$ (gel point). If $\beta < 1$ and $A < o$ one has a class-II-kernel ($\mu=o; \nu=\beta$) where $M_\beta(t)$ remains bounded for all times.

In model (b), $K_{ij} = A(i j^\beta + j i^\beta)/[(i+j)^\beta - i^\beta - j^\beta]$, with homogeneity degree $\lambda = 1$ one finds $M_\beta(t) = M_\beta(o) \exp(At)$, as long as $M_1(t) = 1$. For all values of β this is a class-II-kernel.

In model (c), $K_{ij} = Aij/[(i+j)^\beta - i^\beta - j^\beta]$ with $\lambda = 2-\beta$ one finds $\dot{M}_\beta(t) = M_\beta(o) + \tfrac{1}{2}At$, as long as $M_1(t) = 1$. For $\beta > 1$ and $A > o$ one has $\mu = o$ (class II) and for $\beta < 1$ and $A < o$ one has $\mu = 1-\beta$ (class I). In the latter case the moment $M_\beta(t)$ becomes negative (unphysical) within a finite time (compare the extrapolation of the short time behavior of $M_0(t)$ in fig. 2), which indicates that a gelation transition has occurred before that time.

3c. *Existence and uniqueness*

I review what little is known about existence and uniqueness of solutions to the initial value problem of the coagulation equation for general classes of coagulation kernels:

(i) For kernels K_{ij} bounded from above by Aij a unique solution satisfying the conditions of normalization and positivity exists in a finite time interval [15]. The time interval in which existence has been proved does in general not extend to $t > A^{-1}$. From now on I shall absorb constants such as A in the unit of time.

(ii) For kernels satisfying $K_{ij} \leq i+j$, which includes $\sigma_i \sigma_j \leq (ij)^{1/2}$, White showed [34] existence of global initial solutions, where all moments are bounded on finite time intervals. This excludes gelation within a finite time.

(iii) For kernels satisfying $K_{ij} = \sigma_i \sigma_j$ with $\sigma_k/k \to o$ as $k \to \infty$ Leyvraz and Tschudi [18] have established global existence of nonnegative solutions, but not their uniqueness.

The problem of existence and uniqueness of solutions is perhaps more than a mathematical finesse in the present case. For instance, in the closely related problem of a coagulation kernel with $K_{ij} = a$ and a fragmentation kernel which is constant, Aizenman and Bak [26] have shown that only a unique positive global solution $c_k(t)$ of (1.1) exists in the class of functions with $k^3 c_k(t) \to o$ as $k \to \infty$.

Other mathematical difficulties arise for coagulation kernels with $\nu > 1$ (here again $F_{ij}=0$). There are some numerical and analytic indicates [16] that in the unphysical case, $\nu > 1$, global nor local solutions exist to the initial value problem.

4. OCCURRENCE OF GELATION

4a. *Bounds on the gel point*

In the previous section we have seen that the exactly solved model $K_{ij} = ij$ shows a gelation transition, where the following interesting phenomena occur:

(i) the mean cluster size $s(t)$ diverges as $t \uparrow t_c$
(ii) violation of mass conservation at a finite t_c
(iii) the size distribution decays exponentially (c.q. algebraically) for large k below t_c (c.q. above t_c)
(iv) the generating function of the c_k's, which is initially regular at $x = 0$, develops a branch point singularity after a finite time t_c.

Do other coagulation processes show a similar gelation transition, i.e. which coagulation kernels K_{ij} in Smoluchowski's equation lead to gelation?

As no explicit solutions are known one has to develop other tests to decide on the occurrence of gelation, e.g. one may be able to demonstrate that any of the characteristic phenomena (i-iv) does or does not occur [35]. (In this subsection I will study the possibility of a divergent mean cluster size $s(t)$).

Define some measure for the mean cluster size, such as $s = M_2$ (recall that $M_1 = 1$). If one can construct an upper bound $s^{(2)}(t)$ that remains bounded from above for all finite t, then $s(t) \leq s^{(2)}(t) < \infty$ for all finite t and no gelation occurs (see fig. 3). On the other hand, if one can construct a lower bound $s^{(1)}(t)$ that diverges as $t \uparrow t_1 < \infty$, then $s(t) \geq s^{(1)}(t)$ diverges within a finite time t_c, satisfying $t_c \leq t_1$. Thus gelation does occur.

To illustrate the method I discuss some examples. If one starts below the gel point, supposedly all $c_k(t)$ decay exponentially at large k and all moments $M_\alpha(t) = \Sigma k^\alpha c_k(t)$ are finite. Thus, one can estimate $s(t) = M_2(t)$ from the moment equation (1.3) for $\alpha = 2$. Consider first the example $K_{ij} = i^\mu j^\nu + j^\mu i^\nu$ ($\nu \geq \mu$), for which (1.3) yields:

$$\dot{M}_2 = 2M_{\nu+1} M_{\mu+1} \qquad (4.1)$$

In case $\mu \leq \nu < 1$ it follows from Jensen's inequality that $M_{\nu+1} < M_2^\nu$ (the inequality reserves for $\nu > 1$) and one finds $\dot{M}_2 < 2M_2^\lambda$ with $\lambda = \mu+\nu$ [35]. Using the initial condition, illustrated in fig. 3, one finds the upper bound

$$M_2^{(2)}(t) = [A+2(1-\lambda)t]^{1/(1-\lambda)}$$

If $\lambda = \mu+\nu \leq 1$, then the upper bound is finite for all $t < \infty$ and no gelation occurs. If $\lambda > 1$, then the upper bound diverges as $t \uparrow t_2 = [2(\lambda-1)(M_2(0))^{\lambda-1}]^{-1}$, and gelation may or may not occur.

In the *unphysical* case $\nu > 1$ (see fig. 1) with $\nu \geq \mu \geq 0$ one can construct a diverging lower bound, and thus gelation occurs before some finite t_1. However, there are indications that these unphysical models gel instantaneously ($t_c = 0$) [35,36].

To decide whether gelation does occur for $\lambda > 1$ in the physically relevant case ($\nu \leq 1$)

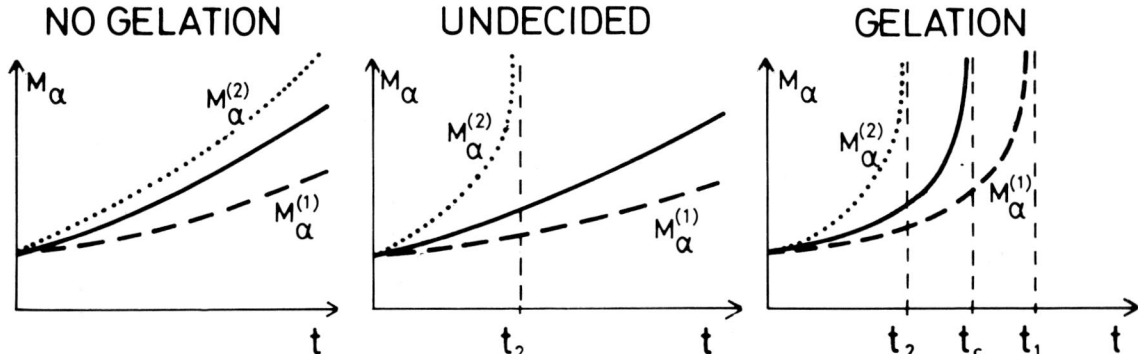

Fig. 3. *Schematic behavior of $M_\alpha(t)$ (solid line) together with an upper bound (dotted line) and a lower bound (dashed line), all having the same initial value. For the upper bound $\alpha = 2$ so that $M_2^{(2)} > M_2$, for the lower bound $\alpha = \mu+1$ so that $M_{\mu+1}^{(1)} < M_{\mu+1} < M_2$, where also $M_{\mu+1}(0) < M_2(0)$. In the gelling case $t_2 < t_c < t_1$; in the undecided case gelation may occur with $t_c > t_2$.*

one has to construct a diverging lower bound. Until now one has only succeeded in doing so for the border line kernels with $\nu = 1$ [36]. Here I will illustrate the method for the example $K_{ij} = ij^\mu + ji^\nu$ ($\mu \leqslant 1$), using the inequalities

$$ij^\mu + ji^\mu > (ij)^\mu (i+j)^{1-\mu}$$

$$(i+j)^\alpha - i^\alpha - j^\alpha > 4(1-2^{1-\alpha}) ij(i+j)^{2-\alpha}$$

valid for $1 < \alpha < 2$. Inserting these inequalities in (4.1) for $\alpha = 1+\mu$ yields $\dot{M}_{\mu+1} > 2(1-2^{-\mu})M_{\mu+1}^2$. For positive μ ($\lambda = 1+\mu > 1$) one finds a diverging lower bound $M_{\mu+1}^{(1)} < M_{\mu+1} < M_2$. Therefore, the above kernel K_{ij} with $o < \mu \leqslant 1$ or $1 < \lambda \leqslant 2$ describes a gelling model with the following estimates for the gel point:

$$[2\mu(M_2(o))^\mu]^{-1} < t_c < [2(1-2^{-\mu})M_{\mu+1}(o)]^{-1} \quad (4.2)$$

So far I have only shown gelation for kernels with $\nu = 1$ and $o < \mu \leqslant 1$. Different arguments are needed for $\nu < 1$.

4b. *Violation of mass conservation*

Here we look for solutions of Smoluchowski's equation with a time dependent sol-mass $M_1(t) = \Sigma\, kc_k(t)$, as a signal of the gelation transition. To illustrate the method I consider the surface interaction model $K_{ij} = (ij)^\omega$ with $\omega \leqslant 1$ [35], so that

$$\dot{c}_k = \tfrac{1}{2} \sum_{i+j=k} (ij)^\omega c_i c_j - k^\omega c_k \sum_{j=1}^\infty j^\omega c_j \quad (4.3)$$

I introduce the generating functions $g(x,t) = \Sigma\, c_k(t) \exp(kx)$ and $f(x,t) = \Sigma\, k^\omega c_k(t) \exp(kx)$, which have the following small-x-behavior:

$$g(x,t) \simeq M_o(t) + xM_1(t) + o(x)$$
$$f(x,t) \simeq M_\omega(t) + o(1) \quad (4.4)$$

It follows from (4.3) that g and f satisfy the relation:

$$\frac{\partial}{\partial t}(g-M_o) = \tfrac{1}{2}(f-M_\omega)^2 \quad (4.5)$$

This relation can be used to test whether the coagulation equation (4.3) admits gelling solutions with a time dependent sol mass, i.e. $\dot{M}_1(t) \neq o$. To that purpose I insert (4.4) into (4.5) and solve for $f(x,t)$. The result is

$$f(x,t) \simeq M_\omega(t) - \left(2x\dot{M}_1(t)\right)^{1/2} \quad (x \to o)$$

This implies for the asymptotic behavior of the size distribution:

$$c_k(t) \simeq \left(-\dot{M}_1(t)/2\pi\right)^{1/2} k^{-\tau} \quad (k \to \infty) \quad (4.6)$$

with $\tau = \omega + 3/2$. The result (4.6) is only consistent if the sol mass $M_1(t) < \infty$, implying that $\omega > 1/2$. Thus, for $1/2 < \omega \leqslant 1$ the coagulation kernel $K_{ij} = (ij)^\omega$ describes a gelling system because there exists a *consistent postgelation solution* (4.6) with an (unknown) time-dependent sol mass $M_1(t)$. In the gel phase $t > t_c$ (t_c is also unknown) the size distribution has an algebraic tail $c_k(t) \sim k^{-\tau}$ with a known exponent $\tau = \omega + 3/2$. From White's theorem [34] in subsection 3c we know already that the kernel $(ij)^\omega$ with $\omega \leqslant 1/2$ ($\lambda \leqslant 1$) does not lead to gelation because $2\sqrt{ij} < i+j$.

Van Dongen and the author [37] have used additional arguments by considering special solutions of the Smoluchowski equation, the so-called similarity solutions, which will be discussed in the next section. There it is found that the mean cluster size behaves as

$$s(t) \sim \begin{cases} (t_o+t)^{1/(1-\lambda)} & (\lambda < 1) \\ (t_c-t)^{-2/(\lambda-1)} & (\lambda > 1) \end{cases},$$

also showing that gelation does occur for $\lambda > 1$ and is absent for $\lambda < 1$.
From these and similar arguments (see section 3b) it seems reasonable to conclude that homogeneous coagulation kernels (2.3) with $1 < \lambda \leqslant 2$ describe a gelling system, as indicated in fig. 1.

5. THE SIZE DISTRIBUTION AT LARGE k AND t

5a. *Similarity solutions*

In coagulation experiments on aerosols and emulsions [38] it has been observed that after a transient period the cluster size distribution approaches a scaling form, as illustrated in fig. 4. Using Smoluchowski's coagulation equation Friedlander [2] formulated the theory of self preserving spectra that gave a satisfactory explanation of the experimental data. The coagulation processes considered were all modeled by kernels belonging to class III. The same theory has been applied in many fields of science [11,40-44] and extended to coagulation models of classes I and II by several authors [21,35,37]. The approach to a scaling form was also seen in recent computer simulations of cluster-cluster aggregation [43-45].

I will outline the theory for kernels of class III. Here one looks for asymptotic solutions of Smoluchowski's equation of the scaling or similarity form

$$c_k(t) \simeq s^{-2}\varphi(k/s) \quad (5.1)$$

where $\varphi(x)$ is a universal function independent

of the initial distribution, and where the mean cluster size $s(t)$ is an increasing function of time that remains bounded for all time (no gelation). This scaling form is supposed to *be valid* in the so-called *scaling limit*, where $k, t \to \infty$ with $k/s(t)$ kept fixed, and the total mass is normalized to unity, $M_1 = \int dx\, x\varphi(x) = 1$.

Thus, one tries to determine the scaling function $\varphi(x)$ by inserting (5.1) into the coagulation equation and taking the scaling limit. The scaling function is found to satisfy the nonlinear integral equation:

$$-wx\varphi'(x) - 2w\varphi(x) = -\varphi(x) \int_0^\infty dy\, K(x,y)\varphi(y) + \tfrac{1}{2} \int_0^x dy\, K(y, x-y)\varphi(x-y)\varphi(y) \quad (5.2)$$

and the mean cluster size satisfies $\dot{s} = ws^\lambda$ or $s(t) \sim t^z$ with $z = 1/(1-\lambda)$ and λ defined in (2.2). The separation constant w can be expressed in moments of $\varphi(x)$ by multiplying (5.2) with x^α (α positive) and integrating over all x. For non-gelling models of classes I and II the two terms on the RHS of (5.2) contain cancelling infinities, which require a more careful formulation [37] of the integral equation. The formulation of the integral equation for $\varphi(x)$, as given by Swift and Friedlander [38] for the class-II-kernel of shear coagulation (see table 1 of section 2) contains "infinities" and is ill-defined. Furthermore, the scaling ansatz should be reformulated for the gelling models of class I. There the size distribution approaches in the scaling limit $(k, s(t) \to \infty$ with k/s kept fixed:

$$c_k(t) \simeq s^{-\tau} \varphi(k/s) \quad (5.3)$$

where τ is an exponent that has still to be determined. In gelling systems Smoluchowski's equation only allows a similarity solution of the form (5.3) in the pre-gelation stage, where the mean cluster size $s(t)$ diverges as $t \uparrow t_c$.

5b. *Asymptotic properties of the solution*

What is known about the integral equation? Its general solution contains two arbitrary (positive) integration constants, since $\bar{\varphi}(x) = a\varphi(bx)$ also satisfies (5.2). For the exactly solved models $K(x,y) = $ constant, $x+y$, and xy the similarity solution is respectively $\varphi(x) \sim e^{-x}$, $x^{-3/2} e^{-x}$ and $x^{-5/2} e^{-x}$. The last solution, referring to the gelling model $K(x,y) = xy$, only applies below the gel point as $t \uparrow t_c$. Numerical solutions of (5.2) have only been considered for some class III-model [39,43].

Analytical information about more general homogeneous coagulation kernels is only available for the small and large x-behavior of the solution, which will be summarized here. The

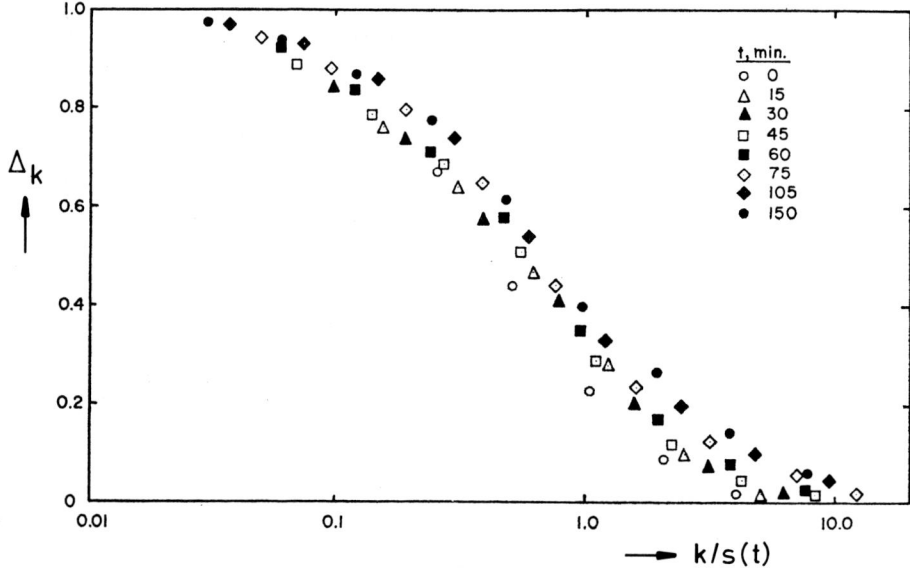

Fig. 4. *Experimental test of the scaling laws for an oil-in-water emulsion [38]. When measured in properly reduced units $k/s(t)$, data taken at different times during the coagulation process fall approximately on a single curve after the first 15 minutes. Here $s(t)$ is the mean volume of an oil drop and $\Delta_k = \int_k^\infty dx\, \varphi(x)$ is the fraction of drops with a volume larger than k.*

mean cluster size increases in non-gelling systems ($\lambda < 1$) [37-43] and gelling systems ($1 < \lambda \leq 2$) [21,35,37] respectively as:

$$s(t) \sim \begin{cases} t^z & (t \to \infty) \\ (t_c - t)^{-1/\sigma} & (t \uparrow t_c) \end{cases} \quad (5.4)$$

with

$$z = 1/(1-\lambda) \qquad (\lambda < 1)$$

$$\sigma = \tfrac{1}{2}(\lambda - 1) \qquad (\lambda > 1)$$

On the borderline ($\lambda=1$) between gelling and non-gelling models the dependence is of exponential form, $s(t) \sim \exp(t^a)$ where $a = \tfrac{1}{2}$ in class I ($\mu > 0$) and $a = 1$ for $\mu = 0$. The *large-x-behavior* of the scaling function in gelling and non-gelling systems with $\lambda < 1$ (see fig. 1) has the form:

$$\varphi(x) \sim x^{-\lambda} e^{-x} \qquad (x \to \infty) \quad (5.5)$$

It is also possible to obtain the next dominant term for large x. On the borderline ($\nu=1$) between the physical and unphysical region the behavior of $\varphi(x)$ at large x can be very different and depends on details of $K(x,y)$ [47]. The *small-x-behavior* of $\varphi(x)$ depends strongly on μ. For class III ($\mu < 0$) the scaling function vanishes exponentially fast

$$\varphi(x) \sim x^{-\tau} \exp(-x^{-|\mu|} + \ldots) \quad (x \to 0) \quad (5.6)$$

where τ depends on the coagulation kernel [37,39,40]. The scaling function for class-III-kernels has a typical bellshape, as it vanishes exponentially fast for small and large x. For instance, $K(x,y) = (xy)^a$ with $a < 0$ gives $\tau = 2$ and $K(xy) = x^a + y^a$ with $a < 0$ gives $\tau = 1$ [37]. For Brownian coagulation in the continuum regime and in the kinetic regime (see first and second example of table 1, section 2) Friedlander c.s. [39] find respectively for $x \to 0$:

$$\varphi(x) \sim x^{-1.06} \exp[-x^{-1/3}]$$

$$\varphi(x) \sim x^{-2} \exp[-x^{-1/2} - ax^{-1/6}]$$

Lushnikov's result [40] for the first kernel with $\tau = 2$ is incorrect.

In class I ($\mu > 0$) one should distinguish gelling ($\lambda > 1$) and non-gelling ($\lambda < 1$) systems, where

$$\varphi(x) \sim x^{-\tau} \qquad (x \to 0) \quad (5.7)$$

with

$$\tau = \begin{cases} \lambda + 1 & (\lambda < 1) \\ \tfrac{1}{2}(\lambda + 3) & (\lambda > 1) \end{cases}$$

The scaling function increases as $t \uparrow 0$ (at least for $\lambda > -1$), as opposed to an exponential decrease in class III. In non-gelling models one could in principle allow $\lambda < -1$, so that $\varphi(x)$ would decrease algebraically as $x \to 0$. A mathematical example could be $K(x,y) = (xy)^\mu (x+y)^{\nu-\mu}$ with $\mu > 0$ and $\lambda = \mu+\nu < -1$. However, we do not know of any physical example in this range of (μ,ν)-values.

I also want to mention the existence of an *exact* unphysical solution to (5.2) for all non-gelling kernels of class I [11]:

$$\varphi(x) = A \, x^{-1-\tau} \quad (5.8)$$

This solution is unphysical because the total mass $\Sigma \, kc_k$ diverges on account of (5.1) and (5.7).

Class II ($\mu=0$) forms the borderline between classes I and III, where there is no clear distinction between the dominant growth mechanisms. The small-x-behavior still has the algebraic form (5.7), but the τ-exponent is expressed in moments of the scaling function and cannot be calculated without solving the integral equation [11,37]. In some cases - for example $K(xy) = x^\lambda + y^\lambda$ and $K(xy) = (x+y)^\lambda$ - it appears possible to construct rather sharp upper- and lower-bounds for the τ-exponent of class-II-kernels [47]. Leyvraz's result [48], $\tau = 1 + \tfrac{1}{2}\lambda$ for $K(xy) = x^\lambda + y^\lambda$ with $0 < \lambda \leq 1$ is not correct [47]. For all gelling kernels, for all non-gelling kernels of class II and III and for the gelling kernels of class I that are sufficiently sharply peaked around the diagonal (a typical example is $K(x,y) \, x^{\lambda+1}\delta(x-y)$) one can determine higher order corrections to the small-x-behavior, that are of algebraic type [37]. For the more common non-gelling kernels of class I, such as $K(x,y) = x^a y^b + y^a x^b$, the assumption of an algebraic correction to (5.7) leads to inconsistencies [37], contrary to the suggestions made in [11], and the analytical structure of the first correction term to the leading behavior (5.7) is not known.

5c. *Further results and problems*

There exists also an extensive literature on coagulation processes in the presence of monomer- or oligomer-sources, where similarity or stationary solutions of Smoluchowski's equation have been studied [5,24,25,49].

In theories on aerosol coagulation it is customary to use the continuous version of the coagulation equation [2]. This equation is invariant under a semi-group of similarity transformations [4]. This symmetry admits an exact (similarity) solution of the form (5.2), provided solutions to the (continous) coagulation equation exist. From a rigorous point of view little is known about the types of kernels for *which positive solutions to (5.2) do exist*.

The difficulties in constructing higher order corrections to (5.7), mentioned at the end of the previous subsection, may be related to the non-existence of physically acceptable scaling solutions of Smoluchowski's equation. One cannot exclude the possibility that the exact unphysical solution (5.8) of Lushnikov and Piskunov [11] is the only solution for certain class-I-models, so that physically acceptable scaling solutions may not exist for such models.

Furthermore, the existence of similarity solutions to Smoluchowski's equation by no means guarantees that the size distribution indeed approaches the scaling form after some transient period has lapsed. For the exactly solved cases of subsection 3a. one can show that the solution $c_k(t)$ in the scaling limit approaches the similarity solution, provided the initial distribution $c_k(o)$ is exponentially bounded. However, Mulholland et al. [50] have shown for the non-gelling model, $K(x,y) = $ constant, that initial $c_k(o)$ with an algebraic tail do not approach to the universal scaling form, $\varphi(x) \sim e^{-x}$, for this model. A similar result has been obtained for the gelling model $K(xy) = xy$ by Ziff et al [16]. These authors showed that $c_k(t)$ for initial distributions with an algebraic tail does approach in the scaling limit to *non-universal* scaling form with exponents explicitly depending on the initial distribution. For the non-gelling model $K(xy) = x+y$ similar results hold. One can apply the mapping (3.5) of the product-kernel on the sum-kernel to conclude from the previous discussion that the initial distribution with $c_k(o) \simeq Ak^{-a}$ ($k \to \infty$, $a > 2$) approaches for $t \to \infty$ and $k \gg e^t$:

$$c_k(t) \simeq Ak^{-a} \exp[(a-1)t]$$

For initial distributions that are exponentially bounded $c_k(o) \leqslant C \exp(-\varepsilon k)$ with $\varepsilon > o$ there are strong indications that the similarity form is approached after a sufficiently long time [46] for non-gelling kernels with $o \leqslant \nu < 1$. For $\nu < o$ and $\nu = 1$ the situation is not entirely clear. In gelling systems with $\nu < 1$ the arguments for a *finite* time t_0, where crossover to universal behavior occurs, do not exclude that $t_0 \geqslant t_c$. In that case the scaling form near the gel point could depend on the initial distributions.

6. SIZE DISTRIBUTION FOR LARGE k

6a. *At infinitesimal times*

The present section concentrates on the asymptotic behavior for large cluster sizes at a fixed value of the time where t is restricted to the sol phase, i.e. well below the gel point in gelling systems. Furthermore, it should be kept in mind that "large" means here: large compared to the mean cluster size s(t), which is itself an increasing function of time.

In the next subsection I shall address the general problem of finite times and general initial distributions. Here I concentrate on the simpler problem of infinitesimal times and monodisperse initial distributions $c_k(o) = \delta_{k1}$, where "large" means large compared to unity.

In statistical mechanical theories of polymerization and percolation [51,52] it is customary to represent the asymptotic form of the size distribution in the sol-phase as

$$c_k \simeq Ak^{-\theta} \exp(-Ck^\zeta) \qquad (k \to \infty) \qquad (6.1)$$

where A and C are positive constants and θ and ζ geometric exponents, characterizing the size distribution.

An expression similar to (6.1) is being used for the size distribution of lattice animals in percolation theories [51]. In statistical mechanical models of lattice percolation the exponents (θ,ζ) are unrelated to the exponents characterizing c_k at the gel point. Furthermore, they are independent of the fraction of bonds already formed, denoted by the variable p, and measured here indirectly by the variable t. By applying renormalization group arguments [52] to these lattice models one finds that the animal exponents (θ,ζ) are determined by a fixed point at $p = o$ (corresponding here to $t = o$ and a monodisperse initial distribution).

Motivated by such results one has calculated [32,53] the exponents (θ,ζ) from a solution of Smoluchowski's equation at short times with a monodisperse initial distribution. One finds that the size distribution behaves as $c_k(t) \simeq N_k t^{k-1}\{1+\mathcal{O}(t)\}$ as $t \to o$, where $N_1 = 1$ and where N_k satisfies the "short time" recursion relation:

$$(k-1)N_k = \tfrac{1}{2} \sum_{i+j=k} K_{ij} N_i N_j \qquad (6.2)$$

The large k-behavior of the solution of (6.2) is determined by the behavior of K_{ij} at large i and j. If one restricts oneself again to kernels K_{ij} that are asymptotically homogeneous of degree λ (see (2.2)), one finds for all kernels with $\nu < 1$ the asymptotic solution [32]:

$$\begin{aligned} N_k &\simeq ak^{-\lambda}R^{-k} \qquad (k \to \infty) \\ a^{-1} &= \tfrac{1}{2} \int_0^1 dx\, K(x,1-x)\,[x(1-x)]^{-\lambda} \end{aligned} \qquad (6.3)$$

Here R is the radius of convergence of the solution of (6.2), which can in general only be determined numerically. Thus, the size distribution has the asymptotic form:

$$c_k(t) \simeq (a/t)k^{-\lambda}(t/R)^k \qquad (k \to \infty) \qquad (6.4)$$

so that $\theta = \lambda$ and $\zeta = 1$ in (6.1).

6b. *At finite times*

6b.1. *Pre-gelation*

In the previous subsection I discussed the speculation that the large-k-result (6.4) – derived from the "short time" recursion (6.2) – should be relevant for the large-k-solution of Smoluchowski's equation at finite times. This speculation turns out to be correct, as has been recently shown [46].

However, one has to impose several restrictions. The first restriction is that the initial size distribution must be bounded by an exponential, $c_k(o) \leq A \exp(-\varepsilon k)$ with $\varepsilon > o$ and $k \to \infty$. If this constraint is not satisfied, e.g. for initial distributions with an algebraic tail, the functional form (6.1) and, in particular, the exponents (ζ, θ) are expected to be non-universal, i.e. depend on $c_k(o)$, as also happens in the scaling limit (see subsection 5c). Secondly, the results are restricted to the sol phase, and thirdly only kernels with $\nu < 1$ are considered, in order to avoid additional complications [46] on the borderline $\nu = 1$.

With these restrictions the asymptotic form of the size distribution at finite times can also be determined from a recursion relation, which is essentially the same as (6.2). One can distinguish transient and universal solutions. At a time t_0, which may be long or short depending on the initial distribution, the transient crosses over to the universal asymptotic form:

$$c_k(t) \simeq a\dot{z}(t) k^{-\lambda} \exp(kz(t)) \quad (6.5)$$

with $k \gg s(t)$, where the constant a and the exponent $\theta = \lambda$ are the same as those obtained from the "short time" recursion relation (6.2). The dependence on the initial conditions enters only through the function $z(t)$, with properties $\dot{z}(t) > o$ and $z(t) < o$, and remains undetermined. A somewhat related result was also obtained by Botet and Jullien [54].

This result is perhaps not so surprising because it refers at any given time to the far out tail of the size-distribution, which is just in its initial stage of development. There $c_k(t)$ has an analytic form, which is for all $k \gg s(t)$ only determined by a single (unknown) function $z(t)$.

I also point out that the universal asymptotic form (6.5) as $t \downarrow o$ reduces to the result (6.4) of subsection 6a [46]. Thus the region of validity of the two limiting solutions: (a) first $t \downarrow o$, next $k \to \infty$, and (b) first $k \to \infty$, next $t \downarrow o$, are overlapping (see upper left corner of the diagram in fig. 5).

One also verifies that (i) the universal asymptotic solution (6.5) for gelling and non-gelling models at fixed t coincides with (ii) the scaling solution (5.1), (5.3) and (5.4) for large scaling argument, $x = k/s(t) \gg 1$. Again

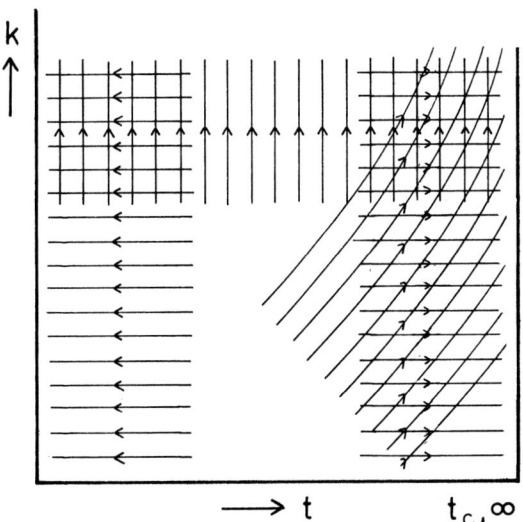

Fig. 5. *Survey of asymptotic solutions $c_k(t)$ in the sol phase (with $t < t_c$ in gelling systems) with their partly overlapping regions of validity. The directions indicate special limits: (i) ↑↑ $k \to \infty$, $t = $ fixed (section 6b.1), (ii) ⇵ $t \to o$, $k = $ fixed (section 6a), (iii) ⇵ $t \to \infty$, $k = $ fixed (section 7) and (iv) ↑↗ $t, k \to \infty$ with $k/s(t) = $ fixed (section 5).*

one has partly overlapping regions of validity of two limiting solutions: (i) $k \to \infty$, t fixed and (ii) first $k \to \infty$, $s(t) \to \infty$ with x fixed, next $x \to \infty$, as illustrated in the upper right corner of the diagram in fig. 5.

Before concluding this section I want to point out the following. In view of the enormous amount of numerical work [55] involved in solving the nonlinear equations of aerosol coagulation (reactor safety, nuclear accidents) analytic information on the structure of the solution at large k should be very useful in improving numerical codes for integrating the coagulation equation for large times and large cluster sizes. The difficulty here lies in estimating the infinite sum in the loss term of Smoluchowski's equation (1.1). After each integration step one could update the results for $z(t)$ and use the analytical formula (6.5) for improved estimates of the infinite sum in (1.1), and thus improving the convergence of the numerical procedures and - hopefully - extend the time range over which the coagulation equation can be integrated.

6b.2. *Post-gelation*

The large-k-result (6.5) at finite times is

exponentially bounded and represents a pre-gelation result. If a gelation transition occurs, post-gelation solutions at large k and finite t with $t \geq t_c$ have an algebraic tail:

$$c_k(t) \simeq B(t)k^{-\tau} \qquad (k \to \infty) \qquad (6.6)$$

This has already been illustrated in (3.4) and (4.6) for the special case $K_{ij} = (ij)^\omega$. For homogeneous coagulation kernels (2.2) with $\lambda > 1$ the coefficient $B(t)$ can be determined by calculating the mass flux $J(L,t)$ from small clusters ($k < L$) to large ones ($k > L$) (see section 3a). The current for the present case is given by replacing kj by $K(k,j)$ on the RHS of (3.2). By inserting (6.6) and requiring that the flux of sol-particles to the gel, $J(\infty,t) = -\dot{M}_1(t)$, be bounded, one finds $\tau = \frac{1}{2}(\lambda+3)$ and

$$B(t) = (-\dot{M}_1(t)/b)^{1/2} \qquad (6.7)$$

with

$$b = \int_0^1 dx \int_{1-x}^\infty dy \, K(x,y) x^{1-\tau} y^{-\tau}$$

The dependence of $\dot{M}_1(t)$ on time remains undetermined. Asymptotic results for the size distribution, covered by the general formula (6.6-7), have been obtained in [21,35,37].

In the next section we shall see that there exists a special post-gel solution for which the sol fraction $M_1(t) = (a+bt)^{-1}$ and that for a general initial distribution one has the asymptotic result $M_1(t) \sim 1/t$ ($t \to \infty$).

The last result refers to the case of a discrete k-variable. If the size-variable k is continuous, the sol-fraction may decay arbitrarily slowly, depending on the initial size distribution $c(k;o)$ at $k \to o$, as shown for the exactly solved case $K(xy) = xy$ in Ref. [17].

7. THE SIZE DISTRIBUTION FOR LARGE t

Two different methods can be used to determine the long time behavior of the solution $c_k(t)$, which lead partly to complementary results. One method uses scaling functions, as was discussed in section 5. The other method uses the "long time" recursion relation (as opposed to the "short time" recursion relation discussed in section 6).

The recursion relation gives the limiting value $\lim c_k(t)/c_1(t) = b_k$ as $t \to \infty$, but not the approach to this value, and is applicable to gelling and non-gelling models of class I only. The scaling function method makes only predictions about large k and applies only to the pre-gelation region. In the present context the latter method is only relevant for non-gelling models. The recursion relation method makes predictions about all k and gives for gelling systems results in the post-gelation region.

I start with the observation [35,47,48,56] that the coagulation equation admits a special solution of the form $c_k(t) = c_1(t)b_k$ ($k=1,2,...$) with $b_1 = 1$, viz.

$$-wb_k = \frac{1}{2}\sum_{i+j=k} K_{ij} b_i b_j - b_k \sum_{j=1}^\infty K_{kj} b_j$$

$$-w = \dot{c}_1(t)/c_1^2(t) \qquad (7.1)$$

where w is a positive separation constant. The monomer concentration behaves therefore as $c_1(t) = 1/(A+wt)$. Since the first equation should also be valid for $k = 1$ it follows that $w = \sum_j K_{1j} b_j$. The unknown w may be eliminated and the "long time" recursion relation becomes:

$$b_k \sum_{j=1}^\infty (K_{kj}-K_{1j})b_j = \frac{1}{2}\sum_{i+j=k} K_{ij} b_i b_j \qquad (7.2)$$

with $b_1 = 1$. The validity of this equation is not restricted to kernels (2.2) that are asymptotically homogeneous, but the solution of (7.2) has only been investigated for such kernels. It turns out [47] that (7.2) has strictly positive solutions for gelling ($\lambda > 1$) and non-gelling systems ($\lambda < 1$) of class I. The solution of (7.2) for class-II-kernels has also been studied [47,48]. For some kernels the coefficients b_k are strictly positive, for others they do not have a definite sign. However, the b_k's have no special significance in class II, whereas in class I they have a universal significance, as I shall discuss below. In class I the large-k-behavior of the solution is given by

$$b_k \sim k^{-\tau} \qquad (k \to \infty)$$

with

$$\tau = \begin{cases} \frac{1}{2}(\lambda+3) & (\lambda > 1) \\ \lambda+1 & (\lambda < 1) \end{cases} \qquad (7.3)$$

Consider the total mass of sol particles $M_1(t) = c_1(t) \sum k b_k$. For gelling systems ($\lambda > 1$) this infinite sum is convergent, implying that the total sol mass decays as $M_1(t) \sim c_1(t) \sim t^{-1}$. Thus, we are necessarily dealing with a post-gelation solution, where the gel points remain undetermined. The recursion relation (7.2) was in fact introduced in [35] to determine the critical exponent τ for the model $K_{ij} = (ij)^\omega$ past the gel point. For non-gelling systems the infinite sum is divergent and the solution $c_k(t) = c_1(t)b_k$ is therefore physically unacceptable. It corresponds in fact to the unphysical scaling solution (5.8) obtained by Lushnikov and Piskunov. However, asymptotic solutions of this form are physically acceptable in non-gelling systems because $c_k(t)/c_1(t)$ may approach b_k non-uniformly in k, so that $\sum k c_k(t) = 1$, but $\sum k b_k$ is divergent.

The special solution determined by the "long time" recursion relation has, however, a *uni-*

versal significance as an asymptotic large-t-solution in non-gelling [47,48] and gelling systems [35,47] of class I, i.e. for general initial distributions the size distribution approaches a limiting value:

$$\lim_{t \to \infty} c_k(t)/c_1(t) = b_k \quad (k=2,3,\ldots) \quad (7.4)$$

which is independent of $c_k(o)$.

By combining (7.1), (7.3) and (7.4) one finds $c_k(t) \sim t^{-1}k^{-\lambda-1}$ for non-gelling class-I-models. The same result follows from (5.1), (5.4) and (5.7) in the scaling limit. Thus, the two limiting solutions: (i) first $t \to \infty$, next $t \to \infty$ and (ii) first $k,t \to \infty$ with $x = k/s(t)$ fixed, next $x \to 0$ have a common region of validity, as indicated on the right of the diagram in fig. 5.

A simple example of a class-I-kernel is $K_{ij} = (ij)^{\omega}$, where the recursion simplifies to

$$p_\omega(k^\omega-1)b_k = \tfrac{1}{2} \sum_{i+j=k} (ij)^\omega b_i b_j$$

with $p_\omega = \Sigma_k k^\omega b_k$. For gelling systems ($\lambda = 2\omega > 1$) the recursion relation was studied in [35,36] and for non-gelling systems ($\lambda = 2\omega \leq 1$) in [48].

In coagulation models of class II (μ=o) the ratio $c_k(t)/c_1(t)$ does approach a constant as $t \to \infty$, but the constant depends on $c_k(o)$ and cannot be determined from the recursion relation (7.2). The monomer concentration, behaving as $c_1(t) \sim t^{-\gamma}$ ($t \to \infty$), decreases in general faster than in class I. The exponent follows from the scaling form (5.1) to be $\gamma = (2-\tau)/(1-\lambda)$. In class-II- kernels τ has been expressed in terms of moments of the scaling function (see below (5.8)), and can only be calculated after solving the integral equation for the scaling function $\varphi(x)$.

For the exactly solved class-II-models one has $c_1(t) \sim t^{-2}$ for K_{ij} = const. and $c_1(t) \sim e^{-t}$ for $K_{ij} = i+j$. One should also notice that the result $c_k(t)/c(t) \sim k^{-1-\omega/2}$ ($k \to \infty$) for $K_{ij} = i^\omega+j^\omega$ [48] are incorrect, as shown in [47].

In coagulation models of class III $c_k(t)/c_1(t) \to \infty$ as $t \to \infty$ for $k = 2,3,\ldots$, and the monomer concentration approaches zero rapidly, in comparison with class I and II. Physically this is understandable, because large clusters grow mainly at the expense of small clusters.

8. CONCLUSION

The analytic information about the asymptotic properties of the solution $c_k(t)$ of the coagulation equation is fairly complete, and best summarized in figs. 1 and 5. For (asymptotically) homogeneous kernels (2.2) of degree λ, fig. 1 shows which models lead to gelation, and fig. 5 shows what type of asymptotic results are known in the sol phase. Similar results for the gel phase are given in sections 6b.2 and 7.

Most results have been obtained by using arguments of self consistency. They may therefore be incomplete and they lack mathematical rigor. Several unresolved problems are mentioned in the different sections, e.g. does the solution of the coagulation equation indeed approach a similarity form, given that a similarity solution exists, or does a similarity solution exist at all for non-gelling kernels of class I.

Until now the integral equation for the scaling function $\varphi(x)$ has only been solved numerically for some class-III-models. It would be interesting to have numerical solutions for more realistic kernels. Also numerical solutions for class-I and II-models would be very interesting. With this information one would be able to carry out more detailed comparison with the wealth of data, available from experiments [38,39,57] and computer simulations of aggregation phenomena [58]. In fact, many properties of the cluster size distribution in real and computer experiments have already been explained on the bases of Smoluchowski's equation [38,39,54,57,58].

REFERENCES

[1] R.L. Drake in *Topics in Current Aerosol Research*, vol. 3, eds. G.M. Hidy and J.R. Brock (Pergamon Press, New York, 1972) part 2.

[2] S.K. Friedlander, *Smoke, Dust and Haze* (Wiley-Interscience, New York, 1972).

[3] R.M. Ziff, in *Kinetics of Aggregation and Gelation*, eds. F. Family and D.P. Landau (Elseviers Sc. Publ. B.V., Amsterdam, 1984).

[4] M.H. Ernst, in *Fundamental Problems in Statistical Mechanics* VI, E.G.D. Cohen, ed. (Elseviers Sc. Publ. B.V., Amsterdam, 1985).

[5] J.D. Klett, J. Atmosph. Sci 32 (1975) 380.

[6] A.A. Lushnikov, (Engl. Transl.), Isv. Ocean. and Atmosph. Phys. 14 (1978) 378.

[7] R.M. Ziff, J. Stat. Phys. 23 (1980) 241.

[8] P.J. Flory, *Principles of Polymer Chemistry* (Cornell Univ. Press, N.Y. 1953).

[9] E.M. Hendriks, M.H. Ernst and R.M. Ziff, J. Stat. Phys. 31 (1983) 519.

[10] P.G.J. van Dongen and M.H. Ernst, Phys. Rev. A32 (1985) 670.

[11] A.A. Lushnikov and V.N. Piskunov (Engl. Transl.) Doklady Phys. Chem. 231 (1976) 1166.

[12] R. Botet, R. Jullien and M. Kolb, J. Phys. A: Math. Gen. 17 (1984) L75.

[13] E.M. Hendriks and M.H. Ernst, J. Colloid Interf. Sci. 97 (1984) 176.

[14] A.M. Golovin, Izv. Geophys. Ser. 1963 (Engl.

Transl.) Bull. Acad. Sci. USSR, Geophys. Ser. No. 5 (1963) 482).
[15] J.B. McLeod, Quart. J. Math. Oxford 13 (1962) 119, 193.
[16] R.M. Ziff, M.H. Ernst and E.M. Hendriks, J. Phys. A: Math. Gen. 16 (1983) 2293.
[17] M.H. Ernst, R.M. Ziff and E.M. Hendriks, J. Colloid Interf. Sci. 97 (1984) 266.
[18] F. Leyvraz and H.R. Tschudi, J. Phys. A: Math, Gen. 14 (1981) 3389.
[19] R.M. Ziff and G. Stell, J. Chem. Phys. 73 (1980) 3492.
[20] J.L. Spouge, J. Phys. A: Math. Gen. 16 (1983) 767; J. Stat. Phys. 31 (1983) 363; P.G.J. van Dongen and M.H. Ernst, J. Phys. A: Math. Gen. 16 (1983) L327.
[21] F. Leyvraz and H.R. Tschudi, J. Phys. A: Math. Gen. 15 (1982) 1951.
[22] P.G.J. van Dongen and M.H. Ernst, J. Phys. A: Math. Gen. 17 (1984) 2281.
[23] A.A. Lushnikov and V.N. Piskunov (Engl. Transl.), Doklady Phys. Chem. 231 (1976) 1403; A.A. Lushnikov, Ya.I. Tokar and M.S. Tsitskishvili (Engl. Transl.) Doklady Phys. Chem. 256 (1981) 1155.
[24] J.G. Crump and J.H. Seinfeld, J. Colloid Interf. Sci. 90 (1982) 469.
[25] W.H. White, J. Colloid Interf. Sci. 87 (1982) 204.
[26] M. Aizenman and T.A. Bak, Commun. Math. Phys. 65 (1979) 203; M.H. Ernst, in *Studies in Statistical Mechanics*, Vol. X, E. Montroll and J.L. Lebowitz eds. (North-Holland, Amsterdam, 1983) chap. 3.
[27] R.C. Srivastava, J. Atmosph. Sci. 39 (1982) 1317.
[28] P.G.J. van Dongen and M.H. Ernst, J. Stat. Phys. 37 (1984) 301.
[29] E.R. Domilovskii, A.A. Lushnikov, and V.N. Piskunov (Engl. Transl.) Doklady Phys. Chem. 243 (1978) 407.
[30] R.W. Samsel and A.S. Perelson, Biophysics J. 37 (1982) 493; 45 (1984) 805; P.G.J. van Dongen and M.H. Ernst, in *Kinetics of Aggregation and Gelation*, F. Family and D.P. Landau, eds. (Elsevier Sci. Publ., Amsterdam, 1984) p. 205.
[31] R.M. Ziff, M.H. Ernst and E.M. Hendriks, J. Colloid Interf. Sci. 100 (1984) 220.
[32] A.A. Lushnikov, J. Colloid. Interf. Sci. 45 (1973) 549; A.A. Lushnikov and V.N. Piskunov (Engl. Transl.), Kolloidn. Zn. 37 (1975) 285.
[33] M.H. Ernst, E.M. Hendriks and R.M. Ziff, Phys. Lett. 92A (1982) 267.
[34] W.H. White, Proc. Am. Math. Soc. 80 (1980) 273.
[35] R.M. Ziff, E.M. Hendriks and M.H. Ernst, Phys. Rev. Lett. 49 (1982) 593; E.M. Hendriks, M.H. Ernst and R.M. Ziff, J. Stat. Phys. 31 (1983) 519.
[36] P.G.J. van Dongen and M.H. Ernst, to be published.
[37] P.G.J. van Dongen and M.H. Ernst, Phys. Rev. Lett. 54 (1985) 1396.
[38] D.L. Swift and S.K. Friedlander, J. Colloid Interf. Sci. 19 (1964) 621.
[39] S.K. Friedlander and C.S. Wang, J. Colloid Interf. Sci. 22 (1966) 126; F.S. Lai, S.K. Friedlander, J. Pich and G.M. Hidy, J. Colloid Interf. Sci. 39 (1972) 395.
[40] A.A. Lushnikov, J. Colloid Interf. Sci. 45 (1973) 549.
[41] K. Binder, Phys. Rev. B15 (1977) 4425; J. Silk and S.D. White, Astrophys. J. 223 (1978) L59.
[42] J.M. Rosen, J. Colloid. Interf. Sci. 99 (1984) 9.
[43] M. Kolb, Phys. Rev. Lett. 53 (1984) 1653.
[44] R.M. Ziff, E.D. McGrady and P. Meakin, J. Chem. Phys. 82 (1985) 5269.
[45] T. Vicsek and F. Family, Phys. Rev. Lett. 52 (1984) 1669; P. Meakin, T. Vicsek and F. Family, Phys. Rev. B31 (1985) 564.
[46] P.G.J. van Dongen and M.H. Ernst, this volume and J. Colloid Interf. Sci., to be published.
[47] P.G.J. van Dongen and M.H. Ernst, Phys. Rev. A32 (1985) 670; J. Phys. A: Math. Gen. 18 (1985) 2779.
[48] F. Leyvraz, Phys. Rev. A29 (1984) 854.
[49] A.A. Lushnikov and V.N. Piskunov (Engl. Transl.), Kolloidn. Zn. 39 (1977) 1076; A.A. Lushnikov and Ya. I. Tokar (Engl. Transl.), Doklady Phys. Chem. 252 (1980) 136.
[50] G.W. Mulholland and H.R. Baum, Phys. Rev. Lett. 45 (1980) 761; G.W. Mulholland, T.G. Lee and H.R. Baum, J. Colloid Interf. Sci. 62 (1977) 406.
[51] D. Stauffer, A. Coniglio and M. Adam, in *Advances in Polymer Science*, vol. 44, ed. K. Dusek (Springer Verlag, Berlin, 1982) p. 103.
[52] H.E. Stanley, P.J. Reynolds, S. Redner and F. Family in *Real Space Renormalization*, ed. T.W. Burkhardt and J.M.J. van Leeuwen (Springer Verlag, Berlin, 1982) ch. 7.
[53] M.H. Ernst, E.M. Hendriks and F. Leyvraz, J. Phys. A: Math. Gen. 17 (1984) 2137.
[54] R. Botet and R. Jullien, J. Phys. A: Math. Gen. 17 (1984) 2517.
[55] S.K. Loyalka, Progress in Nuclear Energy 12 (1983) 1; A. Emani and S.K. Loyalka, Nuclear Technology 52 (1981) 162. F. Gelbard and J.R. Seinfeld, J. Colloid Interf. Sci. 78 (1980) 785; A.F. Middelton and J.R. Brock, J. Colloid. Interf. Sci. 54 (1963) 2.
[56] F. Leyvraz, J. Phys. A: Math. Gen. 16 (1983) 2861.
[57] D. Weitz, J.S. Huang, M.Y. Lin and J. Sung, Phys. Rev. Lett. 53 (1984) 1657; 54 (1985) 1416; J.E. Martin and D.W. Schaefer, Phys. Rev. Lett. 53 (1984) 2457; J. Feder, J. Jøssang and E. Rosenqvist, Phys. Rev. Lett. 53 (1984) 1403.
[58] M. Kolb et al., J. Phys. A 18 (1985) L435.

TAIL DISTRIBUTION FOR LARGE CLUSTERS IN IRREVERSIBLE AGGREGATION

P.G.J. VAN DONGEN and M.H. ERNST

Institute for Theoretical Physics, University of Utrecht, P.O. Box 80.006, 3508 TA Utrecht, The Netherlands

The tail of the cluster size distribution $c_k(t)$ at cluster sizes much larger than the mean cluster size, is determined from Smoluchowski's coagulation equation with rate constants $K(i,j)$. The latter are homogeneous functions of i and j. The tail is found to be exponential, $c_k(t) = A_k(t)\exp(kz(t))$, with $\log A_k(t) = o(k)$ as $k \to \infty$, and $A_k(t)$ may be calculated from a recursion relation. Depending on the initial distribution, $c_k(o)$, and the rate constants, $K(i,j)$, we distinguish universal solutions, coinciding with the scaling solutions as the average cluster size diverges, and non-universal solutions. Our general results have been verified in two exactly soluble models.

1. INTRODUCTION

In this paper we use Smoluchowski's coagulation equation[1] with rate constants $K(i,j)$:

$$\dot{c}_k = \tfrac{1}{2}\sum_{i+j=k} K(i,j)c_i c_j - c_k \sum_{j=1}^{\infty} K(k,j)c_j \quad (1)$$

in order to determine the functional form of the cluster size distribution at <u>large cluster sizes</u>, for different types of initial distributions, $c_k(o)$, and coagulation kernels, $K(i,j)$. "Large" in this context means "much larger than the average cluster size $s(t)$".

The asymptotic properties of the cluster size distribution have been studied extensively in various limits: (i) short-time solutions for monodisperse initial conditions[2], (ii) large-time behavior[3] at fixed k, and (iii) scaling behavior in the limit $k \to \infty$, $s(t) \to \infty$, where $k/s(t)$ remains finite[4]. To complete the available analytic information on the solutions of Smoluchowski's equation (1), we have calculated the asymptotic size dependence of $c_k(t)$ for (fixed) finite time. In gelling systems, where $s(t)$ diverges at a finite time t_c (gelpoint), our considerations are restricted to $t < t_c$. The present asymptotic results coincide on the one hand with the short-time solutions (see (i)) and on the other hand with the scaling solutions (see (iii)), both taken at large cluster sizes ($k \gg s(t)$). This agreement relates and unifies the earlier results.

The structure of the cluster size distribution for large values of k, depends on the behavior of the rate constants $K(i,j)$ at large cluster sizes i and j. As most coagulation kernels used in the literature[1] are homogeneous functions of i and j, at least for large cluster sizes, we restrict ourselves to such kernels, and we characterize $K(i,j)$ by two exponents describing its i- and j- dependence for $j \gg i$:

$$K(ai,aj) = a^\lambda K(i,j) = a^\lambda K(j,i) \quad (2.a)$$
$$K(i,j) \simeq i^\mu j^\nu \quad (j \gg i,\ \lambda = \mu+\nu) \quad (2.b)$$

The reactivity of large clusters should not increase faster than their size, hence $\nu \leq 1$, $\lambda \leq 2$, but no restrictions are imposed on μ. We assume that $K(x,1-x)$ is continuous and positive for all $x \in (0,1)$.

The main result, to be found, is that the cluster size distribution show a universal <u>exponential tail</u>:

$$c_k(t) = A_k(t)\, e^{kz(t)} \quad (z(t) < 0) \quad (3.a)$$

with

$$\lim_{k\to\infty} k^{-1}\log A_k(t) = 0 \quad (3.b)$$

where the large-k-behavior of the prefactor $A_k(t)$ may be calculated from a <u>recursion relation</u>. In solving this recursion relation for large k, we distinguish between kernels with $\nu < 1$ and $\nu = 1$. In either case, we further distinguish <u>universal</u> solutions, that reduce to the scaling solution in the scaling limit, and <u>non-universal</u>, or transient, solutions. Universal solutions have the form $A_k(t) \simeq A(t)k^{-\vartheta}$ ($k \to \infty$), except in certain models with $\nu = 1$, where one finds the universal form $A_k(t) \sim \exp[-\Delta(t)k^\beta]$ as $k \to \infty$. The possibility and the shape of non-universal solutions are determined by the form of the initial distribution $c_k(o)$.

2. THE METHOD

We consider in general solutions of the form

$$c_k(t) = A_k(t)\exp(k^\zeta z(t)) \quad (4.a)$$

with $\zeta > 0$, $\dot{z}(t) < 0$ and

$$\lim_{k \to \infty} k^{-\zeta} \log A_k(t) = 0 \qquad (4.b)$$

If the initial condition is exponentially bounded for some constants $C, \varepsilon > 0$:

$$c_k(o) < C\, e^{-\varepsilon k} \qquad (k=1,2,\ldots) \qquad (5)$$

then one can show[5] that only the assumption $\zeta = 1$, or (3.a,b), leads to consistent solutions. By inserting (3.a,b) into Smoluchowski's equation (1), we find that $z(t)$ and $A_k(t)$ are related through the following equation:

$$\dot{z}kA_k + \dot{A}_k = \tfrac{1}{2} \sum_{i+j=k} K(i,j) A_i A_j - A_k \sum_{j=1}^{\infty} K(k,j) c_j \qquad (6)$$

In solving equation (6) for large k, we distinguish between models with a ν-exponent satisfying $\nu < 1$, and models with $\nu = 1$.

3. SOLUTIONS FOR $\nu < 1$:

Different behavior is found for $\dot{z} \neq 0$ and for $\dot{z} = 0$. If $\dot{z} \neq 0$, then eq. (6) reduces for large k to the recursion relation:

$$\dot{z}kA_k \simeq \tfrac{1}{2} \sum_{i+j=k} K(i,j) A_i A_j \qquad (k \to \infty) \qquad (7)$$

In the derivation of (7) we have used that $\dot{A}_k / A_k = O(k)$ as $k \to \infty$, on account of (3.b), and $\sum K(k,j) c_j \simeq k^{\nu} M_{\mu}$ ($k \to \infty$), where $M_{\alpha}(t) = \sum k^{\alpha} c_k$ represents the α-th moment of $c_k(t)$. The solution of equation (7) satisfying (3.b) has the general form

$$A_k(t) \simeq A(t) k^{-\vartheta} \qquad (k \to \infty) \qquad (8)$$

To determine A and ϑ we substitute the asymptotic form (8) into (7), and equate leading orders in k. This gives, in combination with (3.a):

$$c_k(t) \simeq a\dot{z}(t) k^{-\lambda} e^{kz(t)} \qquad (k \to \infty) \qquad (9.a)$$

where $z < 0$, $\dot{z} > 0$, and the constant a is determined by

$$a^{-1} = \tfrac{1}{2} \int_0^1 dx\, K(x, 1-x) [x(1-x)]^{-\lambda} \qquad (9.b)$$

The integral in (9.b) converges provided $\nu < 1$.

Next we consider the case $\dot{z} = 0$, or $z(t) = z_o$, with $\nu \geq 0$. Then eq. (6) allows for asymptotic solutions of the form:

$$A_k(t) \simeq A_k(o)\, \exp[k^{\beta} \Delta(t)] \qquad (k \to \infty) \qquad (10.a)$$

with $\beta = \nu$ and $\Delta(o) = 0$. The time dependence of $\Delta(t)$ is determined by

$$\dot{\Delta}(t) = \sum_{k=1}^{\infty} k^{\mu} A_k(t) \left(1 - e^{kz_o}\right) > 0 \qquad (10.b)$$

Inspection of (10.a,b) for $\nu > 0$ shows that this solution is possible only if $\dot{\Delta}(t) < \infty$ as $t \downarrow 0$, i.e. if the initial condition satisfies the requirement $-k^{-\nu} \log A_k(o) \geq$ constant as $k \to \infty$. Similarly, for $\nu = 0$, one has the requirement $\sum k^{\mu} A_k(o) < \infty$ in order that $\dot{\Delta}(o) < \infty$. For the ν-values considered ($0 \leq \nu < 1$) one can show[5] that the solution (10.a,b) is a transient, i.e. that it crosses over to the form (9.a,b) in finite time. For this reason we refer to the solution (9) as the universal asymptotic form of $c_k(t)$.

For the case $\dot{z} = 0$ and $\nu < 0$ we infer from (10.a,b) that $A_k(t)/A_k(o) \to 1$ as $k \to \infty$, provided that $\sum k^{\mu} A_k(o) < \infty$. Consequently, the time development of $A_k(t)$ (k fixed) cannot be explained on the basis of the large-k-behavior of the solutions only. Moreover, one cannot decide from the present analysis whether solutions of the form (10) are either transient, or non-universal and long-lived. Next, we consider initial distributions of the form $A_k(o) \simeq Ak^{-\alpha}$ ($k \to \infty$) with $\alpha < 1+\mu$, so that $\sum k^{\mu} A_k(o) = \infty$. If we assume that $A_k(t) \simeq A(t) k^{-\alpha}$, we find from (6) as $k \to \infty$:

$$\dot{A}/A^2 \simeq \tfrac{1}{2} k^{1+\lambda-\alpha} \int_0^1 dx\, K(x, 1-x) [x(1-x)]^{-\alpha} \qquad (k \to \infty) \qquad (11)$$

For α satisfying $1+\lambda < \alpha < 1+\mu$ one finds that $\dot{A} \to 0$, or $A_k(t)/A_k(o) \to 1$ as $k \to \infty$, and one arrives at the same conclusion as for $\alpha > 1+\mu$. For $\alpha < 1+\lambda$, we conclude from (11) that $\dot{A}(o) = \infty$, implying instantaneous crossover to the universal solution. In the special case $\alpha = 1+\lambda$ one finds transient asymptotic behavior, $c_k(t) \simeq A(t) k^{-1-\lambda} \exp(kz_o)$, with $A(t) = 1/(t_o - t)$, crossing over to the universal solutions (9) at $t = t_o$.

4. RESULTS FOR $\nu = 1$:

The calculations for this special case are technically more complicated than for $\nu < 1$. Here we give only the main results. Before doing so we note that the cases $\dot{z} = 0$ and $\dot{z} \neq 0$ need not be distinguished, because eq. (6) contains two terms of $O(k)$. For $\nu = 1$ the asymptotic behavior of $A_k(t)$ depends upon more details of $K(i,j)$ than specified by the leading order (2.b). Therefore, we introduce $Q(i,j)$ which is defined by

$$K(i,j) = (ij)^{\mu} (i+j)^{1-\mu} [1 + Q(i,j)] \qquad (12.a)$$

and assume that the small-x-behavior of $Q(x, 1-x)$ is given by:

$$Q(x, 1-x) \simeq qx^{\rho} + \cdots \qquad (x \downarrow 0) \qquad (12.b)$$

with $\rho > 0$ and q non-vanishing and finite ($-\infty < q < \infty$). The function $Q(i,j)$ is homogeneous, with zero degree of homogeneity, as may be seen from (12.a).

Also for $\nu = 1$ we distinguish universal and non-universal solutions. The criterion for universal behavior is that it reduces to scaling behavior (see below) as the average cluster size $s(t)$ diverges, $s(t) \to \infty$. In either case one finds that the time-development of $z(t)$ is determined by the entire cluster size distribution, as follows:

$$\dot{z}(t) = \sum_{k=1}^{\infty} k^{\mu} A_k(t) (1-e^{kz}) > 0 \qquad (13)$$

However, the asymptotic behavior of $A_k(t)$ as $k \to \infty$, is different for universal and non-universal solutions.

The <u>universal</u> asymptotic behavior of $A_k(t)$ for the various values of ρ and q has been listed in Table I. The large-k-behavior is either of algebraic form, $A_k \simeq Ak^{-\vartheta}$, or of the stretched exponential form:

$$A_k(t) \sim \exp(-\Delta(t) k^{\beta}) \qquad (k \to \infty) \qquad (14)$$

with $0 < \beta < 1$ and $\Delta(t) > 0$. Note that the exponent in (14) is negative, as opposed to (10.a).

A. $\rho>1$ $A_k(t) \simeq A(t)k^{-\vartheta}$
    $\rho=1, q>-2, J(2+\mu)>1$ $1+\mu < \vartheta < m(1)$
    $\rho<1, q>0$ $J(\vartheta)=0$

B. $\rho=1, q=-2, J(2+\mu)<0$ $A_k(t) \simeq A(t)k^{-\vartheta}$
                                       $\vartheta = 2+\mu$

C. $\rho=1, q<-2$ $A_k(t) \simeq A(t)k^{-\vartheta}$
                                       $m(2) < \vartheta < \mu-q$

D. $\rho<1, q<0$ $A_k(t) \sim \exp[-k^{\beta}\Delta(t)]$
                                       $\beta = 1-\rho$

Table I

<u>Algebraic</u> behavior of $A_k(t)$ is found in the classes A, B and C, and stretched exponential behavior is found in class D. In class A, the exponent ϑ is the solution of the <u>transcendental equation</u>:

$$J(\vartheta) = 0 \qquad (15.a)$$

lying in the interval $1+\mu < \vartheta < m(1) \equiv \min\{2+\mu, 1+\mu+\rho\}$. Here $J(\vartheta)$ is defined as

$$J(\vartheta) \equiv \int_0^{\frac{1}{2}} dx \{K(x,1-x)[x(1-x)]^{-\vartheta} - x^{\mu-\vartheta}\} - \int_{\frac{1}{2}}^{\infty} dx\, x^{\mu-\vartheta} \qquad (15.b)$$

In class B one finds a consistent algebraic solution with $\vartheta = 2+\mu$. For the models of class C one finds a solution with ϑ in the interval $m(2) < \vartheta < \mu-q$, where $m(2) \equiv \max\{2+\mu, \mu-q-1\}$.

Since the parameter q may become arbitrarily large and negative in class C, we conclude that for any fixed value of μ, the exponent ϑ may become arbitrarily large, i.e. $\vartheta \to \infty$ as $q \to -\infty$. Finally in class D one finds <u>stretched exponential</u> behavior with $\beta = 1-\rho$. In the special case $\rho=1$, $q=-2$, there are some complications, that will not be discussed here.

Next we consider <u>transient</u> solutions of the form $A_k(t) \simeq F(t) A_k(o)$ $(k \to \infty)$, with $F(o) = 1$. They can be found if $A_k(o) \sim k^{-\vartheta_o}$, with $\vartheta_o > m(3) \equiv \max\{2+\mu, \mu-q\}$, and $\rho \geqslant 1$. The time dependence of $F(t)$ may be expressed in terms of moments of the cluster size distribution. If $k^{\vartheta} A_k(o) \to 0$ as $k \to \infty$, but $k^{m(3)} A_k(o) >$ constant, then the solution crosses over to the universal solution instantaneously. If, on the other hand, the initial distribution is such that $A_k(o) \sim \exp[-D_o k^{\beta}]$ as $k \to \infty$, with $D_o > 0$ and $1 > \beta > m(4) \equiv \max\{0, 1-\rho\}$, then one finds transient behavior of the form $A_k(t) \sim \exp[-D(t)k^{\beta} - \Delta(t)k^{m(4)}]$. Instantaneous cross-over to the universal solution occurs if $\beta \leqslant m(4)$.

5. RELATION WITH SHORT-TIME AND SCALING SOLUTIONS

We discuss the way in which the universal large-k behavior of the cluster size distribution $c_k(t)$ is related to: (1) the short-time solution for monodisperse initial conditions, and (2) the scaling solution. Our arguments are concentrated on the class of models with $\nu < 1$, but the conclusions also hold for models with $\nu = 1$.

The short-time solutions of Smoluchowski's equation for a monodisperse initial distribution, $c_k(o) = \delta_{k1}$, has the form $c_k(t) \simeq N_k t^{k-1}$ as $t \downarrow 0$, where the coefficients N_k satisfy the recursion relation:

$$(k-1) N_k = \frac{1}{2} \sum_{i+j=k} K(i,j) N_i N_j \qquad (16)$$

with $N_1 = 1$. The large-k-behavior of N_k may be determined in a similar manner as that of $A_k(t)$, with the result $N_k \simeq ak^{-\lambda} R^{-k}$, where a is given in (9.b) and R is left undetermined ($0 < R < \infty$). This result for N_k ($k \to \infty$) shows that in the limit $t \downarrow 0$, with $k \gg 1$, $c_k(t)$ has the form (3), with

$$A_k(t) \simeq ak^{-\lambda}/t \qquad (17.a)$$

and

$$z(t) \simeq \log(t/R) \qquad (17.b)$$

The <u>same</u> expressions (17.a,b) are obtained if we take the limit $k \to \infty$ first, i.e. if we start from eqs. (9.a,b), and make the identification $z(t) \simeq \log(t/R)$ as $t \downarrow 0$. We conclude that there exists a common region of validity where the two limiting solutions: (a) first $t \downarrow 0$, next $k \to \infty$, and: (b) first $k \to \infty$, next $t \downarrow 0$, coincide.

Similarly one can show that there exist overlapping regions of validity of the large-k-solution and the scaling solution. The latter applies in the scaling limit, where both the mean cluster size $s(t)$, and the cluster size k are taken to infinity, while the scaling argument $x = k/s(t)$ remains finite. For <u>gelling systems</u> $(1 < \lambda \leq 2)$, the scaling solution has the form

$$c_k(t) \simeq (1/s(t))^\tau \varphi(k/s(t)) \qquad (18.a)$$

$$\dot{s} = ws^{2-\tau+\lambda} \qquad (18.b)$$

where $\tau = (\lambda+3)/2$ and w is a separation constant. The mean cluster size diverges as $t \uparrow t_c$ (gelpoint). The large-x-behavior of the scaling function $\varphi(x)$ in (18.a) is given as:

$$\varphi(x) \simeq w\delta a x^{-\lambda} e^{-\delta x} \qquad (x \to \infty) \qquad (19)$$

with a defined in (9.b). Comparison of (18.a,b), (19) with the large-k-solution (9.a,b) shows that both lead to the same expressions if we identify $z(t)$ as $z(t) \simeq -\delta/s(t)$ as $z \uparrow 0$, or $s \to \infty$. The same arguments apply for <u>non-gelling systems</u> $(\lambda \leq 1)$ where τ in (18.a,b) should be replaced by 2, and $s(t)$ diverges as $t \to \infty$.

6. EXACTLY SOLVABLE CASES

We discuss the large-k-behavior in two exactly solvable models namely $K(i,j) = 2$, which has $\nu = 0$, and $K(i,j) = ij$, which is a model with $\nu = 1$. Detailed calculations are given in reference 5 for $K(i,j) = 2$. For $K(i,j) = ij$, details will be published elsewhere. Here we give only the results.

The large-k-behavior in the model $K(i,j) = 2$ (the value 2 is chosen for convenience) may be expressed in terms of the generating function $v(x)$ of the initial distribution:

$$v(x) \equiv \sum_{k=1}^{\infty} c_k(o) x^k \qquad (20)$$

we distinguish two possibilities: $v(x_c) = \infty$ and $v(x_c) < \infty$, where $x_c \equiv \exp(-z(o))$. (i) If $v(x_c) = \infty$, then we find universal large-k-behavior of the form:

$$c_k(t) \simeq (t^2 v'(x_o) x_o)^{-1} x_o^{-k} \qquad (k \to \infty) \qquad (21.a)$$

where the time dependence of $x_o(t)$ is determined by

$$v(x_o) - v(1) = 1/t \qquad (21.b)$$

One easily verifies that (21) has the form (9.a,b). (ii) If, on the other hand, $v(x_c) < \infty$, or $\sum A_k(o) < \infty$, one finds the following large-k-behavior:

$$c_k(t) \simeq [1-t/t_o]^{-2} c_k(o) \qquad (k \to \infty) \qquad (22)$$

where cross-over occurs at $t_o = [v(x_c)-v(1)]^{-1}$.

Clearly t_o is finite, on account of our assumption (5), or $z(o) < 0$.

In the model $K(i,j) = ij$, the large-k-behavior of $c_k(t)$ may be calculated from the exact solution[6]. In this case the results are expressed in terms of the generating function $u(x) \equiv \sum k c_k(o) e^{kx}$. Again we distinguish the possibilities: (i) $u'(-z_o) = \infty$, and (ii) $u'(-z_o) < \infty$, where $z_o \equiv z(o)$. Case (i), $u'(-z_o) = \infty$, leads to universal asymptotic behavior of the form

$$c_k(t) \simeq [2\pi t^3 u'''(s_c)]^{-\frac{1}{2}} k^{-5/2} e^{kz(t)} \qquad (23)$$

where the time dependence of $s_c(t)$ is implicitly determined by

$$u'(s_c) = 1/t \qquad (24)$$

Furthermore, $z(t)$ is determined by (13) with $\mu=1$, which in this exactly solvable model may be integrated to yield[6]:

$$z(t) = -(s_c + t u(s_c) - t) \qquad (25)$$

In the case (ii), $u'(-z_o) < \infty$, then eqs. (23)-(25) apply only if there exists a root $s_c(t)$ of (24), i.e. if $t > t_o \equiv 1/u'(-z_o)$. For $0 < t < t_o$, there exists transient behavior, depending on the shape of the initial distribution. If we assume that the initial distribution has the form $c_k(o) \simeq A(o) k^{-\vartheta_o} \exp(kz_o)$ as $k \to \infty$, then the transient solution has the form:

$$c_k(t) \simeq A(t) k^{-\vartheta_o} \exp(k z(t)) \qquad (k \to \infty) \qquad (26.a)$$

with

$$z(t) = z_o + [u(-z_o)-1]t \qquad (26.b)$$

and

$$A(t) = A(o) [1-t/t_o]^{-(\vartheta_o-1)} \qquad (26.c)$$

Other initial conditions lead to different transient behavior. E.g. if $c_k(o) \sim \exp[-\Delta(o)k^\beta + kz_o]$, one finds transients, behaving as:

$$c_k(t) \sim \exp[-\Delta(t)k^\beta + k z(t)] \qquad (27.a)$$

with $z(t)$ given by (26.b), and $\Delta(t)$ by

$$\Delta(t) = \Delta(o)[1 - t/t_o]^\beta \qquad (27.b)$$

Thus, the transients cross over to the universal form (23)-(25) within a finite time $t_o=1/u'(-z_o)$, with $t_o<t_c$, where $t_c=1/u'(o)$ is the gelpoint in this model.

We conclude that the asymptotic $(k \to \infty)$ behavior in the models $K(i,j) = 2$ and

$K(i,j) = ij$ is in full agreement with the predictions for models with $\nu < 1$ and $\nu = 1$ respectively.

7. SUMMARY AND CONCLUSIONS

We have shown for general homogeneous kernels with $\nu \leq 1$ that the solution of Smoluchowski's equation has the form $c_k(t) = A_k(t)\exp(kz(t))$, with $k^{-1}\log A_k \to 0$ as $k \to \infty$, where the k-dependence of $A_k(t)$ for k-values, much larger than the average cluster size, may be calculated from a recursion relation. The structure of the solutions of this recursion relation is different for $\nu < 1$ and for $\nu = 1$. Depending on the type of initial size distributions $c_k(o)$ we distinguish universal solutions, and (non-universal) transient solutions. the former reduce to scaling solutions as the average cluster size diverges, $s(t) \to \infty$; the latter cross over to the universal solutions within a finite time. In some exceptional cases ($\nu<0$, $\nu=1$) we cannot exclude that non-universal solutions are long-lived.

The results are as follows. For all kernels with $\nu<1$ one finds universal asymptotic behavior of the form (9.a,b), i.e. $c_k(t) \sim \ell(t)k^{-\lambda}\exp(kz(t))$, provided that the initial distribution satisfies the condition $k^{-\nu}\log A_k(o) \to 0$ $(k \to \infty)$ if $\nu>0$, or $k^{1+\lambda}A_k(o) \to 0$ if $\nu < 0$. If, on the other hand, $k^{-\nu}\log A_k(o) \geqslant$ constant and $0 < \nu < 1$, or $\nu=0$ and $\int k^\mu A_k(o) < \infty$, one finds transients asymptotic solutions $c_k(t) \sim c_k(o)\exp[k^\nu \Delta(t)]$, that cross over to the universal solutions in finite time. For $\nu < 0$, we find asymptotic solutions of the form $A_k(t)/A_k(o) \to 1$ as $k \to \infty$, provided that $A_k(o) = o(k^{-1-\lambda})$. Finally, the class of kernels with $\nu=1$ shows fairly diverse universal and non-universal behavior, depending on the details of the coagulation kernel $K(i,j)$.

Furthermore it was shown that for short times and monodisperse initial conditions, the universal large-k-solution coincides with the short-time solutions at large cluster sizes. Similarly the universal large-k-solution reduces to the scaling solution at large scaling arguments as the average cluster size diverges, $s(t) \to \infty$. Thus, the region of validity of the large-k-solution overlaps with those of the short-time solutions and the scaling solution, both taken at large cluster sizes.

The predictions from the present theory for large cluster sizes have been verified in two exactly soluble models, viz. $K(i,j) = 2$ and $K(i,j) = ij$. We conclude that our present results, taken together with references 2-4 give a fairly complete analytic description of the asymptotic properties of the cluster size distribution if the coagulation kernel $K(i,j)$ is a homogeneous function of i and j, at least for large i and j.

ACKNOWLEDGEMENT

The work of one of us (P.G.J. v. D.) is part of a research program of the Stichting voor Fundamenteel Onderzoek der Materie (FOM), which is financially supported by the Nederlandse Organisatie voor Zuiver-Wetenschappelijk Onderzoek (ZWO).

REFERENCES

1. R.L. Drake, in: Topics in Current Aerosol Research, Vol. 3, eds. G.M. Hidy and J.R. Brock (Pergamon Press, New York, 1972) part 2.
2. A.A. Lushnikov, J. Coll. Interface Sci. 45 (1973) 549; M.H. Ernst, E.M. Hendriks and F. Leyvraz, J. Phys. A: Math. Gen. 17 (1984) 2137.
3. F. Leyvraz, Phys. Rev. A 29 (1984) 854; P.G.J. van Dongen and M.H. Ernst, J. Phys. A: Math. Gen., 18 (1985).
4. S.K. Friedlander and C.S. Wang, J. Coll. Interface Sci. 22 (1966) 126; F. Leyvraz and H.R. Tschudi, J. Phys. A: Math. Gen. 15 (1982) 1951; P.G.J. van Dongen and M.H. Ernst, Phys. Rev. Lett. 54 (1985) 1396.
5. P.G.J. van Dongen and M.H. Ernst, J. Coll. Interface Sci., submitted.
6. R.M. Ziff, M.H. Ernst and E.M. Hendriks, J. Phys. A: Math. Gen. 16 (1983) 2293.

SCALING GENERALIZATION OF THE SMOLUCHOWSKI EQUATION

Zoltán RÁCZ*

Department of Physics, Simon Fraser University, Burnaby, B.C., Canada V5A 1S6

Diffusion-limited cluster-cluster aggregation subject to the condition that single particles are fed into the system at a rate h, while clusters larger than a fixed size are removed is investigated. Fluctuation effects in this system are treated on a phenomenological level by a scaling generalization of the Smoluchowski equation. We find that the cluster size distribution obeys scaling and all the relevant exponents are expressible through a single homogeneity index. This index is determined by arguing that the zero feed-rate process is in one universality class with the diffusive annihilation problem. The scaling theory is checked on the example of one-dimensional diffusive annihilation in the presence of particle sources. We calculate exactly the steady-state particle density $\bar{n}$ and the relaxation time of homogeneous density fluctuation τ. They are found to scale in the $h \to 0$ limit as $\bar{n} \sim h^{1/3}$ and $\tau \sim h^{-2/3}$ in agreement with the scaling theory.

1. INTRODUCTION

A common mechanism of growth is irreversible aggregation. Polymerization, aerosol formation, colloid growth, red-blood-cell aggregation and nucleation at phase transition[1] can all be described to some extent by aggregation models. Most of these models, however, are slight generalizations of Smoluchowski's mean-field approach[2,3] which is based on the assumption that the collisions between the aggregating objects are binary and that the spatial fluctuations in the density of these objects are negligible. Since spatial fluctuations are known to be important at lower dimensions ($d \leq d_c$) one expects that Smoluchowski's theory breaks down for $d \leq d_c$ where the critical dimension d_c is not known accurately for most of the aggregation processes. As one has learned from the theory of critical phenomena, spatial fluctuations can be treated on a phenomenological level by a scaling generalization of mean-field theory. Below we show how this scaling generalization can be carried out for aggregating systems in which particle sources and sinks are present.

2. SCALING PROPERTIES OF THE SMOLUCHOWSKI EQUATION

Smoluchowski's assumption that the clusters coalesce through random binary collisions means that the time evolution of the density of k-particle clusters $n_k(t)$ is determined from the following rate equation

$$\frac{dn_k(t)}{dt} = \frac{1}{2} \sum_{i+j=k} K_{ij} n_i(t) n_j(t) - n_k(t) \sum_{j=1}^{\infty} K_{jk} n_j(t) \quad (1)$$

where the collision kernel K_{ij} takes partial care of such details of the process as how the collision cross-section depends on the size and mobility of the clusters. The usual procedure is to assume[4] that K_{ij} is a homogeneous function of its arguments $K_{bi,bj} = b^\lambda K_{ij}$ and then the degree of homogeneity is used to fit the scaling aspects of the experimental data. This procedure is somewhat reminiscent of the phenomenological method of producing a desired exponent at an equilibrium critical point by introducing appropriately chosen long-range forces.

---
*On leave from the Institute for Theoretical Physics, Eötvös University, H-1088 Budapest, Hungary.

As we shall see, however, one runs into problems with the above approach if sources and sinks are also present in the system (corresponding to the presence of an external field at an equilibrium critical point).

We introduce the simplest kinds of sources and sinks. It is assumed that single particles are produced at a rate h particles per unit volume per unit time and clusters containing more than k_0 particles are instantaneously eliminated from the system. The corresponding Smoluchowski equation is then[5,6]

$$\frac{dn_k(t)}{dt} = h\delta_{k1} + \frac{1}{2} \sum_{i+j=k \leq k_0} K_{ij} n_i(t) n_j(t)$$
$$- n_k(t) \sum_{j=1}^{k_0} K_{jk} n_j(t) . \quad (2)$$

This equation has simple scaling properties with respect to the feed rate h. These scaling properties, however, are determined by the fact that the collision terms are quadratic in the cluster density. Thus K_{ij} cannot be used to fix the corresponding exponents. Indeed, introducing scaled time and scaled cluster-size distribution[7]

$$\tilde{t} = h^{1/2} t \quad \text{and} \quad \tilde{n}_k(\tilde{t}) = n_k(t)/h^{1/2} \quad (3)$$

one can see that h is eliminated from eq. (2) and consequently in the long-time limit (when the initial-condition dependence can be neglected) the solution is written in the form

$$n_k(t,h) \approx h^{1/2} \phi_k(h^{1/2} t) . \quad (4)$$

It follows then that the steady-state ($t \to \infty$) cluster-size distribution $\bar{n}_k$ and the time of relaxation towards the steady state τ_h scale as $\bar{n}_k \sim h^{1/2}$ and $\tau_h \sim h^{-1/2}$ while the relaxation at $h = 0$ is nonexponential $n_k(t,0) \sim t^{-1}$. Monte Carlo simulations show,[8] however, that $\bar{n}_k \sim h^{1/\delta}$, $\tau_h \sim h^{-\Delta}$ and $n_k(t,0) \sim t^{-\zeta}$ and the mean-field values of the exponents obtained from eq. (4) apply only above the critical dimension which seems to be $d_c = 2$ in this case.

3. SCALING GENERALIZATION OF THE SMOLUCHOWSKI EQUATION

In order to develop a theory which might account for the deviations from the mean-field approach, note that eq. (2) can be written in the form

$$\frac{dn_k(t)}{dt} = h\delta_{k1} - G_k(n_1, n_2, \ldots n_{k_0}) \quad (5)$$

where G_k is a homogeneous function of second degree $G_k(\lambda n_1, \lambda n_2, \ldots, \lambda n_{k_0}) = \lambda^2 G_k(n_1, n_2, \ldots n_{k_0})$. Since similarly to free-energy derivatives, G_k is a "generalized force" driving the system towards its steady state, and since scaling is a general feature of the Monte Carlo results,[8] it is quite natural to follow the theory of critical phenomena and postulate a scaling form for G_k, namely

$$G_k(\lambda n_1, \lambda n_2, \ldots, \lambda n_{k_0}) = \lambda^\delta G_k(n_1, n_2, \ldots, n_{k_0}) \quad (6)$$

where the homogeneity degree δ is a free parameter. The analysis of equations (5) and (6) is similar to that of eq. (2). The choice of scaled time and cluster-size distribution which eliminates h from eq. (5) is now $\tilde{t} = h^{1-1/\delta} t$ and $\tilde{n}_k(\tilde{t}) = n_k(t)/h^{1/\delta}$. Thus the solution of eq. (5) for large time is of the form

$$n_k(t,h) \sim h^{1/\delta} \psi(h^{1-1/\delta} t) \quad (7)$$

and one obtains $\bar{n}_k \sim h^{1/\delta}$, $\tau_k \sim h^{-\Delta}$ and $n(t,0) \sim t^{-\zeta}$ with

$$\Delta = 1 - 1/\delta \quad \text{and} \quad \zeta = 1/(\delta-1) \quad (8)$$

i.e. all the exponents are expressible in terms of the homogeneity index δ. The scaling form (7) and the scaling laws (8) have been verified[8] for diffusion-limited aggregation in d = 1,2 and 3. A more stringent test of the theory is provided by the $k_0 = 1$ case which is diffusion-limited annihilation in the presence of particle sources. This problem has been much studied[9-14] for h = 0 and the d = 1 case has been solved exactly.[9] The exact result $n_1(t,0) \sim t^{-1/2}$ implies that if the scaling theory is valid then $\delta = 3$ and $\Delta = 2/3$. Below we calculate δ and Δ directly and verify the validity of the scaling laws (8) for d = 1.

4. ONE-DIMENSIONAL DIFFUSION-LIMITED ANNIHILATION IN THE PRESENCE OF SOURCES

The following model is considered. Particles execute a random walk (with hopping rate Γ per unit time) along a d = 1 lattice and annihilate if they land on the same site simultaneously. Furthermore, single particles are created homogeneously at a rate of Γh per lattice site. Our aim is to calculate both the density of particles $\bar{n}$ in the steady state and the relaxation time τ of the homogeneous density fluctuations. This task is accomplished by mapping the above model onto a generalized kinetic Ising model.

To understand the mapping, consider the d = 1 kinetic Ising model[15] with Glauber's choice of spin-flip rates

$$w_i = \frac{\Gamma}{2}[1 - \frac{\gamma}{2}\sigma_i(\sigma_{i+1} + \sigma_{i-1})], \qquad (9)$$

where $\sigma_i(t) = \pm 1$ is a stochastic spin variable assigned to lattice site i and γ is a parameter related to the temperature of the system. At zero temperature we have $\gamma = 1$ and the i-th spin flips with rate $\Gamma/2$ if $\sigma_{i+1} = -\sigma_{i-1}$ or with rate Γ if $\sigma_{i+1} = \sigma_{i-1} = -\sigma_i$. This means that a domain wall moves to left and right with equal rate $\Gamma/2$ while two neighboring domain walls annihilate each other with rate Γ. Thus identifying the domain walls with particles we see that the zero-temperature limit of the kinetic Ising model describes diffusion-limited annihilation. Single particle sources can now be introduced by allowing spin-flip processes in which all the spins σ_ℓ, $\ell \leq i$ flip simultaneously. This process creates a domain wall and thus a particle between sites i and i+1 (or if there is already a domain wall between those sites then it is destroyed corresponding to the event that creating a particle at an occupied site leads to emptying of that site). If the rate of this process is chosen to be independent of i

$$\tilde{w}_i = \Gamma h \qquad (10)$$

then we have a kinetic Ising model with domain-wall dynamics equivalent to the particle dynamics of the diffusion-limited annihilation model with homogeneously distributed sources.

Since the particle density between sites i and i+1 is given by $n_i = (1-\sigma_i\sigma_{i+1})/2$ all we need to calculate is $\langle\sigma_i\sigma_{i+1}\rangle$ for the kinetic Ising model with spin-flip rates (9) and (10). For translationally invariant initial states $\bar{n} = \langle n_i \rangle$ and the equation for $\langle\sigma_i\sigma_{i+1}\rangle$ involves only the two-point correlation functions $r_\ell = \langle\sigma_i\sigma_{i+\ell}\rangle$. The equation for r_ℓ can be derived following the steps of Glauber's paper[15] and one obtains[16] for $\ell > 0$

$$\Gamma^{-1}\dot{r}_\ell = r_{\ell-1} + r_{\ell+1} - 2(1+h\ell)r_\ell \qquad (11)$$

while for $\ell = 0$ we have $r_0 = \langle\sigma_i^2\rangle = 1$ and $r_\ell = r_{-\ell}$ for $\ell < 0$.

The stationary solution of (11) is given by

$$\bar{r}_k = J_{k+h^{-1}}(h^{-1})/J_{h^{-1}}(h^{-1}) \qquad (12)$$

where $J_\nu(z)$ is the Bessel function[17] of the first kind of order ν. Since $\bar{n} = (1-\bar{r}_1)/2$ we find then in the $h \to 0$ limit

$$\bar{n} \approx 0.46 h^{1/3}, \qquad (13)$$

thus $\delta = 3$ as predicted by scaling.

The relaxation times of homogeneous density fluctuations are found by seeking solutions to (11) in the form $r_k = \bar{r}_k + q_k \exp(-t/\tau)$. Substituting this form into (11) one obtains an infinite set of relaxation times the largest of which scales in the $h \to 0$ limit as

$$\tau_h \approx 0.27 \Gamma^{-1} h^{-2/3} . \quad (14)$$

Therefore the second prediction of the scaling theory $\Delta = 2/3$ is also verified.

4. FINAL REMARKS

Since no new symmetry or conservation law is introduced into the system when going from $k_0 = 1$ to $k_0 = 2,3,\ldots,K < \infty$, one might conjecture that the exponents derived for the diffusion limited annihilation remain unchanged for other diffusion-controlled aggregation processes. Indeed, Monte Carlo simulations in which $\delta = 3$ and $\Delta = 2/3$ have been obtained with good accuracy for $d = 1$ diffusion-limited cluster-cluster aggregation, support the above conjecture.

Diffusion-limited annihilation has been studied in higher dimensions as well[10-14] and the consensus is that $d_c = 2$ above which $\zeta = 1$ while $\zeta = d/2$ for $d \leq 2$. Using scaling and the above argument about universality, one might conjecture that diffusion controlled aggregation problems in the presence of sources and sinks are described by equations (5) and (6) with the scaling exponent

$$\delta = \begin{cases} 2 & d > 2 \\ (d+2)/d & d \leq 2 \end{cases} \quad (15)$$

Since $d_c = 2$ it appears that fluctuations do not play an important role in real aggregating systems. It should be noted, however, that chemical reactions often take place on low-dimensional, fractal-like surfaces and the above theory is directly applicable to those systems.

ACKNOWLEDGEMENTS

I would like to thank M. Plischke for helpful discussions.

REFERENCES

1. Kinetics of Aggregation and Gelation, Edited by F. Family and D.P. Landau (North-Holland, Amsterdam, 1984).

2. M. von Smoluchowski, Phys. Z. 17, 557 (1916).

3. For a recent review see R.M. Ziff, Ref. 1, p. 191 and M.H. Ernst in this volume.

4. P.G.J. van Dongen and M.H. Ernst, Phys. Rev. Lett. 54, 1396 (1985).

5. M. von Smoluchowski, Z. Phys. Chem. 92, 129 (1918).

6. Experiments where the particle sources and sinks play important roles are described in G.J. Madelaine, M.L. Perrin and M. Itoh, J. Aerosol Sci. 12, 202 (1979); L.F. Mocros, J.E. Quon and A.T. Hjelmfelt, J. Colloid Interface Sci. 23, 90 (1967).

7. Z. Rácz, Phys. Rev. A, August 1985.

8. T. Vicsek, P. Meakin and F. Family, Phys. Rev. A, August 1985, L.W. Anacker and R. Kopelman, J. Chem. Phys. 81, 6402 (1984).

9. D.C. Torney and H.M. McConnel, J. Phys. Chem. 87, 1941 (1983).

10. D. Toussaint and F. Wilczek, J. Chem. Phys. 78, 2642 (1983).

11. D.C. Torney, J. Chem. Phys. 79, 3606 (1983).

12. P. Meakin and H.E. Stanley, J. Phys. A17, L173 (1984).

13. K. Kang and S. Redner, Phys. Rev. Lett. 52, 955 (1984).

14. K. Kang and S. Redner, Phys. Rev. A30, 2833 (1984).

15. R.J. Glauber, J. Math. Phys. 4, 294 (1963).

16. Z. Rácz, to be published.

17. Handbook of Mathematical Functions, Ed. by M. Abramowitz and I.A. Stegun (Dover, New York, 1965).

CLUSTERING IN THE UNIVERSE

Francesco LUCCHIN

Dipartimento di Fisica "G. Galilei", Via Marzolo 8, 35100 Padova, Italy.

The Universe is clumpy up to 10^2Mpc, with a clustering fractal dimension $D \cong 1.2$ in a wide range. A short review is given of the clustering and of its cosmological relevance.

1. INTRODUCTION

The galaxy clustering was extensively studied in the last sixty years: the compilation of the Zwicky[1], Lick[2] and Jagellonian[3] galaxy position catalogues was the result of this effort. A great progress in the statistical study of the clustering became then possible; such a work was mainly performed by Peebles and coworkers[4]. Recently a new era was open in the clustering analysis by the advent of greatly improved spectrographs and emulsions which permit to measure the radial velocities v (and, then, the distance d) of faint galaxies (the well known Hubble law $v = H_o d$ relates the velocities of distant galaxies, due to the universal expansion, with d; the Hubble constant H_o is 100h Km/sec/Mpc, with $0.5 \lesssim h \lesssim 1$; 1pc $\cong$ 3.3 light years). In this way it was recently possible to compile redshift-catalogues (the redshift $z \cong v/c$ is a useful distance and look-back time parameter), like the CfA catalogue[5]. Progress in the galaxy clustering analysis are always strictly connected with theoretical progress in the study of the origin of the cosmic structures. In this field a rapid evolution underwent recently mainly due to the application of the "new physics" (GUT's, supersymmetry,..) to the early cosmology (see the texts[6,7]).

2. CLUSTERING IN THE UNIVERSE

2.1. Morphological description

The galaxy distribution is clumpy on all scales. The galaxies are sistems with a mass $10^7 \div 10^{12} M_\odot$ ($M_\odot$ is the solar mass), a size $10 \div 10^2$ Kpc and an enhancement of 10^5 over the average density. Almost half of the galaxies are members of groups or clusters. The groups consist of several tens of galaxies, with a mass $10^{12} \div 10^{14} M_\odot$, a size $10^{-1} \div 10$ Mpc and an overdensity of 10^3. Our Local Group is constituted by about twenty members, the preminent being our galaxy and Andromeda. The clusters are agglomerations of some thousand galaxies, with a mass and a size of the order of $10^{15} M_\odot$ and 10 Mpc respectively and an overdensity of 10^2 or more; the structure is rather lumpy, with low central galaxy density (irregular clusters like Virgo), or smooth and symmetric (regular clusters like Coma). The most important cluster catalogues are the Zwicky[1] and the Abell[8] catalogues.

In the last thirty years the evidence of the existence of superclusters was obtained[9]. The superclusters contain several clusters and have an irregular shape. The mass and size are of the order of $10^{17} M_\odot$ and 10^2 Mpc respectively; the density enhancement is no more than a factor 2÷3 over the average. In the superclusters the galaxy concentrations have often a filamentary or flattened shape. Our local Supercluster, centered on Virgo, is a typical example of a supercluster. Most of the volume of the universe is practically devoid of galaxies. It would be interesting to have catalogues of superclusters and voids (like the large void in Bootes[10]), in order to study their statistical properties;for the moment only a supercluster catalogue with about 20 members exists[11].

In conclusion the matter distribution in the universe, extremely lumpy at small scales, is increasingly homogeneous on scales approaching 10^2 Mpc. On much greater scales the universe is very smooth, as is shown mainly by the high isotropy of the 3°k cosmic background radiation (CBR)[12]: $\Delta T/T \lesssim 10^{-4}$. The CBR is a fossil record of the "last scattering",10^5 years after the "big-bang", before the origin of the cosmic structures.

2.2. The statistical approach

The main work on this subject consisted in the measurement of the low order galaxy and cluster correlation functions on the basis of the position catalogues[1,2,3] containing about 10^6 galaxies. The results of this work[4] were deepened and confirmed by the analysis of the correlation functions on the redshift catalogues[11,13].

The controversial statistical relevance of filaments, sheets and voids in the supercluster morphology and the great debate about different galaxy origin scenarios (some of them seem to produce in a natural way the previous structures[14,15]) stressed the exigence of an alternative and complementary statistics, more sensitive to the topological properties of the galaxy clustering. Among the new proposed approaches, the most important one is the sample percolation test[16] derived from the percolation theory of solid state physics.

2.3. The correlation functions

The galaxy two-point spatial correlation function $\xi(r_{12})$ is defined by the relation

$$\delta P = \bar{n}^2 \left[1+\xi(r_{12})\right] \delta V_1 \delta V_2 \qquad (1)$$

where δP is the joint probability of finding galaxies in volumes δV_1 and δV_2 separated by a distance r_{12}, $\bar{n}$ being the average number density of galaxies. In an analogous way one defines the higher order functions: the three point ζ_{123}, the four-point η_{1234},... correlation functions:

$$\delta P = \bar{n}^3 (1+\xi_{12}+\xi_{23}+\xi_{31}+\zeta_{123})\delta V_1 \delta V_2 \delta V_3 \qquad (2)$$

$$\delta P = \bar{n}^4 (1+\xi_{12}+\xi_{23}+\xi_{34}+\xi_{41}+\xi_{13}+\xi_{23}+\xi_{12}\xi_{34}$$
$$+ \xi_{13}\xi_{24} + \xi_{14}\xi_{23}+\zeta_{123}+\zeta_{124}+\zeta_{234}+\eta_{1234})$$
$$\delta V_1 \delta V_2 \delta V_3 \delta V_4, \qquad (3)$$

....

where $\xi_{ij} \equiv \xi(r_{ij}), \zeta_{ijk} \equiv \zeta(r_{ij},r_{jk},r_{ki}), \ldots$

The observed galaxy two-point correlation function is well approximated by the relation

$$\xi(r) \simeq (r/r_o)^{-\gamma} \qquad (4)$$

in the range $0.1 h^{-1} Mpc < r < 10 h^{-1} Mpc$, with an index $\gamma \simeq 1.8$ and a galaxy correlation length $r_o \simeq 5.5 h^{-1} Mpc$; then $\xi(r)$ becomes negative at $r \gtrsim 30 h^{-1} Mpc$. The data of the three and four point functions are rather well approximated by a suitable symmetric sum of products of two-point functions (the so-called hierarchical form)

$$\zeta_{123} \simeq Q(\xi_{12}\xi_{13}+\xi_{21}\xi_{23}+\xi_{31}\xi_{32}) \qquad (5)$$

$$\eta_{1234} \simeq R_a \left[\xi_{12}\xi_{23}\xi_{34}+cycl.(12 \text{ terms})\right] +$$

$$+R_b\left[\xi_{12}\xi_{13}\xi_{14}+\text{cycl.(4 terms)}\right], \quad (6)$$

in the same range given above, with $Q\cong 1$, $R_a=2.5$ and $R_b\cong 4.3$. A critical analysis of the theoretical implications of the hierarchical form is found in Peebles[4] and Fry[17].

The rich clusters correlate approximately with the same law (4), but with a greater amplitude[18,11,13]: in the range $7h^{-1}$Mpc$<r<150h^{-1}$Mpc the rich cluster two-point correlation function is given by

$$\xi_c(r) \cong (r/r_c)^{-\gamma_c} \quad (7)$$

with a cluster correlation length $r_c \cong 25h^{-1}$Mpc and $\gamma_c \cong 1.8$. There is a great debate about this result[19,20], which implies that galaxies and clusters cannot be both good tracers of the mass distribution: probably the constant slope of galaxy and cluster two-point functions indicates that the galaxy and cluster origin is due to a unique scale-invariant process, the different correlation amplitudes being due to different subsequent processes. Similar scale-free processes are known as fractals[21]: the fractal dimension D corresponding to (4) and (7) is

$$D = 3 - \gamma \cong 1.2 \quad (8)$$

(evidently at scales $r>10^2$Mpc, $D\cong 3$). At present the scale-free processes proposed are[19]: (a) galaxy formation occurs at peaks of the matter density field starting from a primordial scale--free perturbation spectrum ("biased" galaxy formation[22]); (b) galaxies originate from primordial fluctuations created by cosmic strings (see Vilenkin[23]); (c) galaxy formation is driven by percolated explosions of primordial seeds[24,20]. Some geometrical models have been proposed to give the above fractal dimension (power-law cluster model[25]; continuous self-similar hierarchy model[26]; filamentary model[27]).

2.4. The percolation analysis

Given a sample with N galaxies in a cubic volume L^3, one puts a sphere of radius $r=\tilde{r}\bar{d}$ around each galaxy ($\bar{d}=L/N^{1/3}$ is the mean separation between neighbours). The percolation length $L(\tilde{r})\leq L$ is the maximum distance between two galaxies which one can travel remaining in the spheres; when $L(\tilde{r})=L$ the sample is percolated and the corresponding dimensionless quantity $p=\tilde{r}(L)$ is the percolation parameter. A uniform galaxy distribution is characterized by $p=1$; a Poisson distribution with $N>>10^3$ gives[28] $p=0.86$; for an easily percolated distribution (like continuous filaments or sheets) one gets $p<0.86$, while for an uneasily percolated distribution (like isolated clumps) $p>0.86$. The percolation analysis of some toy model distributions is given by Dekel and West[29], who also point out some difficulties of the test. At present only the percolation analysis of the neighborhood of the Local Supercluster was performed[30]: one found $p\cong 0.65$.

3. THE ORIGIN OF THE COSMIC STRUCTURES

3.1. A sketch of the hot big-bang model

The classical hot big-bang cosmology provides a satisfactory description of the evolution of the universe from about the Planck time (when the temperature $T=10^{19}$GeV) up to now (when $T=3°k$) approximately 15 bellions years later[31]. The observative cornerstones of the model are: (a) the isotropic Hubble expansion; (b) the 3°k CBR; (c) the observed abundances of light elements (there is a "miracolous" concordance between the observed and the model predicted abundances). The ultimate theoretical assumptions of the model are General Relativity and the overall isotropy and homogeneity of the hot early universe. Nevertheless for long time several problems plagued the classical model[6]: (a) the baryon asymmetry origin (now observed in the baryon/photon ratio $n_b/n_\gamma \sim 10^{-9}$); (b) the origin of the primordial density fluctuation spectrum (responsible for the observed cosmic structures); (c) the "excessive" flatness of the present universe (roughly, 15 billion years after the Planck era, there is still an innatural balance between expansion and gravitational attraction, implied by the fact that the present density parameter $\Omega_o \equiv 8\pi G \rho_o / 3H_o^2 \cong 1$, $\rho_o \cong 10^{-29} g/cm^3$ being the total present density of the universe); (d) the horizon problem ("ad hoc" assumption of the primordial homogeneity). All these problems are solved in a self-consistent even though, up to now, qualitative way by the application of the "new physics" to the cosmology[6], which yields the baryogenesis[32] and the inflationary[33] scenarios. In particular the "new" early cosmology implies that: (a) Ω_o is very close to one and then, in accordance with some evidences[14], there is a great amount of non clustered "dark" matter (the observed clustering corresponds to $\Omega_{cl} \cong 0.1 \div 0.3$) of non baryonic type (the light elements nucleosynthesis implies for the baryons $\Omega_b \lesssim 0.1$): the non-baryonic component is probably composed by relic weakly interacting massive particle (WIMP's)[14,15]; (b) the primordial perturbation spectrum is adiabatic (the density perturbations are approximately the same for all components: baryons, photons, WIMP's,...) and scale free (the rms gravitational potential depth fluctuations are scale-invariant). Such primordial spectrum with a suitable strength was proposed[34] ten years before the GUT's and the inflationary "revolution" in order to obtain the less unsatisfactory galaxy origin scenarios. Unfortunately up to now the proposed inflationary models predict the requested strength only with some fine tuning of the physics underlying inflation[6,7].

3.2. The galaxy origin scenarios

From a set of initial data, as suggested by the early cosmology, one can construct a detailed scenario of the evolution of the perturbations; the approach is essentially hydrodinamical during the linear phase, when matter and radiation are still coupled, while N-body simulations are often used during the non linear final phases. The theoretical results must finally be compared with the observed properties of the universe; the main observational constraints are: (a) from quasars observations one desumes[35] that galaxies existed at $z \gtrsim 3$; (b) the observed isotropy of the CBR; (c) the clustering properties of the universe; (d) the individual properties of typical galaxies and clusters.

The features of the galaxy origin scenarios depend on the WIMP's nature. At present two main scenarios exist: the "hot" and the "cold" dark matter scenarios[14,15]. In the hot scenario WIMP's decouple "recently" from other particles of the universe, so that at present their average momentum $\bar{p}_o \cong 3°k$ and their number density $n_o \cong n_\gamma$. In such a scenario free streaming destroys any fluctuation smaller than the typical cluster size, so that the first objects, with mass $M \simeq 10^{15} M_\odot$, fragments into galaxies during their initial collapse ("top-down" scenario). A WIMP candidate would be a massive neutrino ($m_\nu \simeq 30 eV$). The cold WIMP's interact more weakly and decouple more early, so that now $p_o \ll 3°k$ and $n_o < n_\gamma$ (but with WIMP mass $m_w > m_\nu$ in order to have $\Omega_o \simeq 1$). In this case free streaming erases fluctuations of no cosmological interest: the first objects of mass about $10^6 M_\odot$ rapidly aggregate into larger structures in a hierarchical way ("bottom-top" scenario). Photinos or axions could be cold WIMP candidates[15,36].

All WIMP's scenario, with some minor strictions on theory parameters, satisfy the CBR isotropy constraint[37]. At present hot models are in trouble because N-body simulations show that galaxies form too late and the clustering is too clumpy [38,14]. The cold scenarios are more promising, even if recent N-body simulations[39] point out some minor difficulties. Other WIMP's models have also been recently proposed[19] to face more involved difficulties of the galaxy origin theory. Such theory, in the modern cosmology, represents the more important and open problem which hopefully will be solved by the interconnected future progresses in cosmology and in particle physics, both theoretical and experimental.

REFERENCES

1. F. Zwicky, E. Herzog, P. Wild, M. Karpowicz and C.T. Kowal, Catalogue of Galaxies and Clusters of Galaxies (Caltech, Pasadena, 1961-68).

2. C.D. Shane and C.A. Wirtanen, Astron. J. 59 (1954) 647; idem, Publ. Lick Obs. 22 (1967).

3. K. Rudniki, T.K. Dworak, P. Flin, B. Baranowski and A. Sendrakowski, Acta Cosmologica 1 (1973) 7.

4. P.J.E. Peebles, The Large-Scale Structure of the Universe (Princeton Univ. Press., Princeton, 1980).

5. J.P. Huchra, M. Davis, D.W. Latham and L.J. Tonry, Astrophys. J. Suppl. 53 (1983) 89.

6. The Very Early Universe, eds. G. Gibbons, S. Hawking and S. Siklov (Cambridge Univ. Press., Cambridge, 1983).

7. Large-Scale Structure of the Universe, Cosmology and Fundamental Physics, eds. G. Setti and L. Van Hove (ESO and CERN, Geneva, 1984).

8. G.O. Abell, Astrophys. J. Suppl. 3 (1958) 211.

9. J.M. Oort, Ann. Rev. Astron. Astrophys. 21 (1983) 373.

10. R.P. Kirshner, A. Oemler, A. Schechter and P.L. Schectman, Astrophys. J. 248 (1981) L57.

11. N.A. Bahcall and R.M. Soneira, Astrophys. J. 270 (1983) 20.

12. J.M. Uson and D.T. Wilkinson, Astrophys. J. 277 (1984) L1.

13. M. Davis and P.J.E. Peebles, Astrophys. J. 267 (1983) 465; T. Shanks, A.J. Bean, G. Efstathiou, R.S. Ellis, R. Fong and B.A. Peterson, Astrophys. J. 274 (1983) 529, A.A. Klypin and A.T. Kopylov, Soviet A.J. Lett. 9 (1983) 41.

14. J.R. Primack, Lectures of the Int. School of Phys. "E. Fermi" Varenna 1984, in print.

15. C.H. Hogan, N. Kaiser, M.S. Turner, N. Vittorio and S.M. White, The formation of structure in the universe, Fermilab-Conf. 85/57-A (1985).

16. Ya. B. Zeldovich, J. Einasto and S. Shandarin, Nature 300 (1982) 407.

17. J.N. Fry, Astrophys. J. 279 (1984) 499.

18. M.G. Hauser and P.J.E. Peebles, Astrophys. J. 185 (1973) 757.

19. D.H. Schramm, Dark matter, tooth fairies, fractals and strings, Fermilab-Conf. 85/42 A (1985).

20. A. Szalay and D.M. Schramm, Are galaxies more strongly correlated than clusters?, Fermilab-Pub - 85/24 A (1985).

21. B. Mandelbrot, The fractal Geometry of Nature (Freeman, S. Francisco, 1982).

22. N. Kaiser, Astrophys. J. 284 (1984) L9.

23. A. Vilenkin, Phys. Reports 121 (1985) 263.

24. B.J. Carr, J.R. Bond and D. Arnett, Astrophys. J. 277 (1984) 445.

25. P.J.E. Peebles, Astron. Astrophys. 32 (1974) 197; J. Mc Clelland and J. Silk, Astrophys. J. 217 (1977) 331.

26. R.B. Maldelbrot, C.R. Acad. Sci., Paris, 280 A (1975) 1551; R.M. Soneira and P.J.E. Peebles, Astron. J. 83 (1978) 845.

27. J.N. Fry, Astrophys. J. 270 (1983) L31.

28. B.I. Shklovski and A.L. Efros, Electronic Properties of Alloyed Semiconductors (Nauka, Moscow, 1979).

29. A. Dekel and M.J. West, Astrophys. J. 228 (1985) 411.

30. J. Einasto, A.A. Klypin, E. Saar and S.F. Shandarin, Mon. Not. Roy. Astron. Soc. 206 (1984) 529.

31. Ya. B. Zeldovich and I.D. Novikov, Relativistic Astrophys, vol. II (The Univ. of Chicago Press, Chicago, 1983).

32. E. Kolb and M. Turner, Ann. Rev. Nucl. and Part. Sci. 33 (1983) 645.

33. A.D. Linde, Rep. Prog. Phys. 47 (1984) 925.

34. Ya. B. Zeldovich, Mon. Not. Roy. Astron. Soc. 160 (1972) 1P.

35. S.M. Faber, Galaxy formation and cosmology, in ref. 8 pp. 187-203.

36. D.N. Schramm, Nucl. Phys. B 252 (1985) 53.

37. S.A. Bonometto, F. Lucchin and R. Valdarnini, Astron. Astrophys. 140 (1984) L27; N. Vittorio and J. Silk, Astrophys. J. 285 (1984) L39; J.R. Bond and G. Efstathiou, Astrophys. J. 285 (1984) L44.

38. C. Frenk, S.D.M. White and M. Davis, Astrophys. J. 271 (1983) 417.

39. M. Davis, G. Efstathiou, C.S. Frenck and S.M. White, Astrophys. J. 293 (1985) 371.

FRACTALS IN PHYSICS
L. Pietronero, E. Tosatti (editors)
© Elsevier Science Publishers B.V., 1986

STOCHASTIC APPROACH TO LARGE SCALE CLUSTERING OF MATTER IN THE UNIVERSE

> *"When a young man in my laboratory uses the word 'universe' I tell him it is time for him to leave".*
> *E. Rutherford* [+]

L. PIETRONERO and R. KUPERS

University of Groningen, Melkweg 1, 9718 EP Groningen, The Netherlands

We formulate a stochastic model that, using suitable assumptions for the merging probabilities, is able to generate self-similar clustering at all scales. This implies that a system with an initial distribution that is homogeneous and random can spontaneously envolve into a fractal distribution with correlation functions described by power laws. This fact may provide a new point of view for the origin of large scale clustering of matter in the universe.

1. INTRODUCTION

The distribution of matter in the universe appears strongly clustered at all length scales, from the galaxy scale up to the present limits of observation[1] (For an up to date discussion see the preceding paper by Lucchin[2]). This implies that the definition of an average density is essentially dependent on the particular volume considered and indicates rather a hierarchical distribution. The observed density correlation function can be well described by a power law relation

$$C(r) = <\rho(r_0)\rho(r_0+r)>_0 \simeq r^{-\gamma} \quad (1)$$

with $\gamma \simeq 1.8$. Since for a fractal[3] we have $C(r) \sim r^{-(d-D)}$ we can interpret Eq.(1) as evidence for a fractal distribution with fractal dimension $D \simeq 1.2$ ($d=3$ is the euclidean dimension of the space). To avoid confusion we note that $\rho(r)$ does not refer to a single element of the system as usually done for the analysis of Cluster-Cluster aggregation models[4,5], but it describes the distribution in space of all the elements of the system considered. In general in the literature the discussion is based on the so-called dimensionless two-point correlation function $\xi(r)$ that is linked to $C(r)$ by the relation[6]

$$\xi(r) = \frac{C(r)}{<\rho>^2} - 1 \quad . \quad (2)$$

This function describes the deviations from the average density $<\rho>$ and for $\xi>>1$ it essentially coincides with $C(r)$. Since for hierarchical distributions the definition of an average density $<\rho>$ depends on the particular volume considered, we prefer to use directly the density correlation function $C(r)$.

This power law behavior, extending over more than three decades in length scale has been extensively studied in the past few years[2]. At the moment, however, there is no satisfactory theoretical explanation[7,8]. Often such a power law is simply assumed in order to compute from it other properties[6-8]. Concerning its origin, however, the most popular view arising from n-body simulations is that these large density fluctuations arise from the amplification of primeval scale-free fluctuations[9-11].

[+] From a citation of J.A. Wheeler, Physics Today, October 1985, p. 66.

The amplification is related to the competition between expansion and gravitational attraction. In this way the question of the origin of these fluctuations is shifted back in time to the properties of the early universe[11]. A basic problem of this approach is that the isotropy of the $3^\circ K$ background radiation imposes severe limits to the size of primeval fluctuations.

The n-body calculations consist of molecular dynamics simulations that include explicitly the expansion and all gravitational interactions. If initially random distributions are used, these calculations do not lead to self-similar hierarchical structures[7,9]. With respect to the real universe they present severe limitations, both in space and time. A single galaxy consists in fact of about 10^{10} stars while the n-body calculations deal with a few thousand particles and the average object generated contains about 30 initial particles[9]. It is clear therefore that there may be a serious scale problem because certain fluctuations may have not yet fully developed on such small scales.

In order to investigate this possibility we consider here simplified stochastic models in which gravity is included only via the probability that, once two objects (galaxies) are close enough, they aggregate irreversibly (merging). No effect of gravity on the trajectories is considered. These drastic simplifications give the advantage that, by choosing appropriate rules for these probabilities, one may actually reproduce the effective behavior of systems of larger scale. In particular this approach allows to control and study in some detail the asymptotic development of fluctuations. This approach is similar in spirit to other stochastic models reported in this volume and aimed at the description of systems whose structure is dominated by fully developed fluctuations. We refer in particular to percolation[12], dielectric breakdown[13], dendritic growth[14], Cluster-Cluster aggregation[5] etc. In analogy with these phenomena, we consider the possibility that scale-invariant fluctuations may be spontaneously generated by a particular dynamical mechanism. In fact we are going to see that a simple model that mimics the effect of the dynamical friction[15] on the aggregation process has indeed such a property: a homogeneous random initial distribution evolves spontaneously into a fractal distribution without including the expansion. The new picture suggested by this result is briefly discussed in Sect. 4.

2. SIMPLE AGGREGATION MODELS

We consider here the evolution of simple aggregation models that start from initially random distributions like the one shown in Fig. 1.

FIGURE 1
Example of a random distribution of 750 points on a two dimensional grid (50x50). The initial configurations we adopt are always of this random type.

All the results reported here refer to a two dimensional grid with periodic boundary conditions but most of these results have also been confirmed by a more realistic dynamics without lattice[16]. At the beginning N particles (typically N∼1000) of equal masses are randomly distributed over the grid (Fig. 1). In addition each particle is assigned a random direction in which it will move. The particles all move one step at a time along linear trajectories. When two of them collide, there is a probability p_a that they aggregate irreversibly and a probability $p_s=1-p_a$ that they scatter. After a collision the outcoming single particle with double mass, or the two initial particles, move again with random directions. This model resembles the Cluster-Cluster aggregation models[4,5] but there are actually two major differences: (i) The particles move along linear trajectories (instead of random walks) between any two encounters; a change of direction can only occur after the encounter with another particle. This provides a relation between space distribution and dynamical properties that is not present in the Cluster-Cluster models. (ii) We neglect the internal structure of single aggregates that are treated as points and consider only their global distribution in space.

Gravity is simulated only through the aggregation probability p_a that is made dependent on the masses of the incoming particles. Typically we have used functions of type

$$p_a \propto (M_i \cdot M_j)^\alpha \qquad (3)$$

with the exponent α ranging from 0 to 2. As a crude approximation to momentum conservation due to the randomness of velocities, particles from a certain mass up are simply stopped. In the simulations presented here, we have stopped all masses which have grown to three initial mass units. This class of models gives rise to distributions of the type shown in Fig. 2 with aggregation of the initial particles into larger masses but without a real clustering. A correlation function analysis of Fig. 2 shows in fact that the system is homogeneous after a certain length-scale and therefore it is not a fractal.

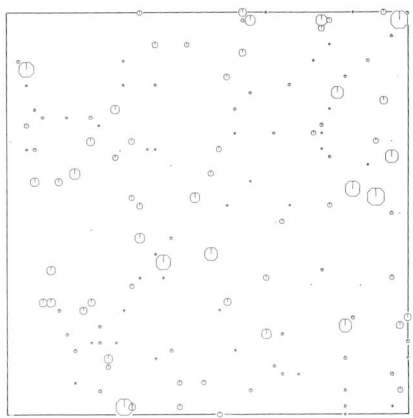

FIGURE 2
Typical final distribution for the class of models described in Sect. 2. The radius of particles gives a measure of their mass. This distribution does not develop hierarchical clustering and it results homogeneous over large length scales.

The reason of this homogeneity is that the probability that a large mass is created at a certain point is independent of its environment. In the next section we describe a mechanism that can provide instead such a dependence.

3. EFFECT OF ENERGY EXCHANGE ON CLUSTERING

Consider two mass points approaching each other from large distance under the effect of their classical potential. If there is no way to dissipate some of their kinetic energy, these two particles will not be able to create a bound state. In most cases this energy dissipation is due to short range scattering with other particles. On cosmic scales, however, the cross section for short range two-body collisions can become extremely small and another mechanism appears to be more important. The thermalization of the orbits of stars into a galaxy is in fact due to the fluctuations of the gravitational field rather than to two-body collisions[17]. This effect suggests that an encounter occurring in the vicinity of other masses (point A in Fig. 3) is more likely to give rise to aggregation than an encounter that occurs in an isolated region (point B in Fig. 3).

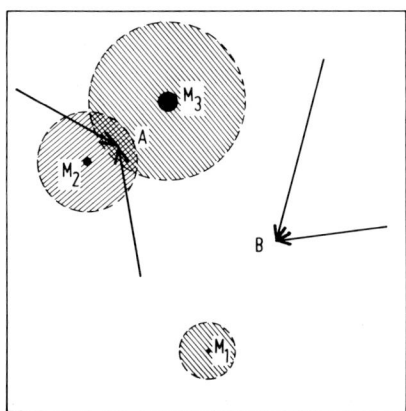

FIGURE 3
Schematic view of how the dynamical friction effect due to the presence of other masses affects the probability of creating a bound state. The encounter B occurs in an empty region of space. The encounter A instead can profit of the presence of M_2 and M_3 in the vicinity to dissipate energy and more likely will lead to a bound state.

This is because the encounter in A has a larger probability to dissipate energy via the neighbouring particles than the one in B. This assumption corresponds to assigning a larger probability to merging when this occurs in a populated region. This hypothesis can be tested by detailed studies of the merging process in various configurations. In order to take into account this effect in our model, we introduce for each mass M_i at position $\vec{r}_i$ a function $f_i(M_i,\vec{r}_i)$ that gives a measure of the influence of this mass on the merging probability. Since there are indications for effects of such nature to be local phenomena[18], we can use functions with a sharp decay as the step function schematically illustrated in Fig. 3 or exponential functions of the type

$$f_i(\vec{r}) = A \exp\left\{-\frac{[\vec{r}-\vec{r}_i]}{M_i^\beta}\right\} . \qquad (4)$$

The total influence of a particular mass distribution is then given by the function $F(\vec{r})=\sum_i f_i(\vec{r})$. In the absence of a detailed theory of this effect, we have related this function to the aggregation probability via a power law

$$p_a(\vec{r}) \sim [F(\vec{r})]^\delta . \qquad (5)$$

It is clear that now the presence of other masses will influence the probability to generate a new large mass at a particular point. This new model gives rise to distributions of the type shown in Fig. 4 with various

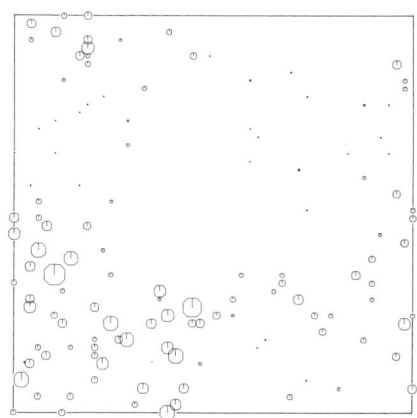

FIGURE 4
Typical final distribution for the stochastic model that includes also the effect of the dynamical friction as described in Sect. 3. We can observe here clustering at all scales and the development of large voids. This distribution is indeed a fractal.

combinations of the parameters β and γ. Typically we have used β= ½ and 1 and δ=1, 2 and 3; more details will be reported elsewhere[19]. This distribution shows development of clustering at all scales and the presence of large voids. Its analysis in terms of the density correlation function C(r) shows the spontaneous development of a fractal distribution starting from a homogeneous one. The value of the fractal dimension D as a function of time is shown in Fig. 5.

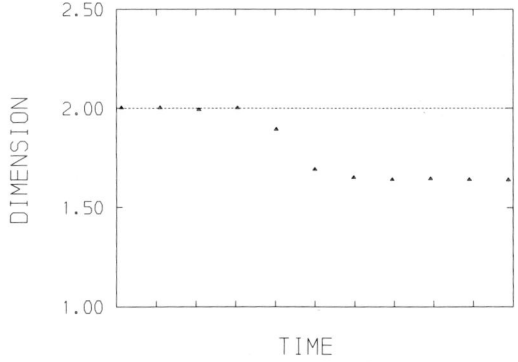

FIGURE 5
Fractal dimension D at various times for the model with dynamical friction whose final distribution is shown in Fig. 4. The fractal dimension is obtained from the analysis of the density correlation function (Eq.(1)).

The asymptotic value of D depends on the particular assumptions used in the model.

We have thus identified a dynamical process that is able to generate spontaneously a fractal distribution starting from a homogeneous one without expansion of the system. One may reasonably expect that the inclusion of the expansion is not going to change this qualitative behavior. This shows that the expansion may not necessarily be an essential feature for the development of a fractal distribution.

4. CONCLUSIONS

The results of the stochastic models we have described, suggest the following description: initially we start with a dense homogeneous system in agreement with the isotropy of the 3°K radiation. Due to the high density, two-body scattering is the main energy loss mechanism and aggregation develops homogeneously as in Fig. 2. Later, due to expansion and aggregation, the two-body cross section is strongly reduced and energy exchange via gravitational fluctuations becomes dominant for energy loss. At this point the system develops large non-homogeneities and spontaneously evolves into a fractal distribution as in Fig. 4 without the need of initial fluctuations of special type. This new picture could in principle resolve a number of long standing problems in this field[6-8,11] but, of course, more work is needed to validate or reject it.

ACKNOWLEDGEMENTS

It is a pleasure to thank S. Taraglio, M. Stiavelli, R. Ruffini, R. Sancisi, C. Evertsz, J. Sellwood and T.S. van Albada, for stimulating discussions.

REFERENCES

1. G. de Vaucouleurs, Science 167 (1970) 1203.

2. F. Lucchin, this volume.

3. B. Mandelbrot, The Fractal Geometry of Nature (W.H. Freeman & Co., New York, 1983).

4. R.M. Ziff in: Kinetics of Aggregation and Gelation, eds. F. Family and P. Landau (Elsevier Publisher, 1984) p. 191.

5. R. Botet, R. Jullien and M. Kolb, this volume.

6. P.J.E. Peebles, The Large-Scale Structure of the Universe (Princeton Univ. Press, 1980).

7. M. Davis and P.J.E. Peebles, Astrophys. J. 267 (1983) 465.

8. F.R. Bouchet and R. Pellat, Astron. Astrophys. 141 (1984) 77.

9. G. Efstathiou and J.W. Eastwood, Mon. Not. R. astr. Soc. 194 (1981) 503.

10. C.S. Frenk, S.D.M. White and M. Davis, Astrophys. J. 271 (1983) 417.
J. Barnes, A. Dekel, G. Efstathiou and C.S. Frenk, Astrophys. J. 295 (1985) 368.
M. Davis, G. Efstathiou, C.S. Frenk and S.D.M. White, Astrophys. J. 292 (1985) 371.

11. G.W. Gibbons, S.W. Hawking and S.T.C. Siklos eds.: The Very Early Universe (Cambridge Univ. Press, 1983).

12. See for example H.E. Stanley and A. Coniglio, this volume.

13. H.J. Wiesmann and L. Pietronero, this volume.

14. See for example P. Meakin, this volume.

15. S. Chandrasekhar, Principles of Stellar Dynamics (Univ. of Chicago Press, 1947).

16. S. Taraglio, thesis (Univ. of Rome, 1985), unpublished.

17. S. Tremaine in: The Structure and Evolution of Normal Galaxies, eds. S.M. Fall and D. Lynden-Bell (Cambridge Univ. Press, 1980) p. 67.

18. S. Tremaine and M.D. Weinberg, Mon. Not. R. astr. Soc. 209 (1984) 729.

19. L. Pietronero and R. Kupers, to be published.

Part VII
DYNAMICAL PROPERTIES OF FRACTAL STRUCTURES

FRACTAL SURFACES AND THE DE GENNES TERMITE MODEL FOR A TWO-COMPONENT RANDOM MATERIAL

H. Eugene STANLEY

Center for Polymer Studies and Department of Physics, Boston University, Boston, Massachusetts 02215, USA

This will be an elementary talk, beginning with a review of the random mixture problem. We will then describe a new way of thinking about this problem: a random walk with two different time scales corresponding to the good and bad conducting regions. The limiting case in which the ratio of the two time scales diverges corresponds to the random superconducting network in which one component is superconducting and the other is normal (de Gennes calls this the termite limit, since the motion of the random walker resembles that of a termite). Finally, we briefly discuss a recent finding that an infinite hierarchy of exponents is needed to adequately characterize the fractal surface of diffusion-limited aggregates.

Before beginning, we note that this talk is based on two recent research projects. The first project, on the "termite" model of electrical conductivity of a two-component random material, was done in collaboration with Armin Bunde, Antonio Coniglio and Daniel Hong (for full details, see Coniglio and Stanley 1984; Bunde, Coniglio, Hong and Stanley 1985; Hong, Stanley, Coniglio and Bunde 1985). This project taught us that the controlling physics underlying the conductivity requires that we distinguish between the screened portions or "invaginations" of the cluster and the unscreened portions or "tips." This led naturally to a second project on fractal surfaces, done in collaboration with Antonio Coniglio, Paul Meakin, and Tom Witten (for details, see Coniglio and Stanley 1984; Meakin, Stanley, Coniglio and Witten 1985, Meakin et al 1986). We benefitted from a rather large number of interactions with others, including Joan Adler, Amnon Aharony, Shlomo Alexander, Pierre-Gilles de Gennes, E. Guyon, Francois Leyvraz, and Dietrich Stauffer. We thank ONR and NSF for financial support.

THE TERMITE PROBLEM

How are the fundamental laws of diffusion and transport modified when the medium in question is a random "AB-mixture" of good and poor conducting regions? This question has received a considerable degree of recent attention for two limiting cases: (i) The random resistor network (RRN)–or pure "ant" limit–for which B, the poor conducting species, has zero conductance, and (ii) The random superconducting network (RSN) or pure "termite" limit, for which A, the good conducting species, has infinite conductance.

The terms "ant" and "termite" arise from the fact that one can replace the conductivity problem with a diffusion problem using the Nernst-Einstein relation. For the RRN limit, no diffusion can occur on the component with zero conductance, so the constrained diffusion problem is rather like an "ant in a labyrinth" (de Gennes 1976). For the RSN limit, the diffusion can occur everywhere since both components conduct, but the fact that the good conductor species has zero resistance means that the diffusion is remarkably different in this region than elsewhere. Some years ago de Gennes (1980) invented the term "termite diffusion" to describe this subtle phenomenon. However to this date there has been no clear statement of exactly how to properly define or measure this phenomenon, in contrast to the "ant" limit where the diffusion is simply constrained to one component. There are many reasons for the current upsurge of interest in this problem.

(i) One reason is that there are many experimental systems that are random and inhomogeneous. For example, a rock is composed of tiny grains of different conductivities (to heat, to fluid flow, to electricity). To the extent that such inhomogeneous materials are also random, we may think of using a site-random description of this material: a "lattice-gas" description. One first coarse grains the material and then assigns to each cell one of two conductivities, σ_a and σ_b. Calculations based upon such a straightforward approach have been usefully compared with a wide range of experiments, from conductivities of thin films of lead depositions on an insulating substrate (roughly the RRN limit) to thin films of superconducting material vacuum deposited on a normal substrate (roughly the RSN limit). Moreover, ionic conductors mixed with a dispersed insulating phase represent random heterogeneous materials, where both limits seem to play an important role.

(ii) A second reason is related, perhaps, to the reason why the Ising model has always been of great interest: it is an extremely simple model that captures the essential physics of a realistic system in nature. The analog of the Ising model for random inhomogeneous materials is a mixture of sites (or bonds) randomly distributed on a lattice. The sites (or bonds) are assumed for simplicity to have only two possible values of the conductance,

$$\sigma = \begin{cases} \sigma_a & [\text{probability } p] \\ \sigma_b & [\text{probability } 1-p]. \end{cases} \quad (1)$$

By convention, we choose $\sigma_a > \sigma_b$, so that the ratio $h = \sigma_b/\sigma_a$ is always less than unity.

Conventionally, one wants to know the *macroscopic* magnetization of an Ising ferromagnet composed of elements (spins) whose *microscopic* property is a two-valued variable. Similarly, we *now* want to know the *macroscopic* conductivity which depends on all possible configurations of the *microscopic* elements (conductors) whose property is again a two-valued quantity (σ_a and σ_b). Just as the magnetization couples to a conjugate field H, the conductivity couples to a conjugate field h.

The two limiting cases mentioned above can now be discussed more precisely: (a) In the RRN limit, the large conductance is set to unity and the small conductance is set to zero. As the percolation threshold p_c is approached from above, the macroscopic conductivity approaches zero with a critical exponent μ,

$$\sum \sim (p - p_c)^\mu. \quad (2a)$$

(b) In the RSN limit, the small conductance is set to unity, and the large conductance is infinite. As the percolation threshold is approached from below, the conductivity diverges to infinity with an exponent $-s$

$$\sum \sim (p_c - p)^{-s}. \quad (2b)$$

The traditional approach to the RRN limit has been to replace Kirchhoff's laws by an equivalent diffusion problem, where the macroscopic conductivity is related to the diffusion constant D by the Nernst-Einstein relation,

$$\sum \sim nD, \quad (3)$$

where n is the density of the charge carriers.

We place a walker on a d-dimensional lattice made of two kinds of bonds, A and B (for illustration: $d = 1$ here, the general-d case is discussed in Hong et al 1985). The walker carries two coins, weighted and unweighted, and a clock. Without loss of generality, let the origin be well inside a high-conductivity A region. At each tick of the clock, the walker tosses the unweighted coin and moves to the left or right depending on the outcome of the coin toss. When the walker comes to a site on the boundary between the A region and the B region, he tosses the other coin that is weighted with probability

$$P_a = f_a/(f_a + f_b) = 1/(1 + h), \quad (4)$$

to stay in the A region, and a probability

$$P_b = f_b/(f_a + f_b) = h/(1 + h), \quad (5)$$

to go outside into the B region. In the event that the walker steps outside the A region, then he must *slow down* by the ratio $f_a/f_b \equiv h^{-1}$. For example, if the conductivity of the B region is 10 times smaller than that of the A region, then f_b is 10 times smaller than f_a ($h = 0.1$) and the walker steps *only* after every 10 ticks of his clock.

Limiting cases of our random walk model are as follows:

(i) $h = 1$. There is no distinction between regions, no reflection on the boundaries ($P_a = P_b$), and no difference in walk speed on and off the A clusters.

(ii) $h \ll 1$. The walker now moves at one step per clock tick when he is on an A cluster, and is almost always reflected when he comes to the boundary. Exremely rarely he passes out of an A region and into a B region, whereupon he walks much, much slower—taking a new step only after his clock has made h^{-1} ticks. Statistically speaking, in a very large time $\gg h^{-1}$, the walker performs $O(f_a)$ moves in the A region and $O(f_b)$ moves in the B region.

Suppose we make a motion picture of the walker's motion. Then we see that the walker is reflected from the walls almost all of the time, and only very rarely—roughly once per h^{-1} trials—will come outside the cluster (see Hong et al 1985). When this does occur, his motion will slow down by a factor of h. If we watch this motion picture, perhaps we become impatient watching the walker in the B region and we speed up the motion picture projector by a factor of $1/h$ so that the walker is now taking *one* step per unit of time while in the B region. Then we are no longer impatient while the walker is in the B region. However, when he finally encounters an A cluster, he moves onto it with a high probability, $1/(1+h)$, and proceeds to move about the A cluster with a motion that is *also sped up by the same factor* $1/h$. Thus the original normally on an A cluster and *extremely slowly* on B clusters has suddenly been transformed into a "termite" who moves normally on B clusters and *extremely fast* on A clusters. *Indeed, the only difference between the two domains, "ant" (RRN domain) and "termite" (RSN domain), is the definition of the time scale.* This simple observation can be formalized in terms of a rigorous transformation (Hong et al 1985). That transformation in turn forms the basis of the scaling laws for the ant and termite limits of the general two-component random mixture.

Thus the two-component random mixture requires for its treatment the understanding of how to handle a diffusion process to which there are two time scales, not one. This problem has not been treated previously and is proving to be quite subtle in many respects. Until quite recently it was widely believed that the physics governing transport near the RSN or "termite" limit was quite different from the physics governing transport near the RRN or "ant" limit. Now we appreciate that these two regions are related by a simple change of time scale (Hong et al 1985; see also Adler et al 1985, Bunde et al 1985c, Leyvraz et al 1985, Sahimi 1985).

SURFACES, INTERFACES AND SCREENING OF FRACTAL STRUCTURES

The next part of this talk is devoted to the subtle and fascinating subject of disordered surfaces. But what do we mean by "the" surface of a fractal object? In fact, we shall see that there are many different surfaces, depending on the physical process in question (Fig. 1). We shall discuss these roughly in order of increasing subtlety.

External Perimeter ("Hull"): d_h

The total number of *external* surface sites, or "hull,"

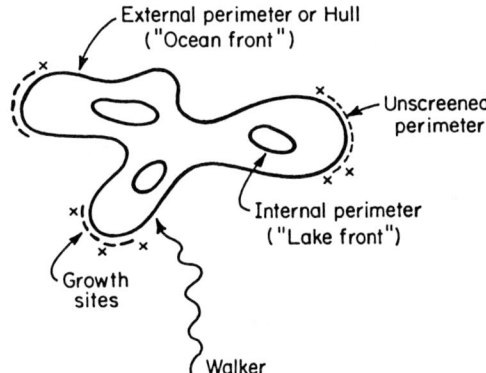

FIG. 1: Schematic illustration of four different fractal surfaces arising in the description of a percolation cluster. (a) The external "oceanfront" perimeter or hull has a fractal dimension d_h. (b) The total perimeter has a fractal dimension d_f, equal to that of the total bulk mass of the cluster. Since $d_f > d_h$, it follows that the internal "lakefront" perimeter must have the same fractal dimension d_f of the total perimeter. (c) The unscreened perimeter where an incoming walker is more likely to hit has fractal dimension d_u (heavy solid lines). (d) The growth sites are those perimeter sites that form the living frontier of the cluster. These have fractal dimension d_g, but the nature of the G-site fractal depends on the actual mechanism of how the percolation cluster grows (see, e.g., Stanley et al 1984; Bunde et al 1985a,b; Herrmann and Stanley 1985).

scales with the caliper diameter or radius of gyration L as

$$N_{\text{hull}} \sim L^{d_{\text{hull}}} \quad [\textit{Fractal Dimension No. 1}]. \quad (6)$$

For $d = 2$ percolation, d_{hull} appears to be about 1.74 ± 0.02 (Sapoval et al 1985), thus motivating the conjecture

$$d_{\text{hull}} = 1 + d_{red} = 7/4, \quad (7)$$

since $d_{red} = 1/\nu = 3/4$ exactly for $d = 2$.

Total Perimeter: d_f

We know that the total number of perimeter sites $N_{perimeter}$ scales in the same fashion as the total number of cluster sites,

$$N_{perimeter} \sim N_f \sim L^{d_f}, \quad [\textit{Fractal Dimension No. 2}], \quad (8)$$

for d-dimensional percolation (Kunz and Souillard 1978) where, for $d = 2$, $d_f = 91/48 = 1.896$. Hence the fact that $d_{\text{hull}} < d_f$ means that the ratio of *external* perimeter sites to *total* perimeter sites approaches zero at the percolation threshold. As the clusters get larger the internal perimeter sites ("lakefront" sites) completely swamp the external perimeter sites ("oceanfront" sites). For $d > 2$, internal perimeter sites are less commonly seen in finite computer simulations since it takes a lot of cluster sites to completely surround a "3-dimensional lake." An open question is the value of d_{hull} for $d > 2$: could it be that $d_{\text{hull}} = d_f$? Work is underway test this possibility.

Unscreened Perimeter: d_u

Coniglio and Stanley (1984) introduced the concept of the "unscreened" perimeter to describe that portion of the hull that is effective in termite motion:

$$N_{\text{unscreened}} \sim L^{d_u} \quad [\textit{Fractal Dimension No. 3}]. \quad (9)$$

Moreover, they showed that the conductivity exponent for the RSN is simply related to d_u. Recalling the Nernst Einstein relation, we have

$$\sum \sim D \sim (R_{\text{cluster}})^2 \tau^{-1}. \quad (10a)$$

Here the jump frequency τ^{-1} scales as the fraction of cluster sites belonging to the unscreened perimeter,

$$\tau^{-1} \sim [N_{\text{unscreened}}/N_f]. \quad (10b)$$

Substituting the Stauffer expression for the mean radius of the finite clusters, R_{cluster}, and the definitions for d_u and d_f into (10), we obtain

$$\sum \sim L^{2-(d-d_f)} L^{d_u - d_f}. \quad (11)$$

Since $\sum \sim \epsilon^{-\tilde{s}}$, we have

$$\tilde{s} = s/\nu = d_u - (d - 2). \quad (12)$$

The conductance between two points scales as $L^{\tilde{\varsigma}_{RSN}}$ where $\tilde{\varsigma}_{RSN} = \tilde{s} + (d - 2)$. From (12) we obtain the extremely simple result that the conductance exponent is identical equal to the fractal dimension of the unscreened perimeter

$$\tilde{\varsigma}_{RSN} = d_u. \quad (13)$$

Let us contrast now the RSN with the RRN, which is well understood. For the RRN, we can also relate the conductance exponent to fractal dimensions characterizing the substrate fractal. The Einstein relation (3) holds, but we must set $n = P_\infty$, the probability that a randomly dropped ant will land on the incipient infinite cluster. Now P_∞ scales as the co-dimension $(d - d_f)$, and D scales as L^2/time. Hence (Gefen et al 1983; Ben-Avraham and Havlin 1982)

$$\sum \sim P_\infty D \sim L^{d_f - d} L^{2 - d_w}. \quad (14)$$

Recalling that $\sum \sim \epsilon^\mu$ for the RRN limit, we have

$$\tilde{\mu} = \mu/\nu = (d - 2) + (d_w - d_f). \quad (15)$$

The conductance between two point scales as $L^{\tilde{\varsigma}_{RRN}}$, where $\tilde{\varsigma}_{RRN} = \tilde{\mu} - (d - 2)$. Hence for the RRN problem, (13) is replaced by

$$\tilde{\varsigma}_{RRN} = d_w - d_f. \quad (16)$$

It is convenient to think of the resistance between two points as the "mass" of 1-ohm resistors that we would place in series between the two points in order to have the same resistance. In this way, we can interpret $\tilde{\varsigma}_{RRN}$ as a proper fractal dimension for some specific fractal object (the set of resistors)

$$N_R \sim L^{d_R} \quad [Fractal\ Dimension\ No.\ 4]. \quad (17)$$

From (13) and (16) it then follows that

$$d_R = \begin{cases} -d_u & [\mathbf{RSN}] \\ d_w - d_f & [\mathbf{RRN}]. \end{cases} \quad (18)$$

From (18) we see that both the RRN and the RSN have resistance exponents d_R that are simply expressed in terms of fractal properties of the substrate. The expressions are completely different, of course, since the mechanism of transport is completely different (Fig. 2). From Fig. 2a, we see that the transport is determined by a "fisherman's net" structure, with a mesh size given by the connectedness length ξ. The strands of the net are made of singly-connected "red" bonds and multiply-connected "blue" bonds, the statistics of which will be described shortly. From Fig. 2b, we see that for the RSN just below p_c transport from one bus bar to the other is determined by the motion of charge carriers from one cluster to another—more precisely *out of* the unscreened perimeter of one cluster and *into* the unscreened perimeter of the next.

Thus the clusters in the RSN limit play the role of the nodes in the RRN limit. As one moves close to p_c in the RRN problem, the critical bonds are the singly-connected "red" bonds (the hottest). As one moves close to p_c in the RSN limit, the critical bonds are those bonds on the lattice which—if occupied—would connect

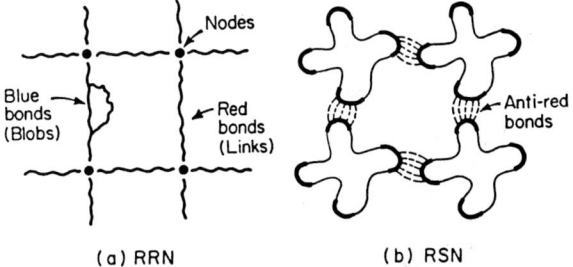

(a) RRN   (b) RSN

FIG. 2: Schematic illustration of the essential features of cluster structure in describing the conductivity of a general two-component random mixture in the limit of (a) the random resistor network, and (b) the random superconducting network. Adapted from Coniglio and Stanley (1984).

two clusters. I have always called these pink (since they are "incipient" red bonds: once occupied they will become red); recently Stauffer suggested the term "anti-red" because in every sense they are the complement of the red bonds. Thus the red bonds control the physics of the RRN problem just above p_c, while the anti-red bonds control the physics of the RSN problem just below p_c. Coniglio (unpublished) has proved that the anti-red bonds have the same fractal dimension as the red bonds,

$$d_{anti-red} = 1/\nu. \quad (19)$$

Thus there is a certain symmetry between the RRN and RSN limits, which in some way should follow directly from the homogeneity theorems mentioned above. Work on this important topic is underway, and perhaps at this meeting some of you can help make progress along these lines.

Can we evaluate the fractal dimensions d_u and d_w appearing in (2) in terms of the fractal dimension d_f of the underlying substrate? Some progress along these lines has been made using arguments that require for their validity certain assumptions. In this section we will review a mean-field type argument (Coniglio and Stanley 1984) that

$$d_u = (d_f - 1) + (d - d_f)/d_w \quad [Coniglio-Stanley]. \quad (20)$$

To this end, we must devise a method of probing the surface of a fractal object. The method we chose (Meakin et al 1985a) was to release random walkers, one at a time. When the random walker touched perimeter site i, a counter on site i was incremented by one unit (N_i becomes $N_i + 1$). After typically a million walkers have been released, statistics were done. Our analysis is based on the idea that only a relatively small fraction of the total perimeter will have a large probability of being contacted. Hence to analyze the distribution function N_i ($i = 1, 2, \ldots, P$—where P is the total number of perimeter sites), we formed the moments μ_j defined through

$$[\mu_j]^j = \frac{N_T^{j+1}}{\sum_i N_i^{j+1}} = \frac{1}{\sum_i P_i^{j+1}} \sim [N_f]^{j\gamma_j}. \quad (21)$$

Here
$$N_T \equiv \sum_i N_i, \quad (22)$$
is the total number of incoming walkers, and
$$P_i = N_i/N_T, \quad (23)$$
is the probability that a given incoming walker will hit site i. The P_i are normalized to unity by virtue of (22).

First we calculated the γ_j for $j = 1-3$, and found that the Coniglio-Stanley mean field relation (20) was satisfied to within the accuracy of the calculations. We did notice a systematic dependence on j, so to test the possibility that the γ_j depend on j we extended the moment calculation to $j = 8$. The hinted dependence from $j = 1-3$ became much clearer (Fig. 3) and so we conclude that there is not a single exponent but rather an entire hierarchy of exponents (Meakin et al 1985a,b). This result has been confirmed by Halsey et al (1985). Why is the Coniglio-Stanley relation wrong? Presumably because it smears out the interface or "active zone" of the fractal into a band, and then assumes that there is an equal probability of capture for all surface sites within this band. In reality, there is a continuous gradation in "temperature," with the outermost tips be-

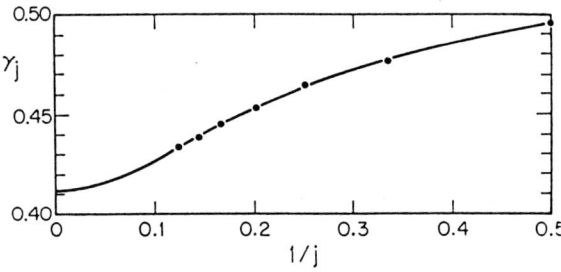

FIG. 3: The exponent γ_j characterizing the behavior of the jth moment of P_i. Here P_i, the basic quantity in surface growth, is the probability that perimeter site i is the next site at which the cluster will grow. In order to see if this apparent hierarchy or "spectrum" of exponents tends in the $j = \infty$ limit toward the expected limit, $1 - 1/d_f$, we have plotted these exponents against $1/j$. From Meakin et al (1985b).

ing immensely hotter and the deepest invaginations being extremely cold. This situation is reminiscent of that found by De Arcangelis et al (1985) for $N(V)$, the distribution of the number of bonds in the backbone across which the voltage drop is V. Here also there is a continuous gradation in temperature from the red bonds (the "hottest" in the sense that the full voltage drop of the entire cluster falls across each red bond) to the very cold bonds arising from the very long loops comprising the blobs.

This discovery of an infinite hierarchy of critical exponents–both in the voltage distribution of the percolation backbone and in DLA–is striking because normally one assumes that two exponents will suffice to describe a critical object. For example, we noted above that $y_h\ (=d_f)$ and $y_T\ (=d_{red})$ were sufficient to describe percolation. However when we "do something" to the fractal, such as put a battery across it or bombard it with random walkers, we introduce a new measure. Instead of each fractal site having weight 1, each site has a weight that depends on what we are doing to the fractal (e.g., each site has a voltage attached to it, or with each site we associate the number of hits on that site). Several groups (Meakin, Stanley, Coniglio and Witten unpublished; Turkevich and Scher 1985; Halsey et al 1985) are currently seeking to understand the meaning of this new measure and what we can learn from this infinite hierarchy of exponents.

In retrospect, we might have anticipated this infinite hierarchy in advance. This is because for two extreme values of j, $j = -1$ and $j = \infty$, exponents differ by more than a factor of two: $\gamma(-1) = 1$ and $\gamma(\infty) = 1 - 1/d_f$. The first result follows immediately from the definition (25) and the fact that the total surface in DLA scales with exponent d_f just like the total mass. The second follows from the recent theorem (Leyvraz 1985) that P_{max} (the maximum value of all the P_i) scales with cluster mass to the exponent $1 - 1/d_f$. This prediction is confirmed by our calculations for $j = 1-8$. Our hierarchy or spectrum of surface fractal dimensions tends clearly toward a number fairly close to the predicted value $1-1/d_f$ (Fig. 3).

FRACTAL GROWTH

How can we characterize the fashion in which a fractal grows? This is the question that we shall address in this final section. It is important to state at the outset an obvious fact: completely different growth mechanisms can lead, eventually, to the same static fractal object (see, e.g., the discussion in Bunde et al 1985a). For now, let us consider one of the simplest kinetic mechanisms of growth, the ant. Instead of dropping the ant onto a pre-formed fractal structure as we did before, we could instead drop the ant onto a Euclidean lattice but give her a set of rules with which she could form the fractal as she moves. This means that the ant would need the four-sided coin that all walkers have (taking a square lattice for now), but she would also need some dynamical mechanism of generating the ultimate static fractal. For percolation fractals, all she needs is a weighted coin so that each site can be blocked with probability $1 - p_c$. The precise mechanics for the ant motion are described in Bunde et al (1985a).

After some time has evolved, the ant is moving around in a rather interesting region of space that is characterized by 3 sorts of sites (the terminology in brackets suggests an epidemic interpretation of this entire growth problem):

(i) cluster sites already visited ["sick"]

(ii) sites already tested and blocked by the second coin ["immune"]

(iii) neighboring sites not yet tested ["growth"].

Thus from the reference frame of the ant, only the growth sites are special in the sense that only these enlarge her territory.

Why are the growth sites interesting? Subjectively, they represent the "open frontier" (Rammal and Toulouse 1983) of the growing fractal. Objectively, this disconnected set of sites has a well-defined fractal dimension d_g in the sense that the number or "mass" of growth sites obeys a scaling relation of the form

$$N_g \sim L^{d_g} \quad [\textit{Fractal Dimension No. 5}]. \quad (24a)$$

Here, as always, L represents the cluster diameter or radius of gyration.

The number of growth sites has been calculated as a function of cluster mass N_f. We predict for the intrinsic exponent

$$N_g \sim (N_f)^x \quad \text{with} \quad x = d_g/d_f. \quad (24b)$$

Stanley et al (1984) obtained the first estimates of x, $x = 0.49$ [$d = 2$ percolation].

We now evaluate d_g for the two popular conjectured relations between d_w and d_f. Above, we discussed the Alexander-Orbach [**AO**] conjecture $d_w = (3/2)d_f$, and we mentioned the Aharony-Stauffer [**AS**] conjecture $d_w = 1 + d_f$. Thus

$$d_g = \begin{cases} d_f/2 = 91/96 = 0.9479 & [\textbf{AO}] \\ d_f - 1 = 43/48 = 0.8958 & [\textbf{AS}]. \end{cases} \quad (25a)$$

For the intrinsic exponent x, we then predict

$$x = \begin{cases} 1/2 = 0.5000 & [\textbf{AO}] \\ 1 - 1/d_f = 0.4725 & [\textbf{AS}]. \end{cases} \quad (25b)$$

Thus we see that the calculated values of x fall in between the AO and AS predictions.

Leyvraz and Stanley (1983) considered the conditions under which the AO conjecture might hold, focussing on the need for complete statistical independence of the increments in N_g. They noted that this statistical independence certainly occurs for the Cayley tree, since it is impossible to have correlations on a loopless fractal. In this fashion, they understand the numerical result that AO holds for the Cayley tree ($d_w = 6$, $d_f = 4$). In the asymptotic limit of a truly huge cluster—say the size of Corsica—they imagined that correlations in growth sites would begin to vanish even though Corsica has loops. To the extent that these correlations eventually drop off, we can expect that in the asymptotic limit the increments in Ng will be statistically independent and hence the distribution will be normal, with $x = d_g/d_f$ exactly 1/2 [AO].

Recently, the subject of growth sites has arisen in various contexts. One concerns a family of epidemic growth models, all of which eventually grow static percolation clusters (Bunde et al 1985a,b; Herrmann and Stanley 1985). Depending on the growth rules, the dynamic critical exponent dg can differ—even though the static critical exponent d_f is the same for all rules. We considered already one such epidemic model above. A second model is a variation of the above model in which the "ant" visit *only* growth sites, never re-visiting cluster sites. Clearly this requires the ant to make long-range jumps. In the simplest version of this model, the ant randomly chooses her next move from among all the existing growth sites, weighting them equally. This model can be simulated very rapidly (at least 100 times faster than the "walking" ant). Had we set $p = 1$, we would have obtained an Eden cluster, with fractal dimension $d_f = d$. When a fraction $1-p$ of the growth sites are poisoned, our flying ant acts rather like a butterfly, choosing carefully to land only on a non-poisoned site. A typical cluster at $p = p_c$ is shown in Fig. 4.

I wish to apologize that time and space considerations do not permit as coherent and complete a description of the subject of fractal materials as I had desired. Since earlier work is reviewed elsewhere (Stanley 1981, 1982a,b,c, 1983, 1984a,b, 1985; Stanley and Coniglio 1983), I've chosen to focus on ideas of recent months, oblivious to the adage that bringing a lecture "up to date" at time $-\tau$ renders it "out of date" at time $+\tau$.

LITERATURE CITED

Adler J, Aharony A and Stauffer D 1985 J Phys A **18** L129

Ben-Avraham D and Havlin S 1982 J Phys A **15** L691

Bunde A, Herrmann HJ, Margolina A and Stanley HE 1985a Phys Rev Lett **55** 653

Bunde A, Herrmann HJ and Stanley HE 1985b J Phys A **18** L523

Bunde A, Coniglio A, Hong DC and Stanley HE 1985c J Phys A **18** L137

Coniglio A and Stanley HE 1984 Phys Rev Lett **52** 1068

deArcangelis L, Redner S and Coniglio A 1985 Phys Rev B **31** 4725

de Gennes PG 1976 La Recherche **7** 919

de Gennes PG 1980 J Phys (Paris) Colloq **41** C3-C17

Gefen Y, Aharony A and Alexander S 1983 Phys Rev Lett **50** 77

Halsey TC, Meakin P and Procaccia I 1985 Phys Rev Lett (submitted)

Herrmann HJ and Stanley HE 1985 Z Phys **60** xxx

Hong DC, Stanley HE, Coniglio A and Bunde A 1985 Phys Rev B **32** xxx

Kunz H and Souillard B 1978 J Stat Phys **19** 77

Leyvraz F 1985 J Phys A **18** xxx

Leyvraz F, Adler J, Aharony A, Bunde A, Coniglio A, Hong DC, Stanley HE and Stauffer D 1985 preprint for J Phys A Letters

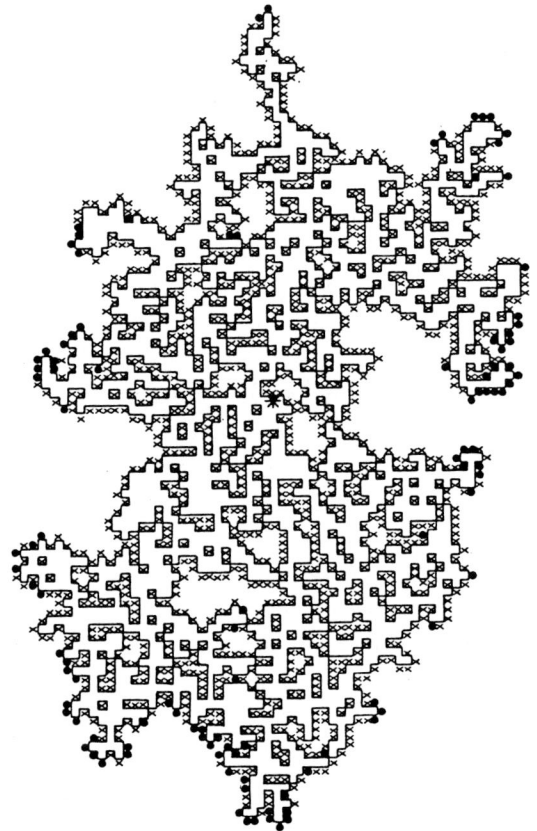

FIG. 4: Typical percolation cluster being grown by the "butterfly" mechanism for the case that each perimeter site has equal weight. This model is equivalent to the Eden model on a "diluted" lattice on which a fraction $1 - p_c$ of the sites have randomly been removed or "poisoned." After Bunde et al (1985a).

Leyvraz F and Stanley HE 1983 Phys Rev Lett **51** 2048

Meakin P, Stanley HE, Coniglio A and Witten TA 1985a Phys Rev A **32** 2364

Meakin P, Stanley HE, Coniglio A and Witten TA 1985b preprint

Rammal R and Toulouse G 1983 J de Physique **44** L13

Stanley HE 1981 in *Int Conf on Disordered Systems and Localization* eds C Castellani, C DiCastro and L Peliti (Springer Verlag, Heidelberg)

Stanley HE 1982a in *Proc NATO Advanced Study Institute on Structural Elements in Statistical Mechanics and Particle Physics* eds K Fredenhagen and J Honerkamp (Plenum Press, New York)

Stanley HE 1982b in *Physics as Natural Philosophy: Festschrift in Honor of Laszlo Tisza* eds A Shimony and and H Feshbach (MIT Press, Cambridge)

Stanley HE 1982c Prog Physics (Beijing) **30** 95 [in Chinese]

Stanley HE 1983 J Phys Soc Japan Suppl **52** 151

Stanley HE 1984a in *Kinetics of Aggregation and Gelation* eds F Family and D Landau (North Holland, Amsterdam)

Stanley HE 1984b J Stat Phys **36** 843

Stanley HE 1985 in *Encyclopedia on Polymer Science* (Wiley, New York)

Stanley HE and Coniglio A 1983 in *Percolation Structures and Processes* eds G Deutscher, R Zallen and J Adler (Adam Hilger, Bristol)

Stanley HE, Majid I, Margolina A and Bunde A 1984 Phys Rev Lett **53** 1706

Turkevich LA and Scher H 1985 Phys Rev Let **55** 1026

FRACTALS IN PHYSICS
L. Pietronero, E. Tosatti (editors)
© Elsevier Science Publishers B.V., 1986

DYNAMICAL PROPERTIES OF RANDOM AND NON-RANDOM FRACTALS

Robin STINCHCOMBE

Theoretical Physics Department, 1 Keble Road, Oxford OX1 3NP, U.K.

Normal and anomalous (critical) dynamics on random and non-random fractals is introduced and approached using scaling methods. The characteristic frequency, dynamic exponent, density of states, spectral dimension and response are discussed for diffusion, phonon and spin wave dynamics on chains, non-random fractals and random fractals (the percolation network). Recent results on anisotropic (including biassed) diffusion on the Sierpinsky gasket fractal, and diffusion on fractal models related to Lévy flights, are also presented.

1. INTRODUCTION

The scale invariance of random fractals[1], like the percolation network at the percolation threshold, and non-random fractals[1] like the Sierpinsky gaskets, gives rise to anomalous dynamical properties. These are here discussed using scaling methods[2] developed from the viewpoints common for critical phenomena.

At a continuous thermal phase transition, the dynamics become anomalous (critical) because the correlation length ξ diverges at the transition. The divergence of a controlling characteristic length and associated anomalous dynamics also occurs in other situations, of which those emphasised here are:

(i) at geometrical phase transitions, of which the most common example is the percolation transition[3] in dilute lattice-based systems (especially dilute magnets[4], which often provide almost ideal examples for experimental investigation);

(ii) on non-random fractals[1], eg. Sierpinsky gaskets, which are of interest in their own right and also as models[5] for naturally occurring random fractals such as the percolation network or its backbone.

In addition, though we shall not discuss this in detail here, the approaches to be introduced can be useful in other situations involving a large length, such as in incommensurate systems (reached as the limit of commensurate systems with longer and longer periodicity lengths) or in quasi scale invariant random systems such as some resins and glasses which appear to have a scaling window in which the anomalous dynamics associated with the so-called fractons may appear[6].

Our discussion of dynamical processes on such random and non-random fractals as those referred to in (i), (ii) above begins with a review of normal and anomalous (critical) dynamics on scale invariant systems, introducing the concepts of scaling and crossover (§2), then proceeds (§3) to a simple illustration of length scaling for dynamics followed by an account of the application of such methods to the dynamics of non-random fractals (§4). In the next section (§5) we go on to discuss the more difficult case of random fractals, especially dilute magnets at the percolation threshold. The last parts of this account present recent work on two special problems:

(a) anisotropic diffusion on a Sierpinsky gasket (§6).

(b) diffusion in a fractal model with flights (§7).

2. NORMAL AND ANOMALOUS (CRITICAL) DYNAMICS

In order to discuss together three basic and important dynamical processes, namely diffusion, spin waves, and lattice vibrations, we introduce[2] a characteristic frequency variable Ω equal, in these respective cases, to $i\omega$ for diffusion, to ω for spin waves, and to ω^2 for lattice vibrations, where ω is the frequency. Then the normal long-wavelength relation between frequency and wavevector k becomes in all these cases

$$\Omega = Dk^2. \qquad (1)$$

This result depends on the wavelength being large compared to any other length in the system, e.g. the lattice spacing (continuum approximation). It clearly breaks down if the characteristic length ξ diverges. This situation, familiar in critical phenomena, can be handled by a scaling hypothesis (usually confirmed by scaling methods) which in the present case becomes[7]

$$\Omega = k^z f(k\xi) \qquad (2)$$

where z is the dynamic exponent. This relation should hold for k, $1/\xi$ both small for any value of the crossover variable $k\xi$ (which measures the competition of the two large lengths ξ and the wavelength, and determines how the characteristic length controls the approach to criticality). It implies the following two asymptotic dependences:

(i) $\Omega \sim k^z$, $k\xi \to \infty$; this is the anomalous (critical) behaviour and occurs provided $f(0)$ is a finite constant;

(ii) $\Omega \sim k^2 \xi^{2-z}$, $k\xi \to 0$; this is the normal behaviour (1) and was recovered by having $f(x) \sim x^{2-z}$, x small, to give the usual k^2 dependence. The associated factor ξ^{2-z} vanishes as $\xi \to \infty$ (since $z > 2$, usually) which is the well-known mode softening effect, overtaken eventually by crossover to region (i).

These effects (mode softening, crossover, anomalous dependence characterised by dynamic exponent z) are to be expected for dynamics at geometrical phase transitions. For fractals, which are strictly self-similar, one is always at the anomalous (critical) limit $k\xi = \infty$.

3. POSITION SPACE LENGTH SCALING FOR DYNAMICS

The dynamic effects just discussed can be treated by length scaling methods[2] since the recursive nature of such methods allows them to probe the diverging controlling length responsible for the effects and to deal with systems (e.g. non-random fractals) constructed recursively. Such methods can lead very directly to scaling behaviour, crossover, and to results for dynamic exponents, etc. In addition, if the scaling techniques are performed in position space, as described below, they can treat the dynamics of non-uniform and random systems such as non-random fractals or the percolation network.

A simple introduction to the technique is provided by the following treatment of dynamics on a chain of lattice sites, with spacing a. Let U_n be the dynamic variable (e.g. the atomic displacement in the case of lattice vibrations) at site n. With the frequency variable Ω defined in the previous section, the equation of motion

$$-\Omega U_n = U_{n+1} - 2U_n + U_{n-1} \qquad (3)$$

applies to all the three cases: diffusion, spin waves, phonons. This equation is trivial to treat using k-space methods, but those will not generalise to non-uniform systems, or cope with scale-invariance, as we require for the treatment of non-random and random fractals. So instead, we use the scaling method which will apply in those more complicated situations. The method (decimation) is to eliminate e.g. every second site ..., n-1, n+1,... from the equations of motion. In place of (3) we then find a similar equation but with $U_{n\pm 1}$ replaced by $U_{n\pm 2}$ and Ω replaced by Ω' where

$$\Omega' = 4\Omega - \Omega^2. \qquad (4)$$

This is therefore the frequency variable for a diluted chain in which the lattice spacing is now

$$a' = 2a \qquad (5)$$

(hereafter referred to as dilatation by a factor b=2). The length scaling transformation equations (4), (5) imply that the relationship,

$\Omega = f(a)$, say, between Ω and a has to be such that

$$f(2a) = 4f(a) - f(a)^2. \quad (6)$$

This has the solution $f(a) = 2(1 - \cos ka)$, where k is an arbitrary constant, clearly identifiable as the wave vector of usual k-space approaches. So the recursion equation (4) is sufficient to give the (usual) full dispersion relation. (4) is actually a special case (with control parameter 4, which makes it here soluble) of the quadratic map discussed by Feigenbaum and others[8]. More complicated cases will be seen to arise in the next example, the Sierpinsky gasket fractal.

The (simplest) triangular Sierpinsky gasket is obtained by dividing a triangle into four equal triangles, discarding the central one and similarly subdividing the remaining ones, and so on indefinitely[1]. Generalisations to d (space) dimensions are obtained by applying a similar recursive process to a hypertetrahedron. The system is non-random, but is obviously non-uniform so k-space methods are inapplicable. Decimation of the sites introduced by the subdivision involved in a given stage of the construction reverses this recursive process, so the length scaling method introduced above is ideally suited to such fractals.

Such a decimation process doubles the site separation (c.f. (5)) and leads to the following exact frequency scaling[2,9,10]

$$\Omega' = \lambda\Omega - \Omega^2 \quad (7)$$

where $\lambda = d+3$ for the general d-dimensional fractal.

The resulting functional equation for the dispersion function f in $\Omega = f(a)$ is not now soluble (for d > 1). Nevertheless the dynamic exponent z can be found, since it describes low frequency dynamics. Here the non-linear term in (7) is negligible, so Ω scales by a factor $\lambda = d+3$ while a doubles and hence

$$\Omega \propto a^z, \quad z = \log_2(d+3). \quad (8)$$

The density of states can also be found from (7) and has a fractal support because all pre-images of the region $\Omega > 4$ scale to unphysical values[11]. By scaling inhomogeneous equations of motion, the Green function scaling can also be obtained[10]. Very recently, the scaling equations for the full frequency and wave-vector dependent response have been obtained and solved numerically[12]. The response function is a power of k times a function (with hierarchical structure) containing the frequency in the combination ω/k^z with z given by (8), as expected from scaling considerations. There is in addition a further periodic dependence on $\log_2 k$ which arises because only the discrete dilatation factors 2^n can be used in decimating the fractal. Details of another recent result, for anisotropic diffusion on the Sierpinsky gasket, are given in §5.

4. DYNAMICS ON RANDOM FRACTALS; SPIN WAVES AT THE PERCOLATION THRESHOLD, ETC.

A brief description is now given of the application of the scaling methods[2] to dynamics on random fractals, taking the special case of the percolation network. Though this is only fractal at the percolation threshold p_c, dynamical processes on it can be treated at general concentration p, allowing a discussion of crossover. In such a case, it is necessary to obtain scaling equations for the concentration p as well as the frequency Ω:

$$p' = R(p) \quad (9)$$
$$\Omega' = S(p,\Omega). \quad (10)$$

Such equations can be obtained by decimation as before, but now considering probability distributions for bond or site variables (depending on the form of dilution) and making cluster approximations[2,10,13].

To be specific, we hereafter discuss the special case of spin waves in one and two dimensional diluted systems, but the conclusions apply also to diffusion and to the simple phonon problem (no shear modes).

In the one-dimensional case, with dilatation factor b=2, the first scaling equation, (9), takes the (exact) form $p' = p^2$. At the fixed points p=0, 1 (empty or pure chain) the second equation, (10), becomes of the form (7) with $\lambda = 3,4$ at p=0,1 respectively[13]. Under iteration of the quadratic map (7) Ω samples only a discrete set of values in the first case (periodic behaviour), or a whole continuous band in the second (chaotic behaviour). However, it can be shown that the sampling density in Ω gives the density of states. This is therefore discrete for $p \sim 0$, corresponding to the levels of (localised) states on small finite clusters, and continuous for p=1 corresponding to the energy band of the pure chain (extended states). Further details are given elsewhere[13]. A complete exact scaling form has been given for the dynamic response function for both the ferromagnetic case[14] and the antiferromagnet[15].

The two-dimensional case has also been treated by scaling methods[10]. One result is that spin waves on the percolation network at p_c have $\omega \propto k^z$ with z = 2.76; this dynamic exponent is obtained from (10) by linearisation at the fixed point $(p,\Omega) = (p_c,0)$ of (9), (10) in the usual way:

$$\left(\frac{\partial S}{\partial \Omega}\right)_{p_c,0} \equiv \lambda = b^z . \qquad (11)$$

The anomalous dynamics associated with this exponent implies an anomalous density of states and hence anomalous thermal properties in Heisenberg layer magnets at p_c. For the pure (ferromagnetic) spin wave system in d=2, the density of states $\rho(\omega)$ is constant at small ω, but at p_c both the quadratic dispersion relation and the dimensionality of the infinite cluster change, leading to a divergent density of states[16]

$$\rho(\omega) \propto \omega^{-\lambda} \qquad (12)$$

where $\lambda = 0.32 \pm 0.01$.

The anomalous spin wave dynamics, characterised by exponent z, is soon to be investigated by neutron scattering[17]. Results are already available from a neutron scattering experiment[18] on Ising dynamics on the percolation network at p_c, yielding a dynamic scaling form for the characteristic relaxation time τ. Recent theoretical work[19,20], however, predicts that this scaling form will break down as the experiment is extended to lower temperatures.

5. ANISOTROPIC DIFFUSION ON SIERPINSKY GASKET FRACTALS

In this, and the next, section recent results on diffusion on fractals are presented. Here we discuss the effect of anisotropy in hopping rates on diffusion on the two-dimensional Sierpinsky gasket[21]. One motivation is to see whether bias causes crossover to new behaviour, but rotational anisotropy is also discussed here.

Coupled non-linear scaling equations for the hopping rates can be obtained by the decimation method introduced §3. These equations have an isotropic fixed point (where all hopping rates are the same). Linearisation about this fixed point yields four eigenvalues, one determining the critical exponent $z = \log_2 5$ for anomalous diffusion (c.f. (8)). The others show that the isotropic fixed point is doubly unstable with respect to bias, with crossover exponent $\phi = 1$, and stable with respect to rotational anisotropy, with a correction-to-scaling exponent $\zeta = \log_2 5/3$. If d and r are measures of bias and rotational anisotropy (both small) we obtain the following exact scaling form for the diffusion length R at long time t:

$$R = t^{1/z} F(dt^{\phi/z}, rt^{-\zeta/z}). \qquad (13)$$

Though the model is probably a very inadequate representation of the percolation network in this context, the result is in qualitative agreement with Monte Carlo simulations[22] of biassed diffusion on the network at p_c in giving a crossover to drift behaviour.

6. DIFFUSION IN A FRACTAL MODEL WITH FLIGHTS

A second problem which has been recently discussed using a fractal model is the effect of arbitrarily long range hoppings ("flights") on diffusion[23]. As in the well-known problem of Lévy flights[24] an anomalous relationship between diffusion length and time will be seen to arise.

The fractal considered is one constructed recursively by replacing a single bond by (r+s) bonds in series of which an inner portion of s bonds is bridged by a single "long" bond (a "flight" of length s). The resulting fractal contains flights of arbitrarily long lengths L distributed according to a probability distribution P(L) whose m^{th} moment is zero or infinite for m negative or positive respectively. As a result[24] the diffusion process will turn out to be non-Gaussian.

The exact scaling relationship between diffusion times t,t' at two successive stages n, n+1 in the recursive construction can be obtained[23] by decimating the equations of motion in such a way as to eliminate the interior sites of the structure that represents a single bond at stage n+1 in terms of bonds at stage n. Linearisation about the long-time fixed point of the full scaling relation yields the following anomalous form for the relationship between diffusion length R and time t, at long times:

$$R \propto t^{1/z} \qquad (14)$$

where

$$z = \ln\left[\frac{(rs + r+s)(r + s + 1)}{(s+1)}\right] / \ln(r+s). \quad (15)$$

This exponent varies with r and s, and its anomalous form results from two features of the fractal system treated: the occurrence of arbitrarily long flights (making the mean square jump displacement infinite) and the fact that these are generated from two hierarchies. This last feature prevents z being simply related to exponents representing the distribution P(L) of flight lengths, as is possible in simpler cases such as the Weierstrass random walk[24].

7. CONCLUSION

This brief summary of basic concepts and some recent work has tried to show that the dynamics of scale-invariant systems can be remarkably rich in new and exotic phenomena, of which the basic ones are the crossover from normal to (anomalous) critical behaviour. Such phenomena are susceptible to length scaling methods which exploit the fundamental scale invariance or hierarchical nature of the systems and lead very directly to the principal features (anomalous exponents, scaling forms, etc). Though some of the systems treated were of a very idealised type (the non-random fractals), they give some insight into the behaviour of the random fractals such as the percolation network at p_c) which are of enormous current experimental interest and which themselves can be treated by the scaling methods described in section 5, where the important special case of dilute Heisenberg spin systems at the percolation threshold was emphasised. It is perhaps remarkable that such systems (and the closely related scale-invariant phonon and diffusion problems) which have simple linear equations of motion can give rise to such rich phenomena, but the scale invariance is the underlying cause. With non-linear dynamic effects also incorporated, the situation becomes even more interesting, allowing now such further possibilities as the breakdown of dynamic scaling (as occurs for example in dilute Ising spin dynamics at p_c). Such non-linear scale-invariant dynamical systems, as well as incommensurate ones, provide many challenges, and the scaling methods need further developments to deal more fully with these.

REFERENCES
1. B. Mandelbrot, Fractals (Freeman, 1977).

2. R.B. Stinchcombe, in: Highlights of Condensed Matter Physics, eds. F. Bassani, F. Fumi and M.P. Tosi (North Holland, 1985); R.B. Stinchcombe, in: Static Critical Phenomena in Inhomogeneous Systems, eds. A. Pękalski and J. Sznajd (Springer, Lecture notes in Physics, Vol. 206, 1984); R.B. Stinchcombe, in: Scaling Phenomena in Disordered Systems, eds. R. Pynn and A. Skjeltorp (Plenum, 1985); and references in these articles.

3. J.W. Essam, Rept. Prog. Phys. 43 (1980) 833.

4. R.B. Stinchcombe, Dilute Magnetism, in: Phase Transitions and Critical Phenomena, Vol. 7, eds. C. Domb and J.L. Lebowitz (Academic Press, 1983).

5. Y. Gefen, B. Mandelbrot and A. Aharony, Phys. Rev. Lett. 45 (1980) 855.

6. R. Orbach, in: Scaling Phenomena in Disordered Systems, eds. R. Pynn and A. Skjeltorp (Plenum, 1985).

7. B.I. Halperin and P.C. Hohenberg, Phys. Rev. 117 (1969) 952.

8. M. Feigenbaum, J. Stat. Phys. 19 (1978) 25; 21 (1979) 669.

9. R. Rammal and G. Toulouse, J. Phys. (Paris) Lett. 44, (1983) L13.

10. C.K. Harris and R.B. Stinchcombe, Phys. Rev. Lett. 50, (1983) 1399.

11. E. Domany, S. Alexander, D. Bensimon and L.P. Kadanoff, Phys. Rev. B28 (1983) 3110.

12. A. Maggs, to be published.

13. R.B. Stinchcombe, Phys. Rev. Lett. 50 (1983) 200.

14. R.B. Stinchcombe and C.K. Harris, J. Phys. A16 (1983) 4083.

15. A.C. Maggs and R.B. Stinchcombe, J. Phys. A17 (1984) 1555.

16. S.J. Lewis and R.B. Stinchcombe, Phys. Rev. Lett. 52 (1984) 102.

17. G. Aeppli, private communication.

18. G. Aeppli, H.J. Guggenheim and Y.J. Uemura, Phys. Rev. Lett. 52 (1984) 942.

19. C. Henley, to be published.

20. C.K. Harris and R.B. Stinchcombe, to be published.

21. R.B. Stinchcombe, J. Phys. A18 (1985) L591.

22. R. Pandey, Phys. Rev. B30 (1984) 489.

23. R.B. Stinchcombe, to be published.

24. E.W. Montroll and M.F. Schlesinger, in: Non-Equilibrium Phenomena II, from Stochastics to hydrodynamics, eds. J.L. Lebowitz and E.W. Montroll (North-Holland, 1984).

THE ELASTIC BEHAVIOR OF FRACTAL STRUCTURES

Itzhak WEBMAN

Serin Physics Laboratory, Rutgers University, Piscataway, NJ 08854 USA

Abstract
A theory for the elastic properties of percolating networks and other fractals is presented. The scaling of the macroscopic elastic moduli with the size is shown to be different than the corresponding scaling of the electrical conductance, leading to a new exponent for the macroscopic moduli near criticality. The theoretical results are discussed in relation to recent experiments and numerical simulations. The corresponding dynamical behavior is characterized by a spectral dimensionality which is smaller than unity, implying a divergence of the density of vibrational modes at low frequencies. As a consequence of this property there are limits to the inherent stability of fractal objects. Tenuous fractal objects are very susceptible to external stresses. The effects of an external stress on the linear elastic response and the main features of the non-linear elasticity are discussed.

I. Linear Elasticity

Small particles may be aggregated to form low density macroscopic materials, characterized by a fractal geometry over a range of length scales [1-4] between the particle size a, and an upper scale ξ. Examples of such materials are gold and silica colloidal aggregates [2,4], and the highly porous metallic composites formed by sintering a powder of submicron silver particles [3]. This paper presents a theory relevant to the elastic properties of these structures, as well as of some other tenuous systems like gels.

Lattice models for the physical properties of disordered fractal materials can be constructed by generating lattice clusters of sites or bonds which possess an appropriate stochastic geometry, together with a definition of a lattice Hamiltonian relevant to the property studied. For concreteness I concentrate mostly on a bond percolation elastic model, which represents a random composite made up of hard particles and very soft particles or voids. An elastic lattice Hamiltonian for such a system has to obey several criteria:

a) Rigid connectedness: Lattice clusters should have finite rigidity. For a percolation model the lattice should have finite macroscopic moduli above the percolation threshold, which vanish as $p \to p_c$.

b) The tensorial aspects of elasticity of thin elements should be properly reproduced.

c) In the absence of significant external interactions, the elastic Hamiltonian describing a system which retains an intrinsic mechanical stability, has to be rotational invariant.

The simplest Hamiltonian which obeys these criteria has the following form for a two dimensional lattice [5]:

$$H = G \sum_{\substack{i,j,k \\ (j,k,nn \text{ of } i)}}^{c} \delta\phi_{jik}^2 + \frac{Q}{a^2} \sum_{\substack{i,j \\ nn}}^{c} (\vec{u}_i - \vec{u}_j)_{\parallel}^2 \quad (1)$$

where $(\vec{u}_i - \vec{u}_j)_{\parallel}$ is the difference of displacements of site i and the site j in the direction parallel to the bond (i,j), and $\delta\phi_{jik}$ is the change in the angle between the bonds (i,j) and (i,k) connected to site i. The summation $\sum^c$ is over lattice sites which belong to the cluster. G and Q are local elastic constants and a is the lattice constant.

In contrast to Eq. (1), a Hamiltonian which contains only the nearest neighbor central force term does not obey the elastic connectedness criterion, and has a rigidity threshold p_r which is higher than p_c [6]. For cubic lattices p_r~1.

The scalar Born Hamiltonian which has been used to represent the elasticity of gels [7], and which leads to an analogy between the elastic moduli and the conductivity, does not possess rotational invariance. It may be an appropriate model for systems, which are mechanically stabilized due to interactions not included in the Hamiltonian of the elastic frame, such as osomotic pressure and excluded volume interactions in gels [8,9].

A basic element which appears in tenuous fractal structures is a thin tortuous chain. The overall elastic response of such a chain can be calculated exactly for a continuum chain as well as for the analogous lattice cluster. The elastic behavior of the chain can be described by a tensorial force constant $\hat{K}$ which depends on both the chain length N and on its geometrical configuration: For long chains, in two dimensions, the change in the elastic energy associated with a change $\delta \vec{R}$ of the end-to end vector is given by:

$$E = \frac{1}{2} \delta\vec{R} \, \hat{K} \, \delta\vec{R} \qquad (2)$$

$$\hat{K} = G \frac{\hat{Z}\hat{S}^{-2}\hat{Z}}{N}$$

Here $\hat{S}^2$ is the tensor of gyration of the chain defined as: $\hat{S}^2 = \frac{1}{N} \begin{pmatrix} \int x^2 ds & 0 \\ 0 & \int y^2 ds \end{pmatrix}$, where the integration is along the chain length ($\hat{S}^2$ is of the order of the squared chain size), and $\hat{Z}$ is a 90° rotation operator. The constant G depends on the width of the chain as: $G \sim a^d$. A generalization of these two dimensional results to higher dimensions is straightforward.

I have omitted a term in $\hat{K}$ due to the central force part of Eq. (1), since it becomes negligible for large N. This result can be generalized for the case of an inhomogeneous chain, in which the local bending strength G(s) varies along the chain. The constant G in Eq. (2) is now replaced by $1/\int^N G^{-1}(s)ds$.

The most striking feature of this result is the dependence of the small strain elastic behavior on the chain configuration. In contrast, for the scalar Born model K ~ 1/N and the elastic behavior depends on the chain length only. A simple example for this difference, is the elasticity of a long thin rod of length ℓ. The tensor K consists of a logitudinal component which scales as ℓ^{-1} and a transverse component which scales as ℓ^{-3}. In the scalar Born model, both components scale as ℓ^{-1}, as does the conductance.

The backbone of a percolating cluster at and above six dimensions contains almost no multiply connected regions up to size ξ. Thus the results for a convoluted chain can be directly applied to obtain the stiffness K(L) of a region of the backbone of size L. Since the Hausdorff dimensionality of the backbone in d=6 is D_B=2, using Eq. 2 with $\hat{S}^{-2} \sim L^{-2}$, and $N(L) \sim L^{D_B}$:

$$K(L) \sim L^{-\zeta_E} \qquad (3)$$

$$\zeta_E = D_B + 2 = 4 \qquad (4)$$

The size scaling can be applied to obtain the exponent τ which describes the behavior of the macroscopic elastic moduli near the percolation threshold:

$$K(p-p_c) = K_o(p-p_c)^\tau \qquad (5)$$

$$\tau = [(d-2) + \zeta_E]\nu = 4$$

Note the difference between this mean field value of $\tau=4$ and the value of the corresponding conductivity exponent: $t=3$.

Generally, the elastic constants of a region of size L of the fractal depend on the structure of the backbone in that region which consists of both singly connected and multiply connected parts. Assuming that the softness of a region is determined by the singly connected bonds and that the multiply connected parts are completely rigid, one can use the above generalized result for an inhomogeneous chain [5], and obtain the following upper bound for the rigidity of a region of size L of the backbone:

$$K(L) < \frac{1}{N_s(L)L^2} \quad (6)$$

$N_s(L)$ is the number of singly connected bonds in a region of size L, $N_s(L) \sim L^{\frac{1}{\nu}}$ [10,11]. These lead to $K(L) \sim L^{-\zeta_E}$, $\zeta_E = 2 + \frac{1}{\nu}$. The following lower bound for the exponent τ results:

$$\tau = d\nu + 1 \quad (7)$$

The values of $\tau = 3.6$ in $d=2$ and $\tau = 3.55$ in $d=3$. These values are vey different from the corresponding values for the conductivity exponent: $t = 1.28$ in $d=2$ and $t = 2.05$ in $d=3$.

Since Ref. 5 was published several numerical calculations of τ were carried out. The values obtained for two dimensional lattices are $\tau = 3.5\pm0.2$ (Ref. 12), $\tau = 3.3\pm0.5$ (Ref. 13,14) and $\tau = 3.5\pm0.4$ (Ref. 15). A recent calculation by Zabolitzky et al. [16] yields $\tau = 3.96\pm0.04$ in $d=2$.

Two recent experiments also obtain values of τ which agree quite well with the theory. Benguigi [17] has studied the elasticity and conductivity of metallic sheets with holes, and obtained for aluminum and copper sheets $\tau = 3.3\pm0.5$, and $\tau = 3.5\pm0.4$ respectively. More recently, Deptuck et al. [3] have studied the Young modulus of porous beams made by sintering submicron silver particles, near the percolation threshold. They obtain $\tau = 3.8\pm0.5$ in good agreement with the theoretical value for three dimensions.

In both experiments the electrical conductivity was also measured, resulting in values of the exponent t in agreement with numerical values, and demonstrating the difference between the two types of critical behavior.

II. Dynamical Properties

The low frequency vibrational modes of a fractal object are determined by the structure of both the backbone and the branches. The backbone determines the elastic constants while the mass contribution is determined by both. Consider a region of size L of fractal structure of Hausdorff dimension D. A dilation of the length by a factor λ gives $K(\lambda L) = \lambda^{-\zeta_E} K(L)$, where ζ_E depends on certain geometrical features of the backbone of the specific fractal structure. The mass in a region of size L scales with L as $M(\lambda L) = \lambda^D M(L)$. From these relations we obtain the scaling property for the vibrational frequencies:

$$\omega(\lambda L) = \lambda^{-(\zeta_E+D)/2} \omega(L) \quad (8)$$

The density of vibrational states is approximately given by $\rho(\omega,L) \sim 1/L^D \Delta\omega$, where $\Delta\omega$ is the spacing between the frequencies of the low vibrational eigenmodes of a structure of size L. The spacing $\Delta\omega$ is of the order of the frequency of the lowest eigenstate, and it scales in the same manner as $\omega(L)$ in Eq. (8). These scaling relations can be combined to yield the exponent $\bar{d}_E$ which describes the low frequency behavior of the density of vibrational modes $N(\omega)$ at low frequencies:

$$N(\omega) \sim \omega^{\tilde{d}_E - 1}$$

$$\tilde{d}_E = \frac{2D}{\zeta_E + D} \quad (9)$$

The mean field (d = 6) value of $\tilde{d}_E$ is 1. For d=2 and d=3, $\tilde{d}_E$ = 0.8 and 0.9, respectively. Thus $N(\omega)$ for a percolating cluster at $p=p_c$ is slightly divergent as $\omega \to 0$. In contrast, $N(\omega) \sim \omega^{\tilde{d}-1}$ with $\tilde{d}$ = 4/3 for the scalar Born model [18]. Similar conclusions [15] can be reached for branched fractals such as diffusion limited aggregates [19].

The density of vibrational states has been numerically calculated by Webman and Grest [15] for percolating clusters and for diffusion limited aggregates. Their numerical results for the elastic spectral dimensionality are $\tilde{d}_E$ = 0.82 ± 0.05 for percolation clusters, and $\tilde{d}_E$ = 0.6 ± 0.05 for DLA. Both values agree well with the prediction of scaling arguments [15].

The most interesting aspect of these results is that an elastic spectral dimensionality that is smaller than unity dominates the low-frequency dynamical behavior of tenuous fractals. The two systems mentioned here, though quite different from each other, show a qualitatively similar behavior, suggesting that this phenomena is rather general.

The divergence of the density of states leads to questions about the mechanical stability of fractal objects. Indeed, our results imply that if the size of the object is sufficiently large so that the frequency of the lowest mode falls below some critical value, the object will not retain its original shape and will become unstable with respect to thermal fluctuations. The argument leads to a criterion for the stability of aggregates which is analogous to the one suggested by Kantor and Witten [20]. It sets an upper limit on the size of nonequilibrium fractal aggregates. On scales larger than this critical size, the aggregate configuration will be determined by relaxation to thermal equilibrium, like in a large branched polymer, resulting in a crossover to a different fractal dimension. As a result the divergence of the density of states will be cut off below a corresponding critical frequency.

III. Elasticity at Large Strains

A tenuous fractal object subjected to a strong external perturbation will become deformed, and over sufficiently large length scales the original geometry will become unstable. An example of such an effect is the coupling to a thermal bath which would make a fractal aggregate become floppy like a polymer in solution over length scales beyond a critical length as discussed above. The change in structure over large length scales implies that the linear elastic behavior of the system will change under fixed external stress. Another, closely related, consequence of the small elastic moduli of fractal networks is the onset of non-linear elastic response at relatively small stress, although the general condition of a strain of order unity for this transition is still valid. Also, in the case of tenuous materials, one can associate the non-linear behavior with the deformation of the system over large length scales, rather than with changes over microscopic length scales, as in the case of dense solids. These features are closely linked to the dependence of the elastic behavior on the geometry discussed in I. Thus, most of the properties discussed here cannot be obtained from the scalar model of elasticity. The discussion in this section also attempts to make a connection between the tensorial elasticity and the scalar elasticity of a network under external strain discussed by Alexander [9].

The present approach is based on the following heuristic picture: As stress is applied to a fractal object deformation occurs mostly on length scales beyond a certain length which decreases with increasing the stress. Little deformation takes place on smaller scales. Thus, the presence of an external stress introduces a new length scale to the problem. A similar idea was invoked by Pincus [21] and de Gennes [22] to study the elasticity of macromolecules. Consider a rigid tortuous thin chain of size L and length N, subject to a stretching force T. The chain will react like a large random spring. The spring constant for linear response depends on the configuration according to Eq. (2): $K(L) \sim 1/NS_\perp^2$, where $S_\perp^2$ is the squared radius of gyration perpendicular to the end-to-end vector. Let the stress be sufficiently large so that the overall strain of the chain is much larger than unity. The chain may then be viewed as a quasi-linear sequence of blobs defined by a subset of points $\{\vec{R}_i\}$ on the chain. One may construct the following "blob Hamiltonian" [23]:

$$H = \sum_i \left(\frac{T}{\chi_\parallel} (\delta \vec{R}_i^\parallel)^2 + \frac{G}{g\chi_\perp^2} (\delta \vec{R}_i^\perp)^2 \right) \quad (10)$$

Here $\chi_\parallel$ and $\chi_\perp$ are the dimensions of the blob parallel and perpendicular to the direction of the stretch T. $\delta \vec{R}_i$ are the displacements of the points R_i, and g is the mean chemical length of each blob. One can simplify and assume $\chi_\parallel \sim \chi_\perp \sim \chi$. G is a local elastic constant for bending which depends on properties such as the width of the chain elements.

The size χ is determined by the condition that the local strain of a blob is unity:

$$\chi \sim \left(\frac{T}{G}\right)^{-\frac{1}{D+1}} \quad (11)$$

Here D is the Hausdorff dimensionality of the chain (which remains unchanged up to scales of order χ). The corresponding force constant of a blob is of the order of:

$$K_B(T) \sim G^{\frac{D+2}{D+1}} T^{\frac{D+2}{D+1}} \quad (12)$$

One may now rewrite Eq. (11) in the form

$$H = \frac{1}{2} \sum_i K_B(T)(\delta \vec{R}_i^{\parallel 2} + \delta \vec{R}_i^{\perp 2}) \quad (13)$$

The effect of the external stress, is thus to transform the elastic energy from the form given by Eq. (2) into a renormalized scalar type energy. The crossover of the overall elasticity of the chain to scalar behavior occurs when the whole chain is a single blob. This condition leads to a crossover stress of the order of

$$\bar{T} \sim GL^{-(D+1)} \quad (14)$$

For external stress smaller than $\bar{T}$, the elasticity remains linear. For fixed $T > \bar{T}$, the behavior for small superimposed strains follows the Hamiltonian given by Eq. (13). Concentrating on the response to a large stress at large extensions, one finds that the length of a strongly stretched chain depends on T in a non-linear manner:

$$R \sim \frac{N}{g} \chi \sim \left(\frac{T}{G}\right)^{\frac{D-1}{D+1}} \quad (15)$$

This picture can be applied to a fractal network, such as a percolating cluster above p_c. Given a fixed external stress T, a crossover from non-scalar to scalar elasticity will take place when the correlation length $\xi = \xi_c$ is of the order of the blob size. The corresponding relation between ξ_c and T is:

$$T \sim G \, \xi_c^{-(D+1)} \quad . \quad (16)$$

where D is the Hausdorff dimensionality of the backbone on scales smaller than ξ. The ξ dependence of a macroscopic linear modulus K is given by:

$$K(\xi) \sim \begin{cases} \xi^{2-d}\xi^{-\zeta_E} & \xi < \xi_c(T) \\ \\ (\frac{T}{G})^{(\zeta_E-\zeta_s)/(1+D)} \xi^{2-d}\xi^{-\zeta_s} & \xi \gg \xi_c(T) \end{cases} \quad (17)$$

where ζ_s is the exponent for the size scaling of a scalar elasticity (and is equal to the corresponding exponent for the conductance). For a percolating cluster, at d<6 the relevant Hausdorff dimensionality, is that of the set of the singly connected bonds, so that $D = \frac{1}{\nu}$, where ν is the correlation length exponent for percolation. In d=6 this value is identical to the Hausdorff dimensionality of the backbone, $D_B = 2$.

Finally, we apply these results to obtain the $p-p_c$ dependence of the linear macroscopic modulus of a percolating system above p_c subject to a fixed stress T. Sufficiently close to p_c, at $p=p_*$ a crossover to scalar elasticity occurs. p_* is given by the relation:

$$p_* - p_c = A (\frac{T}{G})^{\frac{1}{1+\nu}} \quad (18)$$

The behavior of a macroscopic elastic modulus $K(p-p_c)$ in the two regimes is:

$$K(p-p_c) \sim \begin{cases} (p-p_c)^\tau & p > p_* \\ \\ (\frac{T}{G})^{(\tau-t)/(1+\nu)} (p-p_c)^t & p_c < p < p_* \end{cases} \quad (19)$$

Here t is the percolation conductivity exponent. Thus, for a network under fixed external stress, the macroscopic elastic behavior corresponds to the scalar elastic model very close to p_c and crosses over to a tensorial behavior as p increases away from p_c.

Several comments are in order:

This picture of elasticity at large strains is very simplified. At large distortions pieces of the network may collide and entangle with each other, tending to make the system more rigid. In this sense, the above arguments can be expected to hold better for a network which consists mostly of a sparse backbone with few or no dead ends. Additional study is required in order to asses how broad the non-linear regime in various systems is, before the rupture limit of the network is approached. For this, more detailed information on fluctuations in the structure of backbone may be needed.

The term external stress used in this section is meant in the broad sense. It includes any interactions which dilate the original stable configuration of the network [9]. It is clear, however, that in order to crossover to a scalar elastic behavior, this dilation should be quite large on the macroscopic scale. This suggests the possibility that in polymer networks made out of chains of relatively high local rigidity to bending and a large persistence length, a crossover from scalar elasticity to tensorial elasticity may occur at some concentration above the gel point. In this case both exponents, τ and t, could appear in the corresponding regimes. Recent experimental studies of the shear modulus of casein gel seem to confirm this prediction [24].

Acknowledgement: I would like to thank S. Alexander and D.J. Bergman for valuable discussions.

REFERENCES

1. *Kinetics of Agreggation and Gelation*, edited by F. Family and D.P. Landau (North Holland, Amsterdam, 1984).

2. D.AS. Weitz and M. Oliveria, Phys. Rev. Lett. 52, 1433 (1984).

3. D. Deptuck, J.P. Harrison, and P. Zawadski, Phys. Rev. Lett. 54, 913 (1985).

4. D.W. Schaefer, J.E. Martin, P. Wiltzius and D.S. Cannell, Phys. Rev. Lett. 52, 2371 (1984).

5. Y. Kantor and I. Webman, Phys. Rev. Lett. 52, 1891 (1984); I. Webman and Y. Kantor in: Ref. 1.

6. S. Feng and P.N. Sen, Phys. Rev. Lett. 52, 216 (1984).

7. P.G. de Gennes, J. Phys. (Paris) Lett. 37, L1 (1976).

8. S. Alexander, J. Phys. (Paris) 45, 1939 (1984).

9. S. Alexander, In: Proceeding of the Les-Houches Conference on Finely Divided Matter, March, 1985 (to be published).

10. A. Coniglio, Phys. Rev. Lett. 46, 250 (1981), and this conference.

11. R. Pike and E.H. Stanley, J. Phys. A 14, L169 (1981).

12. D.J. Bergman, Phys. Rev. B 31, 1696 (1985).

13. S. Feng, P.N. Sen, B.I. Halperin and C.J. Lobb, Phys. Rev. B 30, 5386 (1984).

14. P.N. Sen, In: Proceeding of the Les-Houches Conference on Finely Divided Matter, March, 1985 (to be published).

15. I. Webman and G.S. Grest, Phys. Rev. B 31, 1689 (1985).

16. J. Zabolitzky, D.J. Bergman and D. Stouffer (unpublished).

17. L. Benguigi, Phys. Rev. Lett. 53, 2028 (1984), and this conference.

18. S. Alexander and R. Orbach, J. Phys. (Paris) 43, L625 (1982).

19. T.A. Witten and L.M. Sander, Phys. Rev. Lett. 47, 1400 (1982).

20. Y. Kantor and T.A. Witten, J. Phys. (Paris) Lett. 45, L675 (1984).

21. P. Pincus, Macromolecules 9, 386 (1976).

22. P.G. de Gennes, *Scaling Concepts in Polymer Physics*, Cornell University Press, 1979.

23. I. Webman (unpublished).

24. M. Tokita, R. Niki and K. Hikichi, J. Chem. Phys. 83, 2583 (1985).

STATIC AND DYNAMIC PROPERTIES OF LOOPLESS AGGREGATES

S. HAVLIN

Department of Physics, Bar-Ilan University, Ramat-Gan, Israel

The static structural and dynamical properties of loopless aggregates are studied in several systems. A cluster growth model for random trees (random aggregates without loops) is presented. In this model the intrinsic (chemical) dimension d_ℓ and the fractal dimension d_f are adjustable. We present a general relationship between the diffusion exponent d_w and the fractal exponent d_f characterizing the trees. These new results were obtained by using a scaling argument and are supported by numerical data. We also present a one-dimensional diffusion model which is analogous to diffusion on loopless aggregates. Solution of the model reproduces the above result for the diffusion exponent. The model is based on parameters characterizing the structure of the tree.

1. INTRODUCTION

The geometrical and topological characteristics of random aggregates are of great interest. Systems such as percolation clusters, lattice animals and diffusion-limited aggregation (DLA) have been extensively studied especially for the purpose of finding those exponents which characterize the physical properties of the systems.[1-12] Considerable interest has recently been shown in the intrinsic properties of random aggregates,[3,5,10,13-17] which were found to be useful in characterizing the dynamics of these systems[3,5,8,14,15].

Two of the intrinsic exponents, d_ℓ and d_ℓ^s can be used to characterize random aggregates. The intrinsic dimension d_ℓ indicates how the mass M scales with the chemical distance ℓ,

$$M \sim \ell^{d_\ell} \quad (1.1)$$

where the chemical distance is the length of the shortest path on the cluster connecting two sites. The exponent d_ℓ^s is the intrinsic dimension of the skeleton.[9] The skeleton of a cluster is defined as the union of all shortest paths from an arbitrary site on the cluster to a shell at chemical distance L, where L << 1. The mass of the skeleton M_s scales with the chemical distance as

$$M_s \sim \ell^{d_\ell^s} \quad (1.2)$$

It has been found[9] that the skeletons of percolation clusters at criticality are chemically linear for all d, namely $d_\ell^s = 1$. This may substantiate the qualitative argument that percolation clusters at criticality are finitely ramified. It was shown recently that for the case of finitely ramified clusters ($d_\ell^s = 1$) for which loops can be neglected (i.e., lattice animals), the diffusion exponents can be expressed in terms of the fractal and intrinsic dimensions as[5]

$$d_w^\ell = d_\ell+1, \quad d_w = d_f(1+1/d_\ell), \quad \bar{\bar{d}} = 2d_\ell/(d_\ell+1) \quad (1.3)$$

where d_w is the diffusion exponent, d_w^ℓ is the chemical diffusion exponent[3] and $\bar{\bar{d}}$ is the fracton dimension. The question arises[8] whether there exist random aggregates or, in particular, trees which are infinitely ramified, that is, trees for which $d_\ell^s > 1$.

In recent works a growth model for trees was presented.[18,19] This growth model enables us to generate trees in any dimension and on a Cayley tree (CT) which can have an arbitrarily chosen value of d_ℓ. We have studied properties of the skeletons of these trees in d = 2 and on CT and find $d_\ell^s > 1$ if d_ℓ is greater than a critical value d_ℓ. This study is presented in Ch. 2.

The analysis of transport properties on fractal aggregates has recently attracted considerable attention.[20-24] In this review, we present two approaches to the study of the transport properties on loopless aggregates. The first approach,[25] presented in Ch. 3, is based on scaling assumptions and on the Einstein relation between diffusion and conductivity. The second approach (Ch. 4) is a phenomenological diffusion model which is based on parameters characterizing the geometrical structure of the tree.[26] In both approaches, we obtain the same general relationships between dynamical diffusion exponents and static exponents for trees

$$d_w^\ell = 2+d_\ell - d_{\tilde{\ell}}^s \qquad d_w = (d_f/d_\ell)(2+d_\ell - d_{\tilde{\ell}}^s)$$
$$\bar{\bar{d}} = 2d_\ell/(2+d_\ell - d_{\tilde{\ell}}^s) \ . \qquad (1.4)$$

Note that for the special case of $d_{\tilde{\ell}}^s = 1$, Eq. (1.4) reduces to Eq. (1.3).

2. CLUSTER GROWTH MODEL FOR TREES

The general model for growing trees with a given intrinsic dimension d_ℓ is as follows. We chose a site on a d-dimensional lattice as the seed of the tree and then select, randomly, B(1) nearest neighbors of the seed to be taken to be blocked. These occupied sites represent the first chemical shell, and their chemical distance to the seed is $\ell = 1$. Similarly, we grow the next shell by examining the non-occupied and non-blocked nearest neighbors of the sites in the first shell. Of the latter, B(2) are then randomly occupied, the sites being chosen with the restriction that a new site can be occupied only if it has but one nearest neighbour already-occupied site. (If the site has more than one nearest-neighbor occupied site, it is considered to be blocked.) This restriction prevents the occurrence of loops in the clusters. Higher-order shells are grown in a similar manner. A similar model without this restriction was studied in Ref. 27. In order to generate figures of predetermined dimensionality, we choose

$$B(\ell) = B_0 \ell^\alpha \qquad (2.1)$$

from which it follows that $M(\ell)$ is given as

$$M(\ell) = \sum_{\ell'=1}^{\ell} B(\ell') \sim \ell^{\alpha+1} = \ell^{d_\ell}, \qquad (2.2)$$

where d_ℓ is the intrinsic dimension of the cluster.

In Fig. 1 we show trees grown in d = 2 with d_ℓ=1.3, 1.5, and 1.8. The fractal dimension d_f of these trees was calculated according to the following scheme. Since $M \sim \ell^{d_\ell} \sim R^{d_f}$, it follows

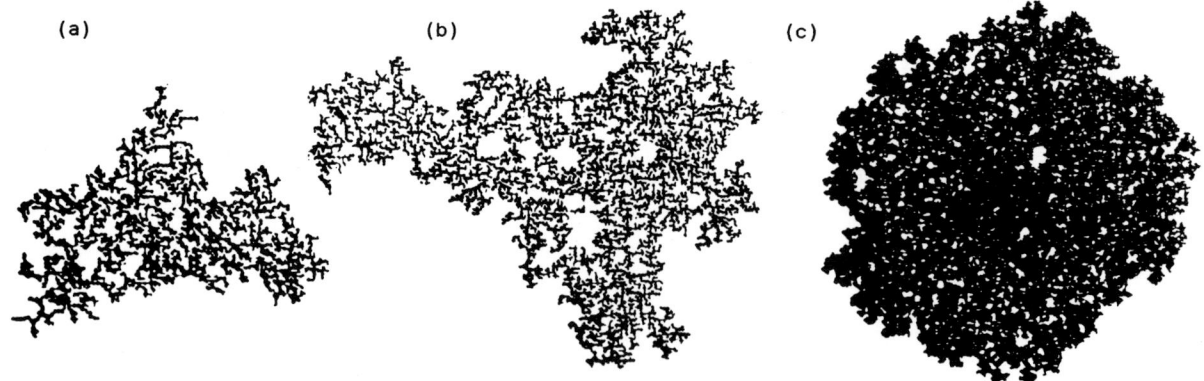

FIGURE 1

Fig. 1: Trees grown by the cluster growth procedure described in the text (L = ℓ_{max} = 300). (a): d_ℓ = 1.3; (b) d_ℓ = 1.5: (c) d_ℓ = 1.8.

that $R\sim \ell^{d\ell/df}$. Thus we numerically evaluate the radius of gyration $R(\ell)$ of the clusters,

$$R(\ell) \sim \ell^{\tilde{\nu}} \qquad (2.3)$$

and use the identity $\tilde{\nu}=d\ell/df$. Results for $\tilde{\nu}$ and df for different values of $d\ell$ are given in Table I.

Table I. <u>Exponents of trees and skeletons</u>

Calculated values for the fractal dimensionality d_f, ratio $\tilde{\nu}=d\ell/df$, and skeleton intrinsic dimensionality d_ℓ^S, for clusters having intrinsic dimensional d_ℓ. The error bars were determined by calculating the extreme slopes when fitting linear curves to the corresponding quantities.

d_ℓ	d_f	$\tilde{\nu}$	d_ℓ^S
1.3	1.71±0.03	0.76±0.02	1.0 +0.03/-0.0
1.5	1.82±0.03	0.82±0.02	1.0 +0.03/-0.0
1.6	1.86±0.03	0.86±0.02	1.0 +0.5/-0.0
1.7	1.91±0.03	0.89±0.02	1.1±0.05
1.8	1.95±0.03	0.92±0.02	1.2±0.05
1.9	2.0 +0/-0.03	0.95±0.02	1.27±0.04
2.0	2.0 +0/-0.03	1 +0.0/-0.02	1.37±0.03

The skeletons of the clusters presented in Fig. 1 are shown in Fig. 2. The structures shown in Figs. 2a and 2b (d_ℓ = 1.3 and 1.5, respectively) do not branch appreciably until ℓ is relatively close to L, whereas in Fig. 2c (d_ℓ = 1.8), we observe that branching occurs for values of $\ell \ll L$. We calculated the intrinsic dimension d_ℓ^S from these and other curves, and the numbers we obtained are presented in Table 1.

It is interesting to compare our results for d_ℓ^S with the result of percolation. Here, for $d_\ell \leq 1.65$, we obtain $d_\ell^S = 1$ as found or percolation clusters at criticality.[9] However, for $d_\ell > 1.65$, we find $d_\ell^S > 1$. The exponent $d_\ell^S - 1$ characterizes the exponent of ramification of the cluster.[12] Thus, for $d_\ell \lesssim 1.65$, these trees are <u>finitely</u> ramified (the ramification exponent is zero), whereas for $d_\ell > 1.65$ the trees are <u>infinitely</u> ramified.

It has recently been shown,[19] using analytical methods, that similar structures generated on a Cayley tree yield $d_\ell^S = 1$ for $d_\ell \leq 2$ and $d_\ell^S = d_\ell^{-1}$ for $d_\ell \geq 2$. These results, together with the results presented in Fig. 3, are consistent with the following relationship between d_ℓ and d_ℓ^S,

$$d_\ell^S = 1 \quad \text{for } d_\ell \leq d_\ell^C$$
$$d_\ell^S = 1 + d_\ell - d_\ell^C \text{ for } d_\ell \geq d_\ell^C, \qquad (2.4)$$

where d_ℓ^C is a critical value.[18] Because Eq. (2.4), with $d_\ell^C = 2$, has been proven

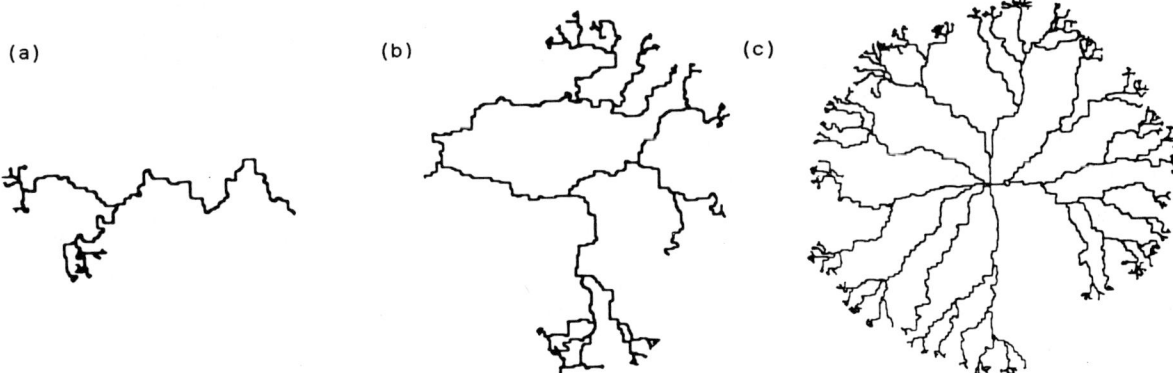

FIGURE 2

Fig. 2: Skeletons of the structures shown in Fig. 1. (a): d_ℓ = 1.3; (b) d_ℓ = 1.5; (c) d_ℓ = 1.8.

analytically for trees grown on a Cayley lattice (which represents growth in high dimensions) and because it seems to hold numerically with $d_\ell^c \sim 1.65$ for $d = 2$, we surmise that it may be valid (with appropriate d_ℓ^c) for trees constructed in any dimension.

We note that $d_\ell^c = 2$ is equal to the value of the intrinsic dimensionality d_ℓ of percolation clusters grown on a Caylet tree. Similarly, the value inferred in the present study, $d_\ell^c = 1.65 \pm 0.05$, is close to the value $d_\ell = 1.64$ which characterizes incipient percolation clusters for $d = 2$.[3] Thus, we conjecture that d_ℓ^c in our tree growth model is equal to d_ℓ for critical percolation clusters generated in the same dimension.

3. SCALING ARGUMENTS FOR TRANSPORT ON TREES

In this chapter we calculate relations between transport exponents for trees using scaling arguments and the Einstein relation[20]

$$d_W = d_f + \bar{\zeta} \qquad (3.1)$$

where $\bar{\zeta}$ is the resistivity exponent which relates the resistivity ρ to the size of the cluster R, by $\rho \sim R^{\bar{\zeta}}$. We define $\rho_{tot}(\ell)$ as the total resistance between a chosen site A on the tree and all sites in the ℓ-th shell surrounding this site, and define the resistivity exponent, $\bar{\zeta}_\ell$, by $\rho_{tot} \sim \ell^{\bar{\zeta}_\ell}$. Let $\rho_1(\ell)$ be the resistance between site A and <u>one</u> site at chemical distance ℓ from A.

Define $B^s(\ell)$ the number of bonds in shell ℓ of the skeleton. Clearly, $B^s(\ell) \sim dM_s(\ell)/d\ell \sim \ell^{d_\ell^s - 1}$. Since there are no loops in the tree (by definition) the quantity $\rho_{tot}(\ell)$ can be related directly to $B^s(\ell)$ by

$$\rho_{tot}(\ell) = \int_\ell^1 \frac{d\ell'}{B^s(\ell')} \sim \frac{1}{\ell^{d_\ell^s - 2}} \qquad (d_\ell^s \leq 2) \qquad (3.2)$$

Also, because of the loopless nature of our aggregates, it follows that $\rho_1(\ell) \sim \ell$ and one finds that

$$\bar{\zeta}_\ell = 2 - d_\ell^s \qquad (3.3)$$

We next make use of the relationships[3]

$$d_W^\ell = \tilde{\nu} d_W, \quad d_\ell = \tilde{\nu} d_f, \quad \bar{\zeta}_\ell = \nu \bar{\zeta} \qquad (3.4)$$

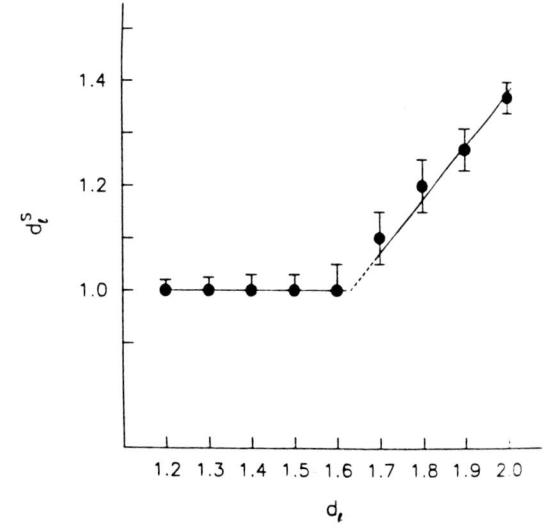

FIGURE 3:
Skeleton intrinsic dimensionality d_ℓ^s for clusters having intrinsic dimensionality d_ℓ. The critical dimensionality is believed to be approximately $d_\ell^c = 1.65 \pm .05$.

and the Einstein relation Eq. (3.1), to obtain the equivalent expression in ℓ-space

$$d_W^\ell = d_\ell + \bar{\zeta}_\ell. \qquad (3.5)$$

By using Eqs. (3.3) and (3.5) and the definition of the fracton dimension $\bar{\bar{d}} = 2d_f/d_W = 2d_\ell/d_W^\ell$, we obtain the general relations given in Eq. (1.4). For the special case of finitely ramified trees, (as lattice animals) for which $d_\ell = 1$, we obtain Eq. (1.3).

4. DIFFUSION MODEL FOR LOOPLESS AGGREGATES

Earlier studies of the anomalous diffusion on general fractal aggregates and treelike structures were usually based on the Einstein relation involving conductivity to calculate the exponents of the anomalous diffusion. In this chapter we introduce a phenomenological approach that allows calculation of several diffusion properties on trees directly from a diffusion model.

Let us therefore consider a loopless tree characterized by a vertex or origin, an infinite branched skeleton, and finite dead ends that branch from the skeleton. For simplicity, the resulting structure will be assumed initially to be discretized in units of $\Delta\ell$. We will later pass to the continuum limit. The random walk is chosen to be restricted to nearest neighbors. Thus a random walker at ℓ can move in a single step only to $\ell \pm \Delta\ell$ on the tree with probabilities denoted by $p_\pm(\ell)$. These will not in general be equal, nor will they sum to one. It will be necessary to define a probability for pausing at any given step equal to $p_0(\ell) = 1 - p_+(\ell) - p_-(\ell)$. In order to derive expressions for these probabilities we use the quantities that characterize the tree $B(\ell)$ and $B^S(\ell)$. The probabilities $p_0(\ell)$ and $p_\pm(\ell)$ are related to these quantities by

$$P_0(\ell) = 1 - P_+(\ell) - P_-(\ell) = 1 - B^S(\ell)/B(\ell) \quad (4.1)$$

$$P_+(\ell)/P_-(\ell) = B^S(\ell+1)/B^S(\ell) \quad (4.2)$$

The first of these relations indicates that the random walker pauses in his progress along the skeleton whenever he finds himself on a dead-end. The second indicates that the relative probabilities of a step in the forward or backward directions along the skeleton depend on the relative number of bonds allowing motion forwards and backwards.

Equations (4.1) and (4.2) together with known scaling properties of $B(\ell) \sim \ell^{d_\ell - 1}$ and $B^S(\ell) \sim \ell^{d_\ell^S - 1}$ at large ℓ, allows us to find asymptotic expressions

$$P_\pm(\ell) = \frac{A}{2\ell^\alpha}\left(1 \pm \frac{B}{2\ell}\right), \quad P_0(\ell) = 1 - \frac{A}{\ell^\alpha} \quad (4.3)$$

where A is a constant related to the proportionality factors in the asymptotic expressions for $M(\ell)$ and $M^S(\ell)$, and the parameters α and B are related to d_ℓ and d_ℓ^S by

$$\alpha = d_\ell - d_\ell^S, \quad B = d_\ell^S - 1 \quad (4.4)$$

Equation (4.4) implies that as the random walker moves further from the origin, he is increasingly likely to remain stationary. This is reasonable because the random walker is increasingly likely to be caught in a dead-end as he moves away from the origin. The terms in the parentheses in Eq. (4.3) stem from the fact that when the skeleton branches, or $d_\ell^S > 1$, the random walk, is biassed in the direction of the more richly branched section.

The assumption that the random walker moves only to nearest neighbors allows one to write a recursion relation for the state probabilities $\{U_n(\ell)\}$ at step n:

$$U_{n+1}(\ell) = p_+(\ell - \Delta\ell)U_n(\ell - \Delta\ell) + p_-(\ell + \Delta\ell)U_n(\ell + \Delta\ell) + p_0(\ell)U_n(\ell) \quad (4.5)$$

Numerical solutions of Eq. (4.5) have shown that the resulting diffusion equation

$$\frac{\partial U}{\partial n} = \frac{A}{2}\frac{\partial^2}{\partial \ell^2}\left(\frac{U}{\ell^\alpha}\right) - \frac{AB}{2}\frac{\partial}{\partial \ell}\left(\frac{U}{\ell^{\alpha+1}}\right) \quad (4.6)$$

leads to results in good agreement with the solutions of the difference equation (4.5)

The solution to Eq. (4.6) that satisfies the initial condition $U(\ell, 0) = \delta(\ell)$ is found to be

$$U_n(\ell) = \frac{(\lambda/n)^{(1+\alpha+B)/(2+\alpha)} \ell^{\alpha+B}}{\Gamma((1+\alpha+B)/(2+\alpha))} \cdot \exp\left(-\frac{\lambda \ell^{2+\alpha}}{n}\right) \quad (4.7)$$

in which $\lambda = 2/[A(2+\alpha)2]$. The expression for $U_n(\ell)$ allows us to deduce relations between the various exponents. The exponent d_W^ℓ is readily found by calculating $\sigma^2(\ell)$ from Eq. (4.7). We find that the time dependence of this parameter is

$$\sigma^2(\ell) \equiv <\ell^2> - <\ell>^2 = Cn^{1/(2+\alpha)} \quad (4.8)$$

where C is a constant. This equation implies that $d_W^\ell = 2 + \alpha = 2 + d_\ell - d_\ell^S$. Furthermore, the behavior of U for large but fixed ℓ and $n \to \infty$ goes like $n^{-(1 + \alpha + B)/(2 + \alpha)}$ which implies that the fracton dimension is

$$\bar{d} = 2(1 + \alpha + B)/(2 + \alpha) = 2d_\ell/(d_\ell-d_\ell^S+2) \quad (4.9)$$

The exponents d_W^ℓ and $\bar{d}$ are in agreement with those derived in Ch. 3.

REFERENCES

1. B.B. Mandelbrot, The Fractal Geometry of Nature (San Francisco: Freeman), 1982.
2. S. Alexander and R. Orbach, J. Phys. (Paris) 43, L625 (1982).
3. S. Havlin and R. Nossal, J. Phys. A: Math. Gen. 17, L427 (1984); S. Havlin, in Kinetics of Aggregation and Gelation, eds. F. Family and D.P. Landau (North-Holland, Amsterdam), in press.
4. T.A. Witten and L.M. Sander, Phys. Rev. Lett. 47, 1400 (1981); P. Meakin, Phys. Rev. Lett. 51, 1119 (1983).
5. S. Havlin, Z.V. Djordjevic, I. Majid, H.E. Stanley and G.H. Weiss, Phys. Rev. Lett. %%, 178 (1984).
6. R. Pike and H.E. Stanley, J. Phys. A: Math. Gen. 14, L169 (1981).
7. M.E. Cates, Phys. Rev. Lett. 53, 926 (1984).
8. T.A. Witten and Y. Kantor Phys. Rev. B30, 4093 (1984).
9. S. Havlin, R. Nossal, B. Trus and G.H. Weiss, J. Phys. A.: Math. Gen. 17, 1957 (1984).
10. P. Meakin, I. Majid, S. Havlin and H.E. Stanley, J. Phys. A 17, L975 (1984).
11. S. Havlin, B. Trus, G.H. Weiss and D. Ben-Avraham, J. of Phys. A 18, L247 (1985).
12. Y. Gefen, A. Aharony, B.B. Mandelbrot and S. Kirkpatrick, Phys. Rev. Lett. 47, 1771 (1981).
13. Z. Alexanderowicz, Phys. Lett. 80A, 284 (1980).
14. H.J. Herrman, D. Hong and H.E. Stanley, J. Phys. A 17, L261 (1984).
15. A.L. Ritzenberg and R.J. Cohen, Phys. Rev. B 30, 4036 (1984).
16. J. Vannimenus, J.P. Nadal and H. Martin, J. Phys. A 17, L351 (1984).
17. Z.V. Djordjevic, S. Havlin, H.E. Stanley and G.H. Weiss, Phys. Rev. B 30, 478 (1984).
18. S. Havlin, R. Nossal and B. Trus, Phys. Rev. A 31, xxx (1985).
19. S. Havlin, J. Keifer, G.H. Weiss, D. Ben-Avraham and Y. Glazer, J. of Stat. Phys. 1985. In press.
20. S. Alexander and R. Orbach, J. Physique 43, L625 (1982).
21. D. Ben-Avraham, S. Havlin, J. Phys. A 15, 2691 (1982); S. Havlin, D. Ben-Avraham and H. Sompolinsky, Phys. Rev. A 27, 1730 (1983).
22. Y. Gefen, A. Aharony and S. Alexander, Phys. Rev. Lett. 50, 77 (1983).
23. R. Rammal and G. Toulouse, J. Physique Lett. 44, L13 (1983).
24. R. Pandy and D. Stauffer, Phys. Rev. Lett. 51, 527 (1983).
25. S. Havlin, R. Nossal, B. Trus and G.H. Weiss, Phys. Rev. B31, 7497 (1985).
26. S. Havlin, B. Trus and G.H. Weiss, J. of Phys. A 18, xxx (1985).
27. D. Hong, S. Havlin and H.E. Stanley, J. of Phys. A 18, xxx (1985).

FRACTALS IN PHYSICS
L. Pietronero, E. Tosatti (editors)
© Elsevier Science Publishers B.V., 1986

THE FRACTAL DIMENSION OF GROWTH PERIMETERS

Alla E. MARGOLINA

E. I. du Pont de Nemours and Company, Central Research and Development, Experimental Station
Wilmington, Delaware 19898

The fractal dimension d_G of cluster perimeters generated by a recently proposed 'butterfly' growth walk is considered. In the long-range limit of the walk on a percolation cluster, d_G appears to be equal to the fractal dimension of the singly connected bonds: $d_G = 1/\nu$. The new relation for chemical dimension d_ℓ is proposed: $d_\ell = d_f/(d_f - d_G)$. In the short-range limit the 'butterfly' walk on a Euclidean lattice appears to be in the same universality class as a random walk. The dynamic aspect of the growth walk is discussed and the continuously tunable spectral dimension is obtained. Both short- and long-range limits of this diffusion process are different from the random diffusion on percolation.

1. INTRODUCTION

Diffusion and transport through random media can be modeled by a random walk on a fractal substrate. One way to study such phenomena is to convert it into a growth problem. Consider a walk that creates its own fractal substrate by choosing the sites it visits according to certain probabilistic rules while walking: the trace left by such a walk forms a fractal cluster of visited sites S. In addition to the usual static and dynamic fractal dimensions[1] (fractal dimension of a substrate d_f, "chemical" dimension d_ℓ, fractal dimension of a minimum path d_{min}, fractal dimension of singly-connected bonds d_{red}, fractal dimension of the walk d_w, spectral dimension d_s) the fractal dimension of the growth perimeter d_G is introduced. Namely, the cluster growth occurs through the sites which are nearest neighbors to visited sites but which were not tested before by a walk. These sites are termed growth sites[2] and the set of these sites we call a growth perimeter G, which scales with the average cluster radius R as

$$G \sim R^{d_G} \sim S^{d_G/d_f} \qquad (1)$$

It is very important to seek relations among the various fractal dimensions, e.g., between static fractal dimensions d_f, d_ℓ and the dynamic d_w, d_G. In this presentation we will discuss a new "butterfly" walk[3] which visits only the growth sites and we will introduce some new relations between the fractal dimensions d_f, d_ℓ, d_{red} and d_G. In the end we will discuss the diffusion aspect of this growth problem and compare the "butterfly" diffusion to the random diffusion on a percolation cluster ("ant").

2. MODEL

Consider spreading of the infection on a square lattice started by an initial sick site [S]. Imagine the infection spreading by an infected "butterfly" who flies from one growth site [G] to another, randomly choosing the next from the probability distribution

$$P(r) \sim 1/r^\alpha \qquad (2)$$

where r is the distance from the most recently added sick particle to the chosen growth site and α is a parameter that governs the effective repulsion or attraction between G sites. The chosen G site is converted into an S site. The typical cluster of S sites with a growth perimeter G is shown on Fig. 1. If this procedure is performed on a percolation cluster, the additional choice has to be made: we convert G sites into S sites with probability p_c or into immune [I] sites with probability $1-p_c$. The sick and immune sites stay as such forever while the growth perimeter G constantly changes its identity: each time the sick site is added all its not yet determined nearest neighbors become G sites. The time t is increased by one only when growth occurs, i.e., when the S site is added; and, therefore, the "butterfly" moves in an artificial "growth" time t=S. After t time steps a large ramified cluster of S sites has been formed. The fractal dimensionality d_f of the grown cluster is the same as for percolation cluster at p_c (d_f = 91/48 in d = 2) since the ultimate connectivity of the S sites is identical to the connectivity of occupied sites in percolation. The main property of interest for this type of a walk is the fractal dimensionality d_G of the growth perimeter G.

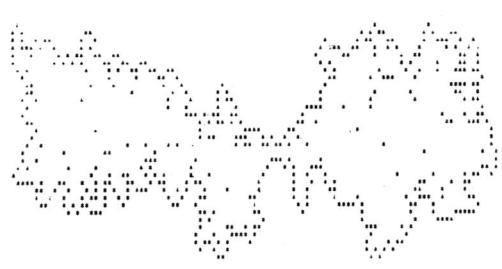

FIGURE 1
The cluster of 1,500 sites formed by an α = -3 "butterfly" on a Euclidean lattice (p=1). Only the growth sites * are shown. The resemblance to a butterfly is arbitrary.

3. THE CROSSOVER FROM THE LONG-RANGE TO THE SHORT-RANGE "BUTTERFLY"

Let us concentrate now on the effect of the tuning parameter α from Eq.(2) on the growth patterns. The case of α = 0 reduces to the Eden model on percolation and G sites are chosen at random (with probability 1/G). One can distinguish between the two well-defined limits in the "butterfly" behavior: the long-range limit when the "butterfly" tends to fly far away (but is limited by the extremities of the cluster) and the short-range limit when the "butterfly" tends to fly close-by (mostly to the nearest neighbors for large positive α). The limiting value α_c can be roughly estimated by considering the mean length of a flight, $\bar{r} = \int dr\, r\, P(r)$, which implies $\alpha_c \simeq 2$. Namely, the long-range behavior is observed for $\alpha < 2$ and it breaks down for $\alpha > 2$. Moreover, we find that for $\alpha > 8$ the short-range behavior establishes and the growth perimeter assumes roughly the same value for all $\alpha > 12$. How is this crossover behavior reflected on growth? The main result is that d_G changes continuously from the smaller long-range value $d_G \simeq 0.76$ ($\alpha < 2$) to the larger short-range value $d_G \simeq 1.04$ ($\alpha > 8$). Therefore, the spatial growth sequence can enhance or hinder the growth of G. We find that the kinetic exponent d_G can be continuously tuned while the static, d_f, does not change. Thus the dynamic universality classes are quite independent of the static ones. The "butterfly" turns out to be the first one-cluster growth model having such a feature and thereby yields insight into a generic feature for growth models.

4. THE FRACTAL DIMENSION IN THE LONG-RANGE LIMIT

A conjecture was proposed[4] for a long-range value of d_G based on the assumption that the scaling form[5] for the finished cluster perimeters P and the perimeters of still growing clusters is the same if the clusters are large enough

$$P = S\frac{1-p_c}{p_c} + AS^\sigma \qquad (3)$$

[here $\sigma = 1/\nu d_f$ and ν is a correlation length exponent; A is a constant]. The first term in

Eq. (3) can be easily identified with the number of blocked sites in a growing cluster of size S. Therefore, the growth perimeter G, which is equal to the number of all nearest neighbors to cluster sites S minus the identified block sites, should be proportional to the "excess" perimeter

$$G \sim S^\sigma \sim R^{1/\nu} \qquad (4)$$

Comparing Eq. (2) and Eq. (4) one gets

$$d_G = 1/\nu = d_{red} \qquad (5)$$

This conjecture is in a very good agreement with the numerical data for the long-range limit in two and three dimensions, and is exact for the Cayley tree. Note that the fractal dimension d_G of the growth perimeter in the long-range limit appears to be the same as the fractal dimension of singly-connected bonds[6] which might lead to some interesting insights into the connectivity dynamics of the growing percolation clusters.

We also suggest[4] a new scaling relation for a "chemical" fractal dimension d_ℓ [defined as $S \sim \ell^{d_\ell}$, where ℓ is a "chemical length"]

$$d_\ell = 1/(1-\sigma) = d_f/(d_f - d_G) \qquad (6)$$

predicted earlier by a numerical observation[7]. Note that since $d_\ell = d_f/d_{min}$, one also gets

$$d_{min} = d_f - d_G \qquad (7)$$

We see that the fractal dimension of the growth perimeter d_G is an important quantity and might reduce the number of "independent" fractal dimensions.

5. THE INTERPRETATION OF THE SHORT-RANGE LIMIT

The short-range limit value $d_G \simeq 1.04 \pm 0.04$ appears to be in a different kinetic universality class than the "ant" for which we find[8] $d_G \simeq 0.93$. Let us see if this statement holds for a short range "butterfly" and a random walk performed on a Euclidean lattice (p=1). For $\alpha >$ 16 the "butterfly" in d=2 is a well defined space-filling walk (see Fig. 2). The resulting behavior turns out to be the same for "butterfly" and random walk growth perimeters: $G \sim S/\ell n\,S$. This makes the discrepancy between the short-range "butterfly" and "ant" dynamic behavior on percolation even more intriguing. At present we do not have a satisfactory explanation of this phenomena.

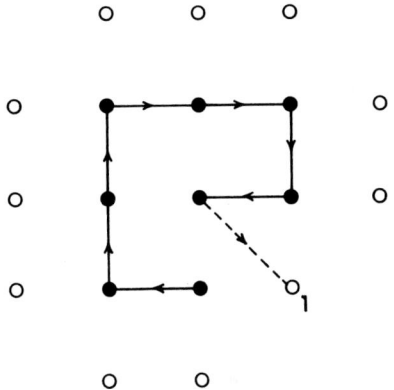

FIGURE 2
Typical 7-step trap for a short-range kinetic walk. "Butterfly" finds a way out by flying to one of the growth sites (o) according to Eq. (2). Growth site 1 is the most probable choice.

6. THE DIFFUSION ASPECT OF THE "BUTTERFLY" MODEL

Let us rescale the "growth" time t = S back to usual $t \sim S^{d_w/d_f}$, thus allowing for revisiting of already visited sites [here d_w is the dimension of the walk defined as $t \sim R^{d_w}$]. It is as if the random walk proceeds normally until it meets a growth site. At this moment the decision to occupy it or not is made according to a "butterfly" probability (Eq. (2)). If the growth site is not occupied, the revisiting continues until the next growth site. I suggest to call this type of a walk a growth-limited diffusion process. If we assume the relation for the "ant"[1]

$$d_w/d_f = 2 - d_G/d_f \qquad (8)$$

to hold for all α, we get a crossover for $d = 2$ percolation in a spectral dimension $d_s = 2d_f/d_w$ from $d_s \approx 1.25$ (long-range) to $d_s \approx 1.38$ (short-range) with the value for the "ant"[8] $d_s \approx 1.32$ in between the two. Thus, the random diffusion process corresponds to a moderately short-range "butterfly" with $\alpha \approx 4$. This is in accordance with our understanding that the random walk must have a finite probability to visit the growth sites that are relatively far away. For d_{min} in the long-range limit I get, combining Eq. (7) and Eq. (8)

$$d_{min} = d_f - d_w \qquad (9)$$

and combining Eq (6) and Eq. (8)

$$d_w/d_f = (1 + 1/d_\ell) \qquad (10)$$

ACKNOWLEDGEMENTS

The collaboration on this work is acknowledged with A. Bunde, H. J. Herrmann, J. Majid and H. E. Stanley. I wish to thank A. Aharony, S. Havlin, F. Leyvraz, R. Rubin and D. Stauffer for many stimulating discussions.

REFERENCES

1. H. E. Stanley, J. Stat. Phys. 36 (1984) 843.
2. F. Leyvraz, and H. E. Stanley, Phys. Rev. Lett. 51 (1983) 2048.
3. A. Bunde, H. J. Herrmann, A. Margolina, and H. E. Stanley, Phys. Res. Lett. 55 (1985) 653.
4. A. Margolina, J. Phys. A18 (1985) L651.
5. D. Stauffer, Phys. Rep. 54 (1979) 1.
6. A. Coniglio, J. Phys. A15 (1982) 3829.
7. S. Havlin, and R. Nossal, J. Phys. A17 (1984) L427.
8. H. E. Stanley, J. Majid, A. Margolina, and A. Bunde, Phys. Rev. Lett. 53 (1984) 1706.

PERCOLATION AND FRACTAL BEHAVIOR IN DISORDERED LATTICES

Panos ARGYRAKIS

Department of Physics, University of Crete, Iraklion, Crete, Greece.

We calculate very accurately the fractal exponents from random walks on randomly occupied 2-dim and 3-dim lattices. Using the efficient algorithm of the cluster-growth-technique we monitor the number of sites visited in an N-step walk and the mean-square displacement for the long time limit. We show that the crossover to the Euclidean behavior occurs very fast in the region above the critical threshold.

1. INTRODUCTION

In a series of papers recently[1-4] random walks on disordered lattices have been shown to be an interesting topic, especially when combined with the fractal properties that such lattices possess. A variety of physical phenomena are related and explored using random walks as the starting point, such as diffusion processes,[6] trapping[1], chemical reaction mechanisms,[5] etc. The idea and methods of percolation are then naturally combined with the above, since the critical percolation threshold provides a good test point for fractal behavior. In this work we summarize the results from random walk calculations at and above the critical point, we give the values of the fractal exponents as derived from S_N, the number-of-sites-visited in an N-step walk. Additionally, we give and discuss the error estimates in these numbers from their statistical analysis. Second, we report our results for the diffusion exponent D, as derived from R_N^2, the mean-square displacement. Last, we give the picture of the crossover from the critical point to the limit of a perfect crystal (no disorder at all).[6]

2. METHOD OF CALCULATIONS

We use Monte-Carlo simulation methods to monitor several random walk properties. The details of the algorithms are described elsewhere.[6] The size of lattices used is 4×10^6 sites, i.e. 2000x2000 for the 2-dim case and 160x160x160 for the 3-dim case. Random walks are monitored up to 2×10^5 steps. Using these numbers it is assured that all calculations are free from revisitation effects, boundary crossings, etc. so that the results reported here contain no approximation or extrapolations. Special attention is paid to the early time behavior, since additional exponents are needed to explain this time limit.[7]

3. RESULTS

It is important to distinguish between the different spectral dimensions reported. The well known formulae are:

$$S_N \sim N^{d_s/2} \quad (1)$$

$$R_N^2 \sim N^{2/D} \quad (2)$$

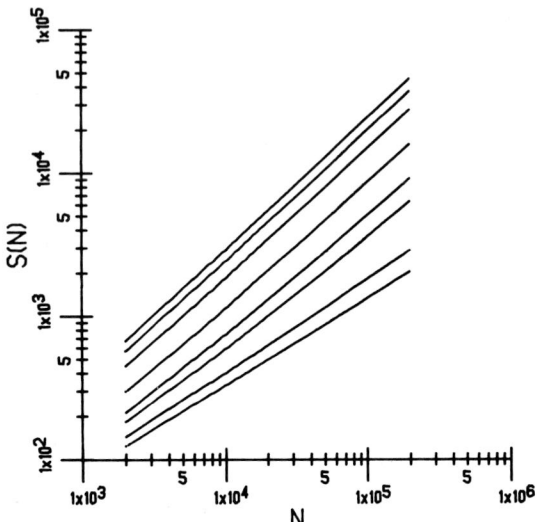

FIGURE 1

The number of sites visited S_N as a function of the number of steps N for 2-dim lattices. These are averages of 1000 realizations for walks that originate on any-size cluster. Top to bottom: p=0.5931, 0.60, 0.63, 0.65 0.70, 0.80, 0.90, and 1.00.

Calculations of random walks at the critical percolation threshold can be done either with the particle originating (a) on the largest percolating cluster, with exponents d_s and D or (b) on any-size cluster, including small finite clusters, with exponents d_s' and D'. For the long time limit it is handy to use the cluster growth technique[6] leading to case (b) and d_s'. But d_s and d_s' are related[4] through:

$$d_s' = d_s \left(2 - \frac{d}{d - \beta/\nu}\right) \qquad (3)$$

where d is the Euclidean dimensionality and β and ν are the static percolation

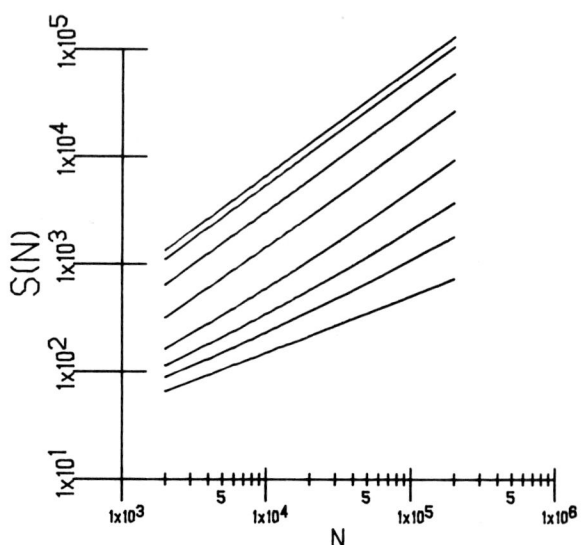

FIGURE 2

Similar to Fig. 1 but for 3-dim lattices. Top to bottom: p=0.3117, 0.32, 0.33 0.35, 0.40, 0.50, 0.75, and 1.00.

exponents. When one of the d_s or d_s' is calculated the other one can easily be deduced. We calculate d_s' from the slope of the lowest curves of Figures 1 and 2, where $\ln S_N$ is plotted vs. $\ln N$, for the 2-dim and 3-dim cases, respectively. In these plots the lowest curves refer to the critical point (i.e. p_c=0.5931 and p_c=0.3117), while the other curves refer to higher occupational probabilities p(see below). The slopes give:

$$d_s' = 1.23 \pm 0.02 \quad \text{(2-dim)} \qquad (4)$$

$$d_s' = 1.06 \pm 0.02 \quad \text{(3-dim)} \qquad (5)$$

Using Eq. 3 and the values of Eq. 4 and 5 we get:

$$d_s = 1.30 \pm 0.02 \quad \text{(2-dim)} \qquad (6)$$

$$d_s = 1.33 \pm 0.02 \quad \text{(3-dim)} \qquad (7)$$

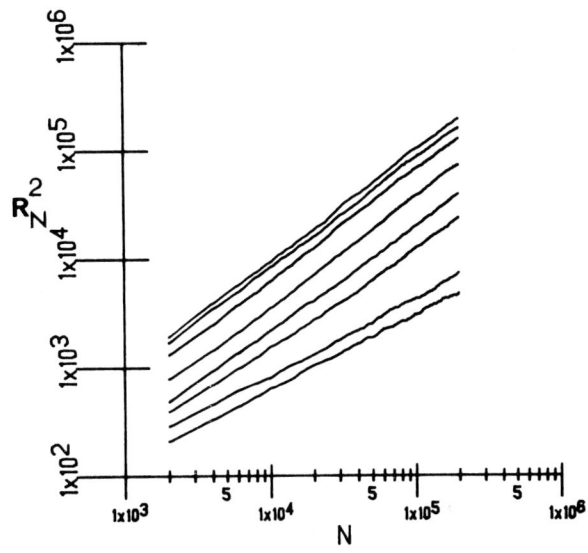

FIGURE 3

The mean-square displacement R_N^2 as a function of the number of steps N for the same p as in Fig. 1 (2-dim case).

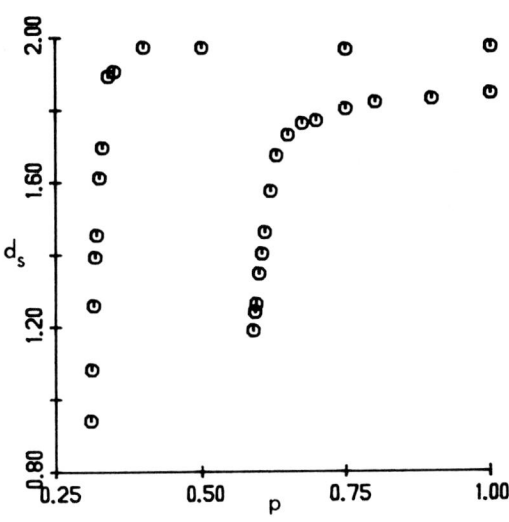

FIGURE 4

The effective d_s as a function of the occupation probability p, for 2-dim (lower) and 3-dim (upper) lattices.

The 2-dim value of d_s, which is calculated here indirectly, is in excellent agreement with previous work[7], where d_s was calculated directly from walks on the largest cluster, but for N up to N=5000 steps. The reported uncertainties of ±0.02 for Eq. 6 and 7 are derived from walks on the largest cluster <u>only</u> (case b) using standard methods of error propagation, i.e. taking into account the standard deviation (σ) of the S_N distribution and the number of realizations. For Eq. 4 and 5 the uncertainty is calculated graphically by considering the extreme-most values of the slopes. This is done so because here the S_N distribution is not smooth as in the previous case but skewed, as it contains a considerable number of S_N values close to zero, which is the contribution from small finite clusters. Thus σ here is very large. Typically, for N=4000 steps:

Case (a) $S_N = 189.8$ $\sigma = 114.0$ (8)

case (b) $S_N = 249.8$ $\sigma = 81.7$ (9)

Figure 3 shows a plot of $\ln R_N^2$ vs. $\ln N$ for the 2-dim walk. From the slope of the lowest curve we derive a value for the D' exponent:

$$D' = 2.89 \pm 0.05 \qquad (10)$$

The uncertainty here is calculated again graphically from Fig. 3, and the D value is in good agreement with previous work[8] for small N (N=300 steps).

Figure 4 shows the effective d_s exponent as a function of the occupation probability p, for the 2-dim and 3-dim lattices. The points are calculated from the slopes of the other curves of Fig. 1 and 2, from the long time limit (N=200000 steps). It is seen that the crossover to Euclidean behavior occurs rather fast and in the region close to the critical point for both dimensionalities. The expected value of 2 is not reached in the 2-dim case because of the additional logarithmic term that is present. From the line shapes of Fig. 4 we expect that scaling will be valid only in this small region of the fast rise, i.e. 0.31-0.35 (3-dim) and 0.59-0.65 (2-dim). This is also verified for both cases.[6]

Summarizing, our results for the spectral dimension d_s as calculated by two different methods using very accurate simulation algorithms agree quite well with the newer Aharony-Stauffer prediction,[9] i.e. for 2-dim lattices d_s is somewhat lower (by about 2%) than the value of 4/3 originally proposed. Our conservative uncertainty for these values is ±0.02, also from two different methods. Our diffusion exponent D' calculated for the long time limit comes out as expected.

REFERENCES

1. A.Blumen, J.Klafter, B.White, and G.Zumofen, Phys. Rev. Lett. 53(1984)1301.

2. R.B.Pandey, D.Stauffer, A.Margolina, and J.G.Zabolitzky, J. Stat. Phys. 34(1984)427.

3. S.Havlin, D.Ben-Avraham, and H.Sompolinski, Phys. Rev. 27A(1983)1730.

4. I.Webman, Phys. Rev. Lett. 52(1984)220.

5. R.Kopelman, P.W.Klymko, J.S.Newhouse, and L.W.Anacker, Phys. Rev. 29B(1984)3747.

6. P.Argyrakis and R.Kopelman, J. Chem. Phys. 81(1984)1015; J. Chem. Phys., In press.

7. A.Keramiotis, P.Argyrakis, and R.Kopelman, Phys. Rev. 31B(1985)4617.

8. D.Ben-Avraham and S.Havlin, J. Phys. A 15(1982)L691; J. Phys. A 15(1982)L311.

9. A.Aharony and D.Stauffer, Phys. Rev. Lett., 52(1984)2368.

FRACTALS IN PHYSICS
L. Pietronero, E. Tosatti (editors)
© Elsevier Science Publishers B.V., 1986

HIERARCHICAL FRACTAL GRAPHS AND WALKS THEREUPON

John MELROSE

Department of Chemistry, Royal Holloway and Bedford New College, University of London, Egham, Surrey, TW20 OEX, U.K.

Hierarchical fractal graphs are lattices which support tractable renormalisation transformations. A Brief discussion will be given of the structure of hierachies. Results for self-avoiding and random walks on hierachies are reported: the bounding of the Ising thermal exponent by the SAW exponent, the convergence of these exponents as D→∞, and proof of fractal Eienstein relation for a simple class of hierarchy. Emphasis is placed on explaining walk properties and anomolies on the basis of hierarchical structure. Monte-Carlo results are presented for random walks on Migdal-Kadanoff hierarchies.

1. INTRODUCTION

Self-similar hierarchical lattices[1,2] are finitly ramified with respect to sites. Generated iteratively, successive generations of lattice will be referred to as n^{th}, $n+1^{th}$ etc. units. The surface sites of units, those that link, some m^{th} unit in an $n^{th} > m^{th}$ unit, will be referred to as nodes; by hierachical definition the number g nodes is fixed independent of m and exact renormalisation can be carried out in a finite parameter space (the Sierpinkski gasket is a three noded hierarchy). Bond-hierachies[3] have just two nodes and are built by bond decoration; this simple class is considered below. Figure 1 shows the n = 4 unit of a Migdal-Kadanoff (diamond) bond-hierarchy.

Many other examples of hierachies can be found in the literature[10]. A number of physical problems have been studied on these lattices[10]. On fractals the analysis of walk problems has revealed anomolous behaviour. Here, after some observations on the general structure of hierachies, results for walks on hierarchies are discussed.

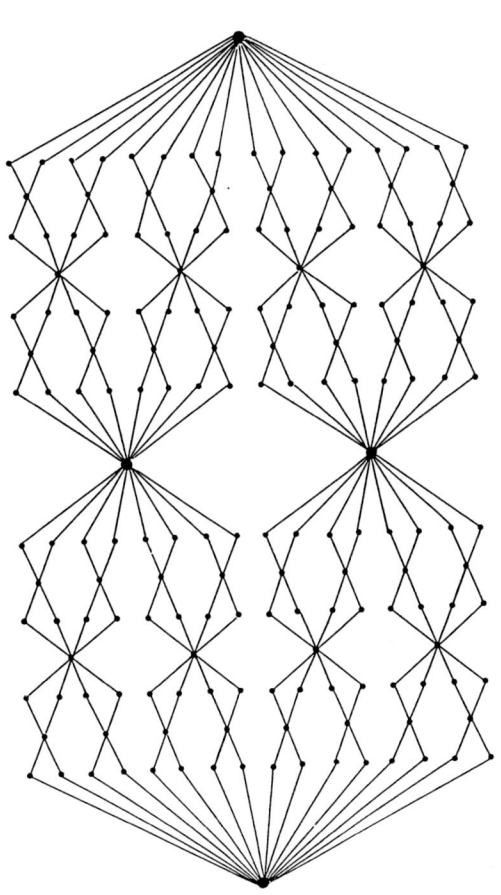

FIGURE 1

Hierarchies may be infinitely ramified with respect to bonds and hence may support phase transitions with Tc>0. Employing an intrinsic metric Melrose[4,5] defined the intrinsic dimension, D, and connectivity, Q, of bond hierarchies: $D = \ln(g)/\ln(b)$ where g is the number of bonds and b is the shortest path between the nodes of a n=1 unit. D characterises how the number of bonds on an nth unit, g^n, grows with its linear scale, b^n, (g=4, b=2 for the example g figure 1).

Infinite bond ramification along with finite site ramification is brought at the expense of an infinite range of site coordinations (see figure 1, this feature is also common on the duals of well known fractal lattices[5]). Let w stand for the coordination of the nodes on the n=1 unit (w=2 for fig. 1) then on an mth unit there are two nodes with coordination w^m and sites with coordinations $k_i w^j$ where $0<j<m-1$ and the k_i are independent of m and are the set of coordinations of sites on the n=1 unit other than the nodes. A site of coordination $k_i w^p$ will be referred to as p^{th} order. A pth order site is surrounded by k_i p^{th} units, $k_i w$ p-1th units etc., and the structure surrounding the site is fixed for this order out to a distance b^p. However, beyond b^p the structure varies from site to site of pth order depending upon location within the larger units.

2. SELF-AVOIDING WALKS

The author[3] examined the saw length exponent, ν_{SAW}, on several bond hierachies and found that the SAW fixed point eigenvalue λ_s, $\nu_{SAW} = \log(b)/\log(\lambda_s)$, is an upper bound on the Ising thermal exponent λ_t. Furthermore by considering families of hierarchy which have D increasing with successive members, it was found[3] that λ_s and λ_t converge as D→∞ if both D→∞ and Q→∞. However the SAW exponent ν_{SAW} does not converge towards the random walk exponent ν_{RW} (see below) as found on Euclidian lattices.

The non-convergence as D→∞ of random walk and SAW exponents can be understood from the structure of the lattices. Consider a high coordination site, the surface of some large unit. From the point of view of a random walk this surface site is similar to a surface of both many bonds and sites on a 'unit' of a regular lattice. However from the point of view of a SAW the single site surface is a strong constraint and the surface of the unit is quite different from that of a 'unit' of a regular lattice.

3. RANDOM WALKS

Renormalisation of random walk generating functions on bond-hierarchies is straight forward to carry out[6]. Recursions relations for the ensemble weight parameter, z, are generating functions for first passage walks across n=1 units. The eigenvalue, λ_z, at the fixed point, $z^*=1$, gives the expected number of steps, λ_z^m, for first passage from one node to the other of an mth unit. The random walk renormalisation is equivalent[6] to that of the gaussian model[8] and harmonic equations of motion with site terms (e.g. masses) scaled according to coordination. Using matrix algebra it can be proved[6] that on the class of bond hierachies the fractal-Einstein relation[7] holds:

$$\lambda_z = g\lambda_r \quad (1)$$

where λ_r is the resistance scaling eigen value.

On fractal lattices one would introduce the spectral dimension[8,9], $F = 2D\nu_{RW}$ where $\nu_{RW} = \ln(\lambda_z)/\ln(b)$. On fractals the number of sites, A(N), within the average spatial extent, R(N), of a walk of N steps obeys[9] $A(N) \sim R(N)^D = N^{F/2}$, where $R(N) \propto N^{\nu_{RW}}$. The range of a random walk, S(N) the distinct number of sites visited in N steps, is argued to obey $S(N) \sim A(N)\{N\}$ for F<2 {F>2}. However in the context of an infinite range of coordinations the intrinsic dimension defined above does not characterise the growth of the hierarchy around an arbitrary site[5] and the relation $A(N) \sim N^{F/2}$ does not hold.

The author has used monte-carlo to examine the range of random walks on the hierarchy of figure 1. The Migdal-Kadanoff hierarchies have a linear nature which allows a simple coding of sites into layers and ease of computation. Figure 2 shows a plot of $\ln(S(N))$ vs $\ln(N)$ for walks from a high order origin. $S(N)$ was averaged over 10^3 walks of 10^3 steps. For $N>50$ the plot was linear and averaging the slope over 10 such plots gave $S(N) \sim N^\Theta$ with $\Theta = 0.865 \pm 0.001$. Statistics were also gathered over 10 randomly chosen 0th order origins on a large cell. In this case the slope of the linear part of the plot was 0.856 ± 0.002. As $F = 2$ for this example the result $S(N) \sim N^{F/2}$ is found not to hold. (Investigation of MKH with larger multiplicity, $M(M=2$ for fig. 1), reveals $\Theta \to 1$ as $M, D \to \infty$.)

Statistics for returns to the origin are also anomolous on hierarchies. For a pth order origin the RG can at most[6] give the generating function for walks that return within b^p (i.e. without passing through any of the other nodes of the k_i surrounding p^{th} units). Consideration of walks that go beyond b^p and return must handle the different possible locations of the pth origin with respect to the larger scale structure, as described at the end of section 1.

Inspection of the linear nature of Figure 1 reveals that on the MKH statistics for returns to a pth order origin within b^p are simply those of the 1-dimensional lattice. For walks that go beyond b^p and return simple heuristic arguments suggest that on the MKH the probability of return does obey $\sim N^{-F/2}$. This is verified by monte-carlo. A ln-ln plot of the probability of return against N for 0th order origins on the example of figure 1 (F=2) has a slope of -0.95 ± 0.06 (statistics were gathered over 10^2 origins with $(3-4) \times 10^3$ walks of 200 steps).

4. CONCLUSIONS

Walks on hierarchical lattices with an infinite range of site coordinations have been shown to have anomalous behaviour over and above that of fractal lattices without this property.

ACKNOWLEDGEMENT

The author would like to thank the S.E.R.C. for a post-doctorial scholarship.

REFERENCES
1. A.N. Berker and S. Ostlund, J.Phys.C: Solid State Phys. 12 (1979) 4961.
2. R.B. Griffiths, Phys.Rev.B. 25 (1982) 5022 (see also Phys.Rev.B. (1984) and references therein).
3. J.R. Melrose, J.Phys.A.: Math.Gen. 18 (1985) L17.
4. J.R. Melrose, J.Phys.A: Math.Gen.16 (1983) 3077.

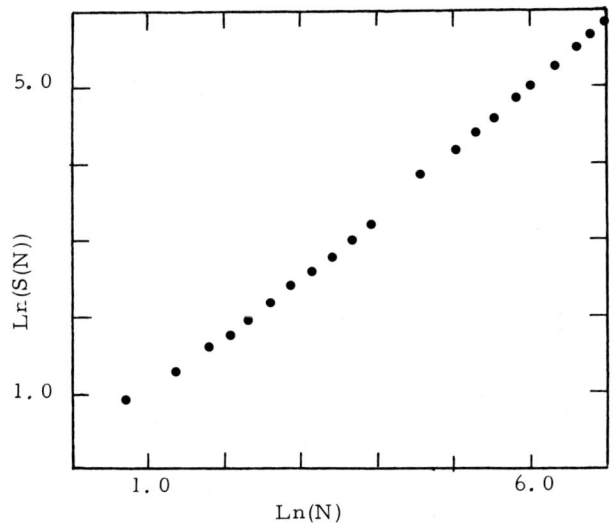

FIGURE 2

5. J.R. Melrose, J.Phys.A: Math.Gen.16 (1983) L407.

6. J.R. Melrose, J.Phys.A: Math.Gen. in print.

7. J.A. Given and B. Mandelbrot, J.Phys.A: Math.Gen.16 (1983) L565.

8. D. Dhar, J.Math.Phys.18 (1977) 577.

9. R. Rammal and G. Toulouse, J.Physique lett. 44 (1983) L13.

10. R.B. Griffiths and M. Kaufman, Phys.Rev.B. (1984) 30 244.

FRACTAL-LIKE EXCITON DYNAMICS: GEOMETRICAL AND ENERGETICAL DISORDER

Raoul KOPELMAN

Department of Chemistry, The University of Michigan, Ann Arbor, Michigan 48109

Fractal-like exciton kinetics were observed in constrained naphthalene samples: 1) Naphthalene embedded in porous glass (vycor) and various porous polymeric (e.g. nylon) membranes. 2) Grain-boundaries inside fresh and annealed naphthalene films formed by rapid, low temperature, vapor deposition. 3) Naphthalene embedded in "cavities" of polymeric glasses (e.g. plexiglass). The anomalous exciton annihilation kinetics gives effective spectral dimensions between one and two. Relative contributions of geometric constrains and energetic strains are obtained.

1. INTRODUCTION

Fractal-like exciton kinetics has been clearly documented for isotopic alloy crystals of naphthalene.[1,2] There the exciton transport is confined to the percolation clusters of the lower energy component (naphthalene molecules) of the binary alloy. Thus an energetic restriction leads to clear-cut fractal spaces (incipient infinite percolation clusters) with a clear-cut fractal random walk exponent[3] that is practically given by the fracton (spectral) dimension[4,5] of 4/3 (corrected for the presence of finite clusters[6,7,8,9]). While the steady-state energy transport experiments[10,11] are dominated by the fractal dimension d_f (i.e., the critical percolation exponents β, γ and δ), the transient exciton fusion kinetics are dominated by the spectral dimension d_s.[1,12,13] We emphasize again that the simple bimodal nature of the excitation energies in these alloys, coupled with a perfect crystal lattice (except for the binary compositional randomness), allowed one to conclude that the fractal-like exciton fusion kinetics are indeed a reflection of a fractal geometry (random clusters). Here we show how the effective spectral dimension, derived from exciton annihilation kinetics, can be used to characterize sample heterogeneities, such as porosity and grain boundaries. Furthermore, we indicate how the contributions of geometric and energetic heterogeneity can be unscrambled.

2. METHODS AND MATERIALS

We have studied triplet exciton annihilation in naphthalene aggregates which are embedded inside pores (of various membranes or vycor glass), cavities (in polymeric glasses) or grain boundaries. The common denominator of all these systems is the non-classical but rather fractal-like annihilation kinetics. The annihilation rate coefficient k is <u>not</u> a constant but rather described by the time-dependent form: $k \sim t^{-h}$, where h is a heterogeneity exponent ($0 < h < 1$) that vanishes ($h = 0$) only for homogeneous samples, giving back the classical form: $k = \text{const}$. We also measure the dependence of h on temperature. The effective spectral dimension of the fractal-like medium is given by: $d_s' = 2(1-h)$.

The fusion rate coefficient k is derived from time-resolved fluorescence (F) and phosphorescence (P) decays:[1,16] $k \sim F/P^2$ (but k can also be extracted from only F or P as a function of time[12,16]). Figure 1 is an example of the results. Table 1 gives the heterogeneity exponent h and the effective spectral dimension d_s' for various samples and temperatures.

Table 1

Sample	Pore Size (μ)	T (K)	Exponent (h)	Spectral dim. (d_s')
Acetate (GA8)	0.2	4	0.16	1.2 - 1.7
Acetate (GA1)	0.5	4	0.47	0.8 - 1.1
Acetate (GA3)	1.2	4	0.44	0.9 - 1.1
Nylon (B214)	0.2	4	0.21	1.2 - 1.6
Porous Glass (Vycor)	0.04	6	0.44	1.1
Plexiglass (PMMA)	?	77	0.1	1.8
Film (Grain Boundaries)	----	6	0.45	1.1
Glass-filter Paper	0.6	6	0.50	0.6 - 1.2
Glass-filter Paper	0.6	80	0.20	1.2 - 1.6
Paper-filter Paper	0.6	6	0.33	0.8 - 1.3
Paper-filter Paper	0.6	80	0.06	1.9

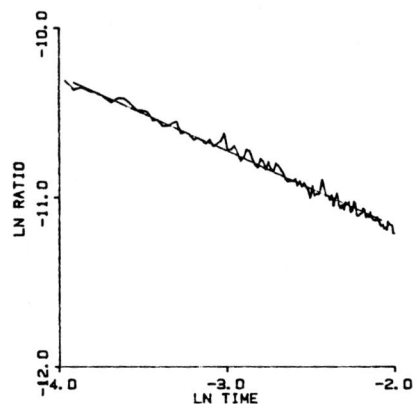 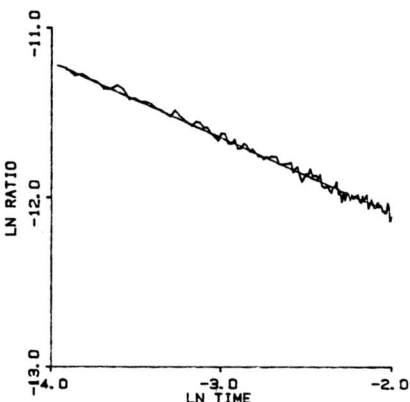

FIGURE 1
Annihilation rate coefficient, $k = F/P^2$, vs. time on a log-log representation for a fresh (left) and annealed (right) film of naphthalene (6K), deposited from the vapor at about 50 K. Slope (h) is 0.44 (left) and 0.45 (right). See Table 1.

3. RESULTS AND DISCUSSION

We see (Table 1) that all samples exhibit a fractal-like (heterogeneous) behavior (h > 0) at low temperatures. There is a reduction in h at higher temperatures. This is consistent with the subordination theorem:[14] $d_s' = \beta d_s$ where β is the parameter characterizing the anomalous hopping time distribution (e.g. for continuous time random walk). We also see that some samples have h → 0 for higher temperatures. This implies that the "fractal-like" effects are entirely due to energy disorder, so that at higher temperatures the energy disorder parameter W is small compared to T. For other samples, the effective geometry may indeed be fractal-like at room temperature. Obviously our d_s' provide only lower limits for the "real" (geometric) d_s. We note that for vycor porous glass (at a low temperature) $d_s' \sim 1.1$, which is consistent with the literature value[15] of $d_f \lesssim 2$ (remembering that for fractals $d_s \leq d_f \leq d$ where d is the embedding Euclidean dimension). We emphasize that in all these samples, and especially in the thin films, most of the bulk might be quite crystalline (i.e., Euclidean), but our slow experimental time scales assure us that only the sluggish kinetics in the fractal-like regions do contribute to our observations.

For the larger pore membranes, at higher temperatures, we observe the same behavior as in deposited films, i.e., no delayed fluorescence on the millisecond time-scale. In short, this approach might characterize crystalline domain sizes. In addition, we have found excellent correlation between the fractal-like kinetics and a number of spectroscopic features: 1) The spectral bandwidths (W); 2) Observation of super-trap (betamethylnaphthalene) emissions; 3) Typical photophysical product (excimer) emissions; 4) Typical photochemical product (radical) emissions.

We have certainly established a correlation between fractal-like exciton annihilation kinetics and the geometrical constrains and/or energetic disorder of a number of samples of pure naphthalene. We have shown an analogous behavior to that of isotopic mixed naphthalene crystals (percolation clusters), in contrast to the behavior of perfect, pure naphthalene crystals. Further work is in progress with the aim to unscramble the contributions of fractal-like geometry from those of energy disorder (fractal-like behavior on the potential energy surfaces or a "fractal" waiting-time distribution of the excitation hopping).

AKNOWLEDGEMENTS

The experimental work is that of L. A. Harmon, S. J. Parus, J. Prasad and E. I. Newhouse. Supported by NSF Grant No. DMR 8303919 and NIH Grant No. 2R01 NS80116-16.

REFERENCES

1. P.W. Klymko and R. Kopelman, J. Phys. Chem. 87(1983) 4565.

2. P. Evesque and J. Duran, J. Chem. Phys. 80 (1984) 3016.

3. A. Aharony and D. Stauffer, Phys. Rev. Lett. 52 (1984) 2368.

4. S. Alexander and R. Orbach, J. Phys. (Paris) Lett. 43 (1982) L625.

5. R. Rammal and G. Toulouse, J. Phys. (Paris) Lett. 44 (1983), L13.

6. P.G. de Gennes, C. R. Acad. Sci. Ser. A296 (1983) 831.

7. D. Ben-Avraham and S. Havlin, J. Phys. A15 (1982) L691.

8. I. Webman, Phys. Rev. Lett. 42 (1984) 220.

9. P. Argyrakis and R. Kopelman, Phys. Rev. B29 (1984) 511.

10. D.C. Ahlgren, Ph.D. Thesis, University of Michigan (1979).

11. J.S. Newhouse, Ph.D. Thesis, University of Michigan (1985).

12. R. Kopelman, P.W. Klymko, J.S. Newhouse and L.W. Anacker, Phys. Rev. B29 (1984) 3747.

13. L.W. Anacker and R. Kopelman, J. Chem. Phys. 81 (1984) 6402.

14. J. Klafter, A. Blumen and G. Zumofen, J. Stat. Phys. 35 (1984) 561.

15. U. Even, K. Rademann, J. Jortner, N. Manor and R. Reisfeld, Phys. Rev. Lett. 42 (1984) 2164.

16. L.A. Harmon, Ph.D. Thesis, University of Michigan (1985).

NYQUIST, DIFFUSION and FLICKER (1/f) NOISE IN FRACTALS AND PERCOLATING NETWORKS

Rammal RAMMAL

Centre de Recherches sur les Très Basses Températures, C.N.R.S., B.P. 166 X, 38042 Grenoble Cédex, France

Recently obtained results for Nyquist noise and for the magnitude of the flicker (1/f) noise in fractal and random resistor networks are reviewed. New results pertaining to continuum percolation models are presented and compared with recent experiments done on metal-insulator mixtures.

1. INTRODUCTION

Statistical self-similarity is emerging as an important concept underlying the behavior of disordered systems. In percolation networks, for example, the notions of fractal, spectral and spreading dimensions have been introduced in order to describe different physical properties of such fractal structures. Intuitively, it is plausible that an infinite number of exponents must be used to characterize a fractal. Recently, this was shown to be the case : the magnitude (not the frequency dependence) of the resistance noise spectrum (flicker 1/f noise) depends actually on a new exponent pertaining to fractal lattices. This exponent, b, can be viewed as a member of an infinite family of exponents which includes the fractal and spectral dimensions. A detailed account of these considerations on flicker noise in fractal and percolation networks has been presented in Refs. 1-5. Accordingly, in what follows we shall limit our exposition to two specific questions : i) the thermal or Nyquist noise in fractal networks and, ii) the flicker noise in continuum percolation models where experimental results are now available.

2. NYQUIST NOISE
2.1. General considerations

In general two types of noise spectrum of a macroscopic sample will be considered. Thermal or Nyquist noise, which is independent of the current intensity across the sample, and flicker (1/f) noise which is proportional to the squared intensity. Microscopic fluctuations of local tensions or local resistances can then be used respectively to model the noise. In this section we shall consider the Nyquist noise only. The assumption that fluctuation-dissipation theorem is valid on a fractal resistor network, leads to the following expression for the power spectrum of the voltage difference V across the sample (Nyquist noise)

$$S_V(\omega) = \lim_{\theta \to \infty} \frac{1}{\theta} \left| \int_0^\theta e^{i\omega t} V(t) dt \right|^2 = 2k_B T/G(L,\omega)$$

where $G(L,\omega)$ denotes the real part of the conductance at length scale L and frequency ω. The fact that $S_V(\omega)$ scales as the inverse of $G(L,\omega)$ can be justified as follows. Assume that on each branch α of the network, which bears a resistor r_α, there is a random tension generator $v_\alpha(t)$ in series with the resistor r_α. The noise generators are defined through their common stationary spectral density $s(\omega)$. It is not difficult to derive the following composition rules for the tension noise :

series rules : $s = \sum_\alpha s_\alpha$, $r = \sum_\alpha r_\alpha$

parallel rules : $s = \sum_\alpha s_\alpha (r/r_\alpha)^2$, $r^{-1} = \sum_\alpha r_\alpha^{-1}$.

Therefore, for identical (r_α, s_α), s assumes the same behavior as the total resistance. It is interesting to compare these composition rules with those derived in Ref. 1 for the resistance fluctuations model.

2.2. Model

In what follow we shall consider a typical model of resistor networks, where the physical content is equivalent to the diffusion process on the considered structure. Nearest neighbour sites of a given lattice are connected by conductors each of which is chosen to have a purely resistive impedance R. Each site not belonging to the electrodes is connected to ground by a circuit consisting of a capacitance C and an emf $U_i e^{i\omega t}$. The emf on the electrodes is equal to the external potential. Noise is introduced into the network in the following way[6]. In each conductor between nn sites k, ℓ there is a random, thermally generated component to the current $J_{k\ell}(t)$ with zero mean and a correlation function
$< J_{k\ell}(t) J_{k\ell}(t') > = (2k_B T/R) \cdot \delta(t-t')$, where T is the temperature. The basic equation of this network is given by the conservation of charge Q_i at easy site i :

$$\frac{dQ_i}{dt} = -\frac{1}{R} \sum_k (V_i - V_k) - \sum_k J_{ik}(t)$$

Here V_i is the voltage at site i :
$Q_i = C(V_i - U_i)$.

2.3. Euclidean networks

Let us consider first the case of an Euclidean lattice, of linear size L, containing L^d sites. The spectra of the source terms are assumed to be white and spatially uncorrelated. Two large planar electrodes, separated by a distance L along the x axis are connected to the network and the voltage power spectrum is measured between these two electrodes. In this case, the expression of $S_v(\omega) \equiv F(\omega)$ is easy to derive and the result can be written as
$F(\omega) = \frac{2k_B T}{G(L)} \psi(\ell(\omega)/L)$. Here, $\ell(\omega) = (2D/\omega)^{1/2}$
refers to the diffusion length, $D = a^2/RC$ is the diffusion constant (a = lattice spacing) and G(L) denotes the macroscopic conductance of the network at $\omega = 0$. This result can also be written as $F(\omega) = (A/DL^{d-2}) \cdot \psi(\ell(\omega)/L)$, where A denotes a constant term and ψ is a scaling function with $\psi(u \ll 1) \sim u$ and $\psi(u \gg 1) \sim 1$. The Nyquist result is then recovered at low frequency : $F(\omega)$ is independent of ω. The crossover to a diffusion regime $F(\omega) \sim \omega^{-1/2}$ occurs at
$\omega = \omega_D \equiv \omega_0 (a/L)^2$, where $\omega_0 = 1/RC$. For $\omega \gg \omega_0$, $\ell(\omega) \ll a$ and $F(\omega) \sim \omega^{-2}$. The physical meaning of these different regimes can be explained as follows.

a) $\omega \ll \omega_D$: in this limit, the fluctuations become correlated across the entire sample. The space-time correlation function of $V(\vec{r},t)$ becomes independent of ω.

b) $\omega_D \ll \omega \ll \omega_0$: $\ell(\omega) \ll L$ and the sample may be devided into correlated regions, each of linear size $\ell(\omega)$. However onlythe two thin plates of width $\ell(\omega)$ located at the electrodes can contribute to $F(\omega)$. Therefore in the previous expression L^{d-2} must be replaced by the conductance of the outer shells $\sim L^{d-1}/\ell(\omega)$ and the above result follows.

c) $\omega \gg \omega_0$: at very high frequency where $\ell(\omega) \ll a$, the capacitor become short-circuit to ground. L^{d-2} must be replaced by L^{d-1}/a as in (b) and $F(\omega)$ becomes:

$$F(\omega) \sim (\frac{\omega_0}{\omega})^2 a/DL^{d-1}$$

It is important to notice that the argument used in cases (b) and (c) can be used in the case where point-like electrodes are used instead of the planar ones. In such a case, the corresponding results can be written without difficilties.

2.4. Fractal networks

Let us consider now the case of a fractal network, where anomalous diffusion is known to occur, without any disorder. The exponent ν_{rw} characterizing this anomalous diffusion is given by the ratio $\tilde{d}/2\bar{d}$ of the spectral ($\tilde{d}$) and the fractal ($\bar{d}$) dimensions[7,8]. Furthermore, the conductance of a sample (linear size L) is given by $G(L) \sim L^{\beta_L}$, where $\beta_L = \bar{d} - 1/\nu_{rw}$. Following the same lines of ideas, the scaling form of $F(\omega)$ is controlled by that of $G(L,\omega)$.

a) At very low frequencies, i.e. $\ell(\omega) \gg L$ ($\ell(\omega) \sim \omega^{-\nu_{rw}}$ denotes the diffusion length), the zero frequency regime must be obeyed and $F(\omega) \sim k_B T . L^{-\beta_L}$. In this limit $\omega \ll \omega_D \equiv \omega_o (a/L)^{1/\nu_{rw}}$, the space-time correlation function of the voltage becomes independent of ω, in contrast with the incorrect statement made in Ref. 9. This result is actually independent of the contact geometry between the electrodes and the sample.

b) $a \ll \ell(\omega) \ll L$: assuming a point-contact geometry between the electrodes (2 points at distance L) and the sample, $F(\omega)$ is given now by the previous expression where L is replaced by $\ell(\omega)$: $F(\omega) \sim k_B T . (\ell(\omega))^{-\beta_L} \omega^{-1+\tilde{d}/2}$ ($\tilde{d} < 2$). In this limit, only regions of linear size $\ell(\omega)$, around the two contact points can contribute to $F(\omega)$.

c) $\ell(\omega) \ll a$: as in case (b), the L dependence is governed by the contact geometry, but the frequency dependence is still given by $F(\omega) \sim (\omega_o/\omega)^2$. It should be noticed that, beside the static limit $\omega \ll \omega_D$, the size dependence of $F(\omega)$ is controlled by the scaling form of the conductance : $G(L,\omega) = L^{\beta_L} f(\frac{\ell(\omega)}{L})$ in the present case, and by the contact geometry. Therefore, the analysis presented here can be extended to more complicated networks of arbitrary impedances in the branches[10].

2.5. Percolation networks

We turn now to the case of the percolation networks at $\Delta p = p - p_c \gtrsim 0$. In this case four length scales are of importance : the lattice spacing a, the diffusion length $\ell(\omega)$, the percolation correlation length $\xi_p = a(\Delta p)^{-\nu_p}$ and the sample linear size L. Different regimes can be classified according to the ratio ξ_p/L. Our discussion will be limited here to the relevant physical regimes.

a) case $\xi_p \ll L$. The sample can be viewed as an Euclidean lattice, with unit cells containing ξ_p^d sites each. Therefore, $a \to \xi_p$, $\omega \to \omega^* = \omega \xi_p^{1/\nu_{rw}}$ and the renormalized diffusion length becomes $\ell^*(\omega) \simeq \xi_p \omega^{*-1/2}$. Here, $\nu_{rw} = \nu_p/(t - \beta_p + 2\nu_p)$.

i) $\ell^*(\omega) \gg \xi_p$. In this limit, two cases must be distinguished. The first corresponds to $\ell^*(\omega) \gg L \gg \xi_p$, i.e. the low frequency regime of Euclidean lattices, and then
$$F(\omega) = \frac{A}{L^{d-2}D(p)} P(p) \sim \frac{A}{L^{d-2}} (\Delta p)^{2\beta_p - t}.$$
Here $P(p) \sim (\Delta p)^{\beta_p}$ denotes the probability of being on the infinite cluster and $D(p) \sim a^2 \omega_o (\Delta p)^{t-\beta_p}$ is the diffusion constant. The second case corresponds to the opposite regime where $\xi_p \ll \ell^*(\omega) \ll L$ and then
$$F(\omega) \sim \frac{A . P(p)}{L^{d-2} D(p)} \cdot \frac{\ell^*(\omega)}{L} .$$

ii) $\ell^*(\omega) \ll \xi_p$. Following the results obtained above, one obtains :
$$F(\omega) = A (\frac{\omega_o}{\omega^*})^2 \frac{\xi_p}{L} \frac{P(p)}{L^{d-2}D(p)}$$

Note that close to p_c, the noise is proportional to $\omega^{-1/2}$ down to a frequency given by $\omega^*/\omega_o \simeq \xi_p/L$, i.e. $\omega \gtrsim \omega_o(a/L)^2 (\Delta p)^{t-\beta_p}$ instead of $\omega_o/(a/L)^2$ in the perfect network ($p = 1$). This noise excess at large ω must be compared with experimental data. Its physical origin is simply the renormalization of the diffusion constant $D(p \gtrsim p_c)/D(p = 1) = (\Delta p)^{t - \beta_p}$ due to the dilution disorder.

b) case $\xi_p \gg L$. Very close to p_c, the correlation length ξ_p becomes larger than L. In this limit, the fractal structure of the infinite cluster becomes relevant. The expression of $F(\omega)$ follows from the results of the previous section.

3. FLICKER (1/f) NOISE

3.1. Resistance fluctuations model

Flicker noise on fractal or random resistor networks can be modeled by a resistance fluctuations model[1-5]. In this model, the resistance of each branch α in the network is assumed to have a small fluctuating part $\delta r_\alpha(t)$. δr_α's are assumed to be time dependent uncorrelated random variables with mean zero and Fourier-Transform covariance :
$<\delta r_\alpha(\omega)\delta r_\beta(-\omega)> = \delta_{\alpha\beta} \cdot \rho_\alpha^2(\omega)$. These fluctuations could be produced by an arbitrary noise mechanism. Assuming that $\rho_\alpha(\omega)$ is the same in all branches, the magnitude of the relative noise for the network can be defined by $S_R/R^2 = <\delta R\delta R>/R^2$. Here R and δR are respectively the overall resistance (one-port configuration) and its time fluctuation. S_R so defined is the relevant quantity for the magnitude of 1/f noise measured under constant external current. In fact, as long as each branch resistance fluctuates independently with the same spectrum, the explicit frequency dependence can be discarded.

For fractal resistor network, of linear size L, S_R has been shown to behaves as a power law: $S_R \sim L^{-b}$ ($L \gg 1$) where b denotes a new exponent pertaining to the fractal structure. Indeed, both b and β_L are associated with the moments of the current distribution $\{i_\alpha\}$ in the network :
$R \sim \Sigma_\alpha i_\alpha^2 \sim L^{-\beta_L}$ and $S_R \sim \Sigma_\alpha i_\alpha^4 \sim L^{-b}$
($-\beta_L \leq b \leq \bar{d}$).
When applied to percolation networks, this result leads to the prediction that near p_c, both R and the relative noise must diverge, but with different exponents. The detailed calculations of b, using various methods, will be found in Refs. 1-5.

3.2. Experimental results

Very recently, a careful study of 1/f noise near p_c, has been performed in 3D carbon-wax mixtures[11]. Close to p_c, both the resistance R and the power spectrum S_R have been shown to diverge in perfect agreement with the above prediction. Using the notations $R \sim (\Delta p)^{-t}$ and $S_R/R^2 \sim (\Delta p)^{-\kappa}$, the measured exponents are $t = 2.3 \pm 0.4$ and $\kappa = 5 \pm 1$. The direct plot of S_R v.s. R leads to $S_R \sim R^Q$ where $Q \equiv 2 + \kappa/t = 3.7 \pm 0.2$ is the noise v.s. resistance exponent. Similar measurement of Q on 2D films have also been performed by two other groups[12,13]. In clumped evaporated gold films[12] subjected to ion milling, a large value Q = 4 has been obtained. Much larger values, $5.4 \leq Q \leq 8.1$ were obtained[13] in metallic films (Al, Cr, In) where the metal was removed by sandblasting. These values of t and κ are actually different from the predictions of lattice percolation theory and must be attributed to continuum percolation corrections. Therefore, as will be shown below, the 3D data seem to provide the first quantitative confirmation of the continuum percolation theory.

3.3. Continuum percolation exponents

First let us mention that the measured exponents are actually outside the bounds found for Q in the lattice percolation theory. Using the known bounds[1-3] for the exponent $b \equiv d - \kappa/\nu = d - t(Q-2)/\nu$:
$-\beta_L \leq b \leq -2\beta_L - 1/\nu$, it is easy to obtain :
$2.82 \leq Q \leq 3.05$ in 2D ($\nu = 4/3$, $-\beta_L = 0.973$) and $2.84 \leq Q \leq 2.85$ in 3D ($\nu = 0.88$, $-\beta_L = 1.16$). Here $\beta_L = d-2-t/\nu$, d is the Euclidean dimension and ν is the correlation-length exponent. The effective medium theory (EMT) gives the value $Q_m = 3$ for Q.

As was pointed out by different authors[14-16], transport exponents such as t and κ, can be modified in continuum percolation models, in contrast with static exponents (e.g. ν). The simplest model is provided by the following probability distribution p(g) of bond conductances in the equivalent lattice model :
$p(g) = (1-p)\delta(g) + p \cdot h(g)$. Here h(g) is a

normalized function. The "Swiss-cheese" class of models is actually a possible realization[14] of p(g), with an anomalous distribution $h(g) \sim g^{-\alpha} (\alpha < 1)$ near g = 0. For this class of models, the conductivity exponent is given by : $t(\alpha) = \nu(d-2) + 1/(1-\alpha)$ for $0 \leq \alpha < 1$ and $t(\alpha) = \nu(d-2)+1$ for $\alpha \leq 0$. This result is implicitly contained in Ref. 15 and coincides with the large d limit[16]. The simplest derivation of this result is probably the following argument. The conductance $g = (\sum_\ell g_\ell^{-1})^{-1}$ of L conductances $\{g_\ell\}$, $1 \leq \ell \leq L$ taken from h(g) and combined in series is given, at $g \sim 0$, by min $\{g_\ell, 1 \leq \ell \leq L\}$. A simple calculation[14] yields $g \sim L^{1/(\alpha-1)}$ ($0 \leq \alpha < 1$), $g \sim L^{-1}$ ($\alpha \leq 0$) and leads to the desired results $-\nu\beta_L = 1/(1-\alpha)$ and 1 respectively. Note that $t(\alpha)$ so obtained differs from its EMT value[16]: $t_m = t(\alpha) - \nu(d-2)$.

The calculation of κ can be carried out similarly[4], using the series composition rule[1] for the relative noise : $s = \sum_\ell s_\ell (g/g_\ell)^2$. For the sake of simplicity we shall consider the class of "Swiss-Cheese" models and assume[17] $g_\ell \sim \delta_\ell^u$, $s_\ell \sim \delta_\ell^{-v}$ where $u = 1/(1-\alpha)$ and v is the exponent relating the relative noise s_ℓ of bond ℓ to the neck width δ_ℓ. Depending on the values of u and v one obtains three distinct relevant cases. a) $\kappa = d\nu + v$ for $u > 1$, $v + 2u > 1$; b) $\kappa = d\nu + (2u + v - 2)$ for $u < 1$, $v + 2u > 1$ and c) $\kappa = d\nu - 1$ for $u < 1$ and $v + 2u < 1$. These values are actually different from the EMT results[17] : $Q_m = 2+v/u$ at $u > 1$ and $Q_m = 3 + (v-1)/u$ at $u < 1$, which are expected to be correct[1-3] far from p_c.

Within the framework of "Swiss-Cheese" models, where $u = d-3/2$ and $v = d-1/2$, one obtains the following results. At d = 2 (case b) : $t = t_m = 1$; $\kappa = 3.16$, $\kappa_m = 2$ and $Q = 5.16$, $Q_m = 4$. At d = 3 (case a) : $t = 2.38$, $t_m = 1.50$; $\kappa = 5.14$, $\kappa_m = 2.5$ and $Q = 4.16$, $Q_m = 3.67$. Clearly the data of Ref. 11 fit nicely with these estimations. The results of Ref. 12 seem to follow the EMT[17] estimations, whereas those of Ref. 13 are definitely larger than the above estimations. Note that, in all cases, an enhancement of Q is obtained. However, a definitive comparison in 2D would require. The simultaneous measurement of both exponents : t and κ. Furthermore, the measurement of S_R at different temperatures would be useful for the identification of the conduction mechanism and the microscopie origin of the observed noise.

REFERENCES

1. R. Rammal, C. Tannous and A.M.S. Tremblay, Phys. Rev. A31 (1985), 2662.

2. R. Rammal, J. Physique Lett. (Paris), 46 (1985), 129.

3. R. Rammal, C. Tannous, P. Breton and A.M.S. Tremblay, Phys. Rev. Lett. 54 (1985), 1718.

4. R. Rammal, Phys. Rev. Lett. "in print".

5. R. Rammal, in : Physics of finely divided matter, Les Houches winter school, Ed. M. Daoud (Springer, Berlin), "in print".

6. M.J. Stephen, J. Phys. C : Sol. State Phys. 11 (1978), L965.

7. S. Alexander and R. Orbach, J. Physique Lett. (Paris), 43 (1982), 623.

8. R. Rammal and G. Toulouse, J. Physique Lett. (Paris), 44 (1983), 13.

9. R. Rammal, J. Physique Lett. (Paris), 45 (1984), 1007.

10. J.M. Luck, J. Phys. A : Math. Gen. 18 (1985), 2061.

11. C.C. Chen and Y.C. Chou, Phys. Rev. Lett. 54 (1985), 2529.

12. R. Koch and R. Laibowitz, Phys. Rev. B, RC, "in print".

13. G. Garfunkel and M.B. Weissman, Phys. Rev. Lett. 55 (1985), 296.

14. B.I. Halperin, S. Feng and P.N. Sen, Phys. Rev. Lett. 54 (1985), 2391.

15. A. Ben-Mizrahi and D.J. Bergman, J. Phys. C : Sol. State Phys. 14 (1981), 909.

16. P.M. Kogut and J.P. Straley, J. Phys. C : Sol. State Phys. 12 (1979), 2151.

17. A.M.S. Tremblay and S. Feng, Phys. Rev. B, RC, "in print".

A REAL-SPACE RENORMALIZATION GROUP APPROACH TO ELECTRICAL AND NOISE PROPERTIES OF PERCOLATION CLUSTERS

J.M. LUCK

Service de Physique Théorique, CEN-Saclay, 91191 Gif-sur-Yvette Cedex, France.

The Migdal-Kadanoff real-space renormalization scheme is used to investigate electrical properties of percolation clusters in two and three dimensions. This method is equivalent to the exact solution of the percolation problem on certain hierarchical lattices.
The static properties of percolation clusters on such lattices are reviewed briefly. We then focus on the complex conductivity of resistor-capacitor mixtures, with emphasis on the critical (at $p \to p_c$) frequency dependence of the loss angle δ, a quantity which has been measured in glass microbeads mixtures and in microemulsions.
The critical amplification of resistor noise (Flicker or $1/f$ noise) is also studied. It is shown to obey universal scaling laws for p close to p_c and low frequency. These laws involve two critical exponents X and Y, which are computed explicitly, and related to the noise dimension b recently introduced by Rammal et al.

1. INTRODUCTION

We present the results of a real-space renormalization group calculation of the frequency dependence of the AC electrical conductivity and flicker noise amplification in random resistor networks near their percolation threshold. The plan of this report will closely follow that of our original publication[1], which the reader is referred to for more detailed derivations, as well as for a complete bibliography. The two - and three - dimensional cases will be treated in parallel.

2. HIERARCHICAL LATTICES AND PERCOLATION

We have chosen to use the Migdal-Kadanoff approximation. This real-space renormalization scheme, which is only approximate for models on regular lattices (i.e. lattices possessing a translation group), becomes *exact* (for non-random models) on certain recursively built *hierarchical* lattices. The lattices we shall use in dimension two and three are generated as indicated on Figures 1 and 2 respectively.

The static properties of the bond percolation problem are very easily determined on these hierarchical lattices. They admit an *exact*

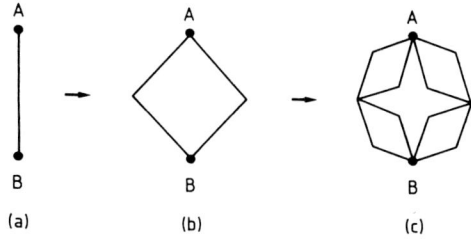

FIGURE 1
The recursive construction of the 2D (diamond) hierarchical lattice.

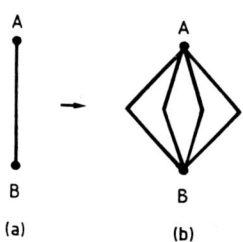

FIGURE 2
Same as Figure 1, for the 3D lattice.

renormalization group transform T acting on the probability p for a bond to be occupied : the lattice at generation N with probability p is equivalent to the same lattice at generation

(N-1) with a *renormalized* probability T(p). The transform T reads:

$$T_{2D}(p) = 1 - (1 - p^2)^2 \quad (1a)$$
$$T_{3D}(p) = 1 - (1 - p^2)^4 \quad (1b)$$

The mapping T has two superstable fixed points at p=1 and p=0, describing percolating and non-percolating pure phases respectively, and one unstable fixed point p_c which corresponds to the percolation threshold. Let $\mu = T'(p_c)$ be the derivative of T at this fixed point. It can be shown that the correlation length exponent ν is related to μ by:

$$\nu = \ln 2 / \ln \mu \quad (2)$$

The numerical values of p_c and the exponents are given in the Table at the end of this report.

3. ELECTRICAL AC CONDUCTIVITY

We consider now the following problem: what is the *macroscopic AC conductivity* Σ of our lattice if each occupied bond is a resistor (impedance R_0) and each empty bond is a capacitor (impedance $(iC_0\omega)^{-1}$). This problem models a wide class of random media such as powders, microemulsions, ... Without giving any detail, let us mention that this problem is still too difficult to be solved in general on hierarchical lattices, since it amounts to solve a non-linear integral equation for the probability distribution of the bond conductance. A truncation of this integral equation is needed: the critical properties of the conductivity $\Sigma(p,\omega)$ are correctly described if the truncation reduces the full problem to a five-dimensional (real) mapping $\overline{T}$. We shall only present here the results of this approach concerning the critical regime: $p \to p_c$ and $\omega/\omega_0 \to 0$, where $\omega_0 = (R_0 C_0)^{-1}$. (The conductivity Σ has a very analogous critical behaviour for $\omega \ll \omega_0$; $p \to p_c$ and for $\omega \gg \omega_0$; $p \to 1-p_c$). The low-frequency critical regime is characterized by the *scaling functions* $\varphi_\pm$:

$$\Sigma = R_0^{-1} |p-p_c|^t \varphi_\pm \left[i\omega/\omega_0 |p-p_c|^{-s-t} \right] \quad (3)$$

where t and s are the usual exponents of the conductor-insulator and normal-supraconductor static (DC) conductivity problems, and where the subscripts ± refer to the sign of $(p-p_c)$. Let us illustrate our results by considering in more detail the *loss angle* δ defined through:

$$\tan \delta = \text{Re } \Sigma / \text{Im } \Sigma \quad (4)$$

which has the advantages of being a dimensionless quantity, and being experimentally measurable. At $p=p_c$ and $\omega \ll \omega_0$, the loss angle assumes the following universal value δ_c:

$$\delta_c = \frac{1}{2} \pi \, s/(s+t) \quad (5)$$

Figures 3 and 4 show plots of tan δ against frequency, for different values of p, both above and below p_c, for the 2D and 3D cases respectively.

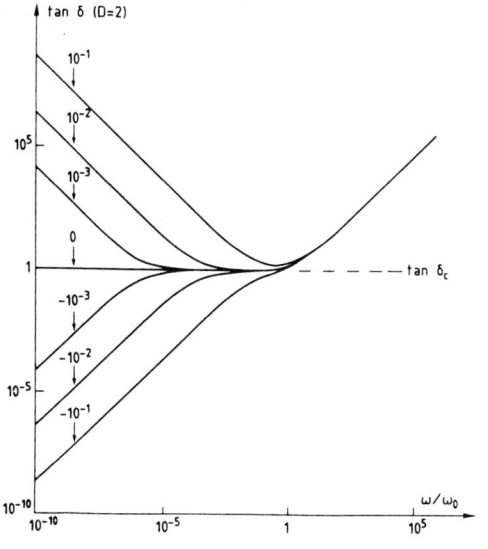

FIGURE 3
Log-log plot of tan δ against ω/ω_0 in the 2D model. Values of $(p-p_c)$ are indicated on the curves.

4. FLICKER NOISE AMPLIFICATION

This section is devoted to the observable noise spectrum of a macroscopic sample of our resistor-capacitor mixture. Two types of noise

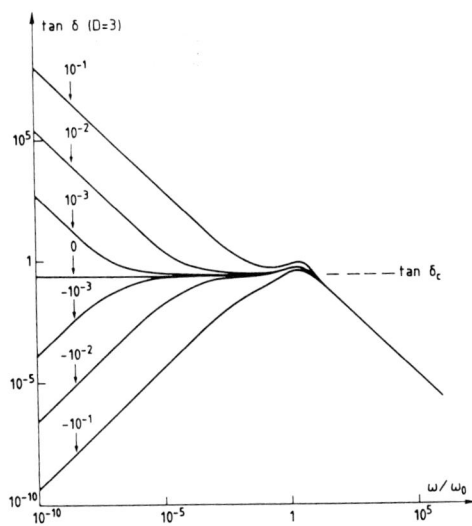

FIGURE 4
Same as Figure 3, for the 3D model.

are usually considered : *thermal* (Nyquist) noise, which is related by Nyquist theorem to the real part of the impedance , and *flicker* (or 1/f) noise. We shall determine the *amplification* of the latter by the structure of percolation clusters near the threshold. We assume that the microscopic resistance of each occupied bond has a small fluctuating part $\Delta_0(t)$:

$$R = R_0 \left[1 + \Delta_0(t) \right] \qquad (6)$$

characterized by its spectral density :

$$S_0(\omega) = |\tilde{\Delta}(\omega)|^2 \qquad (7)$$

By means of the truncation scheme T we have already discussed in Section 3, we can show that the macroscopic noise signal of a large sample of volume (number of bonds) V reads :

$$S_{mac.}(\omega) = S_0(\omega) \, V^{-1} \, G(p,\omega) \qquad (8)$$

where $S_{mac.}(\omega)$ is defined in analogy with $S_0(\omega)$ as being the reduced spectral density of the conductance : $S_{mac.}(\omega) = S_\Sigma / \Sigma^2$. Our method determines the *amplification* (or gain) $G(p,\omega)$ for all values of the parameters. In the critical regime ($(p-p_c)$ and ω/ω_0 both small), the following scaling behavior is obeyed :

$$G(p,\omega) = |p-p_c|^{-X} \chi_\pm \left[\omega/\omega_0 \, |p-p_c|^{-s-t} \right] \qquad (9)$$

The $\omega \to 0$ limit of the amplification therefore diverges as $p \to p_c$ with its own exponent X. At $p=p_c$, the amplification diverges as $\omega \to 0$ as ω^{-Y}. The exponents X and Y are related to the noise exponent b recently introduced by Rammal et. al.[2] through :

$$X = (s + t) \quad Y = (D - b) \, \nu \qquad (10)$$

Figures 5 and 6 show plots of $G(p,\omega)$ against frequency for different values of $p > p_c$ in the 2D and 3D models respectively.

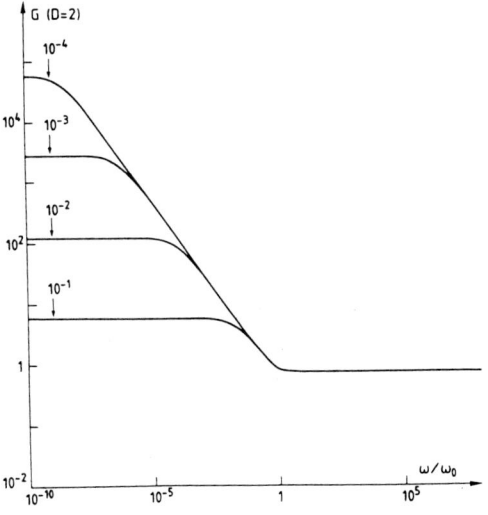

FIGURE 5
Log-log plot of flicker noise amplification $G(p,\omega)$ against ω/ω_0 in the 2D model. Values of $(p-p_c)$ are indicated on the curves.

5. CONCLUSIONS

The following Table presents the numerical values of the percolation threshold and of the exponents we have discussed in this report.

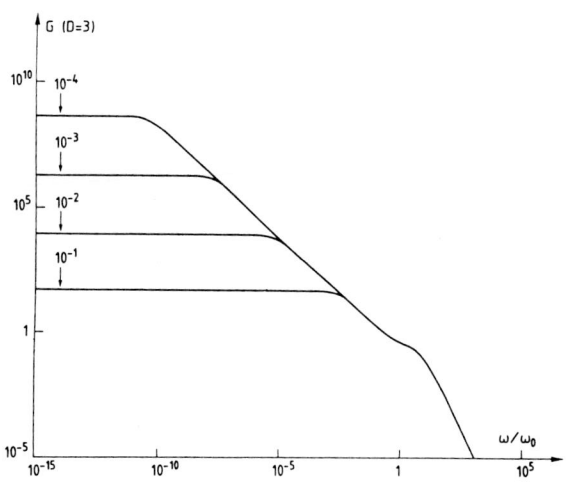

FIGURE 6
Same as Figure 5, for the 3D model.

These numbers are only to be taken as rough estimates, as usual for this type of approach.

The present method could be used to reply on more refined questions concerning electrical properties of percolation clusters, such as the voltage distribution, the transient response to

	D = 2	D = 3
P_c	0.618	0.282
ν	1.635	1.227
s	1.135	0.440
t	1.135	2.243
X	1.339	2.343
Y	0.590	0.874
b	1.181	1.091

arbitrary input signals, surface effects, etc. Let us finally mention that we have also studied, in collaboration with the Marseille group[3], various electrical properties of a deterministic fractal model for percolation clusters.

REFERENCES

1. J.M. Luck, J. Phys. A18 (in press)
2. R. Rammal, C. Tannous, A.M.S. Tremblay, Phys. Rev. A31, 2662
3. J.P. Clerc, G. Giraud, J.M. Laugier, J.M. Luck, J. Phys. A18 (in press)

THEORY OF THE AC RESPONSE OF ROUGH INTERFACES*

S. H. LIU and T. KAPLAN

Solid State Division, Oak Ridge National Laboratory, Oak Ridge, TN 37831, USA

L. J. GRAY

Engineering Physics and Mathematics Division, Oak Ridge National Laboratory, Oak Ridge, TN 37831, USA

It has been known for six decades that the small signal ac impedance of the interface between a blocking electrode and an aqueous or solid electrolyte frequently contains a constant-phase-angle (CPA) element whose impedance has the frequency dependence according to $Z \propto (j\omega)^{-\eta}$, where $0 < \eta < 1$. In recent years it has been shown experimentally that the exponent η is related to the roughness of the interface, with η approaching 1 when the interface is made increasingly smooth. We propose that the CPA originates from the fractal geometry of the rough interface, and derive on the basis of a number of models that $\eta = 3 - d_S$, where d_S is the fractal dimension of the interface.

1. INTRODUCTION

The electrical property of the interface between a metal electrode and an aqueous or solid electrolyte is an important problem in electrochemistry and related technology because it affects the performance of electrochemical devices. In the classical theory an alternating current driven across the interface is expected to encounter an interfacial capacitance in series with the ohmic resistance in the electrolyte. The ohmic resistance in the metal electrode can be ignored. The predicted frequency dependence of the impedance is very simple, i.e. the real part is independent of the frequency and the imaginary part is inversely proportional to the frequency.

It has been known since 1926 that real systems behave quite differently.[1] One finds in nearly all such systems, at least in a limited frequency range, that the frequency dependence is dominated by a term of the form $(j\omega)^{-\eta}$, where $j = \sqrt{-1}$ and the exponent satisfies $0 < \eta < 1$. This added impedance of unknown origin has been called the constant-phase-angle element, or in short, the constant-phase element. In the past decade or so many investigators have established a connection between the CPA element and the roughness of the interface.[2-6] When the interface is made increasingly smooth the value of η approaches unity. De Levie proposed to model the surface roughness by pores whose electrical properties are simulated by transmission lines.[3] However, this type of model always predicts $\eta = 1/2$ unless the resistance and capacitance elements are assumed to have some special spatial distributions.[7]

Recent advances in the theory of disordered systems have made it possible to treat problems involving irregular geometry by statistical approach. Cohen and Tomkiewicz have suggested that a new percolation process takes place at the surface of a semiconductor electrode, and this is the reason for the observed CPA behavior.[8,9] On the other hand, Le Mehaute and

---

*Research sponsored by the Division of Materials Sciences, U. S. Department of Energy under contract DE-AC05-84OR21400 with Martin Marietta Energy Systems, Inc.

Crepy have proposed a connection between the CPA and the fractal nature of the interface.[10] In this paper we carry out rigorous model calculations of the ac response of interfaces which have the fractal geometry. We predict that the CPA exponent $\eta = 3 - d_s$, where $2 < d_s < 3$ is fractal dimension of the interface. Since this result is found from solving different models, one may hope that it is a general relation between η and d_s.

2. FRACTAL NATURE OF REAL INTERFACES

Pfeiffer and co-workers have shown that solid surfaces used in heterogeneous catalysis are naturally rough in the microscopic scale, and the roughness can be described in terms of a fractal dimension between 2 and 3.[11-13] One can measure the fractal dimension by counting the number of adsorbate molecules needed to form a monolayer. The authors discovered that the number of molecules in a monolayer is proportional to the diameter of the molecules raised to the power d, where $2 < d < 3$. This means that the effective area is larger for smaller molecules because they sample more of the surface irregularities. The power d is, by definition, the fractal dimension of the surface.[14,15]

It is possible to determine qualitatively whether a surface has the fractal nature simply by viewing it with an electron microscope under a range of magnifications. A fractal surface has no natural length scale and looks the same under different magnifications. In Fig. 1 we show the electron micrograph of a surface of of Ag β-alumina, which is a much studied solid electrolyte.[6] There are hills and valleys of various sizes, and each hill has smaller hills and valleys of various sizes. In the following sections we will construct models which possess this self-similar nature and demonstrate that self-similarity is the cause of the CPA behavior of the interface.

FIGURE 1
The electron micrograph of a surface of Ag β-Alumina.

3. A PRIMITIVE MODEL

The first successful model which relates the CPA to the fractal geometry of the surface was based on the Cantor bar.[16] We proposed to connect the different stages of a Cantor bar into a contiguous piece shown in Fig. 2. In every new stage a bar is broken into two pieces, each of length 1/a of the original bar, $a > 2$. The fractal dimension of the Cantor bar is $d = \ln 2/\ln a < 1$, and the dimension of the uneven boundary between the black and the white parts in Fig. 2 is easily shown to be $d_s = 2 + d$.[14,15] We associate the black part with the electrolyte side of the interface. The segments represent smooth surfaces perpendicular to the page. This model was inspired by de Levie's observation that polished solid surfaces have

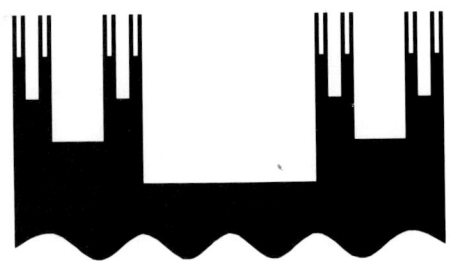

FIGURE 2
The Cantor bar model for a rough metal-electrolyte interface.

grooves with jagged sides.[3] We introduce the fractal geometry into the model by making the grooves self similar.

An ac signal passing from the electrolyte (black) side to the electrode (white) side encounters ohmic resistance in the electrolyte and interfacial capacitance at every segment of the boundary. The equivalent circuit of the interface is shown in Fig. 3. The circuit branches out at every new stage of the Cantor bar. The resistance increases by a factor of a at every branch because the cross sectional area of the branch is reduced by this ratio. The number of interfacial capacitances increases with the number of branches, but the size of the capacitors at every stage is assumed to be the same. This effectively ignores the interfacial capacitance in the dip of the branches. At higher stages the area in the dip becomes very small, and it can be shown to be an irrelevant variable in the asymptotic limit. The common ground represents the electrode.

The input impedance of the network in Fig. 3 can be written as an infinite continued fraction:

$$Z(\omega) = R + \cfrac{1}{j\omega C + \cfrac{2}{aR + \cfrac{1}{j\omega C + \cfrac{2}{a^2 R + \cdots}}}} \quad (1)$$

The function $Z(\omega)$ satisfies the frequency scaling relation:[17]

$$Z\left(\frac{\omega}{a}\right) = R + \frac{aZ(\omega)}{j\omega C Z(\omega) + 2} \quad . \quad (2)$$

In the low frequency limit Eq. (2) reduces to

$$Z\left(\frac{\omega}{a}\right) \cong \frac{a}{2} Z(\omega) \quad . \quad (3)$$

and this relation is satisfied by

$$Z(\omega) = A(j\omega)^{-\eta} \quad , \quad (4)$$

where A is a constant,

$$\eta = 1 - d \quad , \quad (5)$$

and $d = \ln 2/\ln a$ is the fractal dimension of the Cantor bar. Thus the network is a CPA element. Since d lies between 0 and 1, the exponent η is also bounded by 0 and 1. In terms of the fractal dimension of the interface, we find

$$\eta = 3 - d_s \quad . \quad (6)$$

A smooth surface has $d_s = 2$, which gives $\eta = 1$, in qualitative agreement with the experiments.

A better understanding of the CPA behavior can be obtained from studying the frequency dependence of a network which terminates after a finite number of stages. The input impedance is calculated by using the recurrent relation for the continued fraction in Eq. (1).[18] The results for the real part of $Z(\omega)$ are plotted in Fig. 4. At low frequencies the real part of Z reaches a plateau whose height increases by a factor of a/2 for every additional stage. At

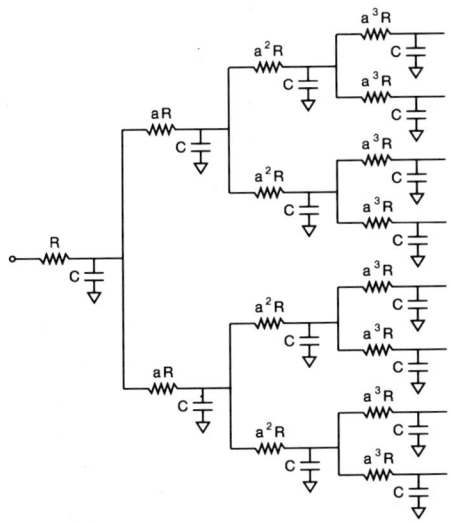

FIGURE 3
The equivalent circuit of the model interface in Fig. 2.

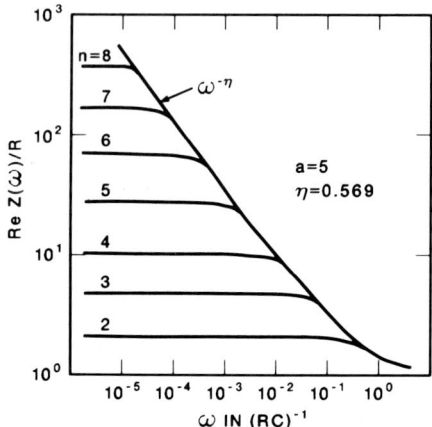

FIGURE 4
The real part of the input impedance of the circuit in Fig. 3 plotted as a function of the frequency. The quantity N denotes the number of stages of a finite network.

high frequencies the impedance has the limiting value R. Between these two limits the system exhibits the CPA property. The imaginary part of Z is inversely proportional to the frequency in the high and low frequency limits and has the CPA behavior in the middle range. These results show that the power-law frequency dependence is the result of competition between resistive and capacitive paths. A signal of lower frequency must propagate further down the network before it crosses the interface through the capacitance, and consequently experiences a higher impedance than a signal of higher frequency. Real surfaces usually are self similar over a finite range of length scales, and this determines the frequency range in which the CPA behavior is evident.

The model can be easily generalized so that it has N branches at every new stage, with $a > N$. The exponent is $\eta = 1 - \ln N/\ln a$. The fractal dimension of the Cantor bar is $d = \ln N/\ln a$, and that of the surface is $d_S = 2 + d$. Thus the relation in Eq. (6) continues to hold.

The result in Eq. (5) is in complete accord with the anomalous diffusion theory for the percolation cluster.[19-22] If we put a resistor R at every bond of a percolation cluster and a capacitor C between every node and a common ground, we obtain a complicated RC network whose ac impedance is a CPA whose exponent is

$$\eta = 1 - d/(2 + \theta) \quad , \qquad (7)$$

where d is the fractal dimension of the percolation cluster and θ is the index for anomalous diffusion. Consider a random walk on a percolation network. The mean square distance travelled by the walker is related to the number of steps by

$$\langle r^2 \rangle \propto N^{2/(2+\theta)} \quad . \qquad (8)$$

The index θ is determined by the geometry of the cluster. If a fixed number of steps is taken in a unit time, Eq. (8) then determines the distance travelled after a given amount of time, or equivalently, the diffusion lengths for signals of different frequencies. For this reason the index θ appears naturally in the CPA exponent.

The formula in Eq. (7) is different from that derived by Gefen et al.[21] These authors applied the dynamical scaling relation to the ac conductivity of the network. On the other hand, Clerc et al. showed that, for Sierpinski gaskets in D dimensional space, the dynamical scaling is satisfied by the total impedance, in a relation similar to Eq. (3).[23] Since in every network a low frequency signal spreads out over a larger portion of the network than a high frequency signal, we argue that the dynamical scaling should be applied to the total impedance of the diffusion volume of a percolation cluster. The result in Eq. (7) follows from this reasoning.

Now consider a random walk on the network in Fig. 3 starting from an arbitrary node on the network. If a step to the left is equally likely as a step to the right, there will be a net drift to the right such that $\langle r \rangle = N/3$. If it is assumed that the probability for a step is inversely proportional to the resistance of the path, there will be a net drift to the left with $\langle r \rangle = -(a-2)N/(a+2)$. In either case we find from Eq. (8) that $\theta = -1$, and this value makes Eq. (5) equivalent to Eq. (6). Thus, the model interface has the anomalous diffusion property similar to all other fractal systems.

4. RANDOM FRACTAL MODELS

Real surfaces do not have the regularity of the Cantor bar. Take for example the surface in Fig. 1. It is self similar in a statistical sense, that is, that the magnification of one part resembles another part of the surface. We have studied a number of random fractal models which approach this realistic property.[17] As a start, we assume that the various stages have the branching number N_1, N_2, N_3, etc. and the reduction ratios a_1, a_2, a_3, etc. The expression for the input impedance is

$$Z(\omega) = R + \cfrac{1}{j\omega C + \cfrac{N_1}{a_1 R + \cfrac{1}{j\omega C + \cfrac{N_2}{a_1 a_2 R + \cdots}}}} \quad (9)$$

The following relation can be proved for sufficiently low frequencies:

$$1/Z(\omega) = (N_1/a_1) 1/Z_1(a_1\omega) , \quad (10)$$

where $Z_1(\omega)$ is the input impedance of each branch of the network after removing the first stage. Assuming that the sets $\{N\}$ and $\{a\}$ are random variables with known distributions, we average over an ensemble of networks to obtain

$$\langle 1/Z(\omega) \rangle = \langle N \rangle \langle 1/aZ_1(a\omega) \rangle . \quad (11)$$

Since the averages of Z and Z_1 have the same power-law frequency dependence, we deduce that

$$\langle N \rangle \langle a^{\eta-1} \rangle = 1 . \quad (12)$$

A nontrivial solution for η exists if

$$\langle a^{-1} \rangle < 1/\langle N \rangle < 1 , \quad (13)$$

which is the condition that the interface represented by the network is a fractal. The fractal dimension of the interface is solved from a similar equation:

$$\langle N \rangle \langle a^{2-d_S} \rangle = 1 . \quad (14)$$

The relation $\eta = 3 - d_S$ is immediately evident. It is interesting to note that the disorder in branching number only enters the solution through its average $\langle N \rangle$, while the exponent depends on the distribution of the reduction ratio a.

In the next step, we randomize the reduction ratio of the branches in the same stage. After a considerable amount of work the results in Eqs. (12-14) are recovered. The details of this work are given in a forthcoming publication.[17] In another publication we will show that the thickness of each stage, which affects both R and C, can be randomized without altering the relation between the CPA exponent and the fractal dimension of the surface.[24]

5. THE CANTOR BLOCK MODEL

The hills and valleys on the surface in Fig. 1 are better simulated by a model which we call the Cantor block. The model is simply a two dimensional Cantor bar constructed by the following algorithm. One starts with a square plate of area w×w and thickness b. In the next step a cross is cut through the center of the plate to divide it into four congruent squares, each of area (w/a)×(w/a), where $a > 2$. The process is repeated ad infinitum. The Cantor block model of the surface is constructed by fusing the various stages together as was done for the Cantor bar model. The fractal dimension of the surface is $d_S = 1 + \ln 4/\ln a$ for $4 > a > 2$. For $a > 4$ we find $d_S = 2$ even though the surface is not smooth. The unevenness diminishes too rapidly with advancing stages that it causes no measurable effect.

The equivalent circuit of the surface has the same structure as the one in Fig. 3 except that there are four branches at every stage, the resistors scale by a^2 at every new stage and the capacitors scale by $1/a$. At low frequencies the scaling relation is found to be

$$Z(\omega/a) = (a^2/4)Z(\omega) , \qquad (15)$$

and this is satisfied by the CPA form of the impedance with

$$\eta = 2 - \ln 4/\ln a = 3 - d_s . \qquad (16)$$

The starting block can be generalized to a rectangle of sides $w_x \times w_y$, and the block can be cut into $N_x N_y$ smaller rectangles of sides $(w_x/a) \times (w_y/a)$. As long as $N_x N_y > a > N_x, N_y$, we obtain $d_s = 1 + \ln(N_x N_y)/\ln a$ and $\eta = 2 - \ln(N_x N_y)/\ln a = 3 - d_s$. Furthermore, the model can be randomized in the same way as the Cantor bar model, namely the branching numbers, the scaling ratio, and the thickness of every stage. We find the remarkable result that the simple relation in Eq. (7) between η and d_s is unaffected as long as the surface remains a fractal, i.e. $d_s > 2$. The detail of this work will be published elsewhere.[24]

6. DISCUSSION

Since the relation between η and d_s in Eq. (6) holds for a variety of models we have studied so far, regular as well as random fractals, we are led to speculate that the relation may be valid in general. This speculation needs to be verified experimentally. Another unknown parameter in the theory is the conversion factor between the length scale and the frequency scale. One way to determine this factor is to manufacture fractal surfaces with known fractal dimensions and measure their impedances in different electrolytes. In this manner we hope that the mystery of the CPA element will be solved eventually.

ACKNOWLEDGMENTS

The authors are indebted to Dr. J. B. Bates for suggesting the investigation and for his continued interest and encouragement. They also wish to thank Dr. J. C. Wang for helpful discussions.

REFERENCES

1. I. Wolfe, Phys. Rev. 27 (1926) 755.
2. P. H. Bottelberghs, Low-Frequency Measurements on Solid Electrolytes and Their Interpretations, in: Solid State Electrolytes, eds. P. Hagenmuller and W. van Gool (Academic Press, New York, 1978) pp. 145-172.
3. R. de Levie, Electrochimica Acta 10 (1965) 113.
4. P. H. Bottelberghs and G. H. J. Broers, J. Electroanal. Chem. 67 (1976) 155.
5. R. D. Armstrong and R. A. Burnham, J. Electroanal. Chem. 72 (1976) 257.
6. J. B. Bates, J. C. Wang and Y. T. Chu, to appear in the Proceedings of the 5th International Conference on Solid State Ionics, Lake Tahoe, USA, 1985.
7. J. C. Wang and J. B. Bates, to appear in the Proceedings of the 5th International Conference on Solid State Ionics, Lake Tahoe, USA, 1985.
8. M. H. Cohen and M. Tomkiewicz, Phys. Rev. B 26 (1982) 7097.
9. J. K. Lyden, M. H. Cohen and M. Tomkiewicz, Phys. Rev. Lett. 47 (1981) 961.
10. A. Le Mehaute and G. Crepy, Solid State Ionics 9 and 10 (1983) 17.
11. P. Pfeiffer and D. Avnir, J. Chem. Phys. 79 (1983) 3558.
12. D. Avnir, D. Farin and P. Pfeiffer, J. Chem. Phys. 79 (1983) 3566.

13. P. Pfeiffer, D. Avnir and D Farin, Surf. Sci. 126 (1983) 569.
14. B. B. Mandelbrot, Fractals: Forms, Chance and Dimension (Freeman, San Francisco, 1977).
15. B. B. Mandelbrot, The Fractal Geometry of Nature (Freeman, San Francisco, 1983).
16. S. H. Liu, Phys. Rev. Lett. 55 (1985) 529.
17. T. Kaplan and L. J. Gray, Phys. Rev. B. (in press).
18. Handbook of Mathematical Functions (National Bureau of Standards, Washington D. C., 1964) p. 19.
19. P. G. de Gennes, Recherche 7 (1976) 919.
20. S. Alexander and R. Orbach, J. Phys. (Paris) Lett. 43 (1982) L-625.
21. Y. Gefen, A. Aharony and S. Alexander, Phys. Rev. Lett. 50 (1983) 77.
22. R. Rammal and G. Toulouse, J. Phys. (Paris) Lett. 44 (1983) L-13.
23. J. P. Clerc, A.-M. S. Tremblay, G. Albinet, and C. D. Mitescu, J. Phys. (Paris) 45 (1984) L-913.
24. T. Kaplan, L. J. Gray, and S. H. Liu, to be published.

Part VIII
HIERARCHICAL AND FRACTAL FEATURES
OF DISORDERED SYSTEMS

FRACTALS IN PHYSICS
L. Pietronero, E. Tosatti (editors)
© Elsevier Science Publishers B.V., 1986

THE NATURE OF TEMPORAL HIERARCHIES UNDERLYING RELAXATION IN DISORDERED SYSTEMS

Michael F. SHLESINGER* and Joseph KLAFTER**

*Physics Division (Code 012), Office of Naval Research, 800 North Quincy Street, Arlington, Virginia 22217-5000
**Corporate Research Science Laboratories, Exxon Research and Engineering Company, Route 22 East, Clinton Township, Annandale, New Jersey 08801

Since 1970 it was realized that dielectric relaxation in many glasses and polymers followed the stretched exponential law. Since then this expression has enjoyed a wide applicability characterizing various types of relaxation (NMR, mechanical, optical correlation, and dielectric) in many complex random materials. Several theories have been proposed to explain this ubiquitious behavior including i) Direct transfer on a fractal; ii) Hierarchially constrained dynamics; iii) Fractal time defect diffusion. While the physics behind these models is radically different we have been able to cast all of these models into the same mathematical form. The underlying unifying concept is a hierarchy of relaxation times broad enough to induce the time t always to appear raised to a non-integer power. This unity holds even though the physical mechanisms may provide for parallel, serial, or some other combination of relaxation paths.

I. INTRODUCTION

Much attention has been focused recently on relaxation phenomena in complex condensed matter systems. This is due to the surprising empirical observation that the decay of correlatation functions for many diverse systems follows the same stretched exponential law,

$$\phi(t) = \exp[(-t/\tau)^\beta], \quad 0<\beta<1 \quad (1)$$

The parameters β and τ depend on the material and can be a function of external variables such as temperature. Equation (1) was first proposed by Kohlrausch[1] in 1864 to describe mechanical creep, and was later used by Williams and Watts[2] in 1970 to describe dielectric relaxation in polymers. The stretched exponential law has since been employed to fit data in other systems such as remanent magnetization in spin glasses[3], and the decay of luminescence in porous glasses[4]. This ubiquitous decay law can be derived from several different physical mechanisms[5-11]. The stretched exponential law has been known in the statistical theory of failure as the Weibull distribution[12]. It is the purpose of this paper to seek connections between three recent different physical models and determine the common feature which is responsible for generating the stretched exponential law.

We will first analyze the Forster direct transfer mechanism[13] which is an example of relaxation via parallel channels, and relate its mathematical structure to the serial hierarchically constrained dynamics model of Palmer, et. al.[10]. The third related theory is the defect diffusion model of Shlesinger and Montroll[9]. All of the above three theories can derive the stretched exponential law in a natural fashion. We show that the underlying reason for this is the existence of scale invariant relaxation rates being generated in each model.

II. THE FORSTER DIRECT TRANSFER MODEL

A. Many Parallel Channels

The origins of this model arose from

studies of an excitation transfer from donor to static defects in various condensed media[13].

Consider the decay law of an initially prepared donor at the origin due to direct energy transfer to a defect at site R_i on a given structure. The relaxation function $\Phi_i(t)$ is the probability that by time t the donor is still excited,

$$\Phi_i(t) = \exp[-tW(R_i)], \quad (2)$$

where $W(R_i)$ is the relaxation rate that depends on the donor-defect. This defines a relaxation time $\tau(R_i) = \frac{1}{W(R_i)}$. For a fixed configuration of many defects at positions R_i excluding the origin one obtains

$$\Phi_{\{R_i\}} = \prod_{i=1}^{\infty}{}' \exp[-t W(R_i)] \quad (3)$$

We now configurationally average Eq. (3) over all possible defect positions. For the use of a substitutional occupancy of sites by the defects with probability p we have[13],

$$\Phi(t) = \prod_i{}' \{1 - p + p\exp[-tW(R_i)]\} \quad (4)$$

which for $p \ll 1$ is

$$\Phi(t) \simeq \exp[-p\sum_i{}' \{1-\exp[-t W(R_i)]\}] \quad (5)$$

By introducing a site density function

$$\rho(R) = \sum_i{}' \delta(R-R_i) \quad (6)$$

we transform the sum to an integral form

$$\Phi(t) = \exp[-p\int dR\rho(R)\{1-\exp[-tW(R)]\}] \quad (7)$$

Two types of isotopic interactions W(R) are assumed:

(a) $W(R) = aR^{-s}$ (8)
(b) $W(R) = B\exp(-\gamma R)$

On regular underlying spatial structures so that $\rho(R) = $ const. we find for case (a)

$$\Phi(t) = \exp[(-t/\tau)^{D/s}] \text{ (in D dimensions)} \quad (9)$$

which is a stretched exponential decay. In Forster's energy transfer problem $s \geq 6$. For case (b)

$$\Phi(t) = (Bt)^{-A} \ln^{D-1}(Bt) \text{ (in D dimensions)} \quad (10)$$

This exhibits an "enhanced" power law decay which takes the form of algebraic decay in one dimension.

The nonexponential decays in Eqs. (9) and (10) are a result of parallel relaxations and a hierarchy of distances. To generalize these results for transfer on a fractal structure of fractal dimension $\bar{d}$, one needs only replace D by $\bar{d}$[14]. If the structure is homogeneous on a spatial scale ξ, but fractal on smaller scales then the relaxation behavior will crossover from fractal to homogeneous when $W(\xi)t \sim 1$[11].

B. ONLY THE FASTEST CHANNEL

The same types of relaxation patterns can be derived when we restrict the transfer to include only the nearest neighbor defects. Then

$$\Phi_{NN} = \int_0^{\infty} f(R) \exp[-tW(R)] dR \quad (11)$$

where f(R) is the probability of having a nearest neighbor at distance R. For a randomly placed defect in one-dimension

$$f(R) = p\exp(-pR) \quad (12)$$

For W(R) in Eq. (8), case (a),

$$\Phi_{NN} = p \int e^{-pR} \exp(-taR^{-s}) \, dR \quad (13)$$

which by the method of steepest descent gives

$$\Phi_{NN} = p e^{-(Ct)^{\frac{1}{1+s}}} \quad (14)$$

again a stretched exponential but with a smaller exponent than for D=1 in Eq. (9), because of truncating the influence of more distant defects. In Eq. (8), case (b) for W(R),

$$\Phi_{NN} = (p/\gamma)(Bt)^{-p/\gamma} \gamma_1(p\gamma, Bt) \quad (15)$$

γ_1 is the incomplete gamma function. At long times Eq. (15) behaves similarly to Eq. (10) for D=1.

The direct transfer decay laws follow a stretched exponential behavior both for the many parallel channels case and for the fastest decay channel when $W(R)=aR^{-s}$. Namely the position dependent relaxation time $\tau(R)$ is scale invariant.

III. HIERARCHICALLY CONSTRAINED DYNAMICS MODEL

Recently Palmer et. al.[10] introduced a model of relaxation which is serial rather than parallel. The hierarchical model supposes that relaxation occurs in stages, and the constraint imposed by a faster degree of freedom must relax before a slower degree of freedom can relax. This implies that the time scale of relaxation on one level is subordinated to the relaxation below.

In one possible realization they[10] considered a system with a discrete series of levels n=0,1,2,... , with the degrees of freedom on level n represented by N_n spins which point either up or down. The spins in level n+1 are only free to change their state when μ_n spins in level n attain one of their 2^{μ_n} possible states. The relaxation time τ_{n+1} of level n+1 is then

$$\tau_{n+1} = 2^{\mu_n} \tau_n \quad (16a)$$

$$= \tau_0 \exp\left(\sum_{k=0}^{n} \tilde{\mu}_k\right) \quad (16b)$$

where $\tilde{\mu}_k = \mu_k \ln 2$. The relaxation function $\phi(t)$ is given by

$$\phi(t) = \sum_{n=0}^{\infty} \omega_n \exp(-t/\tau_n) \quad (17)$$

where $\omega_n = N_n / \sum_{n=0}^{\infty} N_n$ is a weight factor for level n. Note that the hierarchy of relaxation times generated by eq. (16a) is similar to the hierarchy of transition rates discussed in the direct transfer model.

One can now choose specific forms for μ_n and ω_n and caluculate the corresponding relaxation function. The choice $\mu_n = \mu_0$ implies

$$\tau_n = \tau_0 \exp(\tilde{\mu}_0 n) \quad (18)$$

which is essentially case (b) of Eq. (8). Choosing

$$\omega_n = \omega_0 \lambda^{-n} \quad (19)$$

which corresponds to eq. (12), and converting the sum in eq. (17) to an integral yields

$$\phi(t) = \omega_0 \int_0^{\infty} e^{-n\ln\lambda} \exp[-t \exp(-\tilde{\mu}_0 n)/\tau_0] dn \quad (20)$$

which we recognize to be the integral that leads to the algebraic relaxation law of Eq. (15), namely $\phi(t) \sim t^{-(\ln\lambda)/\tilde{\mu}_0}$. This type of relaxation behavior has also been obtained for diffusion in a hierarchial system[15].

The same choice for ω_n, but now coupled with the

$\mu_n = \mu_o/n$ (n≥1, and k≥1)

implies

$$\tau_n = \tau_o \exp\left(\tilde{\mu}_o \sum_{\ell=1}^{n} \ell^{-k}\right)$$
$$\simeq \tau_o n^{\tilde{\mu}_o} \quad \text{(for k=1)} \quad (21)$$

which corresponds to case (a) of Eq. (8). For k=1 this leads to the relaxation integral

$$\phi(t) = \omega_o \int_0^\infty e^{-n \ln\lambda} \exp[-tn^{-\tilde{\mu}_o}/\tau_o]dn \quad (22)$$

which, as in Eq. (13), produces the stretched exponential law with exponent $\beta = 1/(1+\tilde{\mu}_o)$.

Although the physical pictures of relaxing through a serial arrangement of hierarchically constrained levels and relaxing through a direct transfer to a nearest neighbor defect are quite different they both lead to the same relaxation integrals. The weight factor ω_n and the relaxation time τ_n for each level in the hierarchical model correspond to the weight factor $f(R)$ for the defect position, and the transition rate $W(R)$ respectively in the direct transfer model.

IV. DEFECT DIFFUSION MODEL

Glarum[16] in 1960 introduced the concept that migrating defects may trigger the relaxation of frozen dipoles in amorphous materials. It was assumed that the relaxation occurred instantanously when the nearest defect reached the dipole for the first time. This idea was later extended[17] by allowing any defect to trigger the relaxation. In this model a stretched exponential decay with $\beta = 1/2$ was obtained in one dimension and exponential decay in three dimensions. Shlesinger and Montroll[9] showed how these results could be generalized to cover the range $0<\beta<1$ by introducing a hierarchy of waiting times for defects hopping in random environment. Extensions of the above models to fractal, self-similar, geometries were also studied in the framework of the target picture[11a,18]. It has been pointed out[11a] that the defect diffusion model contains parallel decay channels each of which is inherently sequential.

We now recast the defect diffusion model into the mathematical form presented in the last two sections using the approximate arguments of Redner and Kang[19]. We begin with the one dimensional case where a target is located at the origin and defects are randomly distributed around it. Let $f(R_1)$ be the probability of having no defects at distance R_1 from the target. As before for randomly placed defects $f(R_1) = \exp(-pR_1)$. The probability that a defect at R_i has not reached the origin by time t is given by $\exp[-t/4R_i^2]$, where the diffusion constant of the defects has been set equal to unity. The relaxation law for the target is then

$$\phi(R_1,t) \simeq \exp(-pR_1) \prod_{i=1}^{\Pi} \exp[-t/4R_i^2]$$
$$\simeq \exp(-pR_1) \exp\left[-tp\int_{R_1}^{\infty} \frac{dR}{4R^2}\right]$$
$$= \exp\left[-pR_1 + p\frac{t}{4R_1}\right] \quad (23)$$

Averaging over R_i we arrive at

$$\phi(t) = \int_0^\infty \exp(-pR_1) \exp(-pt/4R_1)dR_1 \quad (24)$$

which is again of the form of Eq. (13) in the Forster case and of Eq. (22) in the Palmer et. al. case with $\tau(R)=(4/p)R$ and yields the stretched exponential law with $\beta = 1/2$. This result when generalized to other dimensions gives (9,11a,18,19)

$$\phi(t)=\exp[-pS(t)] \qquad (25)$$

where $S(t)$ is the mean number fo distinct sites visited by a random walker on a lattice after time t. For regular lettices $S(t) \sim t^{1/2}$ (D=1), $S(t) \sim t/\ln t$ (D=2) and $S(t) \sim t$ (D=3). The target relaxation is then a stretched exponential for D=1 with $\beta = 1/2$ and is exponential for D=3. When the defects migrate on fractal, self similar, structures[20]

$$S(t) \sim \begin{array}{ll} t^{\tilde{d}/2} & \tilde{d}<2 \\ t & \tilde{d}>2 \end{array} \qquad (26)$$

When $\tilde{d}$ is the spectral (fracton) dimension[21]. If the structures that carry the defects are fractal only over a given length scale ξ, then we expect a crossover from a stretched exponential to an exponential relaxation at times that correspond to the crossover length.

The above results were extended by incorporating a distribution of waiting times $\psi(t)$ into the motion ofthe defects[9,22]. In the case of a very broad distribution of waiting time whose first moment is infinite[9], $\psi(t) \sim t^{-\alpha-1}$ $0<\alpha<1$, we obtain $S(t) \sim t^{\alpha/2}$ (D=1) which corresponds to Eq. (24) with $\tau(R) \sim R^{2/\alpha-1}$ and $S(t) \sim t^{\alpha}$ (D=3). These give rise, through Eq. (25), to stretched exponential forms with $\beta = \alpha/2$ and $\beta = \alpha$ respectively. The theory of Cohen and Grest [6] for stretched exponential relaxation in glassy materials belongs to same mathematical framework as the theories discussed above. Their relaxation times are scale invariant similar to Eq. (8) case(a) and to Eq. (21) although their physical model is quite different.

IV. CONCLUSIONS

We have demonstrated that a common mathematical framework underlies different physical models which lead in a natural way to the common emperically found stretched exponential relaxation law. The unifying feature of the theories is the generation of a scale invariant distribution of relaxation times. One should be able to differentiate among physical mechanisms underlying the relaxation via the variation behavior f β and τ in Eq. (1) as a function of external variables such as temperature and pressure.

ACKNOWLEDGEMENTS

We thank J. Bendler and A. Blumen for stimulating discussions.

REFERENCES

1. Some history and references are in Montroll, E. W., and Bendler, J. T. (1984) J. Stat. Phys 34, 129.

2. Williams, G. & Watts D. C. (1970) Trans. Faraday Soc. 66, 80.

3. Chamberlin, R. V., Mozurkewich, G., & Orbach, R., (1984) Phys. Rev. Lett. 52, 867.

4. Even, U., Rademann, K., Jortner, J., Manor, N. & Reisfeld, R., (1984) Phys. Rev. Lett. 52 2164.

5. Shore, J. E. & Zwanzig, R. (1975) J. Chem. Phys. 63, 5445.

6. Cohen, M. H. & Grest, G. S. in <u>Structure and Mobility in Molecular and Atomic Glasses</u>, ed. O'Reilly, J. M. and Goldstein M. (Academic Press, N. Y. (1981)) PP. 199.

7. Ngai, K. L. & White, C. T. (1979) Phys. Rev. B20, 2475.

8. Skinner, J. L. (1983) J. Chem. Phys. 79 1955.

9. Shlesinger, M. F. & Montroll, E. W. (1984) Proc. Nat'l. Acad. Sci. (USA) 81, 1280.

10. Palmer, R. G., Stein, D., Abrahams, E. S. Anderson, P. W. (1984) Phys. Rev. Lett. 53, 958.

11. (a) Klafter, J. & Blumen, A. (1985) Chem. Phys. Lett. (in press) and (b) Huber, D. L. (1985) Phys. Rev. B 31, 6070.

12. Cox, D. R. (1962) Renewal Theory (Chapman and Hall, London).

13. (a) Forster, T. (1949) Z. Naturforsch. Teil A4, 321. (b) Blumen, A. (1981) Nuovo Cimento B63, 50.

14. Klafter, J. & Blumen, A. (1984) J. Chem. Phys. 80, 875.

15. Huberman, B. A., & Kerszberg, M., (1985) J. Phys. A. L331.

16. Glarum, S. H. (1960) J. Chem. Phys. 33, 1371.

17. Bordewijk, P. (1975) Chem. Phys. Lett. 32, 592.

18. Zumofen, G., Blumen, A. & Klafter, J. (1985) J. Chem. Phys. 82, 3198.

19. Redner, S. & Kang, K. (1984) J. Phys. A: Math Gen. 17, L451.

20. Ramal, R. & Toulose, G. (1983) J. Phys. (Paris) 44, L13.

21. Alexander, S. & Orbach, R. (1983) J. Phys. (Paris) 43, L625.

FRACTALS IN PHYSICS
L. Pietronero, E. Tosatti (editors)
© Elsevier Science Publishers B.V., 1986

REACTIONS IN DISORDERED MEDIA MODELLED BY FRACTALS

A. BLUMEN[*], J. KLAFTER[§] and G. ZUMOFEN[**]

[*]Max Planck Institut für Polymerforschung, D-6500 Mainz, and Lehrstuhl für Theoretische Chemie, Technische Universität, D-8046 Garching, West Germany

[§]Corporate Research Science Laboratories, Exxon Research and Engineering Co., Clinton Township, Annandale, NJ 08801, USA

[**]Laboratorium für Physikalische Chemie, ETH-Zentrum, CH-8092 Zürich, Switzerland

We study the dynamics of reactions in random systems by using fractal models which account for spatial and temporal disorder. We focus on pseudounimolecular (trapping, target) and on bimolecular reactions and monitor the interplay between the spatial (α) and the temporal (γ) behavior. At long times, most decays show subordination, $\Phi(t) \sim f(t^{\alpha\gamma})$, whereas for trapping the temporal aspect dominates, $\Phi(t) \sim t^{-\gamma}$.

1. INTRODUCTION

Many processes of interest in disordered systems, such as energy transfer and holeburning in condensed molecular media,[1-3] electron and hole transport in amorphous semiconductors,[4,5] particle diffusion in porous materials, and chromatographic separation processes[6] display nonexponential long-time behavior. These decay processes are therefore not characterized by a single relaxation rate. Usual forms of model decay laws, as they appear in the study of relaxation in complex systems are:

a) the Kohlrausch or Williams and Watts stretched exponential:[7-10]

$$\Phi(t) = \exp[-(t/\tau)^{\alpha}] \quad (0<\alpha<1, t>\tau) \quad (1)$$

b) The exponential-logarithmic form of Inokuti and Hirayama[11], and Scher, Lax and Montroll:[4,12]

$$\Phi(t) = \exp[-B\ln^{\beta}(t/\tau)] \quad (\beta \geq 1, t>\tau) \quad (2)$$

and c) the algebraic decay:[5]

$$\Phi(t) \sim (t/\tau)^{-\gamma} \quad (\gamma>0, t>\tau) \quad (3)$$

The above forms were listed in the order of increasingly slower decays.

In this contribution we display ways of modelling the systems mentioned through fractals, such that both the spatial and the temporal disorder aspects are accounted for.[13] As will be shown in the main body of the work, fractal concepts lead readily to decay behaviors of the forms (1) to (3). One should note from the start that spatial inhomogeneities, exemplified for instance by the embedding of molecules of different kinds in a host matrix, or the positioning of atoms in an amorphous semiconductor, determine a distribution of interparticle distances which lead to a wide range of microscopic transfer rates: spatial disorder often implies also temporal disorder, and sometimes even energetic disorder. Evidently, this complex situation is untractable analytically, so that models, such as the fractals presented below, are needed.

2. FRACTALS IN TIME AND SPACE

The basic feature of fractals is their self-similarity,[14] i.e. their invariance (either individually, or in their ensemble) under the group of dilatation operations. Many examples for geometrical (i.e. spatial) fractals are presented in these proceedings, so that we find it advisable to start our considerations with the temporal ones.

As stressed recently, the continuous time-random walk (CTRW) treatment of charge transport in amorphous materials, (such as used by Scher and Montroll[4]) is based on a fractal set of event times.[15,16] Let $\psi(t)$ be the probability density that an event occurs at time t after the previous event has taken place. A simple example is the Poisson process:

$$\psi^P(t) = b \exp(-bt) \quad (4)$$

One constructs now readily a dilatationally symmetric distribution, by taking in the following manner account of events occuring on all time scales:[16]

$$\psi(t) = \frac{1-N}{N} \sum_{j=1}^{\infty} N^j b^j \exp(-tb^j) \quad (5)$$

with N<1. As is evident, the distribution (5) is a normalized sum of Poisson-terms and

$$\psi(bt) = \psi(t)/(Nb) - (1-N) \exp(-tb)/N \quad (6)$$

For later applications we need b<N, so that b<1 and thus at longer times $\psi(bt) \simeq \psi(t)/Nb$. The last expression is equivalent to

$$\psi(t) \sim \frac{1}{t^{1+\gamma}} \quad (7)$$

with $\gamma = \ln N/\ln b$. Equation (7) shows directly the temporal scaling of $\psi(t)$, i.e. its fractal nature in time.

To show that Eq. (5) arises very naturally let us consider the distribution of carrier release times from low-lying traps to the conduction band.[17] This distribution is fundamental in the multiple trapping (MT)-formalism. For activated processes the rates depend exponentially on the energy, so that an equidistant level spacing $E_j = j\Delta E$ leads to rates proportional to $\exp(-E_j/kT) = b^j$, with $b = \exp(-\Delta E/kT)$. Furthermore the density of states in the energy tail is often itself exponential in energy, $\exp(-E_j/kT_0)$ (where we introduced an effective temperature T_0) so that the density of states follows N^j with $N = \exp(-\Delta E/kT_0)$. Thus, in this example $\gamma = \ln N/\ln b = T/T_0$ for $T<T_0$, and one has dispersive transport below T_0.

Scaling also carries over to quantities related to $\psi(t)$. Let $\chi_n(t)$ denote the probability that exactly n events occured in time t. This basic quantity of the CTRW-formalism is simply related to $\psi(t)$ via its Laplace-transform:

$$\mathcal{L}[\chi_n(t)] \equiv \chi_n(u) = [\psi(u)]^n [1-\psi(u)]/u \quad (8)$$

where $\psi(u) = \mathcal{L}[\psi(t)]$.

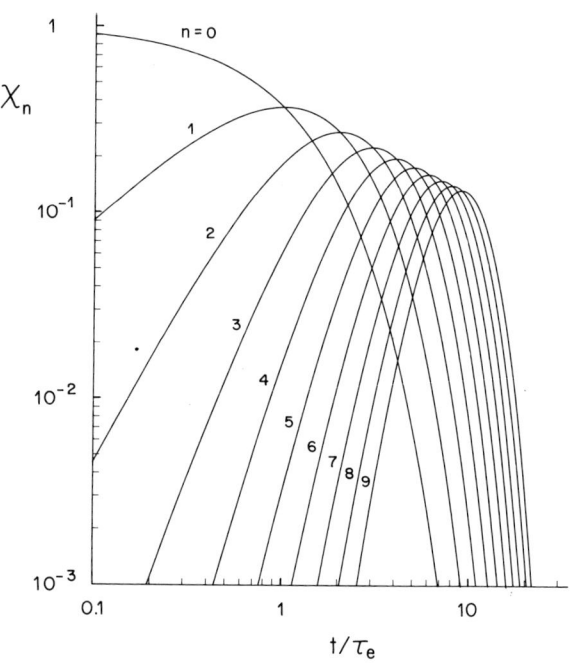

FIGURE 1
Probabilities $\chi_n(t)$ that exactly n steps occurred during t for a Poisson waiting time distribution, $\psi^P(t)$, Eq. (4), and $\psi^P(\tau_e) = 1/e$.

In Fig. 1 we present the set of $\chi_n(t)$ for the Poisson process, Eq. (4). The curves show pronounced maxima which, with increasing n shift to longer times. From the log-log plot no scaling is evident.

To display scaling we have to study $\psi(t)$-forms which behave algebraically at long times. A typical example is the function $\psi_2(t)$, wich belongs to the family of functions defined through:[4,18]

$$\psi_n(t) = c_n a^2 [\exp(a^2 t)] i^n \operatorname{erfc}(at^{1/2}) \quad (9)$$

where the $i^n \operatorname{erfc}(z)$ are repeated integrals of the error function and the c_n are normalization constants. The function $\psi_2(t)$ has no first moment, $\psi_2(u) = (1+u^{1/2}/a)^{-2}$, i.e. $\gamma = 0.5$ in Eq. (7). In Fig. 2 we show the corresponding $\chi_n(t)$. As is evident by inspection, the curves scale very well at long times, and their slope is also given by $\gamma = 0.5$. Indeed, for qualitative arguments one may well approximate the $\chi_n(t)$ through $\chi_0(t)$ in the long-time regime, and we will use this property further in the text.

We now turn our attention to the more common geometrical fractals. Many stochastically disordered structures, like percolation clusters at criticality, aggregates constructed by diffusion-limited growth, linear and branched polymers, epoxy-resins and various porous materials have been characterized as being self-similar under length scaling.[19] As we know, however, fractals need not be stochastic. Well-known examples are the deterministically built Sierpinski-gaskets, whose generators in d-dimensional Euclidean spaces are hypertetrahedrons consisting of d+1 hypertetrahedra of sidelength downscaled by a factor of two.[14] Extensions obtain by changing the generator forms, both by choosing other scaling factors than 2 and by

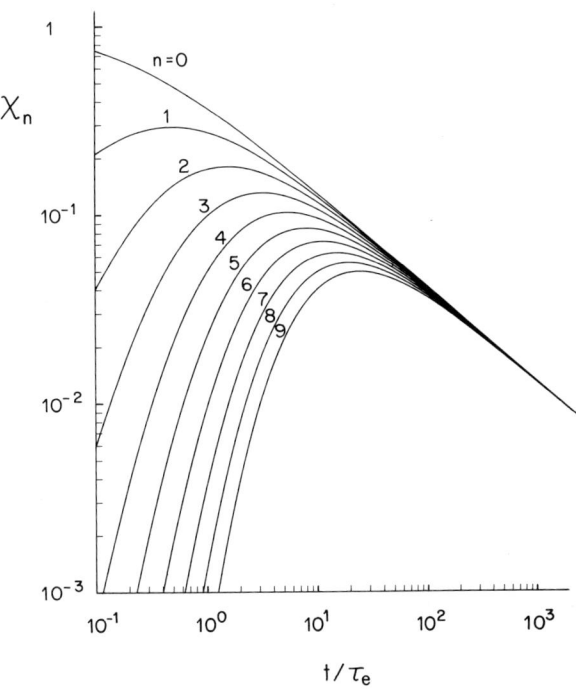

FIGURE 2
Same as in Fig. 1, where the waiting-time distribution $\psi_2(t)$ is given by Eq. (9), and $\psi_2(\tau_e) = 1/e$.

varying the coverage of the original tetrahedron.[20,21] This procedure allows to obtain pseudo-Sierpinski structures with prespecified fractal dimensions. Additonally, one may also obtain deterministic fractals by a direct set-product; thus a Sierpinski gasket multiplied by a linear chain gives rise to a Toblerone-structure.[21,22]

After specifying the fractals in time and in space we are now ready to also connect the two aspects. As already mentioned, spatial disorder is often accompanied by temporal disorder, and both facets appear in the study of dynamical processes, such as transient flow through porous rocks and migration over percolative systems, in which the site energies or the interactions (barriers) are randomly distributed. In previous work we have extended the CTRW to fractal lattices and have studied for several classes of waiting-time distributions $\psi(t)$ the mean squared displacement $<r^2(t)>$ and the decay of the population of the walkers due to trapping. We have shown that the interplay of the two stochastic (spatial and temporal) aspects leads to interesting new behaviors.[13,23]

As an example consider the mean squared displacement of a walker on a geometrical fractal, when the waiting-times between steps are fixed. One has[24]

$$<r_n^2> \sim n^{\tilde{d}/\bar{d}} \qquad (10)$$

where n is the number of steps and $\bar{d}$ and $\tilde{d}$ are the fractal and the spectral (fracton) dimension, respectively. The CTRW analog of Eq. (10) is

$$<r^2(t)> = \sum_{n=0}^{\infty} <r_n^2> \chi_n(t) \qquad (11)$$

where the $\chi_n(t)$ is the probability of having performed exactly n steps during the time t, which probability is given in the Laplace-domain by Eq. (8). In general, Eq. (11) leads to complex expressions. However, for fractals the pattern followed by Eq. (11) is readily obtainable. For $\chi_n(t)$ which scale at long times (see Fig. 2) one has for a fixed t value:

$$\chi_n(t) \simeq \begin{cases} \chi_0(t) & \text{for } n < n_{max}(t) \\ 0 & \text{otherwise} \end{cases} \qquad (12)$$

Using the normalization relation for the $\chi_n(t)$ and Eq. (12) one retrieves the time-dependence of n_{max}:

$$1 = \sum_{n=0}^{\infty} \chi_n(t) \simeq \sum_{n=0}^{n_{max}} \chi_0(t) = \chi_0(t) n_{max}(t) \qquad (13)$$

and hence, for $\chi_0(t) \sim t^{-\gamma}$ one has $n_{max} \sim t^{\gamma}$.

The same argument applied to Eq. (11) gives:

$$<r^2(t)> \sim \sum_{n=0}^{n_{max}} n^{\tilde{d}/\bar{d}} \chi_0(t) \sim \chi_0(t) n_{max}^{1+\tilde{d}/\bar{d}} \qquad (14)$$

which, with help of Eq. (13), is:

$$<r^2(t)> \sim n_{max}^{\tilde{d}/\bar{d}} \sim t^{\gamma(\tilde{d}/\bar{d})} \qquad (15)$$

thus establishing the subordination[13] (i.e. the multiplicative behavior of the exponents) for the two kinds of disorder.

A similar argument applies also to $S(t)$, the mean number of sites visited by the walker in time t. For walks with fixed waiting-times one has $S_n \sim n^{\tilde{d}/2}$ for $\tilde{d} < 2$ and $S_n \sim n$ otherwise. Hence

$$S(t) \sim \begin{cases} t^{\gamma \tilde{d}/2} & \text{for } \tilde{d} < 2 \\ t^{\gamma} & \text{for } \tilde{d} > 2 \end{cases} \qquad (16)$$

i.e. another example for subordination.[13]

3. REACTIONS

In this section we analyze the influence of fractal behavior on several reaction schemes. Here we focus both on pseudounimolecular and on bimolecular-type reactions. An example for the former is the reaction $A + B \to B$, where the A particles are annihilated when encountering a B-particle. Bimolecular decays are exemplified by the $A + A \to 0$ and by the $A + B \to 0$ reactions. In a kinetic scheme the decay of the density of A-particles depends quasiexponentially on time, whereas in the $A + A \to 0$ reaction the long-time behavior follows a $1/t$ law.[25] The latter form also obtains for the $A + B \to 0$ reaction, for an equal number of A and B particles. As we proceed to show, the validity domain of the kinetic scheme (which presupposes a "well-stirred reactor") is limited, and we will display deviations from the decay forms mentioned above.

We start from the pseudounimolecular scheme

A + B → B, and we monitor deviations from exponentiality. As discussed by us in previous works, two special cases are of particular interest, namely those in which one of the species is immobile. The case that A does not move ($D_A = 0$, $D_B \neq 0$) leads to the target-problem, whereas the immobile B-species ($D_A \neq 0$, $D_B = 0$) correspond to the trapping problem.[23]

As shown recently by us for regular, translationally invariant lattices, the target problem admits an exact solution, which is expressible through $S(t)$, the mean number of sites visited.[25,26] For fixed waiting times one defines $H_n(\vec{r})$ to be the probability that a first passage from $\vec{r}$ to $\vec{0}$ occurred in the first n steps. Taking an A-particle to be situated at the origin, and a B-particle at $\vec{r}$ one has for the probability that B did not reach $\vec{0}$ in the first n steps

$$\phi_n(\vec{r}) = 1 - H_n(\vec{r}) \qquad (17)$$

Thus for B-particles which are Poisson-distributed on the lattice the decay of A follows:

$$\Phi_n = \prod_{\vec{r}}{}' \{\sum_j (e^{-p} p^j/j!) [\phi_n(\vec{r})]^j\}$$

$$= \exp[-p\sum_{\vec{r}}{}' H_n(\vec{r})] \qquad (18)$$

where p is the B-particle concentration. Furthermore $H_n(\vec{r})$ is related to S_n, the *mean* number of sites visited in n steps via:

$$\sum_{\vec{r}}{}' H_n(\vec{r}) = S_n - 1 \qquad (19)$$

and therefore:

$$\Phi_n = \exp[-p(S_n - 1)] \qquad (20)$$

Using the $\chi_n(t)$ given in Eq. (8) one obtains for the probability of a visit from $\vec{r}$ to $\vec{0}$ in time t:

$$H(t;\vec{r}) = \sum_{n=0}^{\infty} \chi_n(t) H_n(\vec{r}) \qquad (21)$$

This CTRW-transformation carries through to Eq. (20), as may be readily verified. Hence one has for the target decay in the CTRW-scheme:

$$\Phi(t) \sim \exp[-pS(t)] \qquad (22)$$

For fractal $\psi(t)$, Eq. (7), $S(t)$ is given by Eq. (16). Furthermore the same result, Eq. (22) holds approximately for fractal lattices, since there one has to average over non-equivalent target locations. One should note that the decay form of the target model provides a Kohlrausch-Williams-Watts law, Eq. (1).

Quite different forms obtain in CTRW for the trapping problem. Again we start from the simple random walk with fixed waiting times. Now the A-particle moves, whereas the B are immobile. For a particular realization of the random walk, let R_n denote the *number* of distinct sites visited in n steps. For the same realization of the walk let F_n be the probability that trapping has not yet occurred. If the B-particles are Poisson-distributed, the probability that a site is not occupied by B equals e^{-p} and thus:[25]

$$F_n = e^{-p(R_n-1)} \qquad (23)$$

The measurable survival probability is the average of F_n over all realizations of the walk

$$\tilde{\Phi}_n = \langle e^{-p(R_n-1)} \rangle$$

$$= e^{-p} \exp[\sum_{j=1}^{\infty} K_{j,n}(-p)^j/j!] \qquad (24)$$

where the $K_{j,n}$ are the cumulants of the R_n-distribution. Taking the first two cumulants one has, for instance:

$$\tilde{\Phi}_n \simeq e^{-p(S_n-1)} \cdot e^{p^2 \sigma_n^2/2} \qquad (25)$$

As before, we may revert to the continuous time (CTRW) domain, by use of the $\chi_n(t)$:[27]

$$\tilde{\Phi}(t) = \sum_{n=0}^{\infty} \tilde{\Phi}_n \chi_n(t) \qquad (26)$$

By viewing Eq. (26) as an additional average of Eq. (24) one retrieves in the short-time domain as first term of the cumulant expansion Eq. (22):

$$\tilde{\Phi}(t) \sim \exp[-pS(t)] \qquad (27)$$

On the other hand, the long-time behavior follows from the properties of the fractal $\chi_n(t)$:

$$\tilde{\Phi}(t) \sim \chi_0(t) \sum_{n=0}^{n_{max}} \tilde{\Phi}_n \sim \chi_0(t) <n> \qquad (28)$$

where $<n>$ is the mean number of steps until trapping for a random-walker with fixed waiting times:

$$<n> \equiv \sum_{n=1}^{\infty} n(\tilde{\Phi}_{n-1}-\tilde{\Phi}_n) = \sum_{n=0}^{\infty} \tilde{\Phi}_n \qquad (29)$$

In the last relation of Eq. (28) we made use of the fact that $\tilde{\Phi}_n$ is a decreasing, summable expression, and that for large t, n_{max} is also large, see Eq. (13). The remarkable result of Eq. (28) is that the temporal behavior of $\tilde{\Phi}(t)$ parallels at long times that of $\chi_0(t) = 1 - \int_0^t \psi(t')dt'$ (from Eq. (8)), which gives the probability that no step has occurred until t. This effect appeared in our previous numerical analysis of decay laws and leads to concentration-dependent forms which follow avoided crossing patterns.[28] Interestingly, at long times the dependence on concentration of B-particles and on the lattice enters only through $<n>$. For a simple evaluation for small concentrations:

$$<n> = \sum_{n=0}^{\infty} \tilde{\Phi}_n \sim \sum_{n=0}^{\infty} e^{-p(S_n-1)} \sim \int_0^{\infty} dx \exp(-px^\alpha)$$
$$= p^{-1/\alpha} \Gamma(1+1/\alpha) \qquad (30)$$

In Eq. (30) we made use of $S_n \sim n^\alpha$, as given by Eq. (16) and $\Gamma(x)$ is the Euler-gamma function. We furthermore note that in d=1 the relation $<n> \sim p^{-2}$ is exact,[29] and that it is quite accurately fulfilled in higher dimensions. Putting Eqs. (28),(30),(13) and (16) together one thus has:

$$\tilde{\Phi}(t) \sim \begin{cases} t^{-\gamma}/(p^{2/\tilde{d}}) & \text{for } \tilde{d}<2 \\ t^{-\gamma}/p & \text{for } \tilde{d}>2 \end{cases} \qquad (31)$$

Thus, at long times the dependence of the trapping decay under fractal time is algebraic, as in Eq. (3). The decays of the trapping and of the target problems are thus very different in the long-time regime.

We turn now to the study of the bimolecular reactions and we monitor deviations from the 1/t behavior of the kinetic scheme. We start with A + A → 0. From the previous study of pseudounimolecular reactions, we found that in the short-time-regime the kinetic exponential was modified by the appearance of $S(t)$. Thus one may expect that the A + A → 0 decay will follow a $1/S(t)$ law at longer times.[25,30]

As we have shown for the case of walks with fixed waiting-times, the following form describes the decay well:[25]

$$\Phi_n^{AA} = (1+2pS_n)^{-1} \qquad (32)$$

Here p is the density of A-particles. The long time decay thus follows an algebraic form $\Phi_n^{AA} \sim n^{-\alpha}/p$. Heuristically one may view Eq. (32), the solution of a many-body problem, as being related to the probability of encounter of *two*

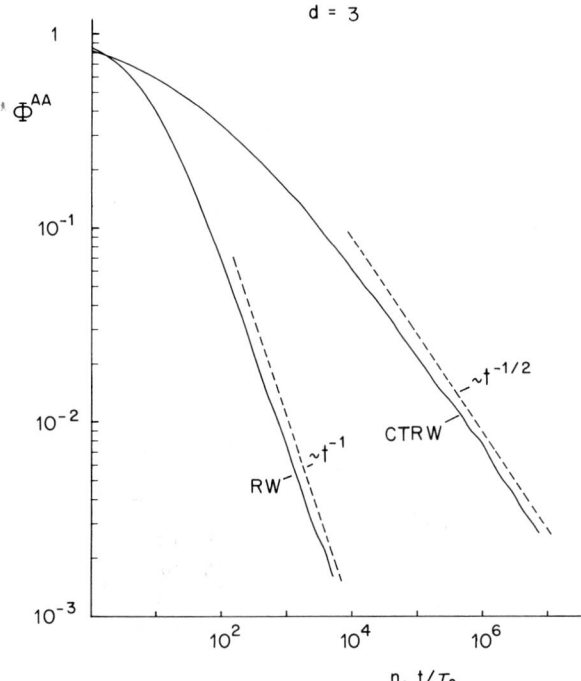

FIGURE 3
Decay law $\Phi^{AA}(t)$ due to the annihilation $A + A \rightarrow 0$ on a simple cubic lattice, d=3, both for a simple RW and for a CTRW with $\psi_2(t)$. The initial particle density is p=0.1. The long-time slopes are indicated as a guide to the eye.

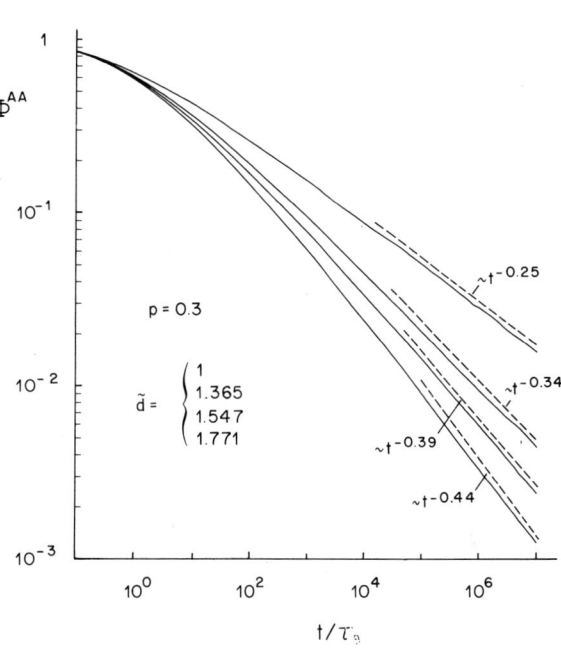

FIGURE 4
Same as in Fig. 3 for CTRW with $\psi_2(t)$ for walkers on several Sierpinski-gaskets and the linear chain.

particles, which itself is expressible, via S_n, by the volume visited by each. Hence, since in the CTRW-scheme the S_n get replaced by $S(t)$, we expect that $\Phi^{AA} \sim [pS(t)]^{-1}$ at longer times.

In order to test this prediction we proceed numerically and analyze the $A + A \rightarrow 0$ reaction under CTRW-conditions. We start from walkers on a simple cubic lattice, d=3, and use $\psi_2(t)$, as given by Eq. (9), and thus $\gamma = 1/2$. In Fig. 3 we present the corresponding decay law and contrast it with the simple RW-results. Whereas at longer times the simple RW-decay follows t^{-1}, for the CTRW with $\psi_2(t)$ we find at longer times a $t^{-1/2}$ dependence, as may be verified by inspection of Fig. 3, in which these asymptotic slopes are also indicated. As a further example we have performed simulations on several Sierpinski-gaskets and on the linear chain for which the numerical findings are presented in Fig. 4. As indicated in the Figure, the long time decay behavior indeed follows the form $t^{-\tilde{d}/4}$ for CTRW with $\psi_2(t)$ instead of $t^{-\tilde{d}/2}$ for the simple RW decay. All findings are consistent with $\Phi^{AA}(t) \sim [S(t)]^{-1}$, i.e.

$$\Phi^{AA}(t) \sim \begin{cases} t^{-\gamma\tilde{d}/2} & \text{for } \tilde{d}<2 \\ t^{-\gamma} & \text{for } \tilde{d}>2 \end{cases} \quad (33)$$

Equation (33) is then another example for subordination.[13]

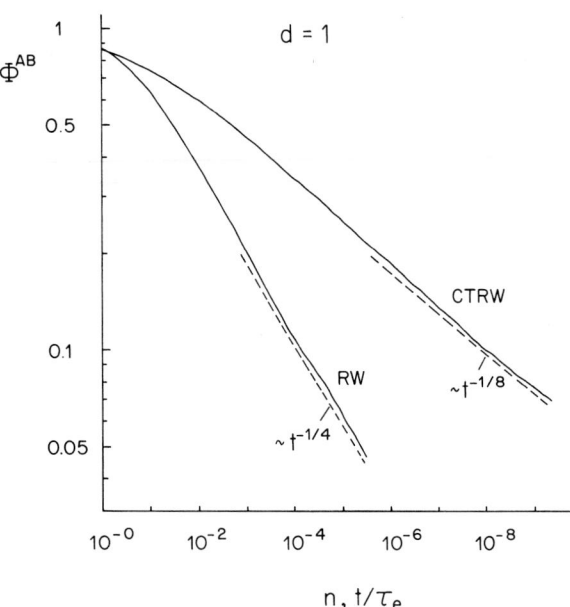

FIGURE 5
Decay law $\Phi^{AB}(t)$ for the strictly bimolecular $A + B \rightarrow 0$, $A(0) = B(0)$, with particles moving according to CTRW with $\psi_2(t)$ on a linear chain. The long-time slopes are also indicated.

As a final example we consider the strictly bimolecular reaction $A + B \rightarrow 0$ for an equal number of A and B particles, $A(0) = B(0)$. In this case it is known[25,31,32] that for A and B particles moving with fixed waiting times the long-time decay does *not* follow n^{-1} but goes rather as $n^{-\tilde{d}/4}$. The marginal dimension of $d=4$ is due to the appearance of large spatial fluctuations.[25,31,32] Since we have already shown that the $n^{-\tilde{d}/4}$ decay is well-obeyed on Sierpinski-gaskets embedded in spaces of different Euclidian dimensions d, it remains to check the influence of the CTRW. As in previous cases, we expect the decay form to subordinate through the time variable to $\psi(t)$, and expect hence:

$$\Phi^{AB}(t) \sim t^{-\gamma \tilde{d}/4} \quad \text{(for } \tilde{d}<4\text{)} \quad (34)$$

In Fig. 5 we present the decay forms for particles moving with $\psi_2(t)$ on a linear chain. As is evident from the Figure the subordination is again well-obeyed.

4. CONCLUSIONS

In this contribution we have shown the interplay of temporal and spatial randomness on pseudounimolecular (target, trapping) and on bimolecular reactions. We have modelled the stochastic aspects through fractals both in time and in space. As we have demonstrated, the classical reaction kinetics is not a good description of the decay since deviations from exponentiality and from the 1/t behavior are quite widespread.

As a summary, we present in Table 1 the decay laws for the reactions considered. In all cases, apart from the very particular trapping problem, the laws show subordination, i.e. the cooperative random aspects lead to a *multiplicative* combination of their scaling coefficients in the final forms. For trapping the temporal aspect is dominant at long times. Although not specifically stressed, the same behavior will obtain for the energetic disorder discussed in Sec. 2., for which the scaling coefficient is proportional to the temperature.

ACKNOWLEDGEMENTS

The authors thank Dr. M.F. Shlesinger for many pleasant, fruitful discussions. The support of the Deutsche Forschungsgemeinschaft and of the Fonds der Chemischen Industrie and a grant of computer time from the computer center of the ETH-Zürich are gratefully acknowledged.

		random walks short times	random walks long times	CTRW long times
$A + B \rightarrow B$	Trapping ($D_A \neq 0$, $D_B = 0$)	$e^{-Ct^{\tilde{d}/2}}$ ($\tilde{d}<2$) e^{-Ct} ($\tilde{d}>2$)	$e^{-Ct^{\tilde{d}/(\tilde{d}+2)}}$	$t^{-\gamma}$ *No* subordination
	Target ($D_A = 0$, $D_B \neq 0$)	$e^{-Ct^{\tilde{d}/2}}$ ($\tilde{d}<2$) e^{-Ct} ($\tilde{d}>2$)	$e^{-Ct^{\tilde{d}/2}}$ ($\tilde{d}<2$) e^{-Ct} ($\tilde{d}>2$)	$e^{-Ct^{\gamma\tilde{d}/2}}$ ($\tilde{d}<2$) $e^{-Ct^{\gamma}}$ ($\tilde{d}>2$) Subordination
$A + A \rightarrow 0$	Bimolecular	$[1+Ct^{\tilde{d}/2}]^{-1}$ ($\tilde{d}<2$) $[1+Ct]^{-1}$ ($\tilde{d}>2$)	$t^{-\tilde{d}/2}$ ($\tilde{d}<2$) t^{-1} ($\tilde{d}>2$)	$t^{-\gamma\tilde{d}/2}$ ($\tilde{d}<2$) $t^{-\gamma}$ ($\tilde{d}>2$) Subordination
$A + B \rightarrow 0$ $A(0) = B(0)$	Bimolecular ($D_A = D_B \neq 0$)	$[1+Ct^{\tilde{d}/2}]^{-1}$ ($\tilde{d}<2$) $[1+Ct]^{-1}$ ($\tilde{d}>2$)	$t^{-\tilde{d}/4}$ ($\tilde{d}<4$) t^{-1} ($\tilde{d}>4$)	$t^{-\gamma\tilde{d}/4}$ ($\tilde{d}<4$) $t^{-\gamma}$ ($\tilde{d}>4$) Subordination

TABLE   Time-dependence of decay forms

REFERENCES

1. J. Klafter and A. Blumen, in: Energy Transfer Processes in Condensed Media, B. di Bartolo ed., Plenum, N.Y., 1984, p. 621.
2. W. Breinl, J. Friedrich and D. Haarer, J. Chem. Phys. 81 (1984) 3915.
3. J. Friedrich and A. Blumen, Phys. Rev. B32 (1985) 1434.
4. H. Scher and E.W. Montroll, Phys. Rev. B12 (1975) 2245.
5. J. Tauc, Semicond. and Semimetals 21B (1984) 299.
6. G.H. Weiss, Separation Sci. 5 (1980) 51.
7. R. Kohlrausch, Ann. Phys. (Leipzig) 12 (1847) 393.
8. G. Williams and D.C. Watts, Trans. Faraday Soc. 66 (1970) 80.
9. K.L. Ngai, Comments Solid State Phys. 9 (1979) 127; 9 (1980) 141.
10. M.F. Shlesinger and E.W. Montroll, Proc. Natl. Acad. Sci. USA 81 (1984) 1280.
11. M. Inokuti and F. Hirayama, J. Chem. Phys. 43 (1965) 1978.
12. H. Scher and M. Lax, Phys. Rev. B7 (1973) 4491; 4502.
13. A. Blumen, J. Klafter, B.S. White and G. Zumofen, Phys. Rev. Lett. 53 (1984) 1301.
14. B.B. Mandelbrot, The Fractal Geometry of Nature, Freeman, San Francisco, 1982.

15. M.F. Shlesinger, J. Stat. Phys. 36 (1984) 639.

16. E.W. Montroll and M.F. Shlesinger, in Nonequilibrium Phenomena II From Stochastics to Hydrodynamics, J.L. Lebowitz and E.W. Montroll eds. North Holland, Amsterdam, 1984, p. 1.

17. G. Pfister and H. Scher, Adv. Phys. 27 (1978) 747.

18. M.F. Shlesinger, J. Stat. Phys. 10 (1974) 421.

19. Contributions to these Proceedings.

20. R. Hilfer and A. Blumen, J. Phys. A17 (1984) L537; L783.

21. R. Hilfer and A. Blumen, these Proceedings.

22. A. Maritan and A.L. Stella, these Proceedings.

23. J. Klafter, A. Blumen and G. Zumofen, J. Stat. Phys. 36 (1984) 561.

24. S. Alexander and R. Orbach, J. Phys. (Paris) Lett. 43 (1982) L625.

25. G. Zumofen, A. Blumen and J. Klafter, J. Chem. Phys. 82 (1985) 3198.

26. A. Blumen, G. Zumofen and J. Klafter, Phys. Rev. B30 (1984) 5379.

27. A. Blumen and G. Zumofen, J. Chem. Phys. 77 (1982) 5127.

28. G. Zumofen, J. Klafter and A. Blumen, J. Chem. Phys. 79 (1983) 5131.

29. E.W. Montroll, J. Phys. Soc. Jpn Suppl. 26 (1969) 6.

30. P.W. Klymko and R. Kopelman, J. Phys. Chem. 87 (1983) 4565.

31. A.A. Ovchinnikov and Ya.B. Zeldovich, Chem. Phys. 28 (1978) 215.

32. D. Toussaint and F. Wilczek, J. Chem. Phys. 78 (1983) 2642.

SELF-SIMILAR TEMPORAL BEHAVIOR OF RANDOM WALKS IN ONE-DIMENSIONAL RANDOM MEDIA

J. BERNASCONI and W.R. SCHNEIDER

Brown Boveri Research Center, CH-5405 Baden, Switzerland

Random walks in one-dimensional random media are analyzed within the framework of a new real-space renormalization procedure. For asymmetric transition probability distributions, the renormalized random walks are shown to approach a directed walk with a limiting distribution of transition times which is evaluated with Monte Carlo methods. Under certain conditions the set of transition times exhibits self-similar clustering with an average fractal dimension $\nu < 1$. It follows that the long time asymptotic behavior of the mean displacement is of the form $x(t) \sim t^\nu F(\ell n t)$, where F is a periodic function.

Random walks in random media are the subject of considerable current interest[1,2]. Corresponding models exhibit a number of remarkable phenomena and are used to study the anomalous transport properties of certain disordered materials. In this paper we are concerned with discrete-time random walks on a one-dimensional lattice with random transition probabilities. If $X_t = n$ denotes the position of the particle at time t, we assume that $X_{t+1} = n \pm 1$ with probabilities p_n and $1-p_n$, respectively. We thus have

$$P_n(t+1) = p_{n-1} P_{n-1}(t) + (1-p_{n+1})P_{n+1}(t) \quad ,$$

where $P_n(t)$ denotes the probability that $X_t = n$, and we shall always assume that $P_n(0) = \delta_{no}$. The transition probabilities p_n are assumed to be independent random variables, identically distributed according to some given probability density $\rho(p)$.

Corresponding models, as well as related continuous-time versions, have been investigated with different methods[1-8], and one has made the remarkable observation that under certain conditions the mean displacement,

$$x(t) = \sum_{n=-\infty}^{\infty} n P_n(t) \quad ,$$

asymptotically increases slower than linearly in time. For some specific types of disorder, an even more surprising phenomenon has been observed[2,3,6]: In addition to the overall sublinear increase, $x(t)$ may exhibit superimposed, non-decaying oscillations as a function of log t. These asymptotic oscillations have been interpreted in terms of a self-similar clustering of waiting times and could be determined exactly for model systems that contain a finite fraction of "diodes"[2,6].

To investigate the detailed asymptotic behavior of $x(t)$ for arbitrary disorder, we propose and analyze a new real-space renormalization procedure for random walks in random media which is more direct and more illustrative than previous renormalization approaches[7-9]. The method is based on a simple decimation of the one-dimensional lattice (see Figure 1), and the renormalized transition probability associated with site 0, p_0', is given by

$$p_0' = \frac{p_0 p_1}{1-p_{-1}(1-p_0)-p_0(1-p_1)} \quad . \quad (1)$$

Equation (1) is obtained by summing over all possible walks that lead from site 0 to site 2 via the intermediate sites -1 and 1. As the num-

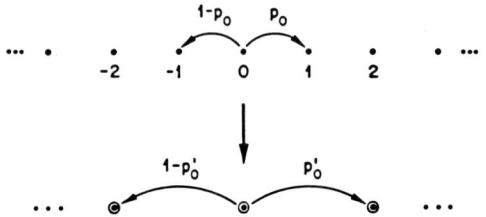

FIGURE 1
Schematic representation of the basic renormalization step.

ber of steps in these walks varies between 2 and infinity, the renormalization in principle leads to a broad distribution of transition times. To simplify the procedure, however, we define a mean transition time by averaging over this distribution after each renormalization step. The corresponding transformation then becomes

$$\tau_0' = \tau_0 + \tau_1 +$$

$$+ \frac{p_{-1}(1-p_0)(\tau_{-1}+\sigma_0)+p_0(1-p_1)(\tau_0+\sigma_1)}{1-p_{-1}(1-p_0)-p_0(1-p_1)} \quad , \quad (2)$$

where τ_n and σ_n denote the mean transition times for jumps to the right and left, respectively.

The original transition probabilities p_n are identically and independently distributed according to a given probability density $\rho(p) = \rho^{(0)}(p)$, and the transition times are all unity, i.e. $\rho_p^{(0)}(\{\tau,\sigma\}) = \delta(\tau-1)\delta(\sigma-1)$. The corresponding probability densities afer k renormalization steps are denoted by $\rho^{(k)}(p)$ and $\rho_p^{(k)}(\{\tau,\sigma\})$, respectively, and we shall be interested in their limiting behavior as $k \to \infty$.

Using Monte Carlo methods, we have analyzed different initial probability densities $\rho^{(0)}(p)$ by performing up to $k = 7$ renormalization steps.

Our results indicate that

$$\rho^{(k)}(p) \xrightarrow[k\to\infty]{} \delta(p-1) \quad (3)$$

for all asymmetric $\rho^{(0)}(p)$ with

$$\int_0^1 dp\, \rho^{(0)}(p) \log \frac{1-p}{p} < 0 \quad . \quad (4)$$

Equation (3) implies that the renormalized random walks approach a limiting walk in which each step is strictly to the right. (If the integral in Eq. (4) is > 0, the limiting walk is directed to the left). With increasing k, the randomness in the transition probabilities thus disappears and is replaced by a random distribution of transition times.

Corresponding explicit results are presented in Figures 2, 3, and 4. They all refer to binary transition probability distributions of the form

$$\rho^{(0)}(p) = x\delta(p-y) + (1-x)\delta(p-(1-y)) \quad . \quad (5)$$

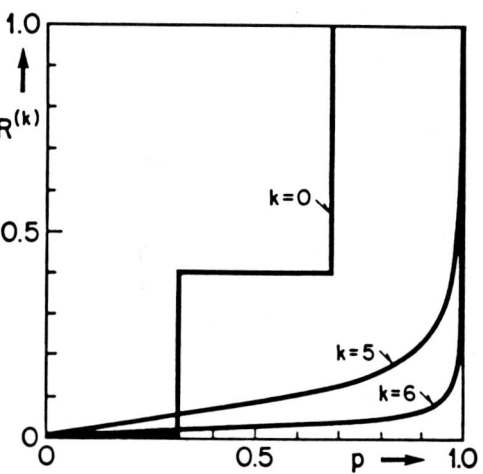

FIGURE 2
Renormalized distribution functions $R^{(k)}(p)$, Eq. (6), for the model of Eq. (5) with $x = 0.4$ and $y = 1/\pi$.

The results of Fig. 2 for the distribution function $R^{(k)}(p)$,

$$R^{(k)}(p) = \int_0^p dp' \, \rho^{(k)}(p') \quad , \tag{6}$$

show that after 6 renormalization steps already about 90% of the transition probabilities are larger than 0.95. As $\rho^{(k)}(p)$ approaches $\delta(p-1)$, the transition times for jumps to the left become irrelevant, and the limiting behavior of the transition time distribution is therefore conveniently studied by evaluating

$$\rho_*^{(k)}(\tau) = \int_0^1 dp \, \rho^{(k)}(p) \int d\sigma \, \rho_p^{(k)}(\{\tau,\sigma\}) \tag{7}$$

for sufficiently large k-values. A corresponding histogram, obtained after k = 5 renormalization steps, is shown in Fig.3. It clearly demonstrates the fractal nature of the transition time distribution, implying[2,6,10] that the set of transition times will exhibit self-similar clustering with a scaling factor $\exp(\beta)$ and an average fractal dimension ν.

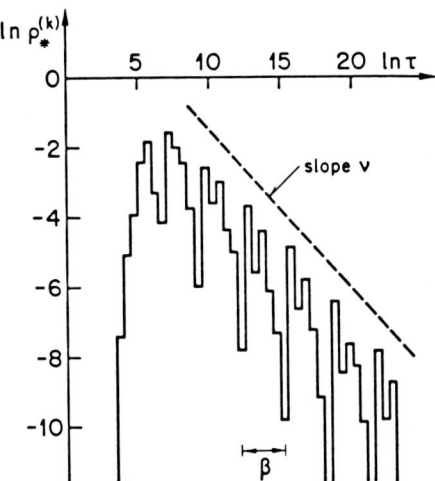

FIGURE 3
Transition time density vs. $\ln \tau$ after k = 5 renormalization steps, for the model of Eq. (5) with x = 0.2 and y = 0.047.

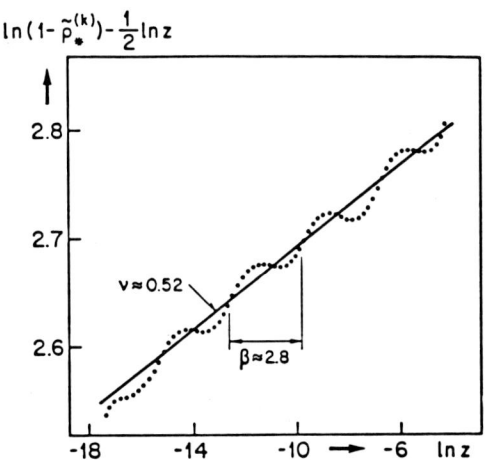

FIGURE 4
Small z behavior of the Laplace transformed transition time density after k = 4 renormalization steps, for the model of Eq. (5) with x = 0.2 and y = 1/17.

Models involving fractal time motion have been used to explain the anomalous behavior of transport and relaxation in certain amorphous materials[10,11]. In these models, however, the fractal form of the transition time distribution is assumed a priori, while in our random walk models it is a consequence of the disorder and emerges naturally through the renormalization process.

Let us now assume that after k renormalizations our random walk is approximately represented by a strictly directed walk with transition time density $\rho_*^{(k)}(\tau)$, and that correlations between successive transition times can be neglected. Then the Laplace transform $\tilde{x}(z)$ of the mean displacement x(t) may be approximated by

$$\tilde{x}(z) = \frac{2^k}{z} \frac{\tilde{\rho}_*^{(k)}(z)}{1-\tilde{\rho}_*^{(k)}(z)} \quad , \tag{8}$$

where $\tilde{\rho}_*^{(k)}(z)$ denotes the Laplace transform of $\rho_*^{(k)}(\tau)$.

If, as in the example of Fig. 4, the small z behavior of $\tilde{\rho}_*^{(k)}(z)$ is of the form

$$1 - \tilde{\rho}_*^{(k)}(z) \sim z^\nu G(\beta^{-1} \ln z) \quad , \qquad (9)$$

where G is periodic with period 1, it follows[2,6] that the mean displacement varies asymptotically as

$$x(t) \sim t^\nu F(\beta^{-1} \ln t) \quad , \qquad (10)$$

where F is also periodic with period 1. Our renormalization procedure thus not only yields the exponent ν, but also the oscillation period β. A comparison with the exact results for "diode"-models[2,6] shows that in general ν and β are predicted rather accurately after only a few renormalization steps. We further note that oscillations are only observed if $\rho^{(o)}(p)$ is discrete and such that $\nu < 1$, in agreement with general expectations[2,4,6].

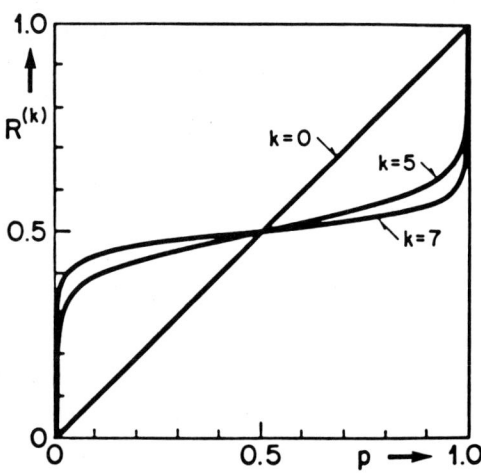

FIGURE 5
Renormalized distribution functions $R^{(k)}(p)$, Eq. (6), for an initially uniform transition probability density.

Finally, we have applied our renormalization procedure to symmetric transition probability distributions, i.e. to $\rho^{(o)}(p)$ for which the integral in Eq. (4) vanishes[12]. Here (see Figure 5 for an example) we find that

$$\rho^{(k)}(p) \xrightarrow[k \to \infty]{} \frac{1}{2}\left[\delta(p) + \delta(p-1)\right] \quad , \qquad (11)$$

and that the corresponding limiting walk exhibits a very strange behavior which is rather hard to analyze quantitatively, but seems to be in agreement with Sinai's[12] predictions.

REFERENCES

1. S. Alexander, J. Bernasconi, W.R. Schneider, and R. Orbach, Rev. Mod. Phys. 53, 175 (1981); and references therein.

2. J. Bernasconi and W.R. Schneider, Helv. Phys. Acta 58, 597 (1985); and references therein.

3. F. Solomon, Ann. Prob. 3, 1 (1975).

4. H. Kesten, M.V. Kozlov, and F. Spitzer, Compositio Math. 30, 145 (1975).

5. B. Derrida and Y. Pomeau, Phys. Rev. Lett. 48, 627 (1982).

6. J. Bernasconi and W.R. Schneider, J. Phys. A: Math. Gen. 15, L729 (1982).

7. J. Machta, Phys. Rev. B24, 5260 (1981); J. Stat. Phys. 30, 305 (1983).

8. R.A. Guyer, Phys. Rev. A29, 2114 (1984).

9. M. Napiórkowski, J. Phys. A: Math. Gen. 16, 3065 (1983).

10. M.F. Shlesinger and B.D. Hughes, Physica 109A, 597 (1981).

11. M.F. Shlesinger, this volume.

12. Ya. G. Sinai, Lecture Notes in Physics (Berlin, Springer) 153, 12 (1982); Theor. Prob. Appl. 27, 256 (1982).

AN OBSERVATION OF SCALING IN TRAPPING REACTIONS

Zoran B. DJORDJEVIĆ

"Boris Kidrič" Institute of Nuclear Sciences, Vinča, Laboratory for Solid State and Radiation Chemistry, P.O.Box 522, 11001 Belgrade, Yugoslavia

We monitored the time dependence of the concentration $C(t)$ of a chemical species which freely diffuses through a solution with randomly distributed and almost static perfect traps. Every contact of a particle and a trap leads to the disappearance of the particle. Concentration $C(t)$ appears to follow a dimension dependent decay law of the form $C(t) \sim \exp(-At - Bt^{d/d+2})$, with $d = 3$ in our experiments.

1. INTRODUCTION

In the last several years we witnessed a considerable growth of interest[1-11] in the diffusion controlled reactions. Most of this interest, at least in the physics community, stems from the realization that long time behavior of concentrations of the involved chemical species (physical particles) follows some anomalous decay laws. The most frequently studied reactions are recombination processes[7], reactions with static traps[1] and scavenger reactions[2,12]. Each of these has its own characteristic behavior, but the physical origin of the anomaly is the same in all cases.

As an example, let us consider a "trapping" reaction in which chemical species P (particles) diffuse through a medium with randomly distributed and static, perfect, traps T. On every contact between a particle P and a trap T, particle disappears. Traps are perfect in the sence of being perfectly absorbing and reaction: P+T → T, can be repeated with every trap "infinitely" many times. It is usually assumed that if concentration of particles and traps is large, or if the reaction vessel is vigorously stirred, one can treat the process as the classical first order reaction and expect to observe a decay of the particle concentration $C(t)$ of the form:

$$C(t) \simeq \exp(-At), \quad (1)$$

where A is a constant proportional to the concentration of traps. However, if concentration of randomly distributed traps is small, there necessarily exist regions of space which are practically trap-free. Particles in these regions need exceedingly long times to reach their traps and consequently the decay of their numbers with time is slower. A formal analysis[13] of this problem demonstrates that particle concentration should decay via a dimension dependent law:

$$C(t) \simeq \exp(-B \cdot t^{d/d+2}), \quad (2)$$

where d is the dimension of the space in which the reaction takes place. The reader will certainly notice the simultaneous appearance of the singular power of time in equation (2) and the large scale fluctuations (inhomogeneities) in the density of traps. The connection is in no way accidental and reflects basically the same physical mechanisms as the ones encountered in the vicinity of the second order phase transitions.

If traps are allowed to move, their motion will in a way average out the effect of the spatial inhomogeneities and bring the system closer to the assumptions which lead to equation (1).

It can be shown[12] (through a non-trivial argument) that in this case, the concentration of particles decays with a combined law:

$$C(t) \sim \exp(-At) \cdot \exp(-Bt^{d/d+2}), \qquad (3)$$

where A is proportional to the difussion constant of traps.

The purpose behind this work was to examine the applicability of the above theoretical prediction to the real experimental systems.

2. EXPERIMENTAL SYSTEM

Trapping reaction we observed is an electron transfer reaction[14] from the methyl viologen, MV^+ (1,1´-dimethyl-4,4´-bipyridine) monocations to the colloidal platinum particles in an aqueous solution. Initially, system contains triply distilled water with small amounts (2×10^{-4} mole/dm^3) of methyl viologen dications, MV^{++}, some 2-propanol (10^{-1} mole/dm^3) and 5×10^{-5} mole/dm^3 of platinum. Colloidal platinum was prepared by the sodium citrate reduction[17] of hexachloroplatinic acid. The average radius of colloidal particles was determined microscopically and equals 3 nm. This corresponds to an average aggregation number of 7.5×10^3.

Pulsed radiolysis of the solution was performed with a 2 MeV electron pulse of the 20 nsec duration.[15] The only active species produced are some $1-3 \times 10^6$ mole/dm^3 of MV^+ monocations. These monocations are our diffusing particles which in contact with colloidal platinum, traps, release one electron each and disappear, i.e. return into the stable dication, MV^{++}, state. On average, every platinum particle can accept close to 10^4 electrons and for all practical purposes can be considered as perfectly absorbing. A relatively large size of platinum particles makes their diffusion constant approximately one order of magnitude smaller than the diffusion constant of MV^+ particles. Before use, the solutions were deaerated by Ar bubbling.

Formally, one represents the above reaction as:

$$m\, MV^+ + (Pt)_{coll} \rightarrow (Pt)^{m-}_{coll} + m\, MV^{++}, \qquad (4)$$

where m is an "arbitrary" integer.

Reaction (4) is followed by the observation of decay of optical absorption of MV^+ monocations at 605 nm ($\epsilon_{605} = 12600$ M^{-1} cm^{-1})[16], a wavelength at which MV^{++} species have a negligable absorbtivity. The concentration C(t) of MV^+ is deduced on the basis of Hook´s law: $I(t) = I_0 \exp(-D\, C(t))$, where I(t) is intensity of light transmitted through solution and D is a constant proportional to the sample thiskness and absorbtivity of MV^+. Measurements of I(t) were performed 1300 to 2000 times every 0.5 micro second.

3. RESULTS AND DISCUSSIONS

Concentration of active, rapidly diffusing particles, MV^+, as a function of time is presented in Fig.1. These particles are created at time zero, during an interval shorter than the time resolution of our measurments. Quantity of our direct interest, logaritham of C(t), is presented in Fig.2. Classical kinetic analysis of the first order reactions suggests that this should be a linear function of time. However, the curvature in Fig.2 is pronounced and we can safely rule out the latter possibility.

In order to examine the applicability of equation 2 and equation 3, we performed the least square fit analysis of the data points with functions of the form: $-At - Bt^C$, where both A and B and exponent C are considered as unknowns. The best results, in terms of the mean square deviations, were achieved with finite values of A and B and a value of exponent C somewhere between 0.5 and 0.6. A conservative estimate of the best fit value of this exponent is $C = 0.56 \pm 0.06$. The best fit values of A and B, given in the caption of Fig.2, reflect the scale of the time axis. One should take a note of the fact that

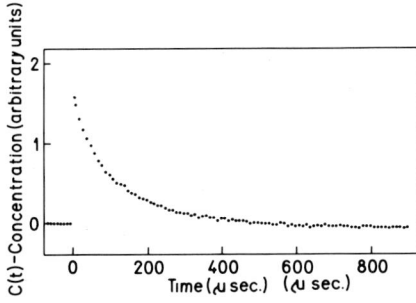

FIGURE 1
Concentration of MV^+ monocations as a function of time.

the singular part of the function $Log(C(t))$ contributes more than 90% of its value, throughout the observed range.

Within the accuracy of our measurements and limitations of our analysis, we may claim that the anomalous decay law of Redner and Kang[12], Eq.3, for a trapping reaction in a system with moving traps, provides a satisfactory description of our three-dimensional kinetics.

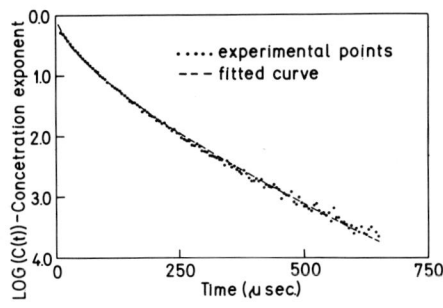

FIGURE 2
Logaritham of the C(t) function of Figure 1. Fitted curve presented by a dashed line is function $Log\, C_f(t) = -1.1825 \times 10^{-3} t - 6.08 \times 10^{-2} \cdot t^{0.6}$.

We are aware of the fact that computer simulation attempts on the verification of the above decay law experience difficulties and observe the non-analytic behavior only at exceedingly long times. However, one should note that in terms of concentrations, our experiments start well beyond the point at which computer simulations stop being practical. For example, we have some 10^{-9} moles of platinum particles per liter of our solution. Translated in the units of the simulation lattice, this number corresponds to, roughly, 1 trap per 10^9 lattice sites. This is obviously a figure too large for an effective computer simulation. Apparently, by achieving such high dilutions we have reached the region where the non-analytic term $t^{d/d+2}$ plays a significant role. Supposedly, the diffusivity of our traps is also sufficiently small (large) to make the observation of both terms in equation 3, possible.

ACKNOWLEDGEMENTS

The author would like to thank his colleagues: Dr. Milica Nenadović and Dr. Olga Mičić for their help in setting these experiments.

REFERENCES

1. B.Ya.Balagurov and V.G.Vaks, Sov. Phys. -JETP 38(1974)968
2. M.F.Shlesinger, J. Chem. Phys. 70(1979)4813
3. M.Bixon and R.Zwanzig, J. Chem. Phys. 75(1981) 2354
4. T.R.Kirkpatrick, J. Chem. Phys. 76(1982)4255
5. P.Grassberger and I.Proccacia, J. Chem. Phys. 77(1982)6281
6. D.F.Calef and J.M.Deutch, Ann. Rev. Phys. Chem. 34(1983)493
7. K.Kang and S.Redner, Phys. Rev. Lett. 52(1984) 955
8. P.Meakin and H.E.Stanley, J. Phys. A. 17(1984) L173
9. L.W.Anacker and R.Kopelman, J. Chem. Phys. 81(1984)6402
10. A.Blumen, G.Zumofen and J.Klafter, Phys. Rev. B. 30(1984)5379
11. S.Havlin, G.H.Weiss, J.E.Keifer and M.Dishon, J. Phys. A. 17(1984)L347

12. S.Redner and K.Kang, J. Phys. A. 17(1984)L451
13. M.D.Donsker and S.R.S.Varadhan, Commun. Pure Appl. Math. 28(1975)525
14. M.S.Matheson, P.C.Lee, D.Meisel and E.Pelizzetti, J. Phys. Chem. 87(1983)394
15. V.Marković, N.Nikolić and O.I.Mičić, Int. J. Rad. Phys. Chem. 6(1974)227
16. M.T.Nenadović, O.I.Mičić and R.R.Adžić, J. Chem. Soc. Faraday Trans. I, 78(1982)1065
17. G.C.Bond, Trans. Faraday Soc. 52(1956)1235

HIERARCHICALLY CONSTRAINED THERMODYNAMICS IN METASTABLE SYSTEMS AND GLASSES

L. PIETRONERO

University of Groningen, Melkweg 1, 9718 EP Groningen, The Netherlands

It has been recently proposed that the configurational degrees of freedom in glasses and metastable systems in general may be coupled in a hierarchical way. We show here how such a picture leads to an effective temperature dependence of the phase space available for thermodynamics. The implications of this result are in agreement with various low temperature properties of glasses.

1. INTRODUCTION

A glass represents a metastable state of a solid without long range order. For practical purposes we can consider it as stable since the relaxation time to the ordered state can be extremely long (sometimes more than the age of our universe). The question of its relative stability may therefore appear as immaterial. On the other hand in view of this fact the formulation of statistical mechanics and thermodynamics for metastable glassy states is a very complex and essentially unsolved problem[1]. The main difficulty lies in the fact that these systems are not ergodic so that the study of equilibrium distributions is not appropriate and the dynamics has to be included explicitely in the theory. Even the nature (thermodynamic or purely dynamic) of the glass-liquid transition is at present not clear[1].

An interesting attempt to include explicitly dynamical effects in the description of the properties of glasses has been recently made by Palmer et al.[2]. These authors consider the effects of hierarchically constrained dynamics for the relaxation of a glass to obtain the Kohlrausch relaxation law: $\exp[-(t/\tau)^\beta]$. The idea is that an atom cannot move until some other atom that blocks its path goes out of the way.

Here we study the effect of such constraints on the thermodynamic properties of glasses and show that <u>the effectively available phase space becomes temperature dependent</u>[3].
This result has consequences on various properties like the line width of impurity levels, the specific heat and the thermal conductivity that compare well with the experimental data. In addition from the present point of view the glass-liquid transition should correspond to a sort of invasion percolation problem in which the accessible degrees of freedom invade the whole phase space.

2. HIERARCHICALLY CONSTRAINED DYNAMICS

In terms of phase space the formation of the glassy state can be viewed as follows. Upon cooling the glass becomes confined to some region in phase space, one out of many equally possible regions, which have become no longer mutually accessible. Accessibility refers to the relaxation time not exceeding the experimental time. More and more degrees of freedom become "frozen" as the temperature is lowered and the system becomes restricted to the vibrational and relaxational degrees of freedom associated with the confined regions. We only restrict our discussion to the configurational degrees of freedom that are the relevant ones for the description of the glassy state. In addition, for simplicity and concreteness we describe them in terms of two-level-systems (TLS)[4].

Consider a glass at low temperature, say $1°K$, the standard description in terms of TLS is that, starting from the local ground state of the system a number of these excitations are available, in the sense that their relaxation times τ_R are shorter than the time of the experiment τ_E. The situation becomes more complex if the dynamics of TLS is subjected to constraints. The type of constraints we intend to discuss is schematically represented in Fig. 1.

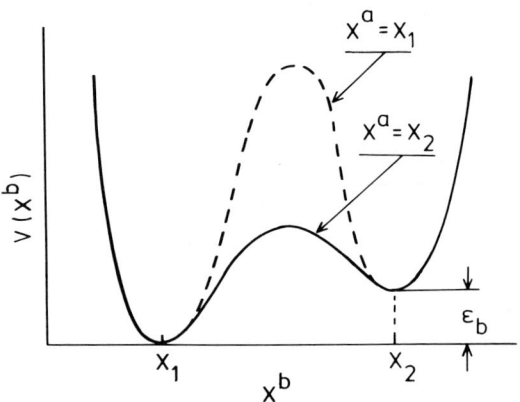

FIGURE 2
Schematic picure of the potential for the TLS indicated by (b) in Fig. 1. If (a) is in 1 the barrier of (b) is very large and the transition is blocked, while when (a) is in 2 the barrier for (b) is lower and the transition is possible. Note that such a coupling is purely dynamical, it does not affect the energies of the local minima.

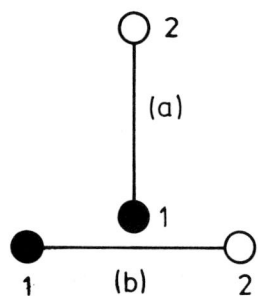

FIGURE 1
Schematic representation of dynamical constraints on a pair of two level systems. When the TLS indicated by (a) is in the local minimum 1 it blocks the path for the relaxation of (b). When it is in position 2 then (b) can also relax.

The configuration of two TLS shown in Fig. 1 contains a free TLS indicated by (a) while the (b) type has a potential barrier that depends on the position of (a) as indicated in Fig. 2. Note that the coupling is purely dynamical in the sense that, as shown in Fig. 2 only the barrier is changed but not the energies of local equilibrium. The constraint is hierarchical in the sense that (a) acts on (b) but (b) does not act on (a) and one can easily generalize this situation to a series of levels as shown in Fig. 3.

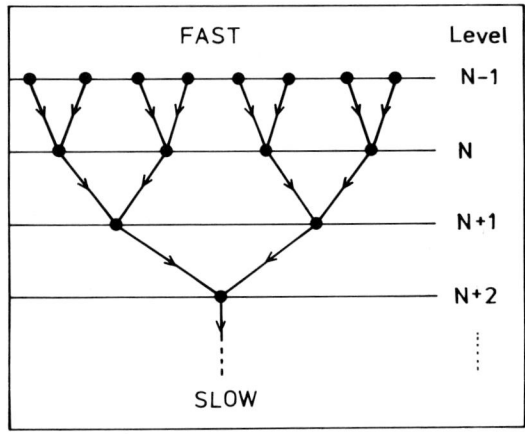

FIGURE 3
Schematic representation of a hierarchy of degrees of freedom. Each arrow represents a constraint of the type discussed in the previous figures. Degrees of freedom go from fast to slow by increasing N.

For simplicity we restrict our discussion here only to a pair of TLS. Suppose $x^{(a)}$ is an accessible degree of freedom. This implies $\tau_E \gg \tau_R^{(a)}$. The question is: under which conditions is $x^{(b)}$ also accessible? Since $x^{(a)}$ is assumed accessible it will reach thermal equilibrium within τ_E and the occupation probability of x_2 is

$$p(x^{(a)} = x_2) = e^{-\beta \epsilon_a}/(1 + e^{-\beta \epsilon_a}) \quad (1)$$

where $\beta = 1/K_B T$ and ϵ_a is the gap of (a). The "effective" experimental time $\tau_E^{(b)}$ available for the dynamics of (b) will be

$$\tau_E^{(b)}(\epsilon_a, T) = \tau_E \, p(x^{(a)} = x_2) =$$
$$= \tau_E \, e^{-\beta \epsilon_a}/(1 + e^{-\beta \epsilon_a}). \quad (2)$$

We can now introduce an "availability function" P that decides whether the degree of freedom (b) is available during the effective experimental time. This function depends also on the properties of (a):

$$P[\tau_E; \tau_R^{(b)}; \epsilon_a] = \begin{cases} 1 & \text{if } \tau_E^{(b)}(\epsilon_a,T) \gg \tau_R^{(b)} \\ 0 & \text{if } \tau_E^{(b)}(\epsilon_a,T) \ll \tau_R^{(b)} \end{cases}. \quad (3)$$

For a particular distribution function $W(\epsilon_a)$ for the gap ϵ_a the probability that $x^{(b)}$ is available is

$$\langle P[\tau_E; \tau_R^{(b)}; \epsilon_a] \rangle_a = \int_0^\infty W(\epsilon_a) P[\tau_E; \tau_R^{(b)}; \epsilon_a] d\epsilon_a. \quad (4)$$

By approximating $1 - e^{-\beta \epsilon_a} \approx 1$ the availability condition is then simply

$$\epsilon_a \ll T \ln[\tau_E/\tau_R^{(b)}] \quad (5)$$

and for a constant distribution of gaps $W(\epsilon_a) \approx W_0$ we obtain

$$\langle P[\tau_E; \tau_R^{(b)}; \epsilon_a] \rangle_a \approx \int_0^{T \ln[\tau_E/\tau_R^{(b)}]} W(\epsilon_a) d\epsilon_a \approx$$

$$\approx T W_0 \ln[\tau_E/\tau_R^{(b)}] . \quad (6)$$

It is then easy to derive the effective density of available degrees of freedom

$$\tilde{\rho}(T) = \rho_a [1 + T(\rho_b/\rho_a) W \ln(\tau_E/\tau_R^{(b)})] \quad (7)$$

where ρ_a and ρ_b are respectively the densities of (a) and (b) type TLS. This result shows how the hierarchical constraints give rise to an available phase space that grows with temperature. It also enlightens the crucial role played by the dynamics in the quasi equilibrium properties of metastable systems.

3. RELATIONS WITH EXPERIMENTS

A number of physical properties become affected by Eq.(7) that in view of the approximations we have made should be considered as a low temperature limit.

The most studied low temperature property of glasses is the specific heat[4]. For this we predict

$$C(T) \approx \rho_0 T + B \rho_0^2 T^2 . \quad (8)$$

A recent analysis of the available low temperature specific heat data suggests exactly such an expression for an optimal fit[5].

Another well known property is the thermal conductivity for which we obtain

$$K(T) \approx T^2(1 - B\rho_0 T) . \quad (9)$$

Such an expression is compatible with the experiments that are usually fitted by a power law with an exponent of about 1.8^4.

These two properties are however restricted to very low temperatures ($T \lesssim 1^\circ K$) where the effect of the extra quadratic term is small.

The situation is different for the homogeneous linewidth of impurity levels[6,7] that can be studied also at intermediate ($T \simeq 10^0 K$) and relatively high ($T \simeq 100^0 K$) temperatures. We obtain for the linewidth

$$T \simeq \rho_0 T + B\rho_2^2 T^2 . \qquad (10)$$

This result is consistent with the fact that the "apparent exponent" systematically increases with temperature for a given material from T at low temperature up to the high temperature T^2 behavior[6,7].

Finally we would like to add some speculative remarks on the implication of the present picture on the glass-liquid transition. Such a transition should correspond to the temperature at which the available degrees of freedom invade the whole phase space. This implies that this transition should be considered closer to some sort of percolation problem (as previously suggested by Grest and Cohen[1] from a rather different point of view) than to a normal equilibrium phase transition.

In order to describe such a phenomenon one cannot restrict to a low temperature perturbation as we have done here but the whole hierarchy of coupled levels as shown in Fig. 3 should be considered. Actually the situation may be even more complex due to the presence of loops in the hierarchy. It is interesting to note that such loops can give rise to hysteresis effects at the transition that are indeed observed[1]. A careful analysis is now needed to substantiate or reject the present speculations.

REFERENCES

1. G.S. Grest and M.H. Cohen, Adv. Chem. Phys. 48 (1981) 55.

2. R.G. Palmer, D.L. Stein, E. Abrahams and P.W. Anderson, Phys. Rev. Lett. 53 (1984) 958.

3. For a more detailed discussion see: L. Pietronero, in print.

4. See for example: W.A. Phillips ed. "Amorphous Solids" (Topics in Current Physics, Springer, Berlin, 1981).

5. C.M. Varma, R.C. Dynes and J.R. Banavar, J. Phys. C 15 (1982) L1221.

6. H.P.H. Thijssen, R.E. van den Berg and S. Völker, Chem. Phys. Lett. 103 (1983) 23; ibid. 97 (1983) 295.

7. J. Hegerty, M.M. Broer, B. Golding, J.R. Simpson and J.B. MacChesney, Phys. Rev. Lett. 51 (1983) 2033.

FRACTAL CLUSTERS AND SCALING IN THE ISING MODEL

J.L. CAMBIER and M. NAUENBERG[*]

Physics Department, University of California, Santa Cruz, CA 95064

Monte Carlo simulations of the Ising model in two and three dimensions show that the clusters of spins are fractals, and that the distribution for large size clusters satisfies scaling properties. The scaling exponents for this distribution function, the surface exponent and the fractal dimension of the clusters are related to the critical exponents of the Ising model.

I. INTRODUCTION

In this lecture we will present some recent Monte-Carlo simulations to study the properties of clusters or droplets of aligned spins in the Ising model and its relation to critical phenomena. Following the droplet model of Fisher[1] and further developments by Domb and collaborators[2,3], who considered the effect of ramified clusters, Binder[4] extended the scaling theory of clusters, and carried out Monte-Carlo simulations to test various conjectured scaling properties. However, after the observation by Muller-Krumbhaar[5] of a percolating cluster in the d = 3 Ising model near but below the critical temperature, doubts have been casted on the usefulness of this approach to describe critical behavior. Proposals for modified clusters have been made[4,6,7] to shift this apparent percolation threshold to the Ising critical point. However these new clusters have surfaces which do not correspond necessarily to the pairs of antialigned spins in the lattice, and therefore do not add up to the internal energy. Consequently these modified clusters do not give a complete description of the critical behavior of the Ising model. More recently, Bruce and collaborators[8,9] have developed a microscopic droplet theory in low dimension which yields scaling exponents for the Ising clusters satisfying relations which differ from those obtained in the phenomenological theory[4]. Furthermore, applying finite size scaling theory and some untested conjectures about clusters, Suzuki[10] obtained a relation between the fractal dimension[11] of the largest cluster and the critical exponents similar to that for percolation clusters[12,13].

In order to clarify some of these issues we have studied the original clusters of aligned spins by Monte Carlo simulations, and evaluated the cluster size distribution, the mean cluster surface, and the cluster radius of gyration as a function of the number of spins for the d = 2 and d = 3 Ising model below but near the Ising critical temperature Tc. We have developed a novel approach to obtain this distribution for large cluster sizes which turns out to be essential to determine its scaling behavior. We review below the scaling theory for the cluster distribution and present the results of our Monte Carlo calculations which are restricted to zero magnetic field.

2. FORMALISM

Below the critical temperature Tc, let $N(n,t,h)$ be the distribution of clusters of n spins aligned in a direction opposite to the magnetic field h at the reduced temperature

---

[*]Supported by a grant from the National Science Foundation

$t = (T_c-T)/T_c$. This distribution is normalized by the condition that the total fraction of overturned spins in the lattice is $\sum_{n=1}^{\infty} n\, N(n,t,h)$ and the magnetization $m(t,h)$ is given exactly by

$$m(t,h) = 1 - 2 \sum_{n=1}^{\infty} n\, N(n,t,h) \qquad (1)$$

Likewise, if we let $s(n,t,h)$ correspond to the mean surface of a cluster of n spins defined as the total number of antialigned spins or broken bonds associated with the boundary of the cluster, the internal energy $e(t,h)$ is

$$e(t,h) = 2 \sum_{n=1}^{\infty} s(n,t,h)\, N(n,t,h) - h\, m(t,h) \qquad (2)$$

Near the critical point, $t = 0$ and $h = 0$, the magnetization and the internal energy satisfy the well-known scaling behavior

$$m(t,h) = t^{\beta} \bar{m}(z) \qquad (3)$$

and

$$e(t,h) = t^{(1-\alpha)} \bar{e}(z) + e_0(t,h) \qquad (4)$$

where $z = h/t^{\phi}$, and α, β and ϕ are the critical exponents. The scaling functions $\bar{m}(z)$ and $\bar{e}(z)$ are regular at $z = 0$, and $e_0(t,h)$ is the nonsingular part of the energy. To obtain this critical behavior from the cluster sums, Eqs. (1) and (2), the cluster distribution $N(n,t,h)$ and the surface $s(n,t,h)$ are assumed to satisfy a scaling dependence on its variables for large cluster size n. Such scaling behavior was introduced by Fisher[1] and discussed more generally by Binder[4] who assumed that for large n and small t and h the distribution $N(n,t,h)$ takes the form

$$N(n,t,h) = n^{-\tau} \bar{N}(tn^{y_T}, hn^{y_H}) \qquad (5)$$

where $\bar{N}(x,y)$ is a scaling function regular at $x = 0$ and $y = 0$, and τ, y_T and y_H are the distribution exponents. For the surface $s(n,t,h)$ we take the simple asymptotic form

$$s(n,t,h) = \bar{S} n^{\sigma} \qquad (6)$$

proposed originally by Fisher[1], where σ is the surface exponent and $\bar{S}$ is a constant, although more generally $\bar{S}$ could satisfy a similar scaling form $\bar{S}(tn^{y_T}, hn^{y_H})$ as in Eq. (5).

We will now show that the asymptotic form, Eqs. (5) and (6), lead to the usual scaling behavior, Eqs. (3) and (4), and derive relations among the new scaling exponents and the well-known critical exponents. Let us assume that the asymptotic scaling expressions, Eqs. (5) and (6) are valid for $n > n_c$, where n_c is a lower bound which need not be determined, and approximate the sums for $n > n_c$ in Eqs. (1) and (2) by integrals over n. For convenience consider the partial sum $M_r(n,t,h)$ for the rth moment of the distribution

$$M_r(n,t,h) = \sum_{n=1}^{n} n^r N(n,t,h) \qquad (7)$$

where r is a constant. This can be written in the form

$$M_r(n,t,h) = \sum_{n=1}^{n_c} n\, N(n,t,h) + \int_{n_c}^{n} dn\, n^{r-\tau} \bar{N}(tn^{y_T}, hn^{y_H}) \qquad (8)$$

where we substituted the asymptotic scaling dependence for $N(n,t,h)$, Eq. 5 under the integral. Changing the variable of integration n to $x = tn^{y_T}$, we have

$$M_r(n,t,h) = \sum_{n=1}^{n_c} n\, N(n,t,h) + \frac{t^u}{y_T} \int_{x_c}^{x} dx\, x^{-u-1} \bar{N}(x, x^{\phi}z) \qquad (9)$$

where $u = (\tau-r-1)/y$, $x = tn^{y_T}$ and $\phi = y_H/y_T$.

For $u > 0$, and $\overline{N}(0,0) \neq 0$ the integral in Eq. (9) diverges as the lower limit of integration goes to zero; we regularize by subtracting $\overline{N}(0,0)$ under the integral to obtain the form

$$M_r(n,t,h) = g_r(t,h) + t^u f_r(x,z) \qquad (10)$$

where

$$f_r(x,z) = \frac{1}{y_T} \int_0^x dx\, x^{-u-1}[\overline{N}(x,x^\phi z) - \overline{N}(0,0)] \qquad (11)$$

and

$$g_r(t,h) = \sum_{n=1}^{n_c} n^r N(n,t,h) - \frac{n^{-uy_T}}{uy_T} \overline{N}(0,0)$$
$$- \int_0^{n_c} dn\, n^{r-\tau}[\overline{N}(tn^{y_T}, hn^{y_H}) - \overline{N}(0,0)] \qquad (12)$$

The last expression determines the nonscaling contribution of small clusters. Although the constant n_c appears in the upper bound of the sum and of the integral in Eq. (12), $g_r(t,h)$ actually does not depend on n_c as can be readily verified by taking the partial derivative with respect to n_c. If $\overline{N}(x,y)$ is regular at $x = 0$, $y = 0$, then the integral in Eq. (12) will also be regular in t and h and the singularities in $M_r(n,t,h)$ are contained in the second term of Eq. (10).

The magnetization $m(t,h)$, Eq. (1), can now be evaluated from Eq. (10) by setting $r = 1$ and taking the limit $n \to \infty$ imposing the constraint $g_1(0,0) = 1/2$. This leads to the scaling form Eq. (3) with

$$\overline{m}(z) = -2M_1(\infty,z) \quad \text{and} \quad \beta = (\tau-2)/y_T \qquad (13)$$

A similar procedure leads to the determination of the internal energy $e(t,h)$ by setting $r = \sigma$ (using Eq. (6)), and we obtain the scaling form Eq. (4), where

$$e(z) = 2M_\sigma(\infty,z) \quad \text{and} \quad \alpha = 1-(\tau-\sigma-1)/y_T \qquad (14)$$

The thermodynamic relation

$$\frac{\partial}{\partial t} m(t,h) = \frac{\partial}{\partial h} e(t,h) \qquad (15)$$

implies the familiar relation for critical exponents $\alpha + \beta + \phi = 2$, and leads to the exponent relation

$$\sigma = 1 + y_T - y_H = 1 - y_T(\phi-1) \qquad (16)$$

3. MONTE CARLO RESULTS AND INTERPRETATION

The Monte Carlo data for the Ising model clusters in the magnetic field was obtained on square and cubic lattices of linear sizes $L = 64$ and 128 for two dimensions and $L = 16$ and 32 for three dimensions. Typical running times were of the order of 10,000 iterations at each temperature. Clusters were analyzed every 50 iterations to obtain the size distribution and the total surface was defined as the number of broken bonds for both outside and inside boundaries.

The results for the surface of the clusters as a function of the number of spins is shown in a logarithmic plot in Fig. (1) for a range of temperatures below the critical temperature in $d = 2$. Similar results are obtained for $d = 3$. In order to show the data for different temperatures, it is shifted vertically by one unit for each decreasing value of the temperature, with the middle temperature shown at the correct scale. The data for $d = 2$ has a much larger spread than for the $d = 3$ case, which gives a nice demonstration of the occurence of a roughening transition for $d = 2$. The mean value of the surface can be fitted to the power law dependence conjectured by Fisher, Eq. 6, provided we restrict the range of sizes of the clusters. For clusters up to medium sizes $n = 200$, and excluding small clusters ($n < 20$), we obtain for the exponents $\sigma_{2d} = 0.68 \pm .04$ and $\sigma_{3d} = 0.88 \pm .04$ for $d = 2$ and $d = 3$ respectively. Our results are in agreement for

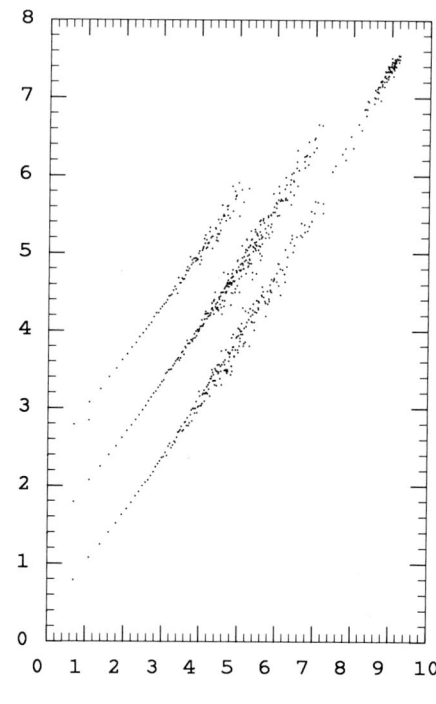

FIGURE 1

$\ln_e$(surface) versus $\ln_e$(size) for $d = 2$. The data by 2 units for $T = 2.15$, and by 1 unit for $T = 2.25$. The highest temperature, $T = 2.35$, is at the bottom and has the correct scale.

$d = 2$ with previous work of Domb et. al.[3], but not with the results of Binder[4].

To obtain the scaling function $\bar{N}(x)$, Eq. (5), by evaluating directly the cluster size distribution $N(n,t,0)$ from Monte Carlo simulations would require an inordinate amount of computer time. We evaluate instead the integrated scaling function $f_\tau(x,0)$, Eqs. (10) and (11), from the partial sum $M_\tau(n,t,0)$, Eq. (7), and determine $\bar{N}(x)$ by differentiation. The result for $d = 3$ is shown in Fig. (2) which demonstrate that scaling is valid for a restricted range of temperatures indicated on these figures. The exponents which give the best scaling fit are $\tau_{2d} = 2.05 \pm .03$ and $y_{2d} = 0.40 \pm .02$ for $d = 2$, and $\tau_{3d} = 2.07 \pm .03$, $y_{3d} = 0.21 \pm .02$ for $d = 3$, where we have assumed the critical temperatures $T_c = 2.25$ and 4.52 for $d = 2$ and $d = 3$ respectively. Substituting these exponents in the scaling relation, Eq. (13) we obtain the critical exponent β for $d = 2,3$ in agreement with the known value. However for $4.3 < T < 4.52$ a percolating cluster appears in $d = 3$, as observed originally by Muller-Krumbhaar[5] and this scaling behavior no longer holds.

Finally we have evaluated the radius of gyration R of the clusters to obtain the fractal dimension d', defined by the relation $R = n^{1/d'}$. We obtain $d'_{2d} = 1.9 \pm .06$ and $d'_{3d} = 2.3 \pm .05$. It should be pointed out that while these results agree approximately with $d' = d - \beta/\nu$, the arguments of Suzuki to derive this relation are not consistent with the behavior of the cluster distributions found in our Monte Carlo simulations.

CONCLUSIONS

We have presented numerical evidence that the cluster size distribution for the Ising model in zero magnetic field satisfies a scaling behavior which is similar in $d = 2$ and $d = 3$ for a narrow temperature range below the critical temperature T_c, and evaluated the critical exponents τ and y_T associated with this distribution in good agreement with the scaling relations for these exponents. For $d = 3$ this scaling behavior occurs below an apparent percolation threshold at $T_p = 4.3$ which, at first sight, appears surprising in view of the current scaling theory of percolation[13] which predicts scaling with respect to T_p. However, we find that the size distribution scales also approximately with the reduced temperature

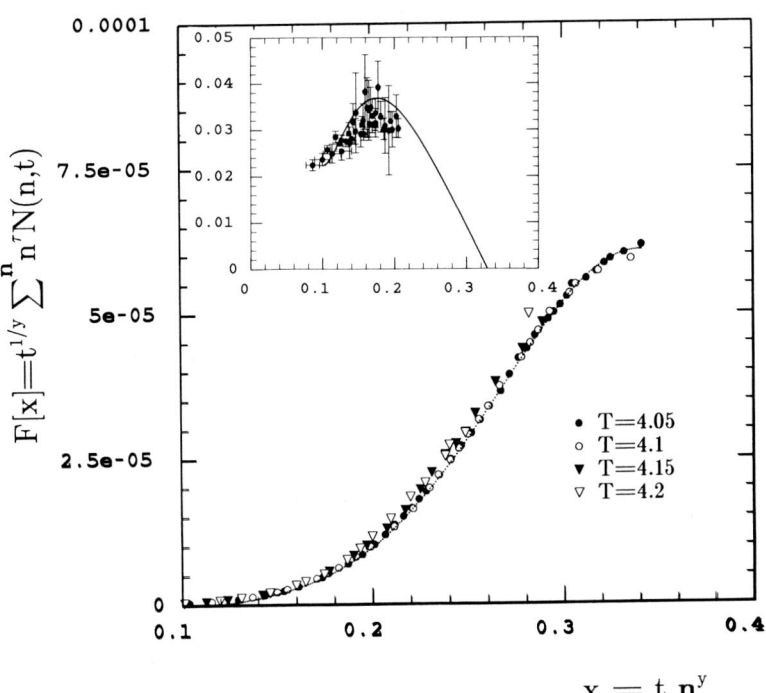

FIGURE 2
Scaling function $f_\tau(x)$ versus $x = tn^{y_T}$ for $d = 3$. The inset shows the scaling function $\overline{N}(x)$ obtained from a polynomial fit to the integrated function $f_\tau(x)$ (full curve). The data points are obtained by measuring directly $\overline{N}(x) = n^\tau N(n,t)$ for medium sized clusters.

$t' = (T_p-T)/T_p$, but with $y'_T \cong 0.4$. It is likely that there exists a more complicated scaling form that accounts for the crossover between these two thresholds. Furthermore, there does not appear to be any simple scaling behavior of the distribution above T_p and below T_c. We believe that finite lattice size effects may distort the size distribution near T_c, and these effects should be investigated further. We have also found that the Ising clusters are fractals, and evaluated the surface exponent which is temperature independent but increases somewhat with the size of the clusters. It should be interesting to find out whether this is due to the fact that the clusters are becoming more ramified, and/or there is an increase of internal boundaries. These results should be extended to study the scaling dependence in an external magnetic field and, of particular current interest, the effects of random field[14] where it is generally believed that the clusters or domains of spins play a crucial role in understanding the thermodynamic behavior of this system. Finally we note that recent efforts to develop a microscopic droplet theory for low dimension[8,9] yield exponent relation which are not in agreement with our results for $d = 2$, and imply incorrectly that the characteristic length of the clusters is the Ising correlation length.

ACKNOWLEDGMENTS

We would like to thank A. Coniglio, W. Klein, J. Cardy and A.P. Young for useful discussions.

REFERENCES

1. M.E. Fisher, Physics 3, 255 (1967).

2. C. Domb, J. Phys. A9 (1976) L141.

3. C. Domb, T. Schneider and E. Stoll, J. Phys. A8 (1975) L90.

4. K. Binder, Annals of Physics 98 (1976) 390.

5. H. Muller-Krumbhaar, Phys. Lett. 48A (1974) 459.

6. A. Coniglio and W. Klein, Phys. A13 (1980) 2775.

7. D.W. Heermann and D. Stauffer, Z. Phys. B44 (1981) 339. D.W. Heermann, A. Coniglio, W. Klein and D. Stauffer, J. Stat. Phys. 36 (1984) 447.

8. A.D. Bruce and D.J. Wallace, J. Phys. A16, (1983) 1721.

9. J.S. Sim and A.D. Bruce, J. Phys. A18 (1985) 1119.

10. M. Suzuki, Prog. Theor. Phys. 69 (1983) 65.

11. B.B. Mandelbrot, The Fractal Geometry of Nature (Freeman, San Francisco, 1982).

12. A. Kapitulnik, A. Aharony, G. Deutscher and D. Stauffer, J. Phys. 16 (1983) L269.

13. D. Stauffer, Phys. Rep. 54 (1979) 1. For a recent review of percolation see J. Kertesz, D. Stauffer and A. Coniglio, Annals of the Israel Phys. Soc. V.5 (1983) 101. Percolation Structures and Processes, Eds. G. Deutscher, R. Zallen and J. Adler.

14. J.L. Cambier and M. Nauenberg (U.C.S.C. preprint 1985).

FRACTALS IN PHYSICS
L. Pietronero, E. Tosatti (editors)
© Elsevier Science Publishers B.V., 1986

DEVIL'S STAIRCASE AND STRANGE ATTRACTOR IN THE ISING MODEL WITH COMPETING INTERACTIONS

M.J. de OLIVEIRA, S.R. SALINAS, and C.S.O. YOKOI

Instituto de Física, Universidade de São Paulo, C.P. 20516, CEP 01498, São Paulo, Brazil

We formulate the Ising model with competing interactions on a Cayley tree, in the infinite-coordination limit, as a two-dimensional nonlinear mapping. The phase diagram displays a Lifshitz point and many modulated phases. We perform calculations to show the existence of a complete devil's staircase at low temperatures. Also, we give strong numerical evidence for the existence of chaotic phases associated with strange attractors.

The anisotropic-next-nearest-neighbor Ising (ANNNI) model, which consists of an Ising model with nearest-neighbor interactions augmented by competing next-nearest-neighbor couplings acting parallel to a single axis direction, is perhaps the simplest nontrivial model displaying a rich phase diagram with a Lifshitz point and many spatially modulated phases.[1-6] There has been a considerable theoretical effort to obtain the structure of the global phase diagram of the ANNNI model in the T-p space, where T is temperature and $p = -J_2/J_1$ is the ratio between the competing exchange interactions. On the basis of numerical meanfield calculations, it has been suggested the existence of an infinite succession of commensurate phases, the so-called devil's staircase, at low temperatures.

In this paper we report calculations for an Ising model on a Cayley tree, in the limit of infinite coordination z, with ferromagnetic nearest-neighbor interactions, $J_1>0$, and antiferromagnetic next-nearest-neighbor interactions, $J_2<0$, along the branches of the tree.[7] As pointed out by Vannimenus[8], who considered this Ising model on a tree of coordination $z=3$, although not entirely similar it is certainly the counterpart on a tree of the ANNNI model.

Initially, we extend the calculations of Vannimenus[8] for a tree of arbitrary coordination z. Then, we use the procedures of Inawashiro, Thompson, and Honda[9] to write a set of three first-order recursion relations. In the infinite-coordination limit, which is defined in the present case by $z \to \infty$, $J_1 \to 0$, and $J_2 \to 0$, with $zJ_1 =$ const and $z^2J_2 =$ const, there remain only two simple coupled first-order recursion relations. We remark that this limit is essential to produce simple equations which are then amenable to some analytical as well as very detailed numerical investigations. The paramagnetic lines in the T-p phase diagram shown in Fig. 1, where $p = -z^2J_2/zJ_1$ and T is the temperature in units of zJ_1/k, can be found

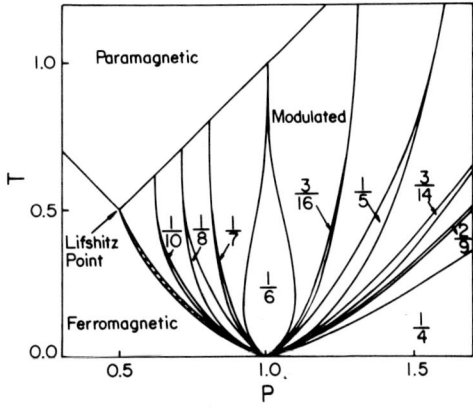

FIGURE 1
Global T-p phase diagram of the Ising model with competing interactions on the Cayley tree. In the modulated region, we show a few commensurate phases, which are indicated by the corresponding principal wave numbers. In the hatched region there is the possibility of coexistence between ferromagnetic and modulated phases.

analytically from the stability of the paramagnetic fixed point. Also, it is easy to find the location of the Lifshitz point and the limit of stability of the ferromagnetic phase. The critical wave number varies smoothly along the paramagnetic-modulated λ line and vanishes at the Lifshitz point. This phase diagram is much richer than the findings of Vannimenus, and the qualitative pattern of the modulated phases agrees with previous work for the ANNNI model.[5,6]

In the present work we perform detailed calculations to obtain, for example, the devil's staircase (principal wave number, q, versus the parameter p) for various temperatures, and the corresponding fractal dimensionalities D (see Fig. 2). At low temperature, since $D<1$, we have a complete devil's staircase, that is, the incommensurate phases occupy a region of zero measure in the phase diagram. However, D increases with temperature, and this indicates that above a certain temperature the incommensurate phases will fill up significant portions of the phase diagram.[10] A quite unexpected finding refers to the fixed orbits in the coexistence region between ferromagnetic and modulated phases. We give strong numerical evidence to show the existence of chaotic trajectories, characterized by a positive Lyapunov exponent. These trajectories exhibit a fractal structure very similar to the strange attractors of Curry and Yorke[11]. Also, the numerical data in this region support the picture of a sequence of bifurcations, that is, of period-doubling modulated phases, which lead to a chaotic state with the universal Feigenbaum exponent δ.[12]

Let us consider the Ising model of Vannimenus, with competing interactions between first and second neighbors along the branches of the Cayley tree (see Fig. 1 in Ref. 8). In order to set up the iterative scheme, we sum successively over spins in the outermost shells (from r=0 to r=N, where r labels the generations of the tree). In the infinite-coordination limit we obtain the

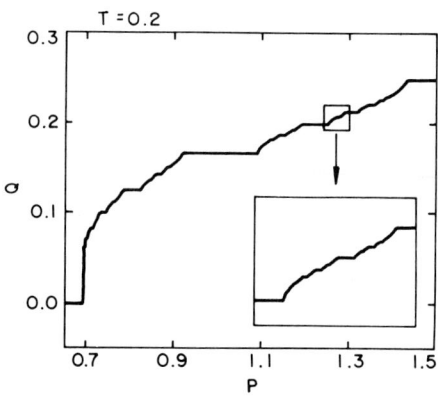

FIGURE 2
Plot of q versus p for $T = 0.2$. The inset shows a detail of the devil's staircase. Between $p = 0.65$ and $p = 1.5$, we estimated the value $D = 0.92$ for the fractal dimension of the staircase.

simple relations $X_r = H/T + (1/T)\tanh X_{r-1} + Y_{r-1}$ and $Y_r = -(p/T)\tanh X_{r-1}$, where H is the applied field and X_r may be interpreted as an effective field per site induced in the new rth shell. Also, it is convenient to define an effective magnetization per spin in the rth shell, $M_r = \tanh X_r$, and write the second-order difference equation

$$M_r = \tanh[T^{-1}(M_{r-1} - pM_{r-2} + H)] . \quad (1)$$

The phase boundaries between the modulated phases shown in Fig. 1 have been obtained from a numerical analysis of Eq.(1) in zero field. The equilibrium configuration for given values of T and p is found by repeated iterations of this recursion relation, starting from the initial condition $M_0 = M_1 = 1$. The magnetization flows to one of the following attractors: (i) a trivial fixed point $M^* = 0$ (which corresponds to the paramagnetic phase); (ii) a nontrivial fixed point $M^* \neq 0$ (the ferromagnetic phase); (iii) a well defined periodic cycle (which corresponds to a commensurate phase with a given period); (iv) a one-dimensional orbit (associated with an incommensurate phase); and (v) a strange attractor with a fractal character. For the periodic orbits (commensurate phases) the Lyapunov exponent is negative, but it vanishes, within the numerical precision, for the one-dimensional attractors. For the fractal attractors, on the other hand, we calculated positive Lyapunov exponents. In the hatched region of the phase diagram the magnetization flows either to the ferromagnetic or to the modulated fixed point, depending on the initial conditions. The ferromagnetic-modulated transition is thus of first order, with the possibility of coexistence between different phases. Also, the strange attractors were found in this region, although they are still present even beyond the line of stability of the ferromagnetic phase.

The most remarkable aspect of our calculations is shown in Figs. 3(a)-3(d), which were obtained for $T = 0.2$, $p = 0.69662$, and the initial conditions $M_0 = M_1 = 1$. The fractal nature of the strange attractor is clearly seen in this sequence of pictures. The Lyapunov exponent

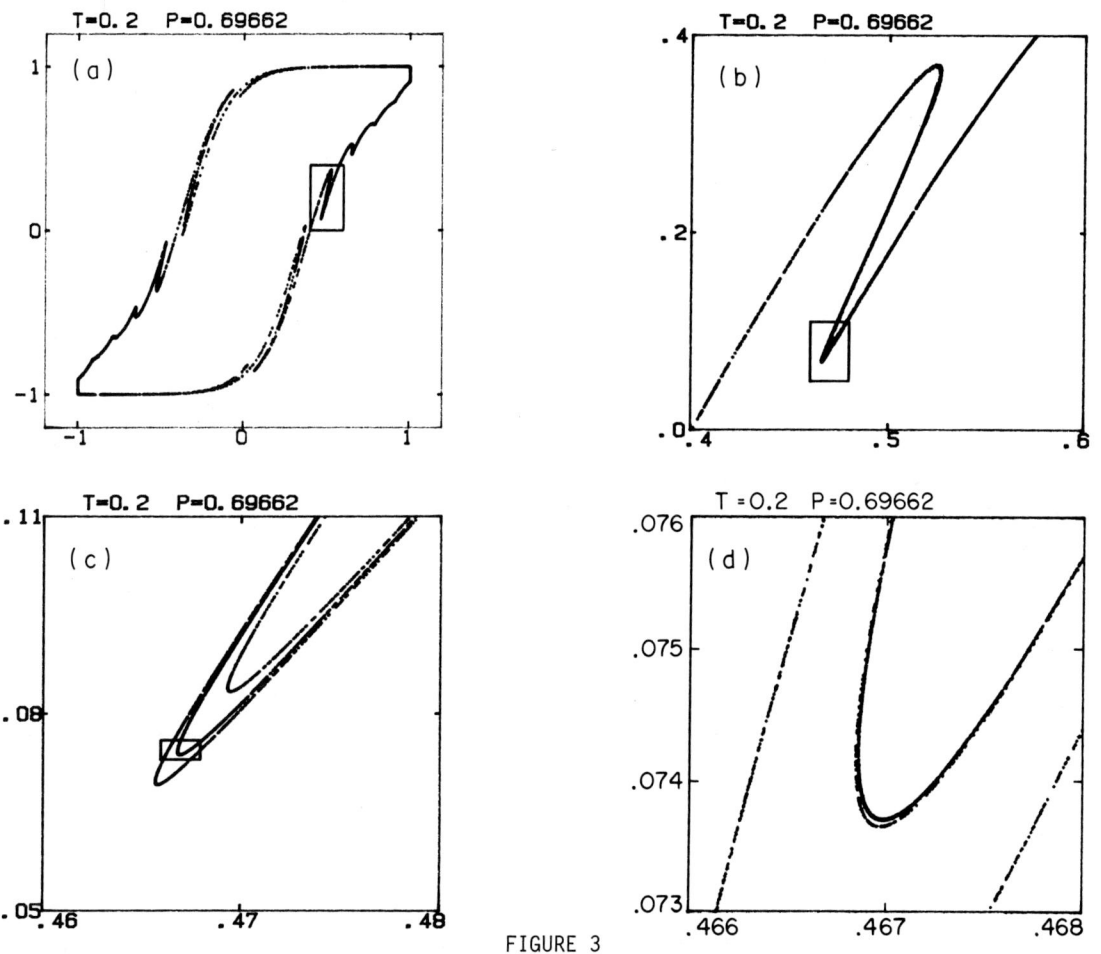

FIGURE 3
(a) Plot of the attractor associated with the mapping given by Eq.(1), for $T = 0.2$, $p = 0.69662$, and the initial conditions $M_0 = M_1 = 1$. The first 10^4 iterations are discarded; the subsequent 5000 iterations are plotted. (b) Enlargement of the box shown in (a). 3000 out of a total of 49439 iterations are plotted; the initial 10^4 iterations are discarded. (c) Enlargement of the box in (b). 2000 out of about 10^5 iterations are plotted; the initial 10^4 iterations are discarded. (d) Enlargement of the box in (c). This graph corresponds to a plot of 1000 iterations out of a total of about 10^6. This sequence of pictures gives some evidence for the self-similarity of the strange attractor.

of this orbit is positive ($\lambda=0.12$), and the Fourier analysis of the magnetization per spin for successive iterations yields a typical noise spectrum. Also, we analyzed the route to chaos. Within the numerical accuracy of the calculations, we were able to find a sequence of nine bifurcations (that is, of period-doubling commensurate phases), from which we estimated the Feigenbaum exponent $\delta=4.669\pm0.001$. (See Table I).

TABLE I
Bifurcations, for $T=0.2$, as a function of p.

n	p_n	$\delta_n = (p_{n+1}-p_n)/(p_{n+2}-p_{n+1})$
1	0.69595967544	4.4934
2	0.696021075024	4.6345
3	0.6960347394183	4.6609
4	0.6960376878381	4.6675
5	0.69603832041215	4.6688
6	0.6960384559389	4.6691
7	0.69603848496684	4.668
8	0.6960384911838	
9	0.6960384925155	

In conclusion, the infinite-coordination limit of the Ising model with competing interactions on a Cayley tree has been used to obtain a simple nonlinear two-dimensional mapping. The global phase diagram displays the same features of the mean-field phase diagram of the ANNNI model. The numerical investigations of the mapping established the existence of a complete devil's staircase at low temperatures, and the appearance of chaotic phases, in the core of the tree, for a given set of boundary conditions.

REFERENCES

1. R.J. Elliot, Phys. Rev. 124 (1961) 346.
2. R.M. Hornreich, M. Luban, and S. Shtrikman, Phys. Rev. Lett. 35 (1975) 1678.
3. S. Redner and H.E. Stanley, Phys. Rev. B16 (1977) 4901.
4. W. Selke, Z. Phys. B29 (1978) 133; W. Selke and M.E. Fisher, Phys. Rev. B20 (1979) 257.
5. P. Bak and J. von Boehm, Phys. Rev. B21 (1980) 5297.
6. M.E. Fisher and W. Selke, Phys. Rev. Lett. 44 (1980) 1502.
7. A preliminary account of these calculations has been published by C.S.O. Yokoi, M.J. de Oliveira, and S.R. Salinas, Phys. Rev. Lett. 54 (1985) 163.
8. J. Vannimenus, Z. Phys. B43 (1981) 141.
9. S. Inawashiro, C.J. Thompson, and G. Honda, J. Stat. Phys. 33 (1983) 419.
10. A similar dependence of the fractal dimensionality upon temperature has been found, in the case of the chiral Potts model on a Cayley tree, by C.S.O. Yokoi and M.J. de Oliveira, J. Phys. A18 (1985) L153.
11. H. Curry and A. Yorke, in: The Structure of Attractors in Dynamical Systems, eds. N.G. Markley, J.C. Martin, and W. Perrizo, Lecture Notes in Mathematics, Vol. 668 (Springer, Berlin, 1978), p. 48.
12. M.J. Feigenbaum, Physica (Utrecht) 7D (1983) 16.

FRACTALS IN PHYSICS
L. Pietronero, E. Tosatti (editors)
© Elsevier Science Publishers B.V., 1986

PROBLEMS ABOUT THE SELF-SIMILAR STRUCTURE OF WAVEFUNCTIONS IN DISORDERED SYSTEMS

A.P. SIEBESMA and L. PIETRONERO

University of Groningen, Melkweg 1, 9718 EP Groningen, The Netherlands

We discuss some problems connected with the recently proposed existence of fractal features in the internal structure of localized wavefunctions. Our analysis, even if not conclusive, points rather against this possibility.

1. INTRODUCTION

Recently Soukoulis and Economou[1] have considered the possibility that wavefunctions localized by disorder may manifest fractal features in the structure of their internal fluctuations. The probability distribution corresponding to an Anderson localized wavefunction decays exponentially after the localization length λ. This length, however, can be sensibly larger than the lattice spacing a, so that there may be some decades in the length scale in which self-similar fluctuations may develop. The above authors[1] have analyzed various wavefunctions in this regime and concluded that fractal behavior is indeed present with the fractal dimension D dependent on the amount of disorder. In view of the restrictions of the numerical analysis the data are best characterized in one dimension, because in two and three dimensions the available length-scale is much shorter.

In order to extend these studies we started by repeating their one dimensional calculations. The result was that even if at first sight we obtained a rather similar overall picture, a more careful analysis of the data revealed the presence of a small but persistent negative curvature of the slopes in a log-log plot. This implies that the observed deviations from D=1 (non fractal behavior) may be due just to finite size effects rather than to self-similar properties. In view of this problem we have decided then to consider idealized examples of self-similar normalized probability distributions constructed in a deterministic way. Such an analysis shows that these systems cannot be considered as homogeneous fractals because scale invariance can be present only with respect to particular origins. It is then hard to imagine that probability distributions corresponding to real localized wavefunctions may possess the properties of homogeneous fractals as implied by the analysis based on the density density correlation function[1]. In our opinion therefore, the evidence for fractal properties of localized wavefunctions is rather questionable.

2. ANALYSIS OF LOCALIZED WAVEFUNCTIONS IN ONE DIMENSION

We discuss in this section the method of analysis of Soukoulis and Economou[1] and point out some inconsistencies that raise serious questions about the validity of the approach.

The idea is that self-similar fluctuations may develop for length scales bounded by the lattice spacing a and the localization length λ. In numerical calculations this interval is largest for one dimension, that is the case we consider here. For the rest we adopt the standard Anderson tight binding model with diagonal disorder characterized by a rectangular distribution of total width W, identical therefore to the case studied by Soukoulis and Economou[1]. Some eigenvectors of this Hamiltonian are computed with the method of inverse iteration for a system of 5000 lattice sites with periodic boundary conditions. A typical probability distribution corresponding to a wavefunction of energy $E \approx 0$ (band center) and for $W=0.5$ is shown in Fig. 1.

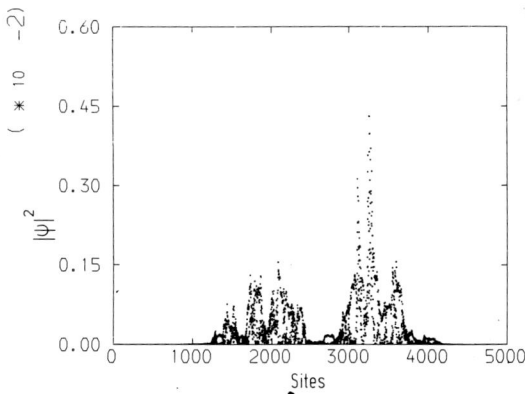

FIGURE 1
Example of the probability distribution corresponding to a localized wavefunction in one dimension. The disorder is diagonal with rectangular distribution of total width $W=0.5$.

The conjecture is that, if the fluctuations of this probability distribution are self similar (fractal), then the distribution can be described (on average) by a power law $\rho(x) = |\psi(x)|^2 \approx x^{-\alpha}$ for $a < x < \lambda$. For $x \gtrsim \lambda$ ($\lambda \approx 400a$ in the case of Fig. 1) the exponential envelope function should then take over. We can then define a fractal dimension $D = 1 - \alpha$ via the relations

$$\int_0^L \rho(x) dx \simeq L^{(1-\alpha)} \equiv L^D \quad . \quad (1)$$

An analysis of probability distributions like that of Fig. 1 from the point of view of Eq.(1) implies the choice of a center for the distribution and gives rise to very large fluctuations that could be eliminated only averaging over an extremely large number of samples. To avoid these problems Soukoulis and Economou[1] propose to consider the correlation function that is an appropriate quantity for homogeneous fractals[2]. This gives for a fractal

$$I(L) = \int_0^L <\rho(x_0)\rho(x_0+x)>_0 \, dx \simeq L^D \quad (2)$$

where the average $<...>_0$ is taken over x_0. With this definition even a single wavefunction provides a rather smooth relation between $\ln I(L)$ and $\ln L$ as shown in Fig. 2 for the distribution of Fig. 1.

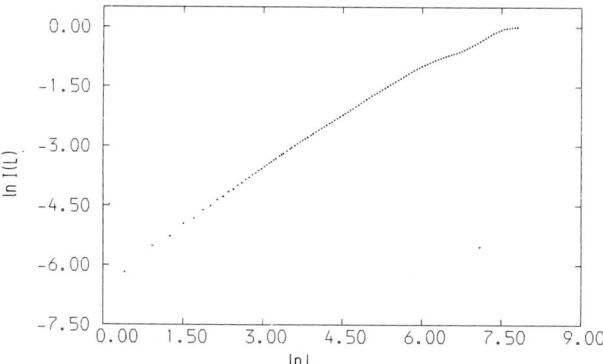

FIGURE 2
Correlation function (see Eq.(2)) of the probability distribution of Fig. 1. A linear behavior with slope different from one would provide evidence for fractal features. Note the slight curvature that is analyzed in detail in Table I and that in our opinion does not support a fractal interpretation.

If such a relation is linear then its slope provides the fractal dimension D. At first glance the relation of Fig. 2 looks reasonably linear, on the other hand its slope is also reasonably close to $D=1$ that corresponds to non fractal behavior. In order to analyze this question carefully we start from $\ln L_{min} = 1.7$ corresponding to $L_{min} \approx 5$ lattice sites and perform a linear fit in an interval up to a $\ln L_{max}$. The resulting apparent fractal

dimension "D" is then analyzed as a function
of the upper limit of this interval. For a
fractal structure one expects a well defined
"D" with random statistical fluctuations
followed by an abrupt change occurring when
L_{max} approaches the localization length λ.

TABLE I
Analysis of the slope of the curve of Fig. 2
as a function of the size of the interval
considered. The gradual variation of
apparent fractal dimension "D" points towards
a finite size effect rather than a true
fractal behavior.

L_{max}	$\ln L_{max}$	"D"
20	3	0.971
33	3.5	0.958
55	4	0.941
90	4.5	0.918
148	5	0.902
245	5.5	0.892
403	6	0.880
665	6.5	0.859

In table I we observe instead a gradual change
of "D" from 1 to about .86, that is very close
to the value of Soukoulis and Economou (D=.87)
for this type of disorder. This systematic
change of the slope provides in our opinion a
strong argument against the evidence of fractal features. We have also performed the same
analysis over many other wavefunctions
(about 50) and found exactly the same behavior
with the same type of curvature. This means
that averaging over many functions (that we
also did) does not improve the situation. This
result indicates therefore that the apparent
change in slope (with respect to D=1) may be
just a finite size effect. Considering a
larger value of W that should correspond to
smaller values of D[1] also does not make the
problem more clear because in this case the
interval of possible fractal behavior is
restricted by the shorter localization length.

Facing these problems we have decided to
consider a more basic question: can a localized
wavefunction manifest the properties of a
homogeneous fractal? We discuss this question
in the next section by studying idealized
deterministic distributions.

3. IDEAL SELF-SIMILAR NORMALIZED DISTRIBUTIONS
In order to elucidate the nature of a
possible self-similar structure in localized
wavefunctions we consider here a deterministic
construction of self-similar probability distributions that preserve normalization. This
can be obtained as a generalization of the
Cantor set construction that gives different
weights to the points of the Cantor set[2]. We
start with a Cantor set $C_k(\lambda,\eta)$ where λ is
the number of intervals, η their length and
k is the order of iteration. Define $U_{k=1}(x)$
by giving a weight to each of the λ intervals
of $C_1(\lambda,\eta)$:

$$U_{k=1}(x) = \begin{cases} w_1 & \text{if } x \in [0,\eta] \\ w_2 & \text{if } x \in [\frac{1-\eta}{\lambda-1}; \frac{(\lambda-1)\eta+1-\eta}{\lambda-1}] \\ \vdots \\ w_\lambda & \text{if } x \in [1-\eta,1] \\ 0 & \text{otherwise} \end{cases} \quad (3)$$

and, because of normalization

$$\sum_{i=1}^{\lambda} W_i \eta = 1 \quad . \quad (4)$$

Within the next iteration we proceed in the
same way with the same weight distribution.
For simplicity we consider the specific case
$\lambda=3$, $\eta=\frac{1}{4}$ and $W(w_1,w_2,w_3)=(w,2w,w)$ as shown
in Fig. 3.

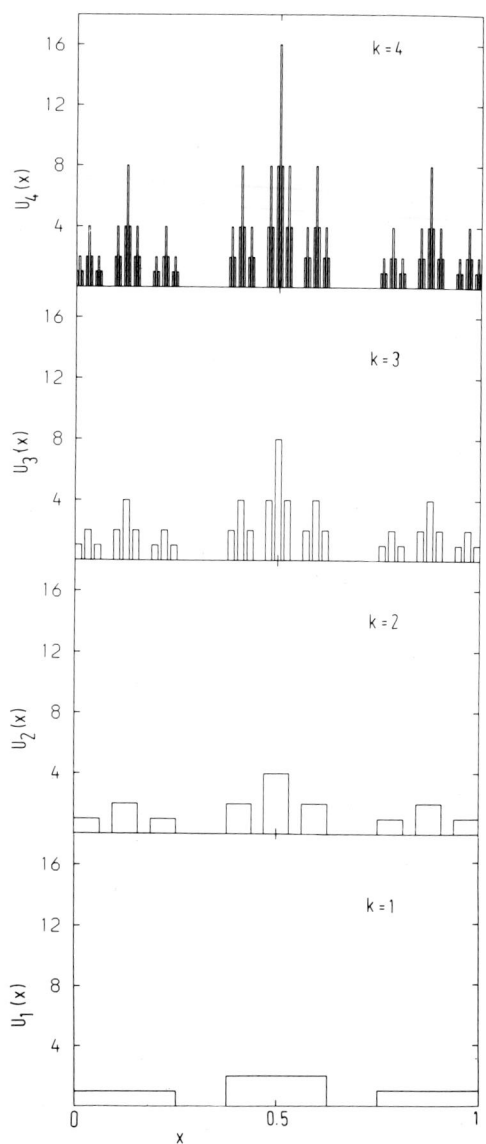

FIGURE 3
Example of the generalization of the Cantor set constructed to mimic the possible self-similar features of localized wavefunctions. See text for a detailed description.

Because of normalization the relation between η and w is simply $4w\eta=1$. We have then the scaling property

$$U_{k+1}(\eta x) = \frac{1}{4\eta} U_k(x) \quad . \quad (5)$$

For small x (note that the origin of our coordinates is at the edges of Fig. 3), $U(x)$ can be described by the smooth function

$$\bar{U}(x) \sim x^{(D-1)} \quad (x \to 0) \quad (6)$$

where $D=-\ln 4/\ln \eta = 1$ for $\eta=\frac{1}{4}$. These well defined scaling properties have been derived for a particular origin. We now show that by changing the origin one obtains different scaling properties. Consider the translation $x \to x-\frac{1}{2}$ so that the origin lies in the center of the distribution. In this case we recover the problem studied by Schilling[3]. The scaling relations are now given by

$$U_{k+1}(\eta x) = \frac{1}{2\eta} U_k(x) \quad , \quad (7)$$

so that $D=-\ln 2/\ln \eta = 0.5$ (for $\eta=\frac{1}{4}$) which is a factor of two smaller than the previous case. This origin dependence is due to the introduction of different weights. We thus conclude that the normal Cantor set is homogeneous while the generalized is not. If now we consider a localized wavefunction to be a generalized Cantor set with stochastic features it is hard to see how it could possess the properties of a homogeneous fractal. For one idealized case an integration reproduces the two different fractal dimensions obtained from scaling only if taken from the two origins $x_0=0$ and $x_0=\frac{1}{2}$. When x_0 is taken to be on a local peak there is no well defined power law. The use of the correlation function as given by Eq.(2) produces a sort of average fractal dimension that lies in between the two values described above, but it is not clear how to interpret this number that has no counterpart in the scaling properties we have used to generate the system. Other methods of analysis lead, when averaged, to different fractal dimensions[4].

4. SUMMARY

In conclusion we have seen that the correlation function analysis as given by Eq.(2) applied to localized wavefunctions is far from providing a clear proof of fractal behavior. In addition we have constructed generalizations of the Cantor set to mimic the possible scaling properties of localized wavefunctions. From the idealized examples it is hard to see how the real wavefunctions may actually possess the properties of homogeneous fractals. Our analysis therefore, even if not conclusive, points against the presence of self-similar properties for localized wavefunctions.

REFERENCES

1. C.M. Soukoulis and E.N. Economou, Phys. Rev. Lett. 52 (1984) 565.

2. B. Mandelbrot, The Fractal Geometry of Nature (W.H. Freeman & Co., New York, 1983).

3. R. Schilling, Phys. Rev. Lett. 53 (1984) 2258.

4. A.P. Siebesma and L. Pietronero, unpublished.

Part IX
CHAOS, TURBULENCE AND RELATED TOPICS

CIRCLE MAPS IN THE COMPLEX PLANE

Predrag CVITANOVIĆ

Institute of Theoretical Physics, Chalmers University of Technology, S-412 96, Goteborg, Sweden

Mogens H. JENSEN, Leo P. KADANOFF

The James Franck and Enrico Fermi Institutes, University of Chicago, Chicago, Illinois 60637, USA

and

Itamar PROCACCIA

Department of Chemical Physics, The Weizmann Institute of Science, Rehovot 76100, Israel

Circle maps of polynomial, exponential, and rational polynomial types are studied numerically in the complex plane. The golden mean universality for real circle maps does not extend into the complex plane.

1. INTRODUCTION

The discovery of the period-doubling universality in one-dimensional iterations[1],[2] has prompted a search for universal scalings in other low dimensional dynamical systems (the theory and the experimental observations of period doublings are reviewed in refs. [3],[4],[5]). One class of such problems in which the universality ideas have had some success are the transitions to chaos for diffeomorphisms on the circle (circle maps). Maps of this type model a variety of physical systems: we refer the reader to refs. [6],[7] for a discussion of their physical applications.

The first example of a universal scaling for the critical circle maps was discovered in a study of mappings with the golden mean winding number[8],[9],[10]. An elegant formulation of such asymptotically universal self-similarities is afforded by the unstable manifold equations[11],[12],[13],[14],[3].

For the circle maps the unstable manifold equation is given by (see for example refs. [12],[15]):

$$g_p(z) = \alpha g_{1+p/\delta}(\alpha g_{1+1/\delta+p/\delta^2}(z/\alpha^2)) \quad (1.1)$$

However, numerical solutions of this equation are made difficult by subtle convergence problems. To best of our knowledge, only two successful numerical solutions of the unstable manifold equations are extant[16]. These convergence problems, as well as the interest in understanding other universalities associated with the circle maps[15], has motivated us to investigate the structure of the complexified circle maps. Such investigations have previously yielded new insights into universal scaling laws[13],[14],[17],[18], as well as much beautiful mathematics (see for example refs. [19],[20],[21],[22],[23],[24]).

The reader is referred to refs. [6],[15] and [25] for an introduction to the circle maps. In this note we concentrate only on some general properties of complexified circle maps.

2. THE CUBIC CIRCLE MAP

As the first example of a complexified circle map, consider the critical cubic map

$$z_{n+1} = \Omega + 4z_n^3, \quad \text{Re}(z + 1/2) \bmod 1 \quad (2.1)$$

(a cubic circle map is critical if it has a cubic inflection point $z'=z''=0$). This map is periodic along the real axis, and discontinuous at $z = \pm 1/2 + iy$ for all $y \neq 0$. This is the crudest example of the polynomial approximations to circle maps of kind used in the numerical solutions of the unstable manifold equation (1.1).

The Mandelbrot set (the set of all parameter values for which the iterates of the critical point tend to infinity) for a polynomial cubic map is plotted in fig. 2.1.

The boundary of the large central component (parameter values for which the iterates of the critical point converge to a stable fixed point) is given by the cardioid of parameter values for which the fixed point is marginally stable, $|dz'/dz| = 1$. The z map is monotone for z real and therefore cannot exhibit period doubling along the real axis; however, bifurcations to period 2 occur on the imaginary axis, generating the pair of 2-cycle "hearts". One of these

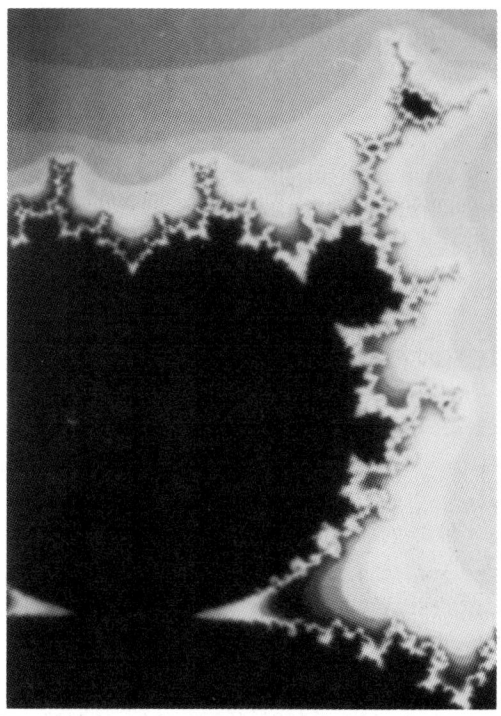

FIGURE 2.2 An enlargement of the cycle-2 "heart" from fig. 2.1.

hearts is plotted in fig. 2.2. A bifurcation to a stable periodic orbit takes place at every rational value of dz'/dz phase along the cardioid boundary. This gives rise to a set composed of self-similar "hearts", analogous to the Mandelbrot set for the quadratic map $z \to c+z^2$. Just as in the quadratic case[14], each component is centered on a superstable m/n cycle, and one expects universal scaling laws for the infinite sequences of bifurcations.

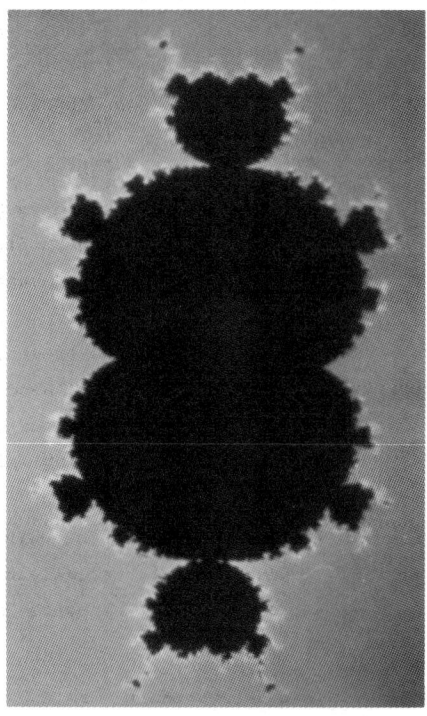

FIGURE 2.1 The Mandelbrot set for the critical polynomial cubic mapping $z \to c + 4z^3$.

The shape of the Mandelbrot set for the circle maps will be described in the next section.

The basin_of_attraction (the set of all initial values of z whose iterates do not escape to infinity) of the superstable fixed point for the cubic circle map (2.1) is plotted in fig. 2.3.

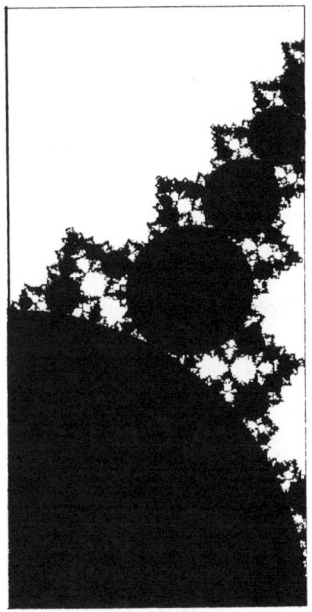

FIGURE 2.3 The upper right quadrant of the basin of attraction of the superstable fixed point for the cubic map (2.1) ($\Omega = 0$). The remaining quadrants are obtained by reflection about the x and y axes.

The general shape of the basins of attraction for the complex circle maps is easy to understand. The map (2.1) has a boundary of marginal stability $|dz'/dz| = 1$ on the circle of radius $|z| = 1/2$. This defines the central disk. The periodicity condition ($Re(z+1/2)$ mod 1) generates an infinity of such disks, one for every integer value of Re z. The remaining self-similar structure arises from this periodicity; it is generated by all preimages of the $|z| = 1/2$ disks along the real axis. z^3 maps initial points with phase $\exp(\pm 2i\pi/3)$, $\exp(\pm i\pi/3)$ onto the real axis. This gives rise to the four rays of smaller disks at $\pm 60°$, $\pm 120°$. The higher iterates of (2.1) image similarly the other sequences of disks visible in fig. 2.3 onto the real axis. (The abrupt truncation close to Im z = .95 is an artifact of the Re z modularity condition in (2.1).)

3. THE STANDARD CIRCLE MAP

As the second example of a complexified circle map we take the sine map

$$z_{n+1} = \Omega + z_n - \frac{k}{2\pi} \sin(2\pi z_n) , \quad \mod 1 \quad (3.1)$$

(k=1 for the critical case). The basins of attraction have the same basic structure as in the previous example; the superstable fixed point basin of attraction is given in fig. 3.1, and the superstable 3-cycle basin of attraction is given in fig. 3.2 as typical examples.

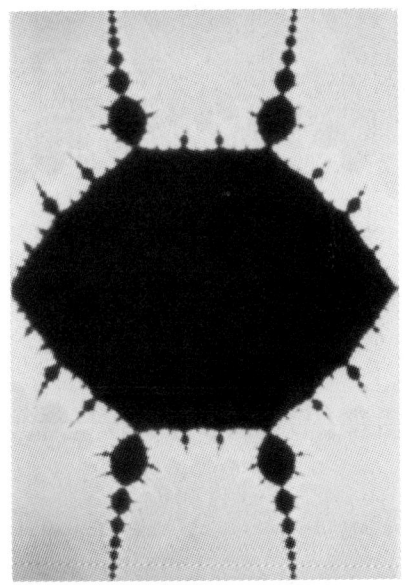

FIGURE 3.1 The basin of attraction for the superstable fixed point of the sine map (3.1).

The complexified sine map is an example of an exponential map[26]: with substitutions

$$u = \exp(2\pi i z) \qquad c = \exp(2\pi i \Omega)$$

(3.1) becomes

$$u' = cue^{-k(u-1/u)/2} \qquad (3.2)$$

(here k=1). Unlike the polynomial maps of type (2.1), or the rational maps which we shall discuss in the next section, the exponential maps have basins of attraction which extend to infinity and are dense everywhere in the complex plane. For example, the image of $z = \pm 1/4 + iy$, y large, lies close to the real axis in fig. 3.2, and in the same way any ray of disks which is a preimage of the real axis extends to infinity.

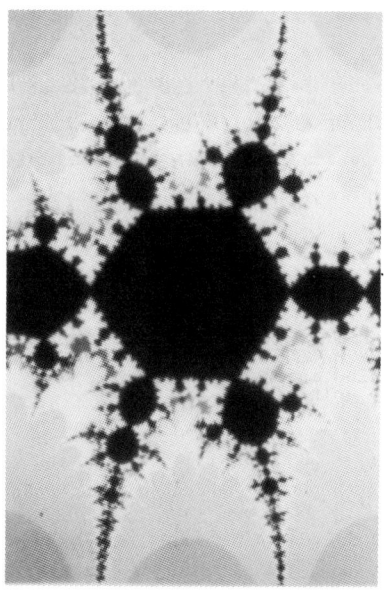

FIGURE 3.2 The basin of attraction for the superstable 1/3 cycle of the sine map (3.1).

For the same reasons the Mandelbrot set for the sine map (3.1), plotted in fig. 3.3, is dense everywhere over the entire complex plane.

The general structure of the Mandelbrot sets for the critical circle maps is illustrated by fig. 3.3 . Along the real axis there is a mode-locking interval for every rational number (the "devil's staircase"[6]). As the map has a cubic inflection, each mode-locking interval extends in the complex plane into a "hearts" set of fig. 2.1. By crossing from the central component into one of the hearts, one drives the mapping through a period n-tupling without changing its winding number (for example, the 1/2 winding number period-doubles to 2/4). Hence the Mandelbrot set for a circle map is constructed from two basic building blocks: first, the map has a cubic inflection, so the complex period n-tuplings are characterized by the self-similar hearts of fig. 2.1; second, the map is periodic, and that gives rise to the infinite sequences of copies of the basic hearts set along the real axis, and along the rays in the complex plane.

Similarly, the basins of attraction for the critical circle maps are constructed from the corresponding building blocks: first, a basin of attraction for the critical cubic

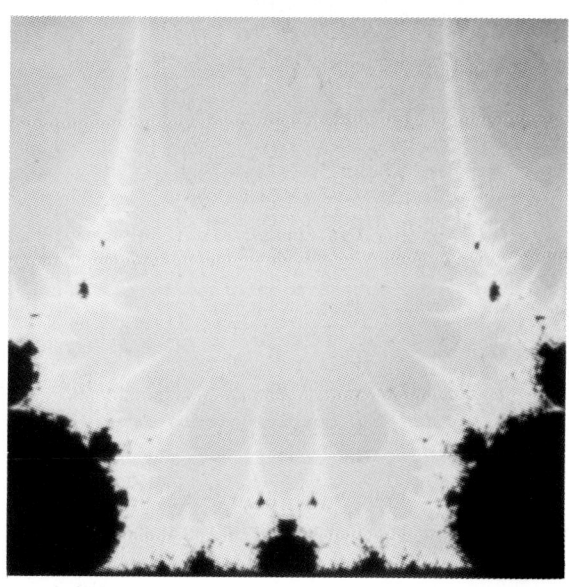

FIGURE 3.3 The Mandelbrot set for the sine map (3.1). The lower complex halfplane is obtained by the reflection about the real axis. The black region corresponds to values of parameter for which the critical point does not iterate away to infinity. The different shades of gray are an indication of the escape velocity; the longer the number of iterations needed to reach a cutoff, the lighter the shading. The large "hearts" set on the ends of the interval corresponds to the fixed point, the hearts set in the middle corresponds to the orbits with the winding number 1/2, and so on.

polynomial map; second, from the infinity of its preimages generated by the periodicity condition. This is illustrated in fig. 3.4 for the superstable period-doubled cycle with winding number 0/1. The triangular arrangement of the central 4 disks is typical of the basins of attraction for period-2 polynomial cubic maps; the remainder of the set is generated by the periodicity along the real axis, and its preimages in the complex plane.

series expansion of the exponentials in (3.2). It is a critical circle map with a cubic inflection point at u=1:

$$u' \simeq c(1 - r^3/2 - r^5/4 + r^6/8 + \ldots) \quad (4.2)$$
$$r = 2\pi i z$$

The Mandelbrot set for this map is given in fig. 4.1.

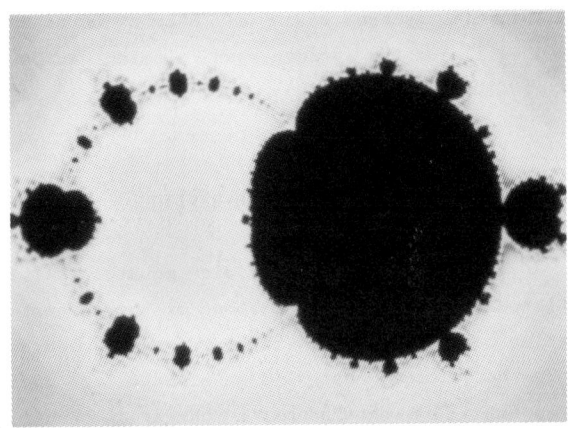

FIGURE 4.1 The Mandelbrot set for the rational map (4.1). The fixed point hearts set is centered on c=1, the 1/2 winding on c=-1. The remaining infinity of other mode lockings lie on the unit circle.

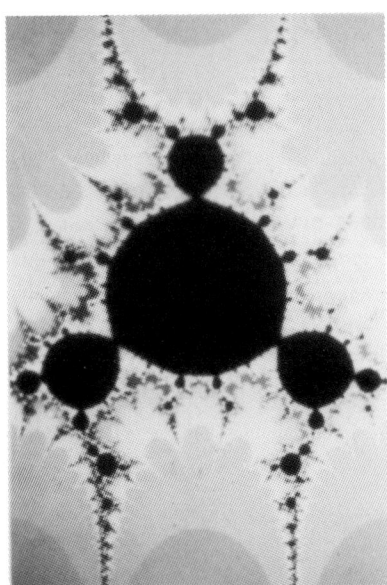

FIGURE 3.4 The basin of attraction for the superstable period-doubled cycle of (3.1) with winding number 0/1. The central part of the basin of attraction is the same as for the period-2 polynomial cubic maps. The remainder of the set are replicas generated by the periodicity along the real axis, and the preimages in the complex plane.

4. A RATIONAL POLYNOMIAL MAP

As the third example of a circle map we take the rational polynomial map

$$u' = cu(3-ku)/(3-k/u) \quad (4.1)$$

(k=1 here). The form of this map is motivated by the lowest order truncation of the power

FIGURE 4.2 The basin of attraction of the rational polynomial map (4.1) for the parameter value corresponding to the golden-mean winding number.

The rational maps differ from the exponential maps in one important aspect; their Mandelbrot sets and their basins of attraction are finite in extent, because for large $|u|$ the rational maps behave like polynomials. The preimages of the real axis of the circle map (more precisely, as (4.1) is an exponentiated circle map, the preimages of the $|u|=1$ unit circle) are themselves closed loops. This is illustrated by the basin of attraction of the map (4.1) for the golden-mean winding number, fig. 4.2.

5. THE GOLDEN MEAN UNSTABLE MANIFOLD

The unstable manifold equation (1.1) states that in the neighborhood of the parameter value corresponding to the golden-mean winding, the Mandelbrot set is self-similar under rescaling by the Shenker's universal scaling number δ. We have investigated this numerically by blowing up the neighborhood of the parameter value corresponding to the golden mean winding, and comparing the sizes of the hearts sets corresponding to the successive ratios of Fibonacci numbers. In this neighborhood the Mandelbrot set does indeed scale with the same universal factor as in the real case. However, the golden mean universality does not generalize to the complex plane in several important ways.

The first non-universality is familiar from the theory of polynomial iterations [27], [24]; the longer the cycle, the "hairier" is the exterior of the hearts set. This can be seen by comparing fig. 2.1, the fixed point hearts set, with fig. 5.1.

The other non-universal feature of the complex circle maps is the fact that the exteriors of hearts sets for different maps remain unmistakably distinguishable, regardless of the degree of magnification of the the golden

FIGURE 5.1 The Mandelbrot set fig. 3.3 for the sine map (3.1) in the neighborhood of the parameter value corresponding to the golden mean winding. The largest hearts set corresponds to the 55/89 Fibonacci numbers ratio. While the successive Fibonacci ratios' hearts sets do scale by Shenker's universal scaling, their exteriors contain more and more "hair".

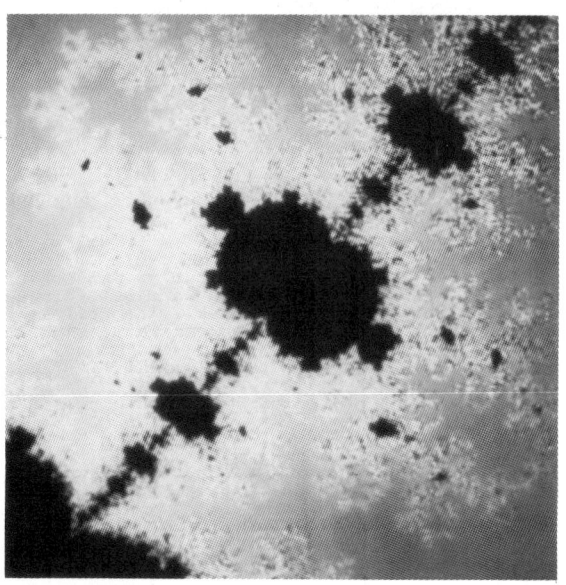

FIGURE 5.2 The Mandelbrot set fig. 4.1 for the rational polynomial map (4.1) in the neighborhood of the parameter value corresponding to the golden mean winding. The largest hearts set corresponds to the 21/34 Fibonacci numbers ratio. While the successive Fibonacci ratios' hearts sets do scale by Shenker's universal scaling, their exteriors are different from those for the exponential maps, such as fig. 5.1.

mean winding neighborhood. This can be seen by comparing fig. 5.1 with fig. 5.2. While all the external rays for the sine map go to infinity, the exteriors of hearts sets for the rational polynomial map are decorated by looplike preimages of the unit circle. Numerically, these loop decorations scale by the same universal Shenker's number as the interiors of the hearts sets, so we can always determine the type of the starting approximation to the unstable manifold, regardless of the degree of magnification. We conclude that the unstable manifold equation (1.1) has no unique analytic continuation into the complex plane.

ACKNOWLEDGEMENTS

We are grateful to J.H. Hubbard and B. Branner for many inspiring discussions. P.C. thanks J.H. Hubbard and the Department of Mathematics, Cornell University, for the access to the Mathematics VAX, and B. Wittner for the kind assistance with the two-dimensional iteration programs, with which most of the graphics in this paper were generated. P.C. is grateful to M.J. Feigenbaum for the hospitality at the Laboratory of Solid State Physics, and acknowledges support from DOE contract no. DE-AC02-83-ER13044. I.P. acknowledges partial support by the Minerva Foundation, Munich, Germany. M.H.J. and L.K. work has been supported in part by the U.S. Office of Naval Research and by the Materials Research Laboratory.

REFERENCES

1. M.J.Feigenbaum, J.Stat.Phys.$\underline{19}$, 25 (1978), reprinted in ref. [4].

2. M.J.Feigenbaum, J.Stat.Phys.$\underline{21}$, 669 (1979), reprinted in refs. [3],[4].

3. P. Cvitanović, ed., Universality in chaos (Hilger, Bristol, 1984).

4. Bai-Lin Hao, ed., Chaos (World Scientific, Singapore, 1984).

5. P. Collet and J.-P. Eckmann, Iterated maps on interval as dynamical systems (Birkhauser, Boston, 1980).

6. M.H. Jensen, P. Bak and T. Bohr, Phys Rev. $\underline{A30}$, 1960 (1984).

7. T. Bohr, M.H. Jensen and P. Bak, Phys Rev. $\underline{A30}$, 1970 (1984).

8. S.J. Shenker, Physica $\underline{5D}$, 405 (1982), reprinted in ref. [3].

9. M.J. Feigenbaum, L.P. Kadanoff and S.J. Shenker, Physica $\underline{5D}$, 370 (1982), reprinted in ref. [4].

10. S. Ostlund, D. Rand, J. Sethna and E.D. Siggia, Physica $\underline{D8}$, 303 (1983).

11. H. Daido, Phys. Lett. $\underline{83A}$, 246 (1981); $\underline{86A}$, 259 (1981).

12. H. Daido, Prog. Theor. Phys. $\underline{67}$, 1698 (1982).

13. A.I. Golberg, Ya.G. Sinai and K.M. Khanin, Usp. Mat. Nauk $\underline{38}$, 159 (1983).

14. P. Cvitanović and J. Myrheim, Nordita preprint 84/5 (1984), submitted to Comm. Math. Phys. (Feb. 1984).

15. P. Cvitanović, M.H. Jensen, L.P. Kadanoff and I. Procaccia, Phys. Rev. Lett. $\underline{55}$, 343 (1985).

16. G. Gunarathne (unpublished); M.J. Feigenbaum (unpublished).

17. N.S. Manton and M. Nauenberg, Commun. Math. Phys. $\underline{89}$, 555 (1983).

18. M. Widom, Commun. Math. Phys. $\underline{92}$, 121 (1983).

19. A. Douady and J.H. Hubbard, C.R. Acad. Sci. Paris $\underline{294}$, 123 (1982).

20. P. Fatou, Bull. Soc. Math. France $\underline{47}$, 161 (1919); ibid. $\underline{48}$, 33 and 208 (1920).

21. G. Julia, J. Math. Pures et Appl. $\underline{4}$, 47(1918)

22. B.B. Mandelbrot, Ann. N.Y. Acad. Sci. $\underline{357}$, 249 (1980).

23. B.B. Mandelbrot, The Fractal Geometry of Nature, (Freeman, San Francisco, 1982).

24. John Milnor, IHES preprint (May 1985).

25. P. Cvitanović, B. Shraiman and B. Söderberg, Phys. Scripta (to appear).

26. Similar maps have been studied by R. Devaney (unpublished) and J.H. Hubbard (private communication).

27. A. Douady and J.H. Hubbard (private communication); B.B. Mandelbrot (unpublished conjecture).

FRACTALS IN PHYSICS
L. Pietronero, E. Tosatti (editors)
© Elsevier Science Publishers B.V., 1986

FRACTAL MODELS FOR TWO- AND THREE-DIMENSIONAL TURBULENCE

G. PALADIN and A. VULPIANI

Dipartimento di Fisica, Universitá "La Sapienza", Piazzale Aldo Moro 2, I-00185 Roma, Italy, and GNSM-CNR, Unitá di Roma

Energy dissipation in three-dimensional turbulent flows is believed to be concentrated on a set with non integer fractal dimension. Experimental data for the moments of velocity fluctuations seem however to indicate that small scales properties can not be described by an usual homogeneous fractal generated by a set of rules which relate its statistical properties over a certain scale of length to those of a larger scale of lengths. We introduce "inhomogeneous fractals" where those rules at each step in length scale are drawn at random according to a certain probability distribution. The energy transfer is thus described by means of a random fragmentation model assuming no correlation among different steps of the process.

This model can not be extended to the two-dimensional case by considering a local cascade of enstrophy instead of energy as exact results involve that intermittency does not affect power laws of the structure functions. We introduce therefore a new model where the fragmentation mechanism is forbidden whenever the density of enstrophy dissipation is greater than a critical value.

We can reproduce a kind of intermittent behaviour but without corrections to the two dimensional Kolmogorov power law and thus find that zero enstrophy dissipation regions as well as strong vorticity gradients regions in R^2 have fractal dimension identical to the topological one.

1. INTRODUCTION

Fractal objects have been recently introduced in the analysis of a wide class of phenomena in physics and natural sciences[1].

Geometric aspects of turbulence particularly need an investigation with "selfsimilar" rather than "euclidean" techniques.

Homogeneous fractals (i.e. with a global exact dilatational invariance) are usually considered while we want to introduce simple models where selfsimilarity holds only in average in order to reproduce realistic intermittent behaviours in 2- and 3--dimensional turbulence. We shall namely use inhomogeneous fractals, generated by rules that, at each scale of lengths, are not fixed but are given at random according to a certain probability distribution.

Let us clarify our terminology by an example: a 2 dimensional Sierpinski carpet obtained by iterating a fragmentation process. At each step i' a square of scale ℓ_i originates n^2 of scales $\ell_{i+1} = \ell_i/n$ of which only a fraction β are "full" (see Fig. 1 where n = 2).

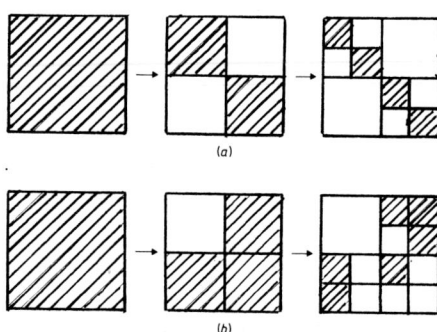

FIGURE 1
Schematic view of a homogeneous fractal (case a) compared with an inhomogeneous one (case b). The dashed areas are the zones active during the fragmentation process.

The carpet is called homogeneous if the number of boxes is fixed (i.e. one has $\beta n^2 = n^{D_F}$); inhomogeneous if this number is not constant i.e. if β is a random variable depending on the particular square and step considered.

The fractal dimension, even if well defined (for the carpet $D_F = 2 + \log_n \{\beta\}$) (where $\{\ \}$ indicates the average on the probability distribution of β) can not characterize the inhomogeneity, of course. On the other hand the fragmentation is determined by the probability distribution of the β's which is in principle known from the moments $\{\beta^q\}$. These moments can be related in a natural way to the structure function exponents in 3-dimensional turbulence. This is no longer possible in 2 dimensions where fragmentation does not lead to a selfsimilar structure.

II. 3 DIMENSIONAL INTERMITTENCY

The Navier-Stokes equations:

(1) $\partial_t \underline{u} + (\underline{u} \cdot \underline{\nabla}) \underline{u} = -\frac{1}{\rho} \underline{\nabla} p + \nu \Delta \underline{u}$

are formally invariant under the scaling transformations:

(2) $\underline{r} \to \lambda \underline{r}, \quad \underline{u} \to \lambda^h \underline{u}, \quad t \to \lambda^{1-h} t, \quad \nu \to \lambda^{h+1} \nu$

Note that the mean energy dissipation $\bar{\epsilon} = \nu \overline{(\nabla \underline{u})^2}$ changes as follows under the scaling transformation (2):

(3) $\bar{\epsilon} \to \lambda^{3h-1} \bar{\epsilon}$

The celebrated Kolmogorov's laws[2] can be obtained by imposing the invariance of $\bar{\epsilon}$ under the scaling transformations (2): h = 1/3: this implies that:

(4) $\lim_{\Delta x \to 0} \frac{\Delta u}{\Delta x^{1/3}} = \lim_{\Delta x \to 0} \frac{|\underline{u}(\underline{x} + \Delta \underline{x}) - \underline{u}(\underline{x})|}{\Delta x^{1/3}} \neq 0$

In eq. (4) $\Delta X \to 0$ means $\Delta X \sim$ dissipation length, so in the limit of infinite Reynolds number the velocity gradients are singular. Note that in the Kolmogorov theory there is the implicit assumption that the set of singular points (those where (4) holds) is space filling.

One can think (as Landau[3] first remarked) that the energy dissipation $\epsilon(\underline{x}) = 1/2 \cdot \nu \cdot \Sigma_{ij} (\partial_i u_j + \partial_j u_i)^2$ presents strong fluctuations. This happens for example if the set of points where the velocity gradients are singular has a non integer fractal dimension.

An approach to the problem using homogeneous fractal has been proposed with the so called β-model[4]. Briefly, in this approach eq. (4) is replaced by

$$(5) \quad \lim_{\Delta x \to 0} \frac{\Delta u}{\Delta x^h} \neq 0; \quad h = \frac{D_F - 2}{3}$$

and the set of points where (5) holds has a fractal dimension $D_F < 3$. Practically in this model the energy dissipation $\varepsilon(\underline{x})$ is uniformly distributed on a homogeneous fractal object. In the β-model one obtains for the fluctuation of velocities:

$$(6) \quad <|\Delta u(\ell)|^P> \sim \ell^{\zeta_p}, \quad \zeta_p = \frac{D_F - 2}{3} p + (3 - D_F)$$

In Fig. (2) we report data on ζ_p for various experimental tests[5].

A linear fit is consistent only for $p \lesssim 7$ while for greater values of p ζ_p has a tendency to behave in a nonlinear way.

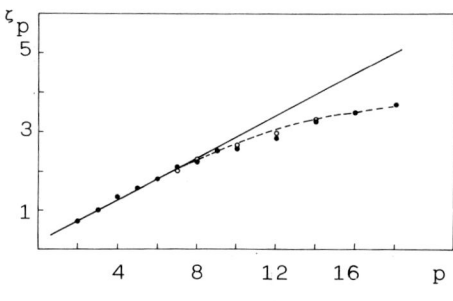

FIGURE 2

ζ_p versus p. Dots and circles represent experimental data ref. (5). Solid line is the β-model with $D_F = 2.83$. Dashed line refers to equations (13) and (14) with x = .125.

Therefore one concludes that the region containing energy dissipation is not a homogeneous fractal and there are many singularity structures with different values of h. Recently[6,7] a multifractal model has been proposed in order to include more complicated structures.

Let us define S(h) as the set of points for which:

$$(7) \quad \lim_{x \to y} \frac{|\underline{u}(\underline{x}) - \underline{u}(\underline{y})|}{|\underline{x} - \underline{y}|^h} \neq 0$$

and let us indicate d(h) the fractal dimension of S(h). Because the probability to belong to S(h) at scale ℓ is proportional to $\ell^{3-d(h)}$ one has

$$(8) \quad <|\Delta u(\ell)|^P> \sim \int d\mu(h) \, \ell^{ph + 3 - d(h)} \sim \ell^{\zeta_p}$$

With a saddle point technique we obtain for ζ_p

$$(9) \quad \zeta_p = \min_h \{ph + 3 - d(h)\}$$

Physically eq. (5) means that for a given value of p, ζ_p depends on a particular value of h. Hence the kind of instabilities which are needed to set up the sets S(h) are picked out by the moments of velocity differences. Fig. 2 can then be interpreted as the evidence for different instability mechanisms acting on the flow to select the probability distribution of energy transfer and dissipation.

We now introduce a generalization[7] of the β-model in order to obtain an inhomogeneous fractal object with different kind of singularities, i.e. different h, for velocity gradients. Let us consider the scales $\ell_n = 2^{-n}\ell_o$

where ℓ_o is the scale on which energy is injected. If at scale ℓ_n there are N_n active eddies, each eddy $\ell_n(K)$ generates eddies of size ℓ_{n+1} (k labels the "father" eddy, $k = 1, \ldots N_n$). Because the rate of energy transfer is constant among $\ell_n(K)$ and $\ell_{n+1}(K)$:

$$(10) \quad V_n^3(K)/\ell_n(K) = \beta_{n+1}(K) V_{n+1}^3(K)/\ell_{n+1}(K)$$

Where, as in the standard β-model, $V_n(k)$ is the velocity difference in the active eddy between two points at distance ℓ_n and $\beta_{n+1}(k)$ is the percentage of volume occupied by the active eddies of scale ℓ_{n+1} generated by the eddy $\ell_n(k)$. Eq. (10) implies that the velocity difference V_n in an eddy generated by a particular set of fragmentation $\beta_1,\ldots,\beta_n$ is

$$(11) \quad V_n \sim \ell_n^{1/3} (\prod_{i=1}^{n} \beta_i)^{-1/3}$$

from (11) it follows that:

$$(12) \quad <|\Delta u(\ell_n)|^p> = \int \prod_{i=1}^{n} d\beta_i \; p(\beta_1 \ldots \beta_n) \; \beta_i |V_n|^p$$

Because we assume that there are no correlations among different steps of fragmentation i.e.: $p(\beta_1 \ldots \beta_n) = \prod_{i=1}^{n} P(\beta_i)$, one obtains for the exponent ζ_p:

$$(13) \quad \zeta_p = p/3 - \ln_2 \{\beta^{(1-p/3)}\}$$

where $\{\ldots\}$ means average on the distribution $P(\beta)$. Note that if β is a constant $(2^{(D_F - 3)})$ one obtains the results of the standard β-model (i.e. eq. (7)). The knowledge of the probability distribution $P(\beta)$ is connected to the understanding of the nature of singularities of the Navier-Stokes equations. Fig. 2 shows that the simple form:

$$(14) \quad P(\beta) = x\delta(\beta - .5) + (1-x)\delta(\beta - 1)$$

with $x = .125$ leads to a good fit of the available experimental data, x being the only free parameter. There is no deep reason to choose the form (14) for $P(\beta)$. We have assumed in (14) following a simple phenomenological idea, that an active eddy can generate either velocity sheets ($\beta = .5$) or space filling Kolmogorov-like eddies ($\beta = 1$).

Note that the fractal object generated with the above rules has no global dilatation invariance properties; one can nervertheless compute the fractal dimension D_F defined by

$$(15) \quad <N_n> \sim \ell_n^{-D_F}$$

where N_n is the number of active eddies at the n-th step of fragmentation. It is easy to show, with a simple calculation[7] that:

$$(16) \quad D_F = 3 + \ln_2 \{\beta\} = 3 - \zeta_o$$

We remark that the fractal dimension D_F defined by eq. (15) is different from the (D^*) often used in the experimental papers and computed by the energy dissipation correlation. In terms of our ζ_p, $D^* = 1 + \zeta_6$. It is easy to see that in the general case $D_F > D^*$; $D_F = D^*$ only holds for the standard β-model (i.e. homogeneous fractal). For example $D_F = 2.91$ and $D^* = 2.83$ with the fit given by eq. (14) and $x = .125$; this is an indirect check of the multifractal nature of fully developed turbulence.

III. 2 DIMENSIONAL INTERMITTENCY

The random β-model described in the previous section can not be easily extended to the 2-dimensional turbulence by considering a local cascade of enstrophy instead of energy.

If we assume that the fragmentation process is not intermittent (i.e. Kolmogorov-like, space filling), the usual power law for the structure function is obtained in the inertial range [9]

(17)
$$< |\Delta u(\ell)|^p > \sim \ell^{\zeta_p}$$
$$\zeta_p = p$$

Let us show that the intermittent correction to (17) can not be found by the β-models machinery.

Let us recall that a central role is played by the assumption that the enstrophy transfer rate is constant from which it follows by dimensional counting that:

(18) $\zeta_3 = 3$

Moreover it is possible to show that ζ_p is convex in p [10]. There is an exact result for the 2-dimensional Euler equation [9] which can be extended to the viscous case without any difficulties:

(19) $|\Delta \underline{u}(r)| < $ constant $r |\log(r)|$

which implies $\zeta_p > p$.

This last inequality is consistent with (18) and the convexity of ζ_p only if $\zeta_p = p$.

It might seem rather surprising that intermittency does not affect the structure function exponents. There are nevertheless some numerical evidences of this odd behaviour which is quite peculiar of 2-dimensional turbulence.

We need therefore to modify the random β-model in order to reproduce at least the known phenomenological aspects.

As we have postulated a constant enstrophy transfer rate, formula (10) becomes, in any case:

(20) $$\frac{v_n^3(K)}{\ell_n^3(K)} = \beta_{n+1}(K) \frac{v_{n+1}^3(K)}{\ell_{n+1}^3(K)}$$

The inequality (19) can be satisfied by imposing a further constraint:

(21) $\beta_{n+1} = 1$ if $\frac{v_n^3}{\ell_n^3} > n_{max}$

Equation (21) has modified the old Bernoullian process of fragmentation into a Markovian one as the steps of the cascade are no longer independent.

This Markovian random β-model has an intermittent behaviour even if $\zeta_p = p$.

Indeed there are regions in $\mathbb{R}^2$ of zero enstrophy dissipation with fractal dimension $D_F = 2$ as well as the active eddies with strong vorticity gradient.

Roughly speaking, we have no more a multifractal but a "checkers"-like structure.

The fragmentation is space filling on scales small enough but however larger than the viscosity dissipation ones and the intermittency should be seen a somewhat "macroscopic" phenomenon in 2 dimensions.

Recent numerical results[11] show that the scenario for 2 dimensional turbulence is in qualitative agreement with our model. The enstrophy cascade indeed seems to be inhibited in some coherent structure which dominate the energy spectrum for all scales.

REFERENCES

1) B. Mandelbrot, The fractal geometry of nature (Freeman and Company, S. Francisco, 1982).

2) A.N. Kolmogorov, C.R. Acad. Sci. USSR 30, (1941).

3) L.D. Landau and E.M. Lifchitz, Mecanique des Fluides (Moscow M.I.R. ed. 1971).

4) U. Frisch, P. Sulem and M. Nelkin, J. Fluid. Mech. 87 (1978), 719.

5) F. Anselmet, Y. Gagne, E.J. Hopfinger and R.A. Antonia, preprint Institute de Mecanique de Grenoble (1983).

6) U. Frisch and G. Parisi, Varenna Summer School LXXXVIII (1983).

7) R. Benzi, G. Paladin, G. Parisi and A. Vulpiani, J. Phys. A. (Math. Gen.) 17 (1984) 3521.

8) G. Paladin and A. Vulpiani, Lett. Nuovo Cimento 41 (1984) 82.

9) H.A. Rose and P. Sulem, J. Physique 39 (1978) 441.

10) W. Feller, An introduction to probability theory and its applications, vol. 2 (Wiley, New York 1971).

11) R. Benzi, G. Paladin, S. Paternello, P. Santangelo and A. Vulpiani to be published 1985.

FRACTALS IN PHYSICS
L. Pietronero, E. Tosatti (editors)
© Elsevier Science Publishers B.V., 1986

NUMERICAL INVESTIGATION OF NONUNIFORM FRACTALS

Remo BADII* and Antonio POLITI+

Physik-Institut der Universität, Schönberggasse 9, 8001 Zürich, Switzerland*
Istituto Nazionale di Ottica, Largo E. Fermi 6, 50125 Firenze, Italy+

Besides the concept of dimension, which gives a first average characterization of fractals, a second relevant quantity, the nonuniformity factor, is here introduced and discussed. A theoretical calculation is performed for the binary Cantor set in order to point out the intuitive meaning of nonuniformity as the spread among the different contraction rates of lengths. The Sinai map is then investigated in detail, as a prototype for a map with non-constant Jacobian, both in the invertible and noninvertible parameter region.

1. INTRODUCTION

Recently, there has been a great deal of interest in the study of fractal[1] sets occurring in many areas of physics. In order to characterize them, many different definitions of dimension-like quantities have been introduced[2-4]. They can be roughly classified into two groups, the first one deriving from purely geometrical requests, the second being related to information theory. The definitions of the first group usually seem to give identical results for physical systems, while the differences among quantities belonging to the second one are a measure of the degree of "nonuniformity"[4-6] of the fractal under consideration.

In order to better understand and quantify this concept, a continuous infinity of dimensions (called dimension function) has been recently introduced[4] through the moments of the distribution $P(\delta,n)$ of nearest-neighbour (nn) distances δ among n points on the fractal. As the dimension function DF can be proved to yield, in suitable points, quantities as self-similarity dimension (SSD), information dimension and all other Renyi dimensions[4b], its evaluation provides an useful tool for the study of nonuniformity in physical systems.

2. DIMENSION FUNCTION AND NONUNIFORMITY

For convenience, we recall here the definition of Renyi dimension of order q. Assuming a partition of the phase space with boxes of size ε, and defining $P_i(\varepsilon)$ as the probability of the i-th box, Renyi introduced[2] generalized entropies $K_q(\varepsilon)$, $q = 0,1,2,\ldots,n$, as

$$K_q(\varepsilon) = \{\ln \sum_i P_i(\varepsilon)\} / (1-q) \quad (1)$$

Consequently, the Renyi dimensions are given by

$$D_q = -\lim_{\varepsilon \to 0} K_q(\varepsilon) / \ln \varepsilon \quad (2)$$

In particular, D_0 is the SSD (usually evaluated by means of the box-counting algorithm), D_1 is called information dimension[3a] and D_2 is the correlation integral exponent[3b]. These quantities satisfy the relation $D_q > D_p$ (when p>q), the equality being obtained in the case of uniform sets, i.e. such that the probability P_i scales as $P_i(\varepsilon) \approx \varepsilon^D$, where D indicates any dimension[6]. As the direct evaluation of the D_q's is, in general, impractical for computer memory reasons and lack of statistical convergence, we followed a different approach. Consider a reference point $\underline{x}$ in a d-dimensional Euclidean space, plus (n-1) others, all of them chosen at random, with respect to the natural measure, on the set*. We then define $\delta(n)$ as the distance between $\underline{x}$ and its nearest-neighbour $\underline{y}$ among the $(n-1)$ other points. Evidently, $\delta(n)$ is a nonincreasing function of n and, in general, some average over all points $\underline{x}$ will behave as $<\delta(n)> \approx n^{-1/D}$. To be more specific, we introduce the probability distribution $P(\delta,n)$ of nn distances among n points. Now, following Ref. 4, we compute the moments of $P(\delta,n)$ as

$$<\delta^\gamma(n)> = M_\gamma(n) = \int_0^\infty P(\delta,n)d\delta = K\, n^{-\gamma/D(\gamma)} \qquad (3)$$

where $D(\gamma)$ is a γ-dependent definition of dimension hereafter called dimension function (DF). The prefactor K depends on both γ and n: However, its dependence on γ is, by definition, irrelevant, while the dependence on n reduces, in a large class of systems, to an unessential periodicity in $\ln(n)$[8]. It has been shown[4] that, whenever $\gamma = D(\gamma)$, i.e. a fixed-point relation is satisfied, this value of the DF coincides with the SSD, at least in the case of self-similar[6] fractals. Moreover, for the same class of systems, it is possible to prove the general relation $D\{\gamma = (1-q)D_q\} = D_q$, where D_q indicates the order-q Renyi dimension. For $q = 0$, the fixed point relation is recovered. Therefore, the evaluation of moments (3) allows the determination of any D_q by means of a recursive method: An initial value of γ is chosen to obtain a first estimate of the desired dimension which, in turn, is used as a new input until a satisfactory accuracy is reached. However, this is not in general necessary, as the estimation of $D(\gamma)$ in some prefixed points gives the same information. Anyway, in order to check the stability of the method, it is necessary to compute the slope of the DF in the fixed points. In Ref. 4b, it has been shown that $D'(\gamma)$ is always bounded between 0 and 1 and, for uniform fractals, is equal to 0 and all dimensions coincide. Moreover, the slope in the fixed point $\lambda = D'(D_0)$ is related to a suitable entropy[4] and can be used to sinthetize the degree of nonuniformity of fractals, as it grows towards 1 for increasing nonuniformity: hence, this quantity has been called "uniformity factor". It is then worthwhile to examine some properties of the nn-distribution $P(\delta,n)$. First, note that it is smooth, contrary to the distribution of points in the Euclidean space containing the fractal. Secondly, for some physical systems, single points are not accessible but $P(\delta,n)$ can be evaluated, either analitically or numerically interpreting it in the following way: We consider a covering of the fractal by n unequal balls of diameters $\delta_i (i=1,\ldots,n)$ and then refine it by taking more balls of smaller size. For percolating systems, for example, $P(\delta,n)$ would represent the probability of finding a cluster of size δ in a box (of size L) containing n clusters: Then, one should consider a larger box (i.e. larger n) and renormalize it to the scale L, in such a way that the δ's would shrink to zero with increasing n. Finally, notice that, in general, $P(\delta,n)$ will spread when n grows: Actually, in Ref. 4b it has been shown that the relative r.m.s. $\Delta(n)$ of the distribution $P(\delta,n)$ behaves as $\Delta(n) = (M_{2D_0} - M_{D_0}^2)^{1/2}/M_{D_0} \approx n^\lambda$, for small λ: This clarifies the meaning of nonuniformity as the spread among the different contraction rates of lengths, when n goes to infinity. For uniform sets (i.e. $\lambda = 0$) all distances decrease with the same speed. The DF is, indeed, bounded by two

---

*In the case of dynamical systems, points are generated in a sequential way. For fractals such as percolating lattices or polymer aggregates, they can be picked up according to any rule, in order to get information on the structure: Following different rules corresponds to assigning different probability weights to the various regions of the object.

horizontal asymptotes corresponding to the fastest ($\gamma = -\infty$) and the slowest ($\gamma = \infty$) contraction rates.

3. FRACTAL MEASURES

Let us now investigate the shape of the function $P(\delta,n)$ in a simple case: The uniform Cantor set, generated by keeping two segments of lenght α and deleting the middle part of the unit segment. Also, the same probability is assigned to the two parts. For this set, the following expression for $P(\delta,n)$ has been obtained[4b]

$$P(\delta,n) = 2D_0 n(2\delta)^{D_0-1} \exp\{-n(2\delta)^{D_0}\}. \quad (4)$$

In order to see how nonuniformity manifests itself in the shape of the probability $P(\delta,n)$, let us consider a binary Cantor set[3a], that is the set of points x ($0 < x < 1$) such that their binary expansion contains zeros and ones with probabilities p_1 and p_2 ($p_1 + p_2 = 1$), respectively. This is an example of fractal measure[9] or "non-lacunar fractal"[1]: The whole unit segment is filled (i.e. $D_0 = 1$), while Hausdorff[1] and information dimension coincide $\{D_1 = -(p_1 \ln(p_1) + p_2 \ln(p_2))/\ln 2 < 1\}$. We are interested in the evaluation of the integral probability $S(\delta,n)$ for a reference point x to have a nn, chosen among (n-1) points, within a distance δ, $S(\delta,n) = \int_0^\delta P(y,n)dy$. In Ref. 4b, we obtained the following expression for the complementary distribution $S = 1-\overline{S}$

$$\overline{S} \propto [-\log 2\delta]^{1/2} \int_{-\log p_1}^{-\log p_2} \exp[-n(2\delta)^D]$$
$$\exp\left[\frac{\log 2\delta}{p_1 p_2 \log^2(p_2/p_1)} (D+D_1)^2\right] dD \quad (5)$$

where log indicates the base-of-two logarithm and $p_2 < p_1$. The first term in the integral is the result one would obtain for a uniform ternary Cantor set of dimension D[4b]. Therefore, the whole expression for S can be interpreted as a superposition of distributions of uniform Cantor sets with different dimensions. Note that D ranges between $-\log p_1$ and $-\log p_2$ which coincide with the asymptotes of the DF $D(-\infty)$ and $D(\infty)$. Moreover, the weight function is a Gaussian centered at the information dimension which, hence, appears to play a special role. Similar considerations should apply also to fractals embedded in higher dimensional spaces: Indeed, some plots of $P(\delta,n)$ reported in Ref. 4b suggest that deviations from the form (4) are an indication of nonuniformity.

4. LYAPUNOV DIMENSION IN NONINVERTIBLE MAPS

As a more physical example of fractal measure, we consider the Sinai map[3c,10]

$$\begin{aligned} x' &= x + y + g\cos(2\pi y) \quad \text{mod } 1 \\ y' &= x + 2y \quad \text{mod } 1 \end{aligned} \quad (6)$$

whose solution covers the whole unit square for sufficiently small g. As, however, the sum of Lyapunov exponents is always negative, an element of area will shrink to zero in time: This means that the probability distribution will be highly peaked, that is, the attractor is a fractal measure.

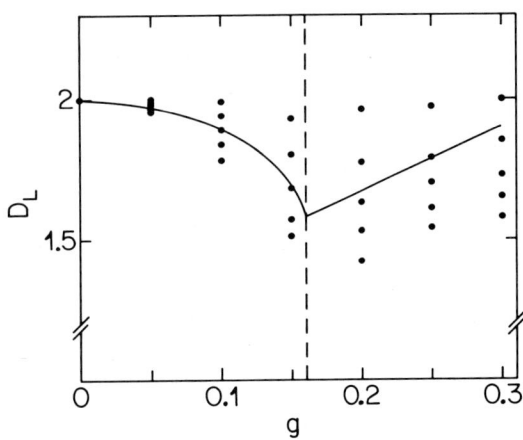

Lyapunov dimension D_L (full line) and Dimension Function $D(\gamma)$ (dots) versus g. The DF is computed for 5 different values of γ (-2, -1, 0, 1, 2).

Another reason to expect nonuniformity for this transformation is the nonconstancy of the jacobian $J = 1 + 2\pi g \sin(2\pi y)$. Infact, as shown by Grassberger and Procaccia[11], fluctuations in the point-like value of Lyapunov exponents are strong in these systems and are responsible for the difference among the various dimensions. Moreover, the information dimension D_1 of the attractor can be computed in terms of the Lyapunov exponents as $D_L = 1 + \lambda_1/|\lambda_2| \geq D_1$. The equality holds for two-dimensional invertible maps[12] while, for more general systems, D_L provides a good approximation to D_1. It is, therefore, interesting to investigate the behaviour of this "Lyapunov dimension" as a function of the parameter g. In Figure 1, we display D_L (as a full line) and five values of the dimension function $D(\gamma)$ (the central one corresponding to the information dimension), versus g. At the point $g = 1/(2\pi)$ (indicated by a vertical dashed line), the map becomes noninvertible and, for larger g-values, D_L gives only an upper bound to $D_1 = D_{(0)}$, as expected. Moreover, the system is rather nonuniform already at $g = 0.1$ and the highest spread among the various dimensions is obtained at the transition point $g = 1/(2\pi)$. It is, however, evident that nonuniformity changes rather smoothly with g and does not have a relation with the noninvertibility of the map.

REFERENCES

1. B.B. Mandelbrot, The Fractal Geometry of Nature (Freeman, San Francisco, 1983).
2. A. Renyi, Probability Theory (North-Holland, Amsterdam, 1970).
3. a) J.D. Farmer, Z. Naturforsch. 37A (1982) 1304,
   b) P. Grassberger and I. Procaccia, Phys. Rev. Lett. 50 (1983) 346,
   c) J.D. Farmer, E. Ott and J.A. Yorke, Physica 7D (1983) 153,
   d) Y. Termonia and Z. Alexandrovitch, Phys. Rev. Lett. 51 (1983) 1265,
   e) J. Guckenheimer and G. Buzyna, Phys. Rev. Lett. 51 (1983) 1438.
4. R. Badii and A. Politi,
   a) Phys. Rev. Lett. 52 (1984) 1661,
   b) J. Stat. Phys. 40 (1985) 725.
5. P. Grassberger, Phys. Lett. 97A (1983) 227.
6. H.G.E. Hentschel and I. Procaccia, Physica 8D (1983) 435.
7. C. Tricot, Math. Proc. Camb. Phil. Soc. 91 (1982) 57.
8. R. Badii and A. Politi, Phys. Lett. 104A (1984) 303.
9. B.B. Mandelbrot, Multiplicative Chaos and Fractals, this volume.
10. Ya. Sinai, Russ. Math. Surveys 4 (1972) 21.
11. P. Grassberger and I. Procaccia, Physica 13D (1984) 34.
12. L.S. Young, J. Ergodic Theory and Dynam. Sys. 2 (1982) 109.

FRACTALS IN PHYSICS
L. Pietronero, E. Tosatti (editors)
© Elsevier Science Publishers B.V., 1986

GENERALISED SCALE INVARIANCE AND ANISOTROPIC INHOMOGENEOUS FRACTALS IN TURBULENCE[1]

Daniel SCHERTZER and Shaun LOVEJOY[2]

EERM/CRMD, Météorologie Nationale, 2 avenue Rapp, 75007 Paris, France

A generalisation of scaling is presented to deal with anisotropy and (multidimensional) intermittency. Implications, especially for meteorological fields, are discussed.

1. INTRODUCTION

Many geophysical fields are extremely variable over a wide range of time and space scales. The variability of the atmosphere is large over at least 9 orders of magnitude ($\sim$ 1mm to $\sim$ 1000 km) and creates strongly intermittent and anisotropic structures : the energy spectrum ($E(k)$) of the horizontal wind in the horizontal is $\sim k^{-5/3}$ whereas it is (roughly) the much steeper $\sim k^{-11/5}$ in the vertical. This difference is the spectral counterpart of the (large) vertical stratification.

For both analysing and simulating these structures, it is necessary to generalise both the notion of scale invariance and intermittency, through the introduction of anisotropic metrics and dimensions, and scale invariant measures characterised by multiple (fractal) dimensions. Interesting consequences are that multidimensionality is directly connected with the divergence of high statistical moments of average cascade quantities, multiplicative processes and new questions on detectability and predictability.

2. GENERALISED SCALE INVARIANCE (G.S.I.)

2.1 Motivations

To avoid the untenable dichotomy 2D/3D for large/small scales, we have proposed an alternative scaling model[1-3] (see also [4-5] for non-mathematical reviews) of atmospheric dynamics: the anisotropy introduced by gravity via the buoyancy force results in a differential stratification and a consequent modification of the effective dimension of space (from the isotropic value D=3 to 23/9=2.5555...).

In order to take into account this and other effects such as the differential rotation introduced by the Coriolis force, a general formalism of scaling is required. The fundamental problem is that of finding a family of "balls" representing the statistical properties of eddies at different scales, via (mathematical) random measures, such as the flux of energy through structures of a given scale.

2.2 Generalised notion of scale

Close examination of the phenomenology of

---

[1] partially supported by ATP-RA (85-2601) of CNRS.
[2] present address : Dpt. of Physics, McGill University, University ST., 3600 University St., Montreal, Que. H3A 2T8, Canada

turbulent cascades outlined the basic properties associated with the notion of scale and leads[6] to the following abstract definition in terms of a group (the "scaling group") of operators T_λ acting on a topological space M:

(i) T_λ is a multiplicative group ($\lambda \in R_+^*$) of transformations from M to M:

(1) $T_{\lambda \lambda'} = T_\lambda \circ T_{\lambda'}$ $\forall \lambda, \lambda' \in R_+^*$

(in particular: $T_1 = 1$ = the identity and $T_\lambda^{-1} = T_{\lambda^{-1}}$)

(ii) there exists a family $\mathcal{B}_1$ of "balls" (open sub-sets of M) such that $\mathcal{B} = T_\lambda \mathcal{B}_1$ is a basis for the topology of M

(iii) there exists an increasing function ϕ from $\mathcal{B}$ to R_+, bounded on $\mathcal{B}_1$ and which factorizes in ($D \in R_+$):

(2) $T_\lambda \phi = \lambda^D \phi$, $\forall \lambda$

(T_λ is naturally defined by: $T_\lambda \phi(B) = \phi(T_\lambda B)$, $\forall \lambda, B$)

Note the expression λ^D results from the group property of T_λ since it would be implied by the assumption of the existence of a continous function $g(\lambda)$ in (2).

As is easily shown, in case of a metric space, D plays the role of a dimension and ϕ can be taken as the radius of the balls defined by the distance, T_λ is the usual dilatation in case of isotropy. More generally we can use the measurability property of the balls. For instance on R^d, we can take ϕ as the Lebesgue (d-volume) measure, by supposing that the B's are Lebesgue measurable, and D equals d if the balls are the usual spheres or cubes. This is no longer true with strongly anisotropic balls (such as self-affine, but not self-similar, ellipsoïds). Even more anisotropic (and/or irregular) balls can be dealt with: (anisotropic) fractal sets. In all these cases ϕ can be taken as the measure which is finite and positive on the balls (Lebesgue or Haussdorff) and the scale is given by $\phi^{1/D}$.

2.3 Linear GSI case

The group T_λ is generated by a (bounded) linear application G according to:

(3) $T = \exp(G\log \lambda) = \sum_{n=0}^{\infty} (\log)^n G^n/n!$

The following conditions[6] are necessary and sufficient to obtain a scaling group:

a) measurable case: $D_{el} = \text{Trace}(G) > 0$ D_{el} can be considered as the effective dimension of the space, or its elliptical dimension [1-3]. Non-linear examples are given in [6].

b) metric case: $\inf \text{Re}\, \sigma(G) \geq 1$ where $\sigma(G)$ is the spectrum of G. If the unit ball is defined by the ellipsoïd generated by a symmetric operator A, the following condition is obtained:

(5) $\inf \sigma(\text{sym}(AG)) \geq 1$

(Sym (AG) denotes the symmetric part of AG)

Particularly simple examples of linear GSI are obtained by the use of quaternions[6], and used to exploit[7] the FSP model[8-9] to give examples of (mono-dimensional) fields respecting linear metric GSI.

3. GSI AND MULTIDIMENSIONAL INTERMITTENCY

3.1 Introduction

Usual stochastic processes (such as Brownian motion) are obtained by the (weighted) addition independent identically distributed (i.i.d.) random variables (e.g. integrals of white noise). Conversely, the multiplicative group T_λ suggests that in GSI the most natural type of process to use are those obtained by multiplication, corresponding also to the non-linear breaking of eddies.

The former case is mono-dimensional, while the latter generally leads to multiple dimensions. Many efforts have been made to relate the most obvious aspect of intermittency-its "spottiness"[10]- to a turbulent support with a single fractal dimension[11,12]. However, as pointed out in[1,2], phenomenological models of intermittency[13,14,11] lead, generally to multiple dimensions, corresponding to the different (tensorial) powers of the measure of the flux of energy. Indeed, a sequence of dimensions is easily obtained[2,3,15,16] by considering the divergence of high moments of the density ε_A of this flux with respect to different D_A-dimensional Hausdorff measures:

(6) $<\varepsilon^h_A> = \infty$ for $D_A < C(h) = D_{el} - D(h)$

$C(h) = \log_\lambda <W^h> /(h-1)$

where W is the random variable which distributes the density during a step of the cascade. Note that the condition stated in Eq.6 corresponds to the one of non-intersection of sets A and S(h) of co-dimension C(h) (since, usualy for sets A, B: $D(A \cap B) = D(A) - C(B)$, D and C indicating the dimension and co-dimension of the referenced sets). Increasing h corresponds to studying the more intense regions. C(h) is an increasing function of h, or the most intense regions are the most sparsely distributed.

3.2 GSI and multiplicative processes-multiplicative chaos

Instead of adding random increments of finer and finer resolution along the cascade (as in the FSP[8-9]), one may multiply by random increments of finer and finer resolution. This multiplicative procedure corresponds to the non-linear break-up of eddies into sub-eddies (Mandelbrot's cascade model of intermittency on a rigid grid corresponds to a discrete product).

The limit of such processes represents a mathematical problem -called multiplicative chaos- where some results have been recently obtained[17]. Nevertheless, due to the multiplicative property of both T and the way the process is constructed, we may introduce the co-dimension function C(h) (f being a multiplicative density increment):

(8) $<T_\lambda^{-1} f^h> = \lambda^{(h-1)C(h)} <f^h>$

and generalise thus ealier results (Eq.6).

3.3. Implications of multidimensionality

We introduce the "structure integral" S(h, A), instead of the classical structure function, to study the behavior of a stochastic measure m:

(9) $S(h, A) = <m^h(A)> / <m(A)>^h$

where A is a D_A-dimensional measuring-set (e.g.: on which the averages are taken). Generalised scale invariance implies :

(10) $S(h, T_\lambda A) = \lambda^{-p(h,D_A)} S(h, A)$

For the simple case where the phenomenon is mono-dimensional with dimension D_s (co-dimension = $D_{el} - D_s = C_s$) $p(h, D_A)$ is linear in h:

(11) $p(h, D_A) = C_s \cdot (h - 1)$, if: $D_A > C_s$

In other cases, C_s has to be replaced by the co-dimension function C(h) (hence $p(h, D_A)$ is no longer linear in h) and the (physical) measures become sensitively dependent on D_A (especialy due to the the condition of intersection). Such a study on radar-data[18] (with $D_A = 1, 1.5, 2, 3, 4$), supports multidimensional behaviour for the rain.

A dimensional detectability condition results from the condition of non-degeneracy of the statistics (i.e.: non-zero and finite) that is (according to Eq.6 or 11) the intersection of the set of observation (A) and the support sets S(h). Thus scale resolution is not sufficient to estimate the detectability of the phenomena, e.g. the most intense phenomena will be lost with a sparse set of observations. It has been shown[19] that indeed ground networks have lower dimensions than 2 (e.g. the world meteorological surface network has D_A 1.75). Due to non-linear interactions, it turns out to raise new questions on predictability, which until now has been studied only in terms of scales.

4. CONCLUSION

Motivated by the strong anisotropy and intermittency of the atmosphere, we have developed a formalism called generalised scale invariance. The formalism is based on two sets of elements and may be regarded as an extension of earlier work on cascade processes (especially[10,3]).

The first is a group of general scale changing operators, whereas the second are the intermittent measures invariant under the operators. In a turbulent cascade, the scale changing operator transforms eddies into sub-eddies, while leaving the physically significant energy flux invariant (here represented by a scaling measure). It may be worth noting that explicit geometry is not always required, since measurable properties are sufficient.

We stressed that multidimensionality is theoretically the rule for multiplicative processes, and such a behaviour has been tested directly on radar determined rain field[18]. It raises new questions on detectability and predictability of turbulent phenomena because of dimensional dependance.

ACKNOWLEDGEMENTS

We acknowledge fruitful discussions with G. Austin, D. Lilly, R. Cahalan, P. Muller.

REFERENCES

1. D. Schertzer and S. Lovejoy, preprint vol., IUTAM symp. on turbulence and chaotic phenomenon in fluids, Kyoto, Japan (5-9/09/1983) pp 141-144.
2. D. Schertzer and S. Lovejoy, on the dimension of atmospheric motions in: Turbulent and chaotic Phenomena in Fluids, ed. T. Tatsumi, (North-Holland, 1984), pp.505-512
3. D. Schertzer and S. Lovejoy, The dimension and intermittency of atmospheric dynamics, in Turbulent Shear Flow 4, B. Launder Ed. (New York, Springer ,1985) pp.7-33
4. D. Schertzer and S. Lovejoy, Sciences et Techniques (1984) 69.
5. S. Lovejoy and D. Schertzer, AMS Bulletin 67 (1985).
6. D. Schertzer and S. Lovejoy, PCH Journal 6, 5/6 (1985) 623.
7. S. Lovejoy and D. Schertzer, Wat. Resour. Res. (1985) 21, 8, 1233.
8. B. Mandelbrot, Fractal Sum of Pulses (available from the author).
9. S. Lovejoy and B. Mandelbrot, Tellus 37A (1985) 209.
10. G.I. Batchelor and A.A. Townsend, Proc. Roy. Soc., A199 (1949) 238.
11. B.B. Mandelbrot, J. Fluid Mech. 62 (1974) 331.
12. U. Frisch, P.L. Sulem and M. Nelkin, J. Fluid Mech. 87 (1978) 719.
13 E.A. Novikov and R. Stewart, Izv. Akad. Nauk. SSSR Ser. Geofiz. 3 (1964) 408.
14 A.M. Yaglom, Sov. Phys. Dokl. 2 (1966) 26.
15. D. Schertzer and S. Lovejoy, Note CRMD N° 69 (Met. Nat., Paris).
16. B.B. Mandelbrot, J. Stat. Phys. 34 (1984) 895.
17 J.P. Kahane : Multiplicative Chaos (in preparation).
18. S. Lovejoy and D. Schertzer, Extreme variability, scaling and fractals in remote sensing, in: Digital image processing in remote sensing. P.J. Muller Ed. (Taylor and Francis, London, 1985) Ch.14
19. S. Lovejoy, D. Schertzer and P. Ladoy, Nature (in press).

ANALYSIS OF THE FRACTAL SHAPE OF SEVERE CONVECTIVE CLOUDS

Franz S. RYS

Fritz-Haber-Institut der Max-Planck-Gesellschaft, Faradayweg 4-6, D-1000 Berlin 33, West Germany
and
A. WALDVOGEL

Institut für Atmosphärenphysik, ETZ-Zürich, Hönggerberg, CH-8093 Zürich, Switzerland

An analysis of recent radar data on hail clouds yields, for larger clouds, a fractal shape with a fractal (Hausdorff) dimension D_1 = 1.36 + 0.1 in agreement with an earlier analysis of cloud shapes by Lovejoy[9] and the theory of relative turbulent diffusion[10]. Moreover, a particle drift, present in strong winds as well as in the center of hail clouds is presumably responsible for a rather sharp cross-over to a smooth behaviour (D_1 = 1) showing the existence of an apparent characteristic length.

1. INTRODUCION

Since 1977, when the mathematical notion of the Haudorff dimension was renamed as the fractal dimension and introduced in the physical literature[1], it has been shown to be an important geometrical concept to quantify disordered structures of simple models and physical systems in general. In fact, an increasing number of articles on experimental systems (such as polymer solutions or melts[2], adsorbate domains[3], electric discharges[4], etc.) and numerical simulations (e.g. in aggregate systems[5]) prove the usefulness of this notion.

The fractal dimension D characterises any self-similar system: Upon changing the linear dimensions by a scale factor f, a fractal quantity (such as a contour or a surface) changes by the factor f^D, for any value of f. The value of the fractal surface dimension D_s is comprised between 2 and 3 (where D_s = 2 describes a smooth surface) whereas for the dimensions of a contour line, $1 \leq D_1 \leq 2$ (D_1 = 1 for smooth lines). A planar section of a fractal surface has a dimensionality $D_1 = D_s - 1$ in general (if pathological cases are disregarded[6]). Up to date, there are only a few rather academic models known which allow an exact evaluation of their fractal dimension and hence the usefulness of the latter stems mainly from numerous experimental observations[7].

In this work we show that severe convective clouds which occur in hail storms have also a fractal shape. An analysis of radar echo data from these clouds yields:

(1) a non-trivial fractal behaviour of larger cloud sections, and

(2) the existence of an apparent characteristic length.

Surprisingly, the fractal dimension D of the perimeter of the cloud cross-section has, within error bars, the same value of D≃4/3 as for rain and cloud areas up to very large sizes of 10^6 km^2 which were detected from infrared satellite- and radar-determined geometries. On the other hand, small hail clouds have a smooth shape. Also, an apparent smoothening of the cloud shape is observed for hail storms with strong wind strengths causing a fast displacement of the hail center with velocities up to 15 m/sec. Although there is so far no theoretical understanding of this observation, we conjecture that the presence of strong winds smoothens the cloud shape in a similar way as an *anisotropic* diffusion constant smoothens the shape of diffusion-limited agglomerates. Analogous anisotropy effects are observed in electrical discharges in presence of uniaxial fields[12].

2. METHOD

The experimental data have been assembled during the hail suppression experiment "Großversuch IV" in the years 1976-1982. The radar information is available as a plan position indication (PPI), i.e. a XY-plot of nearly horizontal contours of constant radar reflectivity. The PPI's were obtained after a full azimuthal revolution of the antenna at a fixed elevation angle of 5.5 degrees. The measuring time was 1 minute for 1 PPI and the radar resolution was 1 degree in azimuth, 0.3 km in distance and 1 dBZ in reflectivity.

The latter is defined by:

$$Z = \int N(d) \, d^6 \, d(d) \qquad (1)$$

where $N(d)$ denotes the Rayleigh distribution of diameters of the hydrometeors (i.e. Rayleigh scatterers such as rain drops and hail stones) per m^3 and diameter interval d and $d + d(d)$ (in units of mm^6/m^3, i.e. $10^{-18} \, m^3$). Conveniently, the quantity

$$I = 10 \log Z \text{ (in units of dBZ)} \qquad (2a)$$

is used. For a Poisson distribution $N(d) \sim \exp(-\lambda d)$, which describes well the observations, it follows:

$$I = \text{const} - 70 \log \lambda \qquad (2b)$$

i.e. the dBZ values are monotonously increasing with the particle size $\lambda d(\text{median}) = \text{const}$. Details of the radar data processing etc. have been described recently[8].

A typical PPI is shown in Fig. 1, where contours for $I = 45, 55, 60$ and 65 dBZ are plotted. In particular, the 65 dBZ curve represents a heavy rain area (with a precipitation rate $R = 500$ mm/h) including hail precipitation.

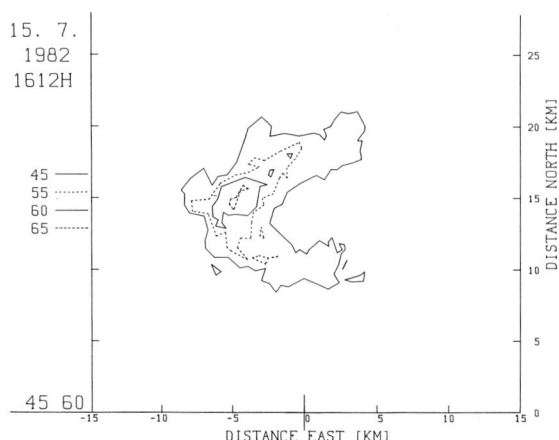

FIGURE 1
Example of a horizontal cross section of constant radar reflectivity surfaces of an instant hail cloud shape, characterised by their dBZ values (see text).

The fractal shape of the observed hail clouds manifests itself in the fractal properties of the contours with constant dBZ values. The area F within a contour depends fractally on the contour length (i.e. the perimeter) U as

$$F = F_0 \, U^E \qquad (3)$$

where the exponent E is related to the fractal dimension D_l of the contour through

$$E = 2/D_l \qquad (4)$$

and, correspondingly, to the fractal dimension $D_s = D_l + 1$ of the surface of constant dBZ values and, thus, of the hail cloud.

3. RESULTS

In Fig. 2, a log-log plot of the area F vs. the perimeter U of 45 dBZ contours are shown. F and U were taken in intervals of 1 minute during the typical time evolution of a hail storm. Similar plots have been obtained for the dBZ values 55, 60 and 65. Smaller contours lie on a linear plot corresponding to a power law (eq. 3) behaviour with a slope E = 2 (i.e. D_1 = 1). Hence, these contours are *smooth*. The points for larger contours lie on a different line with a slope E = 1.5, i.e., D = 1.33 describes the *fractal* structure of the larger contours.

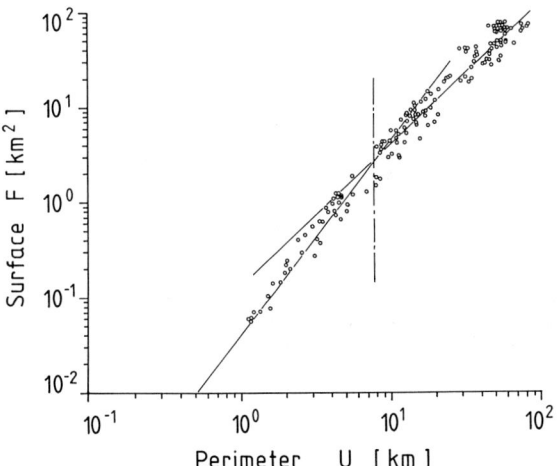

FIGURE 2
log-log plot of the surface (F)-perimeter (U) relation for the (constant) dBZ value 45. Every point corresponds to a particular time interval (of 1 minute) during the temporal evolution of one hail storm event. From the linear fits (shown as straight lines) of 24 different plots the results of eq. (5) and (6) are obtained.

From the evaluation of 24 hail events we have obtained:

(1) For smaller values of U,

$$D_1^< = 1.0 \pm 0.1 \; (D_s^< = 2.0 \pm 0.1) \quad (U<U_o) \quad (5)$$

describes a smooth behaviour.

(2) For larger values of U,

$$D_1^> = 1.36 \pm 0.1 \; (D_s^> = 2.36 \pm 0.04) \quad (U>U_o) \quad (6)$$

describes a fractal shape.

(3) The rather sharp cross-over around $U=U_o$ indicates the existence of a characteristic length

$$L = U_o/\pi = 3 \pm 1 \text{ km}.$$

4) In the presence of strong wind strengths during the hail storm the cross-over value U_o is shifted towards *higher* values and larger contours appear smooth.

4. DISCUSSION

1) The fractal dimension of larger contours $U>U_o$ (eq. 6) is, within error bars, equal to the value 4/3 found recently by Lovejoy[9]. His value D=1.35±0.05 is valid in the range $10<U<10^4$ km. An attempt to interpret the fractal dimension D of cloud shapes in the frame of the relative turbulent diffusion theory has been published recently[10].

2) The observed cross-over at U_o indicates the existence of a chracteristic length L in the hail cloud problem. Its value is independent of the dBZ value, i.e. of the particle size but does vary with the wind strength. Note that the value of L is one order of magnitude *larger* than the resolution of the radar (of ∼0.3 km) and hence, it is unlikely that the cross-over at L is caused by the finite value of the latter. Instead, the dependence of L on the wind strength suggests, that the cross-over to a smooth shape reflects the presence of a particle drift (or convective flow). It is well established that particularly strong vertical up- and down winds are present in the center of thunderstorms and hail clouds. This would explain the smoothness of the smaller contours near the cloud center.

Consequently, we conjecture that the smoothening (i) of the smaller clouds, and, (ii) of larger cloud contours in the presence of strong winds is caused by the convective flow. At large drift velocities the particles are expected to follow linear trajectories (instead of Brownian ones at zero drift). This is a possible mechanism for the smoothening of the cloud shapes. A similar smoothening of the corresponding patterns has been observed recently in the diffusion-limited aggregation problem[11] and for electric discharges in strong homogeneous external fields[12].

A generalisation of the theory of relative turbulent diffusion to include a non-zero particle drift should describe this phenomenon quantitatively. Interestingly, the value of U_o at low wind strengths corresponds to the lower end of the range of contour lengths observed in Ref. 9. From this fact as well as from the equality of the fractal dimensions of severe convective (hail) clouds, light rain and fair weather clouds, we conclude that our findings describe the fractal nature of clouds in presence of winds in general.

3) These observations could be the basis of a useful phenomenological method for determining cloud areas of strong winds by measuring the corresponding fractal cloud dimensions.

Note that our data were sampled during the whole hail storm and therefore, temporal variations of the wind strengths could be an important source of deviations from the mean values of our plot (Fig. 2). Therefore, more observations on various scales at well-defined wind strengths are needed to improve the statistics. Moreover, a quantitative theoretical description is to be developed in order to give a quantitative description of cloud shapes in the presence of winds in general.

ACKNOWLEDGEMENTS

One of us (F.S.R.) acknowledges the financial support by the Deutsche Forschungsgemeinschaft under contract Sfb-6-TP A7/2. It is a pleasure to thank Jean-Pierre Eckmann, Max Kolb, Peter Pfeiffer, Luciano Pietronero and Hans-Jürg Wiesmann for various informative discussions.

REFERENCES

1. B.B. Mandelbrot, Fractals (Freeman, San Francisco, 1977).

2. P.J. de Gennes, Scaling Concepts in Solids (Cornell University Press, Ithaca, N.Y. 1979).

3. W.T. Elam, S.A. Wolf, J. Sprague, D.U. Gubser, D. von Vechten, G.L. Barz, jr. and P. Meakin, Phys. Rev. Lett. 54 (1985) 701.

4. L. Niemeier, L. Pietronero and H.J. Wiesmann, Phys. Rev. Lett. 52 (1984) 1033.

5. T.A. Witten and L.M. Sander, Phys. Rev. Lett. 47 (1981) 1400.

6. P. Mattila, preprint.

7. cf., e.g., Proc. 3rd Conf. on Fractals in the Physical Sciences, (NBS Gaithersburg, MD, USA; November 1983).

8. A. Waldvogel and W. Schmid, J. Appl. Meteorology 21 (1982) 1228.

9. S. Lovejoy, Science 216 (1982) 186; See also: Proc. 20th Conf. on Radar Meteorology (Boston, MS, USA, November 1981).

10. H.G.E. Hentschel and I. Procaccia, Phys. Rev. A29 (1984) 1461.

11. P. Meakin, Phys. Rev. B28 (1983) 5221.

12. H.J. Wiesmann, private communication (1985).

FRACTALS IN PHYSICS
L. Pietronero, E. Tosatti (editors)
© Elsevier Science Publishers B.V., 1986

NESTED CELLULAR AUTOMATA: CONTINUOUS ASPECTS OF DISCRETE SYSTEMS

Uwe QUASTHOFF

Karl-Marx-Universität Leipzig, Sektion Mathematik, DDR-7010 Leipzig, DDR

Cellular automata by definition consist of discrete cells containing these automata and work in discrete time. To model continuous processes, we describe a nesting procedure which replaces the cells by smaller and smaller ones so that one can have arbitrary small cells and arbitrary small time units while the macroscopic structure is still the same. We investigate automata with simple neighborhood structure whether they admit such a nesting procedure or not.

1. INTRODUCTION

Figure 1 shows Pascals triangle modulo 2 up to order 128 with a dot indicating a 1 and a blank for a 0. While our picture consists of a finite number of discrete cells it is obvious that there is a much finer fractal structure giving the same macroscopic picture.

Fig. 1

Figure 1 can be thought as a result of the following cellular automata: To get the state of a cell at time t+1 just add the states of this cell and its right neighbor at time t modulo 2. The resulting state is shown below the original states in the next line. The figure starts with just one cell in state 1.

Figure 2 shows the same cellular automata but with random starting configuration. In this picture it is not clear whether one can have a much finer structure with the same transition rule giving the same macroscopic structure. While pictures like these appear also for other transition rules, the nesting process is possible only for the described automaton. We give a full classi-

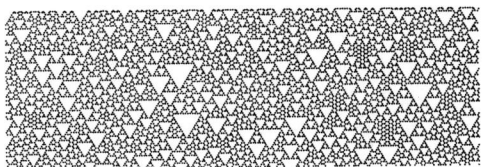

Fig. 2

fication of the cellular automata with two states and (one- or twosided) next neighbor interaction and a nesting function of the same type. The modulo 2 example is generalized to more states, larger neighborhoods and higher dimensions. Applications to model continuous state spaces are discussed. For the proofs of the theorems we refer to Quasthoff[1].

2. THE NESTING PROCESS

We will consider cellular automata

with the cells placed in a d dimensional lattice isomorphic to $\mathbb{Z}^d$.
For this and the next section we will assume d=1. The state space Y of a cell is assumed to be finite. The transition function T of a cell gives the state of this cell at time t+1 depending on the states of the cells in a neighborhood at time t. For simplicity we will assume that the neighborhood U(x) of a cell x will consist of several adjacent cells containing x, i.e. $U(x)=\{x-k, x-k+1, \ldots, x+l\}$ for some natural numbers k,l. More complicated neighborhood structures are investigated in Quasthoff[1].

The nesting process goes as follows. We replace each cell (simultaniously) by m smaller cells with with the same transition function T. Call m the order of the nesting process. The neighborhood of a small cell is just a condensed copy of the original neighborhood. A nesting function is a function $N: Y^m \to Y$ which assigns to a large cell a state depending on the states of the smaller cells inside (some ordering of these cells is always assumed).

<u>Definition</u>: A pair (T,N) of a transition function T and a nesting function N is called compatible if there is a natural number n such that
$$T \bullet N = N \bullet T^n.$$
log n is called growth rate.

The existence of a nesting function N for a given transition function T such that (T,N) is compatible has several consequences. We can iterate the nesting process by considering the smaller cells as new large cells and construct smaller and smaller cells. A very small cell constructed in the k-th step is just the n^t-th part of the original cell and the time unit of this small cell is just the n^k-th part of the time unit of the original cell. So we can approximate processes continuous both in time and space by such nested cellular automata. On the other hand, we can interpret the nesting function $N: Y^m \to Y$ as a map of an $\#(Y)^n$ state space ($\#(Y)$ means the number of elements of Y) on a $\#(Y)$ state space. Combining these two points of view one can even assume that the smaller automata have more states so that we can approximate continuous state spaces, too. In this case we need a different transition function for the automata with larger state space. An example is given in section 5.

For the rest of the paper, we will assume n=k+l+1, i.e. the number of elements of the neighborhood of any point equals the number of small cells which replace one large cell.

3. COMPLETE ANALYSIS OF TWO STATE AUTOMATA WITH ONE- AND TWOSIDED NEXT NEIGHBOR INTERACTION

For onesided next neighbor interaction we may assume $U(x)=\{x, x+1\}$, in the case of twosided interaction we take $U(x)=\{x-1, x, x+1\}$.

To give names to the different transition functions we follow Wolfram[3]. Read the pair or triple s of states in in a neighborhood in their natural order as binary number b(s) and let T(s) be the transformed state. Then give the transition function T the name $\sum T(s) \cdot 2^{b(s)}$ where summation is taken over all 2^2 or 2^3 possible couples of states, respectively.

For example, for onesided interaction we get the names 6, 10, and 12 for the transition function of the introduction, the left shift and the identity, respectively. The nesting functions are numbered in the same way. We will consider only transition functions and nesting functions with even names because we want the global state with all automata in state 0 to be stable.

We have the following result.

Theorem 1: i) In the case of onesided next neighbor interaction, the only compatible pairs are (6,6), (6,10), (6,12), (8,8), (10,n), (12,n), and (14,14) with n=2,4,...,14.

ii) In the twosided case all compatible pairs are (128,128), (136,128); (170,n), (192,128), (204,n), (238,254), (240,n), (252,254), and (254,254) where n=2,4,...,254.

So a transition function is compatible with all nesting functions iff it is the identity or a shift.

All other transition functions except 6 in i) are of class 1 (see Wolfram[4]), i.e. they evolve for a.e. initial state to a unique homogeneous state and show simple behaviour. The only remaining transition function is the mod 2 example of the introduction. It is of class 3 (i.e. leads to "chaotic" patterns) and it is the only function leaving the global state with all automata in state 1 not invariant.

4. DIFFERENT GROWTH RATES AND HIGHER DIMENSIONS

To construct a compatible pair with arbitrary high growth rate log n we use a nesting process of order n. Then if T is the left shift (that means the state at point x and time t+1 equals the state at x+1 and t) and N is any nesting function of order n, then (T,N) is a compatible pair with $T \cdot N = N \cdot T^n$, i.e. we have growth rate n. The same result is true for higher dimensions.

To give a less trivial example we generalize the mod 2 transformation. Let $Y=\{0,1,\ldots,p-1\}$ for some odd number p, $U_p(x)=\{x-\frac{p-1}{2},\ldots,x+\frac{p-1}{2}\}$ and let T_p be the function summing up its p arguments (the states of a neighborhood) modulo p. Define a nesting function N_p of order p also by summation modulo p. For p=2 and a slightly different neighborhood the pair (T_2, N_2) is the pair (6,6) and compatible by Theorem 1. More generally we have:

Theorem 2: i) For every prime number p, we have the compatible pair (T_p, N_p) satisfying $T_p \cdot N_p = N_p \cdot T_p^p$.

ii) Let N_p' be a nesting function of order p with $N_p'(s_1,\ldots,s_p) = s_q$ for some q with $1 \leq q \leq p$. Then the pair (T_p, N_p') is also compatible with growth rate p.

These examples can be generalized to higher dimension in the following way. Take $\tilde{U}_p = U_p \times \ldots \times U_p$ (d factors) as d dimensional neighborhood. In the nesting process each cell is of the form of a d dimensional cube and will be replaced by p^d small cubes. Let $\tilde{T}_p$, $\tilde{N}_p$ as before be functions adding their arguments modulo p. Then $(\tilde{T}_p, \tilde{N}_p)$ is compatible.

5. A MODEL FOR CONTINUOUS STATE SPACE

There is a second point of view concerning the nesting procedure. After replacing a large cell by m small ones we can interpret the m-tupel of states inside the large cell as one complicated state of this cell. Instead of getting more cells we construct a new state space for each cell. This induces a new transition function which can have a simpler neighborhood structure as can be seen in the following example. Again the process can be iterated.

Let us start with the transition function T_5 of Theorem 2. Replacing each cell by 5 smaller cells of 5 possible states is the same as enlarging the state space of the large cell from 5 to 5^5 possible states. Let $V: Y^5 \to \{0, 1, \ldots, 5^5 - 1\}$ be the function $V(s_1, \ldots, s_5) = 5^4 s_1 + 5^3 s_2 + \ldots + s_5$ reading five adjacent states as a number to the base 5. A new transition function T' on these many state automata is induced by T_5 via $T' = V T_5 V^{-1}$. In contrast to T_5, the function T' has next neighbor interaction.

Let N be the nesting function $N(s_1, \ldots, s_5) = s_1$. Then (T_5, N) is compatible by Theorem 2. From $T_5 \cdot N = N \cdot T_5^5$ one has $VNV^{-1} T'^5 = T' \cdot UNU^{-1} = UTNU^{-1}$, where UNU^{-1} is the function sending a number x to the largest multiple of 5^4 not greater than x. So the transformed nesting function UNU^{-1} reduces the state space in a way similar to rounding.

REFERENCES

1. U. Quasthoff, Nested cellular automata, preprint 1985.
2. S.J. Wilson, Discrete Appl. Math. 8 (1984) 91.
3. S. Wolfram, Rev. Mod. Phys. 55 (1983) 601.
4. S. Wolfram, Physica 10D (1984) 1.

THE HYPERBOLIC HELIX HYPOTHESIS: STAPLETON'S FRACTAL MEASURE ON THE HYDROPHOBIC FREE ENERGY MODE DISTRIBUTIONS OF ALLOSTERIC PROTEINS

Arnold J. MANDELL

Laboratory of Biological Dynamics and Theoretical Medicine (M-003), University of California, San Diego, La Jolla, California 92093, USA

I. Background

Poincaré's geometric-topological approach to the problem of the number, n, of periodic and near-periodic families, Fn, of orbits, $z(n)$, on R^2, led to the conjecture that $nFn = \infty$. His initial work, using continuations of analytic series, suggested that the measure m on parameter space (r, λ) of each z_i, $m_i(r, \lambda)z_i \sim \log nFn$. The properties of this distribution of orbits, $Pz(nFn)$, its dynamical origins and significance for the behavior of proteins is the focus of our program of research.

Birkhoff[1] studied bounded, linear, invertible transformations of the annulus, $\phi^t{}_z : R^2 \to R^2$ which conserved measure, $\phi^t{}_z = e^{At}$, det $A = 1$. The simultaneous motions of the paired images of a single orbit along the identified inner and outer boundaries travel in opposite directions. $\phi^t{}_z$ is a set of mirror-symmetric representations of a one-way transformation. We shall see that two strong symmetry conditions, enthalpy-entropy compensation and the stereoisomerism restriction of polypeptide-protein information exchange, make these orbits paired geodesics with eigenvalues that are symmetrically separated from the unit circle. R^2 is a Banach space $E = E^s \otimes E^u$, $[\phi^t{}_u(z)] \subset E^s$, $[\phi^t{}_v(z)] \subset E^u$, representing a decomposition of the tangent space of the manifold of E into stable and unstable foliations[2]. $p_u \in \phi^t{}_u(z)$ and $p_v \in \phi^t{}_v(z)$ are mirror images of a hyperbolic fixed point, a whirlpool-like generator of nFn which along with det A = 1 constitutes the biologically oriented reconstruction of Birkhoff's proof of Poincaré's last geometric theorem: Any one-to-one transformation of an annulus admitting an invariant integral and rotating the boundaries in opposite directions possesses at least two distinct invariant points with symmetrically opposite signed indices. At the dissipative limit of an annular construction, one might think of the superimposition of the stable, $\lambda -$, and unstable, $\lambda +$, manifolds of a circle (spiral) in resonance. This uniform hyperbolicity condition on fixed points and orbits in R^2 achieves straightforward expression in the thermodynamic properties of proteins.

Hydrophobic free energy, the surface tension between elliptical proteins and their aqueous surround, is decomposed into: the x-ray crystallographically observable, enthalpy-containing structure, the stable foliation $\phi^t{}_u$, calculated by measurements of the protein's surface at 25 to 35 cal per mol per $Å^2$ [3]; the "icebergs" in the surrounding solvent and the invisible and inferred destabilizing loss of polypeptide chain entropy with folding, "entropic strain," the unstable foliation, $\phi^t{}_v$, estimated at 2 to 3 kcal per mol per residue[4]; and a very small, usually passive, H^+ bond, solvent-self variation involving 3 or so out of a possible 600 bonds, $\epsilon_u \sim 15$ kcal per mol of protein. The hyperbolic second law for an average protein and its halo of enthalpy-entropy com-

---

Written while the author was a Visiting Fellow at the Institut des Hautes Etudes Scientifiques, Bures-sur-Yvette, France. Numerical analyses by analog and digital computer were supported by U.S. Army Research Contract DAAG-20-83-K-0069 and Department of Defence Instrumentation Grant DAAG-29-84-G-0072.

pensation in the surrounding solvent in kcal/mol[5] is:

$$\phi^t{}_u - \alpha\phi^t{}_v \pm \epsilon_u = 0 \qquad (1a)$$
$$350 - 350 \pm 15 = 0 \qquad (1b)$$

This enthalpy-entropy compensation,composed of large terms of opposite sign, characteristic of water itself and all organic compounds in aqueous or related solution[6], is the hidden symmetry that reduces the problem of protein structure and dynamics to a two-dimensional phase space where the Poincaré-Birkhoff theorem is valid. We seek the value of α, an exponent that symmetrically scales the power and mode ratios of the unseen complex molecular and solvent entropy: $\phi^t{}_v(z) \sim e^{\alpha^{-1}\omega^\alpha}$.

An important step was taken in this direction by the Stapleton group[7,8] who deduced a value for α^{-1} as the Fourier dimension of a density of states mode spectrum using a transformation of the temperature dependence of the electron spin relaxation rate of low-spin Fe^{3+} of ferric proteins. Here, the enthalpy-entropy symmetry is seen as a reciprocal relation between the real and imaginary parts of the eigenvalues of the distribution of the complex entropy $p(z) \sim e^{\alpha^{-1}\omega^\alpha}$ and the value found was 1.65 ± 0.04. We note with respect to the ubiquitous "1/f noise" of membrane protein ion-conductance fluctuations and Stapleton's protein power law spectrum that $e^{-\alpha^{-1}-\alpha} = 1$ iff $\alpha = 0.618...$ Equivalent values were obtained on the spatial dimension $\dim_x$ using stroboscopic snapshots of the protein dynamics from x-ray crystallographic maps on $\mathbf{R}^2$; $\alpha \sim \dim_x$.

This ratio between sequential modes agreed with the integer-valued common ones found in our Fourier transformations of the amino acid sequences in hydrophobic free energies of over 100 polypeptides and proteins[9] using their ethanol-water partition energies[10]: 2, 3, 5, 7-8, 11, and 13; $<z_i / z_{i+1}> \sim \alpha$. As a reflection of the aggregate dynamics of the brain's membrane ion-conductance proteins, α is seen in the ratio of the fundamental frequency bands of the time-varying oscillations of the electroencephalogram: $\Delta = 2$-3; $\theta = 5$-7, $\alpha = 11$-13; $\beta = 17$-19, along with its characteristic z^α power law.

II. A Family of Mirror-Image, Recurrent, Hydrophobic, Hyperbolic Geodesics, **F**n

Poincaré resolved the behavior of one parameter transformations of the plane in the neighborhood of an equilibrium point into the generic family of sinks, sources,and saddles. However, degeneracies of solutions not removable by small perturbations (they reappear in the same small neighborhood) have positive measure in co-dimension two. Representing the enthalpy with u, entropy v, nonlinear hydrophobic coupling $r \sim$ (average hydrophobicity per residue in kcal/mol, $<u>$), and hydrophobic mode relation parameter λ, we numerically studied the Cartwright-Littlewood equation[11] in $(u + v) \equiv \underline{u} = f(r,\lambda)$:

$$\underline{\ddot{u}} - r(1 - \underline{u}^2)\underline{\dot{u}} + \underline{u} = r\lambda \cos \lambda t \qquad (2)$$

Differentiating and rescaling $(\hat{v} / r \rightarrow v)$ with the implicit variable time step, $\Delta t = \theta/n$, results in the a nonlinear equation for the enthalpy and entropy:

$$\phi^t{}_u = v - r(u - u^3/3) \qquad (3a)$$
$$\phi^t{}_v = -u/r + \lambda \cos \lambda t \qquad (3b)$$

Figure 1a portrays the resulting map of the annulus as in the Poincaré-Birkhoff theorem, a set of mirror-image geodesics and hyperbolic fixed points. The latent left-right, down-up mirror symmetry is seen with a small change in r (Fig. 1b). Whereas a parameter region of degenerate geodesics suggests the presence of chaotic mathematical solutions and sensitivity to initial conditions, in the hyperbolic helix of the hydrophobic free energy, near-periodic modes of proteins, the geodesic family, $\phi^t{}_v \in \mathbf{F}n$ is stereoisomerically inverted and exists only in the solvent mirror. The scaling mode spectrum of Stapleton is not the 1/f noise of chaos but the discrete modes of the universal spectrum with a power law and low order frequency peak sequence that go like

α^{-1}, α. As we shall see below, these Fibonacci modes of the Farey sequences inscribe a partition of the hyperbolic unit circle as a perfect set [12,13].

Using the zeta function, $\varsigma(z)$, the distribution of hyperbolic geodesics on $\mathbf{R}^2$ was proven to be analogous to those of the primes, Φn [14]. We are currently investigating this theorem in the context of (3) and conjecture that the distribution of the orbits of $z_i \in \phi^t{}_v$ goes like that of Φn with low order n:

$$Pz(n\mathbf{F}n) \sim \varsigma(z) = (1 - e^{-(z+n)\alpha(z)}) \quad (4)$$

Another approach we are taking is the Fourier-Laplace transformation of $Cu(\tau)$ as a time-dependent decay in the correlation function:

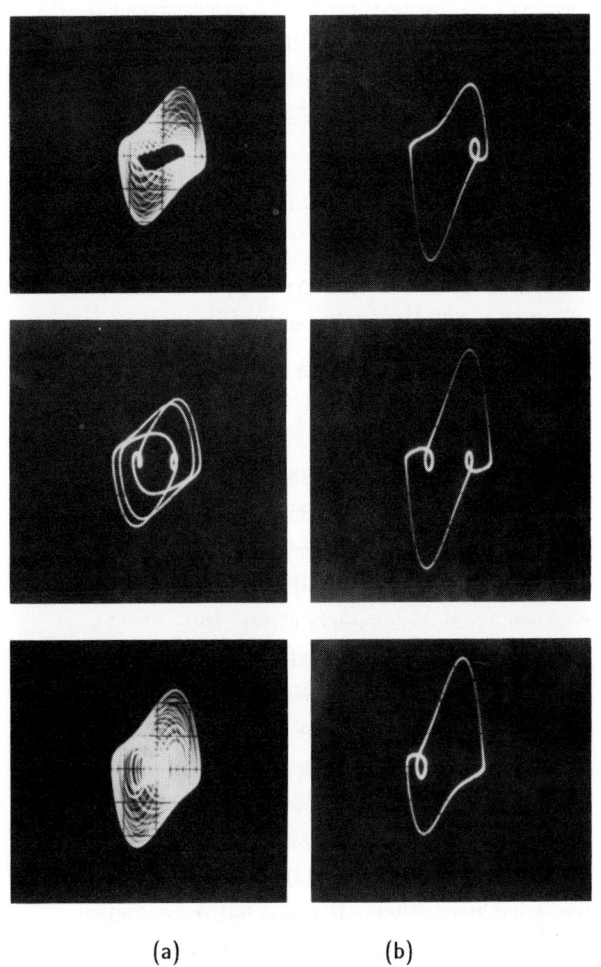

(a)　　　　　(b)

Figure 1

$$Cu(\tau) = \lim_{t \to \infty} 1/t \int_0^t z_i \cdot z_j \, dt \quad (5a)$$

$$- \lim_{t \to \infty} 1/t \int_0^t z_i \, dt \int_0^t z_j \, dt$$

$$C(z) = \frac{z_i \cdot z_j}{2 - e^z - e^{\alpha z}} + f(z) \quad (5b)$$

We can generate the Stapleton mode spectrum directly by expanding the enthalpic component of the hydrophobic free energy, u, in a trigonometric series, $|\phi^t{}_u(z)|^\alpha$:

$$\phi^t{}_u = (\Gamma \alpha + 1)^{-1} \int_x^\infty z(u)^{-(\alpha+1)} e^z \, dz(u) \quad (6a)$$

$$\Gamma(\alpha + 1) = \int_0^\infty (\phi^t{}_u)^\alpha e^{-u} \, du \quad (6b)$$

and predict the Levy distribution of convolutionally stable, identically distributed hydrophobic free energy modes with scaling and characteristic exponents α^{-1}, α [15]:

$$\ln Pz = i\delta z - \alpha^{-1} |z|^{-\alpha} (1 + i\beta z) \quad (7)$$

in which δ locates the spectrum and β describes its skew.

If p/q and r/s are successive terms in a Farey series of order n, the irreducible fractions between 0 and 1 whose denominators do not exceed n, the geodesic families, $\mathbf{F}n = f(r, \lambda)$ can be defined by the matrix A of map $\phi^t{}_u = e^{At}$, det $A = 1$, norm $(\lambda-, \lambda+) \neq 1$ as, for example,

$$\mathbf{F}n = 3 \quad (8)$$

$$\begin{pmatrix} q & s \\ p & r \end{pmatrix} \to \begin{pmatrix} 3 & 2 \\ 1 & 1 \end{pmatrix} \begin{pmatrix} 2 & 3 \\ 3 & 2 \end{pmatrix} \begin{pmatrix} 3 & 1 \\ 2 & 1 \end{pmatrix}$$

The matrix A at the strong coupling fixed point $\mathbf{F}n = 1$, when exponentiated

$$\begin{pmatrix} 1 & 1 \\ 1 & 0 \end{pmatrix}^n \to \begin{pmatrix} 1 & 1 \\ 1 & 2 \end{pmatrix} \begin{pmatrix} 1 & 2 \\ 2 & 3 \end{pmatrix} \begin{pmatrix} 2 & 3 \\ 3 & 5 \end{pmatrix} \begin{pmatrix} 3 & 5 \\ 5 & 8 \end{pmatrix} \cdots \quad (9)$$

accumulate orbits like the elements of Tr A, $\sim \alpha$.

Consistent with a range of helical pitches found in x-ray studies of proteins, the range of values $\alpha = (1/3 - 2/3)$, reported recently by the Stapleton group in studies of over 70 proteins[16], can be understood by a transformation of ϕ^t_u as a two-parameter (λ, r) map of the annulus to dynamics on elliptical curve **C**, parameterized by a harmonic triangle, Δu, along the smallest and Δv oriented along the largest diameter. $\Delta u / \Delta v$ determines whether a Taylor-trigonometric expansion of a perturbation of **C** will converge to both Δu and Δv or lose its hyperbolic character by always converging on Δu. Now, let Δu and Δv be complex numbers, $\Delta u \equiv z_u$, $\Delta v \equiv z_v$, with Im $z \equiv n/k$, $k \equiv$ circuits of **C** ($k < n/2$) and prime to n) and Re z as the moduli of z. The question becomes: What is the limit on the moduli of their differences, mod $|z_v - z_u|$, in order *not* to converge smoothly to z_u upon perturbation? Using both geometric and analytic formalisms, Cayley[17] proved

$$\text{mod}\,|z_v - z_u| = \alpha \, \text{mod}\, z_u \quad (10)$$

The overlapping space of mirror-symmetric geodesic degeneracy, the place where the hydrophobic hyperbolic helix lives, is in the range of values found for α by the Stapleton group, two thirds from each mode in the middle third of the classic ternary dissection set of Cantor in complex space. The Farey-Fibonacci hydrophobic free energy mode partition and the spectral and fractal dimensional isometry relate the Salem perfect set of unicity in a trigonometric series representing a distribution of infinitely many independent Bernoulli distribution functions[18], the Levy convolutionally stable mode distribution[15], the critical modulus scaling of the Cartwright-Littlewood equation for degenerate solutions[11], and the universal spectrum[12,13] to the hydrophobic mode power law on the density of state spectrum in some of Stapleton's proteins[7]. Elsewhere we have argued that this zone between topological conjugacy (Hölder continuity) and Lipschitz differentiability marks the neighborhood of incipient loss of the scaling power spectrum at the onset of macromolecular phase transitions. The $e^{\alpha^{-1}\omega^\alpha}$ spectrum may be a signatory property of allosteric proteins with the specialized capacity for these low energy conformational transitions. Only 3.7 kcal/mol is required for the transconformation of hemoglobin (Stapleton $\dim_x = 1.65$) between its low and high oxygen affinity states. Myoglobin (Stapleton $\dim_x = 1.67$) has similar dynamical properties. In contrast, trypsin (Stapleton $\dim_x = 1.34$) has no allosteric properties.

III. Optical Isomerism and Solvent Entropy

Since $\phi^t_u - \phi^t_v \sim 0$, $(\lambda-) \in \phi^t_u$ and $(\lambda-) \in \phi^t_v$ describe a mirror symmetric exponential divergence of z_u and z_v (Fig. 1a,b). We can represent these maps as paired spiral orbits with increasing (decreasing) radii r rotating with a constant unit of angle t in radians. γ represents the constant angle made at the points that δr, the change in the length of sequential radii, and the tangents to the curve there intersect. Since $\pm (\delta r/r) \sim e^{(\lambda+,-)}$

$$\frac{\delta r}{r\delta t} \sim \cot \gamma \quad (11a)$$

$$\ln r \sim t \cot \alpha + \ln r_0 \quad (11b)$$

$$\ln \lambda(\lambda^{-1}) \sim t(t^{-1}) \cot \gamma \quad (11c)$$

A curve in which the tangent at any point makes a constant angle with the radius drawn to that point from a fixed point is a near-periodic two-parameter (t, γ) expanding (contracting) transformation of the plane, is self-similar across scale, has dilation symmetry and sequential moduli of mod $|z_i - z_j| = \alpha z_j$. Because its inverse with respect to the pole is an equal spiral, this is the stereoisomeric solvent dual of the hydrophobic hyperbolic helix. We note again that $\phi^t_u(z)$ is the protein enthalpic observable and $\phi^t_v(z)$, the inferred and mirror symmetric solvent entropy. The change in the solvent is passive with respect to its induced stereoisomerism because pure water has a two-fold axis of symmetry, S_2, so a rotation by $180° = 360°/2$ takes it back to an indistinguishable condition. Optically active absorption bands are often the result of symmetric groups perturbed into asymmetry by their molecular environment[19].

If a molecule has a helical structure its polarizability, as measured by the optical rotation of light that has been polarized into handedness, depends upon whether or not the field of the probing radiation follows the direction of the spiral. Molecules having a helical structure are not superimposable on their mirror image. $\phi^t{}_u(z)$ and $\phi^t{}_v(z)$ are mirror symmetric clockwise $(+)$ or counter clockwise $(-)$ rotators of polarized light. The optical activity of proteins is quantified as rotation per unit mass normalized as percent helix and are dependent on both protein hydrophobicity and the surrounding solvent structure. Increasing average hyperbolic-hydrophobic "pressure", $<u>$, with 2-chloroethanol in the solvent can increase a protein's percent helix 3-4 fold, whereas destroying the water structure with a chaolytic agent such as urea eliminates all optical activity of macromolecules. With $<u>$ representing the average hydrophobic free energy per amino acid residue of the protein in kcal/mol, z_i, the protein's hydrophobic free energy modes, an interesting relationship may exist between optical activity and Stapleton's exponents. We use two of the most thoroughly studied protein enzymes as examples.

Table 1

Protein	$<u>$	z_i	% Helix	dim_x
Lysozyme	1.15	2.2, 3.3, 5, 7, 13.3	29	1.69
Ribonuclease	1.01	5.7, 10.1	16	1.33

The hyperbolic helix hypothesis can be tested in part by correlating the proteins' specific optical activity under standardized conditions with their values in Stapleton dim_x as in (11). $\alpha = f(\gamma)$. This work is in progress.

The solvent shadow of the enantiomorphic complement of $\phi^t{}_u$, the $\phi^t{}_v$, exists in the normally two-fold symmetric supply of aqueous entropy. This entropic pool has $(+)$ and $(-)$ room for the necessarily stereospecific entropy exchanges of substrates, hormones, transmitters, and drugs with helical receptor and enzyme proteins. A well studied counter example of this kind of interaction involves the bias of the solvent entropic environment using very hydrophobic, stereochemically specific "hosts" such as the cyclodextrins or crown ethers to optically resolve by crowding out of solution the $(+)$ or $(-)$ member of racemic mixtures of amino acids or amine salts[20]. These processes are quantitatively dependent on $<u>$ [21].

Since the entropic state of the gel-like fluid compartments are global properties of biological organisms, it is possible that some generalized disorders of protein function may result from crowding the solvent's S_2 condition on the entropy-enthalpy plane, R^2, into an optically active, mirror symmetric bias as in Fig. 1b. As a hypothetical and medically counter-intuitive example: high levels of non-relevant stereochemically specific protein antibody titers could optically bias the solvent entropy so as to reduce the optically active room for the stereochemically-specific entropy-enthalpy information exchange between the protein antigen of a new, invading pathogenic agent and the membrane protein responsible for triggering the defensive cascade in the lymphatic tissue. Hyperfunctioning antibody generating systems and immunological paralysis with vulnerability to usually harmless organisms could paradoxically co-exist.

Bifurcation from divergent flow to periodicity results from the loss of the hyperbolic stability of mathematical solutions: a transition from mirror image geodesics to a mode locking pattern. We postulate that this state in membrane receptor-proteins generates stereotypic behavior and loss of regulatory sensitivity in cells. Fig. 2 compares the power spectrum of a time series of samples of growth hormone released from pure cell type perifusion systems of normal and tumor pituitary cells[22]. A loss of the hyperbolic continuous spectrum and a pattern of non-responsive, mode-locked periodicity is seen in the tumor cell line.

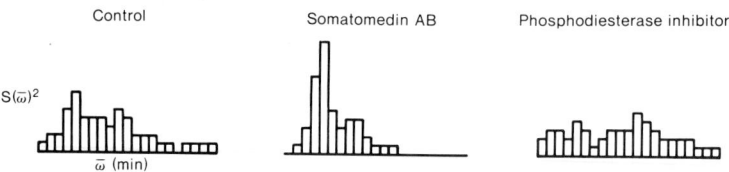
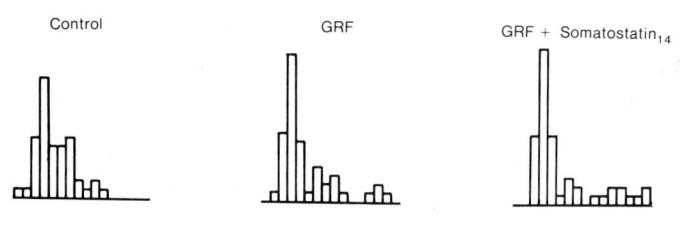

Figure 2

References

1. G.D. Birkhoff, Acta Math. 43 (1920) 1.
2. S. Smale, Bull. Am. Math. Soc. 73 (1967) 747.
3. J.A. Reynolds et al., Proc. Natl. Acad. Sci. USA 71 (1974) 2925.
4. C. Chothia, Nature 248 (1974) 338.
5. P.L. Privalov and N.N. Khechinashvili, J. Mol. Biol. 86 (1974) 665.
6. C.N. Hinshelwood, J. Chem. Soc. 538 (1937) 858.
7. H.J. Stapleton et al., Phys. Rev. Lett. 45 (1980) 1456.
8. J.P. Allen et al., Biophys. J. 38 (1982) 299.
9. A.J. Mandell, Ann. Rev. Pharmacol. Toxicol. 24 (1984) 237.
10. Y. Nozaki and C. Tanford, J. Biol. Chem. 246 (1971) 2211.
11. M.L. Cartwright and J.E. Littlewood, J. Lond. Math. Soc. 20 (1945) 180.
12. S.J. Shenker, Physica 5D (1982) 405.
13. D. Rand et al., Phys. Rev. Lett. 49 (1982) 132.
14. W. Parry and M. Pollicott, Ann. Math. 118 (1983) 573.
15. B.V. Gnedenko and A.N. Kolmorogov, Limit Distributions for Sums of Independent Random Variables (Addison-Wesley, Reading, MA, 1968).
16. G.C. Wagner et al., J. Am. Chem. Soc. (1985) in press.
17. C. Cayley, Quart. J. Pure Appl. Math. 16 (1879) 179.
18. R. Salem, Trans. Am. Math. Soc. 54 (1943) 218.
19. P. Crabbé, Optical Rotary Dispersion and Circular Dichroism in Organic Chemistry (Holden-Day, San Francisco, 1965).
20. E.P. Kyba et al., J. Am. Chem. Soc. 99 (1977) 2564.
21. T. Sugimoto and N. Baba, Isr. J. Chem. 18 (1979) 214.
22. R. Guillemin et al., in Synergetics of the Brain, eds. E. Basar, H. Flohr, H. Haken, and A.J. Mandell (Springer-Verlag, Berlin, 1983) pp. 155-162.

AUTHOR INDEX

ALEXANDROWICZ, Z., 125
ALLAIN, C., 61, 283
ARGYRAKIS, P., 361

BADII, R., 453
BALL, R.C., 231, 237
BERNASCONI, J., 409
BLUMEN, A., 33, 399
BOTET, R., 251, 255
BRADY, R.M., 231, 237

CAMBIER, J.L., 421
CAPPELLI, A., 111
CAPRILE, B., 279
CHERNOUTSAN, A., 115
CHHABRA, A., 129
CLOITRE, M., 61, 283
CONIGLIO, A., 97, 165
CVITANOVIĆ, P., 439

DACCORD, G., 193
DE OLIVEIRA, M.J., 427
DEKEYSER, R., 101
DEWAR, R., 145
DJORDJEVIĆ, Z.B., 413

ERNST, M.H., 289, 303
EVERTSZ, C., 87, 159

FLORES, F., 177

GONZALEZ, A.E., 133
GOUYET, J.F., 137
GRASSBERGER, P., 273
GRAY, L.J., 383
GUINEA, F., 177

HARRIS, C.K., 141, 145
HAVLIN, S., 351
HERRMANN, H.J., 129
HILFER, R., 33

JAKEMAN, E., 55
JAN, N., 97
JENSEN, M.H., 439
JULLIEN, R., 251, 255

KADANOFF, L.P., 439
KAPLAN, T., 383
KEEFER, K.D., 39
KLAFTER, J., 393, 399
KOLB, M., 255, 259, 263, 267,
KOPELMAN, R., 369
KUPERS, R., 319

LANDAU, D.P., 129
LEVI, A.C., 279
LIGGIERI, L., 279
LIU, S.H., 383
LIVI, R., 111
LOUIS, E., 177
LOVEJOY, S., 457
LUCCHIN, F., 313
LUCK, J.M., 379
LUNG, C.W., 189
LYKLEMA, J.W., 87, 93

MAJID, I., 97
MANDELBROT, B.B., 3, 17, 21
MANDELL, A.J., 469
MARGOLINA, A.E., 357
MARITAN, A., 101, 107, 111
MEAKIN, P., 205, 213
MELROSE, J., 29, 365
MILOŠEVIĆ, S., 115
MURAT, M., 169

NAUENBERG, M., 421
NITTMANN, J., 193

PALADIN, G., 447
PELITI, L., 73, 83, 269
PFEIFER, P., 47
PIETRONERO, L., 83, 151, 159, 319
    417, 431
PLISCHKE, M., 217
POLITI, A., 453
PROCACCIA, I., 439

QUASTHOFF, U., 465

RÁCZ, Z., 217, 309
RAMMAL, R., 373
ROSSI, G., 231, 237
ROSSO, M., 137
RUFFO, S., 111
RYS, F.S., 461

SALINAS, S.R., 427
SANDER, L.M., 241
SAPOVAL, B., 137
SATPATHY, S., 173
SCHAEFER, D.W., 39
SCHER, H., 223
SCHERTZER, D., 457
SCHNEIDER, W.R., 409
SHLESINGER, M.F., 393
SIEBESMA, A.P., 431
SOLLA, S.A., 185

STANLEY, H.E., 97, 193, 327
STELLA, A.L., 101, 107
STINCHCOMBE, R., 337

TAKAYASU, H., 181
THOMPSON, B.R., 231, 237
TSALLIS, C., 65
TURKEVICH, L.A., 223

VAN DONGEN, P.G.J., 303
VICSEK, T., 213, 247
VULPIANI, A., 447

WALDVOGEL, A., 461
WEBMAN, I., 343
WIESMANN, H.J., 151, 159

YANG, Y.S., 119
YOKOI, C.S.O., 427

ZHANG, Y.-C., 269
ZUMOFEN, G., 399

MELEGA